Biorefinery Production Technologies for Chemicals and Energy

Scrivener Publishing
100 Cummings Center, Suite 541J
Beverly, MA 01915-6106

Publishers at Scrivener
Martin Scrivener (martin@scrivenerpublishing.com)
Phillip Carmical (pcarmical@scrivenerpublishing.com)

Biorefinery Production Technologies for Chemicals and Energy

Edited by

Arindam Kuila and Mainak Mukhopadhyay

This edition first published 2020 by John Wiley & Sons, Inc., 111 River Street, Hoboken, NJ 07030, USA and Scrivener Publishing LLC, 100 Cummings Center, Suite 541J, Beverly, MA 01915, USA

For more information about Scrivener publications please visit www.scrivenerpublishing.com.

Wiley Global Headquarters
111 River Street, Hoboken, NJ 07030, USA

For details of our global editorial offices, customer services, and more information about Wiley prod-ucts visit us at www.wiley.com.

Library of Congress Cataloging-in-Publication Data

ISBN 978-1-119-59142-9

Cover image: Pixabay.Com
Cover design by Russell Richardson

Set in size of 11pt and Minion Pro by Manila Typesetting Company, Makati, Philippines

Printed in the USA

Contents

Preface

Today, the world's major economies are totally dependent on fossil fuels and other nonrenewable energy resources, and once substrate components are depleted, they will be gone forever. This has become an issue since, due to an ever-increasing world population, the use of technological devices in our everyday lives has been on the rise. In turn, the ever-increasing use of these devices has led to a rise in the use of fossil energy sources, resulting in a hike in global warming brought about by their consumption. Therefore, the world has been shifting toward sustainable development, including the generation of energy and industrially valuable chemicals from renewable substrates. Consequently, there is an eagerness to find alternative sources that are easily available and produce zero pollution. For example, in the search for possible fuel replacements, scientists hit upon the concept of producing fuel from lignocellulosic biomass, which is easily available in many forms, easily processed, cost-effective, and nonpolluting. However, whatever the source, biorefineries will play an important role in the integration of the conversion process with high-end equipment facilities for the generation of energy, fuel, and chemicals. With that in mind, this book was written as an invaluable resource for those interested in all aspects of biorefinery—from basic principles to end results.

The book is divided into four parts. The first part, "Basic Principles of Biorefinery," covers the concept of biorefinery, its application in industrial bioprocessing, the utilization of biomass for biorefinery application, and its future prospects and economic performance. The second part, "Biorefinery for Production of Chemicals," covers the production of bioactive compounds, gallic acid, C4, C5, and C6 compounds, etc., from a variety of substrates. The third part, "Biorefinery for Production of Alternative Fuel and Energy," covers sustainable production of bioethanol, biodiesel, and biogas from different types of substrates. The last part of this book discusses sequential utilization of wheat straw, material balance, and biorefinery approach.

Dr. Mohd Asyraf Kassim (Universiti Sains Malaysia), Dr. Massimo Lucarini (CREA Research Centre for Food and Nutrition, Italy), Gunjan Mukherjee (Chandigarh University, India), Dr. Sachin A. Mandavgane (VNIT, Nagpur, India), and Dr. Ashok Pandey (Indian Institute of Toxicology Research, Lucknow, India) are just a few of the eminent scientists involved in the writing of this book. The book will be a useful resource for researchers and students in the areas of environmental biotechnology, bioprocess engineering, renewable energy, chemical engineering, etc.

Last but not least, we are grateful to the production team of Wiley-Scrivener Publishing for encouraging and extending their full cooperation and help for a timely completion of this book.

Dr. Arindam Kuila
Banasthali Vidyapith (Deemed University), Rajasthan-304022, India
Email id: arindammcb@gmail.com

Dr. Mainak Mukhopadhyay
JIS University, Kolkata-700109, India
Email id: mainak.mukhopadhyay12@gmail.com

Part 1

BIOREFINERY BASIC PRINCIPLES

1

Principles of Sustainable Biorefinery

Samakshi Verma and Arindam Kuila*

Department of Bioscience and Biotechnology, Banasthali Vidyapith, Vidyapith, Rajasthan, India

Abstract

Sustainable biorefineries have a critical role to play in our common future. The need to provide more goods using renewable resources, combined with advances in science and technology, has provided a receptive environment for biorefinery systems development. Biorefinery offers the promise of using fewer non-renewable resources, reducing CO_2 emissions, creating new employment, and spurring innovation using clean and efficient technologies. Lessons are being learned from the establishment of first-generation biofuels operations. The factors that are key to answering the question of biorefinery sustainability include: the type of feedstock, the conversion technologies and their respective conversion and energy efficiencies, the types of products (including co-products) that are manufactured, and what products are substituted by the bioproducts. The BIOPOL review of eight existing biorefineries indicates that new efficient biorefineries can revitalize existing industries and promote regional development, especially in the R&D area. Establishment can be facilitated if existing facilities are used, if there is at leastone product which is immediately marketable, and if supportive policies are in place. Economic, environmental, and social dimensions need to be evaluated in an integrated sustainability assessment. Sustainability principles, criteria, and indicators are emerging for bioenergy, biofuels, and bioproducts. Practical assessment methodologies, including data systems, are critical for both sustainable design and to assure consumers, investors, and governments that they are doing the "right thing" by purchasing a certain bioproducts. If designed using life cycle thinking, biorefineries can be profitable, socially responsible, and produce goods with less environmental impact than conventional products ... and potentially even be restorative!

Keywords: Biorefineries, environmental impacts, sustainable development

1.1 Introduction

Sustainable development is a positive socio-economic modification which allows ongoing and upcoming generations to encounter their requirements. In order to do this, development has to assist the natural and social systems as it is based on them. All the three dimensions (economic, environmental, and social) of development are interconnected with each other and schematically, the place where they meet is known as sustainability (viable). Sustainability is not static; it is highly vigorous because it varies with geographical surroundings and develops over time [1].

**Corresponding author*: arindammcb@gmail.com

Arindam Kuila and Mainak Mukhopadhyay (eds.) Biorefinery Production Technologies for Chemicals and Energy, (3–14) © 2020 Scrivener Publishing LLC

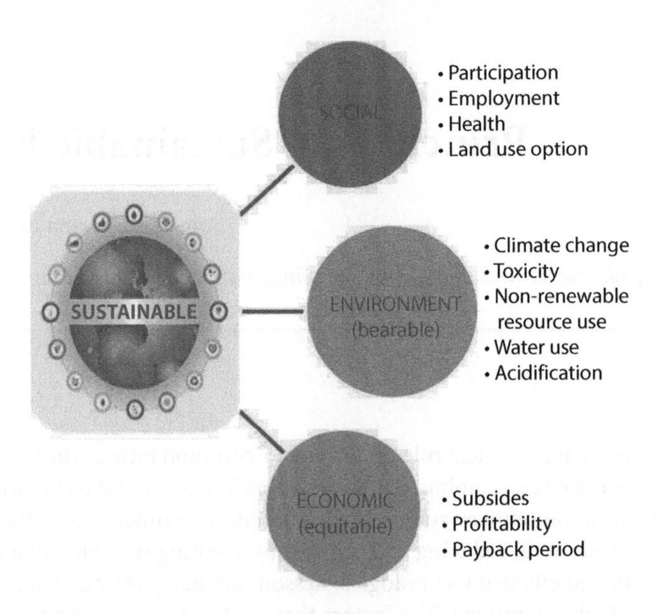

Figure 1.1 Sustainable development.

Because of increased production and exploitation of fossil fuel energy and use of certain materials and chemicals, it causes huge environmental destruction which takes us to unavoidable termination that there should be the development of new production systems. These new systems of production should work on reducing pollutants or harmful materials and will produce safe and eco-friendly products within the green and sustainable supply chain. A renewable and constant supply having low carbon cost is required to do this process. Biomass is the only source of such renewable feedstock globally [2].

There are several factors that are responsible for determining environmental sustainability which are following: types of products that are manufactured, feedstock types; the technologies with their energy and conversion efficiencies; which type of product is exchanged by bioproducts; how emissions are dispensed to these products; and how these bioproducts are utilized and discarded with the end of life [1].

There is a need of developments in sustainable production for commercialization, co-products, and energy carriers that are being developed from biomass by using biorefinery. There is a rapid development of technology among these areas. It is understood that biomass has all the elements which are present in fossil resources, whereas in different integration due to which we are able to conclude that ongoing and upcoming technologies will head toward the future depending upon low carbon, renewable, and sustainable economies [2]. In our common upcoming time, there will be an important role to play by sustainable biorefinery. As there is no such development within the biorefinery products and technologies, so there will be immense possibilities for great significance. Superior knowledge about upcoming climatic modification footprints, sustainability affairs and inventive technologies will lead us to select more informed methods which will help in developing our renewable resources so that they can fulfill the requirements of society as well as nature on "spaceship Earth" [1].

Necessity to provide more goods and servicesby utilizing renewable sources in combination with advances in science and technology together have provided a providential environment for the development of biorefinery systems. A more bio-based economy has proposed the assurance to manufacture and preserve new employment, minimize CO_2 emissions, exploit lesser non-renewable resources, and motivate novelty by utilizing sterilized and more productive technologies. Developed biorefinery should be capable of furnishing future products [1]. A brilliant estimation of such products under various technology frameworks is required within the next decades so that sustainable designs must be consciously formulated and evaluated. Sustainability is not just about renewability or about ecosystem or about GHG emissions because human health effects, economical, environmental, and social work are all need to be directed. If we move rapidly without even knowing the full outcome of any new development, then accidental effects can evolvedas the land utilization problem has been demonstrated. Eventually, public favor for developing biorefinery, i.e., acquiring bioproducts, biofuels and bioenergy will be vanished if there is any variability regarding their faithful subscription toward society and nature. It is mandatory to address the awaking concerns in an understanding manner [1].

1.2 Biorefinery

According to the NREL (National Renewable Energy Laboratory), a process that involves various equipments and biomass conversion methods in order to produce transportation biofuels, biomaterials, biochemicals, and power and heat is known as biorefinery [3].

Biorefineries are defined by International Energy Agency (IEA) Bioenergy Task 42 as the sustainable handling of biomass into a spectrum of marketable products and energy. It incorporates the process of sustainability, multiple products, and system integration. According to the literature, biorefineries are often classified by predominant conversion technology or feedstocks. If the number of platforms and/or products increases, then biorefinery complexity also increases [1].

An integrated biorefinery is planned in order to provide sustainable supply of biofuels and to produce fine and large amount of biochemicals (e.g., ether, glycerol, methanol, and syngas) with least waste generation. Depending upon the various conversion technologies utilized, i.e., biological conversion, chemical conversion, and thermo chemical conversion, biorefineries are further divided into three major categories [4].

All over the world biorefineries are appearing in a variety of many sizes, forms, or configurations. Their development rely on the present infrastructure, accessibility of biomass raw material, the requirement of given products, and to get knowledge about how some policies like public acceptance, scale-up facilities, and the level of expenditure in research can help in alteration toward a more productive and greener compound economy [1].

According to Wellisch *et al.* [1] choosing a feedstock will have some social, environmental, and cost implications. There are majorly three types of biomass feedstocks through which a biorefinery can be served and they are given below.

1. Primary biomass feedstocks: This type of feedstocks contains biomass which is immediately directly collected from water bodies, agricultural area, or forest [1].

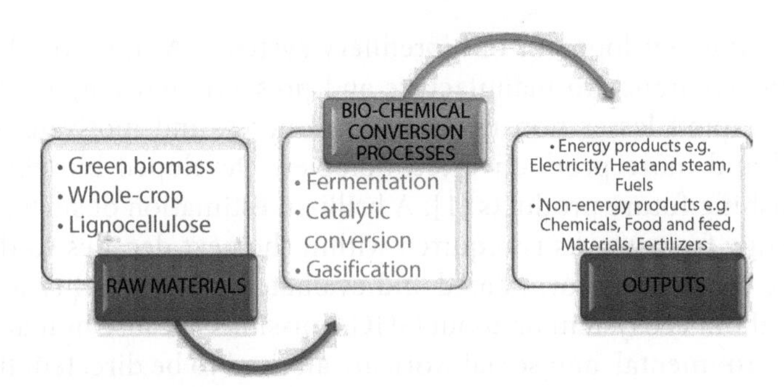

Figure 1.2 General concept of biorefinery.

2. Secondary biomass feedstocks: These are basically process sediments, i.e., black liquor or saw mill remnants which are released by the forest products industry such as lignin is released from a lignocellulosic biorefinery or food processing wastes is released from agricultural food industry [1].
3. Tertiary biomass feedstocks: These are mostly post-consumer remnants or wastes, i.e., wastewater, waste greases, and municipal solid waste [1].

Biorefineries are established by the exploitation of secondary (vs. primary) feedstocks, modification of existing facilities (vs. green field construction); at least one product is manufactured with benevolent renewable energy approaches and an existing market. Product-driven biorefineries are generally more complex systems than first-generation biofuels biorefineries [3].

1.3 Conversion Technologies of Biorefineries

In each category, there are various types of conversion technologies which are given below in a tabular format, i.e., Table 1.1 [5]. Choosing a correct conversion technology is not worthless, and this selection of a conversion method is based upon various factors such as the process productivity, the attribute and accessibility of biomass feedstocks, environmental and economical needs, the types of necessary products, and several other facets associated with the complete biorefinery supply chain [4].

There are typically five stages of biorefinery supply chain which are the following [4].

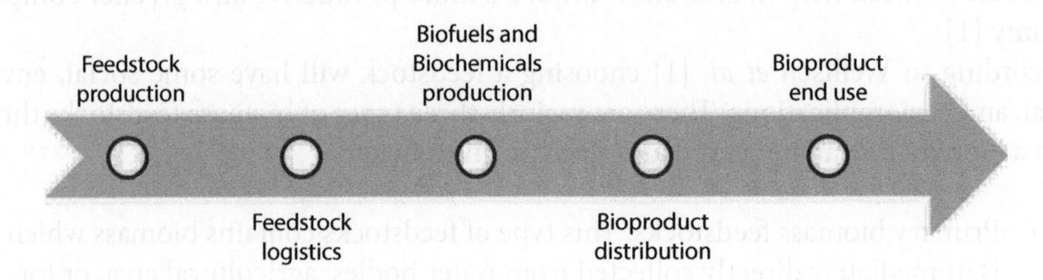

Figure 1.3 Five stages of biorefinery supply chain.

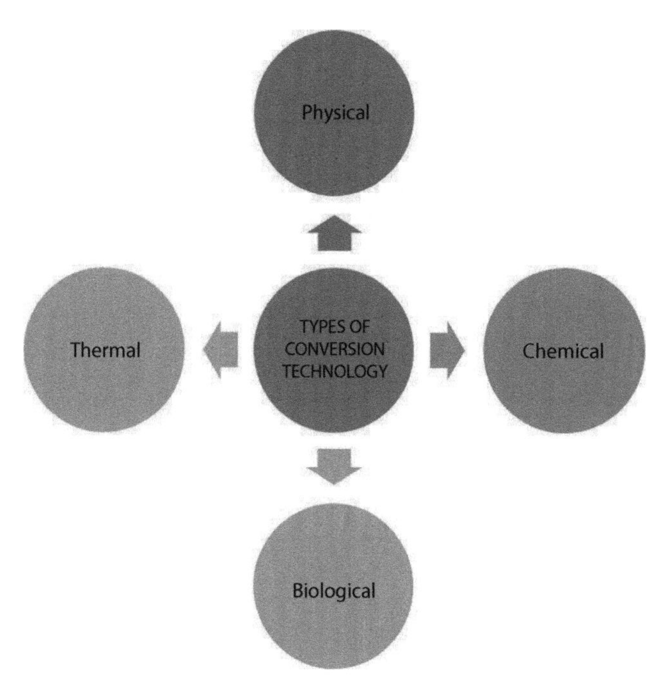

Figure 1.4 Four types of conversion technology in each category of biorefinery.

Different stages of biorefinery supply chain are associated with each other throughout the complete process via some important factors such as handling, storage, and transport. Additionally, there are some major key issues regarding specific stages of biorefinery supply chain [6].

Achieving sustainability, selecting a crop, and enhancing plantation efficiency are some major issues related to feedstock producers. Majorly, there are issues associated with the scale of operation within the preprocessing that can be approached as an opportunity either for on-farm added values to biomass or for a new, isolated industry. Scaling up of biofuels technologies is a major issue in conversion of biomass. Affordability, availability, and ease of use of the biofuel/bioenergy products are some major issues among consumers [4].

As we know, biorefinery conversion technology metamorphoses available biomass feedstocks in many other forms into various energy products which comprises of power and heat; gaseous fuels, i.e., syngas and biogas; and liquid fuels, i.e., alcohols, bio-oil and biodiesel. Any capable biorefinery system will become complex due to this great heterogeneity of products, biomass feedstocks, and conversion technologies. Additionally, different anxiety and norms could be directed at various biorefinery supply chains, and the decision maker will always requires a substantially ideal design solution, that will ultimately increases the complexity in the design and in the evaluation tasks also [4]. There are basically four types of conversion technology in each category of biorefinery which are shown in Figure 1.4.

1.4 Some Outlooks Toward Biorefinery Technologies

According to Kamm and Kamm [7], in research and development, there are basically three biorefinery systems that are involved nowadays: i) Whole-crop biorefinery (WCB) utilizes

Table 1.1 Conversion technology in each category of biorefinery.

Conversion category	Conversion technology	Type of resource	Examples of resources	Products	End use
Physical	Mechanical Briquetting of biomass Torrefaction Grinding Distillation Liquid-liquid extraction Absorption	Both liquid and solid biomass	Oil seeds, Agricultural, forestry residues, and other waste biomass materials	Biogas, By-products, and Liquid fuels (bioethanol and biodiesel)	Electricity, Heat, and Transport fuel
Chemical	Hydrolysis Solvent extraction Polymerization Supercritical conversion Transesterification	Both liquid and solid biomass	Oil or fat waste, Rape seeds, and Soybeans	Liquid fuels (biodiesel)	Electricity, Heat and Transport fuel
Biological	Enzymatic hydrolysis Anaerobic Digestion Growing of biomass Fermentation	Basically solid biomass but in case of anaerobic digestion, wet biomass	Wheat Colza, Maize, Manure, Miscanthus, Potatoes, Sewage sludge, and Sugar	Biogas, By-products, and Liquid fuels (bioethanol)	Electricity, Heat, and Transport fuel
Thermal	Pyrolysis Combustion Gasification Liquefaction	Basically solid biomass	Agricultural residues, Chicken litter, Wood logs, chips, and pellets	Heat, Product gas, Pyrolysis oil, and By-products (product gas, char)	Electricity, Heat and Transport fuel

Source: - E4Tech.com, http://www.e-sources.com//biomass_converion.htm (accessed June 30, 2012).

feedstocks, i.e., maize or cereals; ii) Green biorefinery (GB) utilizes naturally wet biomass, i.e., immature cereal, clover, lucerne, or green grass; iii) Lignocellulose feedstock (LCF) biorefinery utilizes naturally dry feedstocks, i.e., wastes and cellulose-containing biomass.

Physical methods are used to separate the biomass containing precursors within the first step of process. After that, the by-products and the main products are exposed to chemical or microbiological methods. Now, the main and by-products are left over with follow-up products which can further be treated or move into a conventional refinery. Hence, biorefinery is of dual significance because it undergoes biological genesis of respective feedstocks, and on the other hand, there is an uplifting in the biological traits of selected treatment and processing methods [7].

1.5 Principles of Sustainable Biorefineries

According to the IEA Bioenergy Task 42, sustainable handling of biomass into a spectrum of valuable products and energy is known as biorefining [8]. Although a very small amount of general list of biorefineries have been distinctly identified till now and several other kinds of biorefineries are still in growing phase, and it is acknowledged that all types of biorefineries will be developed and handled in a sustainable manner according to the social, economic, and environmental (SEE) aspects. Taking this in consideration, a biorefinery system should follow the given basic sustainability principles.

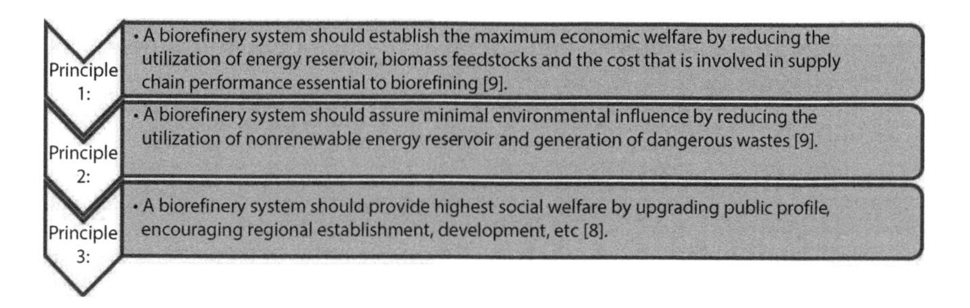

Principle 1:
• A biorefinery system should establish the maximum economic welfare by reducing the utilization of energy reservoir, biomass feedstocks and the cost that is involved in supply chain performance essential to biorefining [9].
Principle 2:
• A biorefinery system should assure minimal environmental influence by reducing the utilization of nonrenewable energy reservoir and generation of dangerous wastes [9].
Principle 3:
• A biorefinery system should provide highest social welfare by upgrading public profile, encouraging regional establishment, development, etc [8].

According to the SCM (Supply Chain Management), principles 4–7 have been allotted to biorefinery systems by the IEA Bioenergy Task 42 [8].

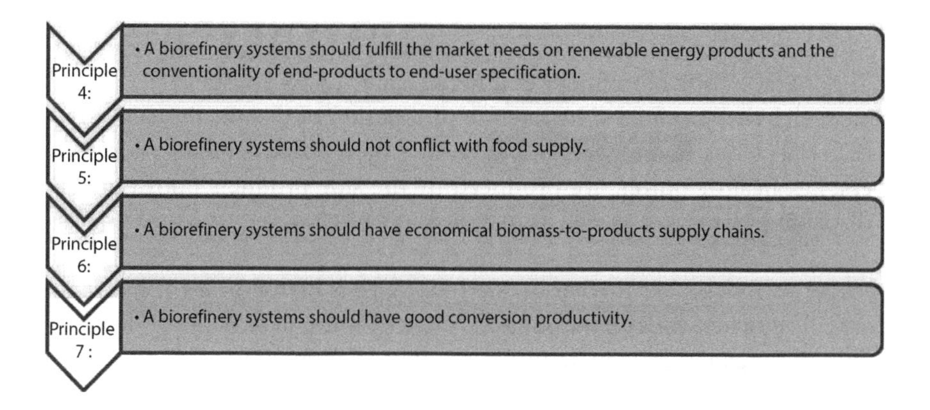

Principle 4:
• A biorefinery systems should fulfill the market needs on renewable energy products and the conventionality of end-products to end-user specification.
Principle 5:
• A biorefinery systems should not conflict with food supply.
Principle 6:
• A biorefinery systems should have economical biomass-to-products supply chains.
Principle 7:
• A biorefinery systems should have good conversion productivity.

These seven principles of sustainable biorefinery system are predominating instructions for the analysis and design of a biorefinery. There is a requirement of getting further knowledge if they are implemented to specific cases. For example, principle 2 can be explained as a necessity on life cycle GHG radiation depletion or no threat to air, soil, water, etc. Additionally, other principles could also be acquired for biorefineries as an essential accretion, i.e., the principles of sustainable development [10] and 12 other principles of green chemistry [11].

1.6 Advantages of Biorefineries

There was a saying that "if we have energy than we have everything" is not properly right. If we have energy, then other requirements of humans can be resolved by producing sustainable biomass feedstock. Biorefineries convey these demands as well as it also address the SEE requirements of our society. They will significantly provide employment and rural development, accompanied by reduced production costs with the development of emerging technologies and economies of scale with reasonably low carbon costs [2]. Biorefineries are strengthened by manufacturing co-products and energy carriers due to which this system becomes more economic. There is a variation in feedstocks that can be geographically based, manufacturing a diversity of valuable products which discovers this process as the substantial nominee in upcoming sustainable developments. Biorefineries remit matters of sustainability from all point—social, economic, and environmental. This technology is based upon the combination of the science, agro-engineering, marketing disciplines, and chemistry which demands a new prototype in sustainable development [2].

1.7 Classification of Biorefineries

Nowadays, classification of biorefineries is based upon the types of feedstocks utilized, technological (implementation) status, or major type of conversion technology applied. According to the literature, there is a variety of terms explaining biorefineries—see Figure 1.5.

There is a more relevant classification of biorefinery system given by the IEA Bioenergy Task 42. This classification depends upon the simplified description of full supply chains from biomass to end product. Approximately, biorefineries can be categorized in major two types:

Energy-driven Biorefinery: Production of biofuels or energy is the main target. The characteristic biorefinery enumerate value to co-products.

Product-driven Biorefinery: In general, production of chemicals, feed, food, or materials is the main target of biorefinery method. Secondary energy carriers, i.e., heat or power, are produced by the side products both utilized for dispersing in market as well as in-house advantages.

IEA Bioenergy Task 42 has further divided the various biorefineries. This type of recommended classification depends upon the latest ruling operator in biorefinery development, i.e., effective and well-organized manufacturing of transportation biofuels, so that the sharing of

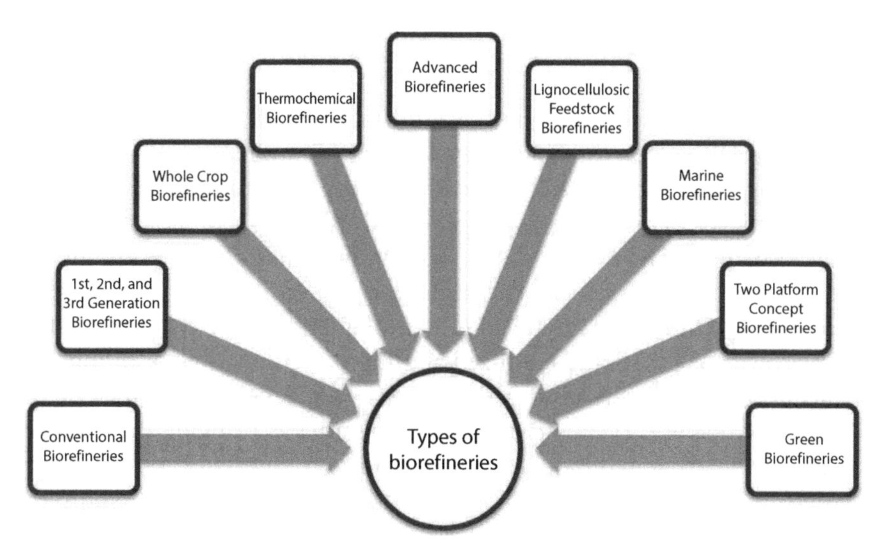

Figure 1.5 Varieties of biorefineries.

biofuels can be extended in the transportation area. In order to identify, classify, and describe the various biorefineries, this classification approach involves four major following features:

1. There are four principal conversion technologies, i.e., biochemical (e.g., enzymatic conversion and fermentation), chemical (e.g., acid hydrolysis, esterification, and synthesis), mechanical processes (e.g., size reduction, pressing, and fractionation), and thermo-chemical (e.g., pryolysis and gasification).
2. Products (e.g., chemicals, feed, food, and materials) and energy (e.g., bioethanol, biodiesel, and synthetic biofuels) are two main biorefinery product groups.
3. Biomass residues from agriculture, forestry, industry, and trade (e.g., bark, straw, used cooking oils, waste streams from biomass processing, and wood chips from forest residues) and energy crops from agriculture (e.g., short rotation forestry and starch crops) are two major feedstock groups within this type of classification.
4. Several biorefineries and their processes are linked with each other through an intermediates called as platforms (e.g., biogas, C5/C6 sugars, and syngas). Biorefinery systems complexity will increases with the increase in number of platforms.

- Classification of biorefineries is done by stating the involved feedstocks, platforms, products, and, if mandatory, the processes. Some examples of classifications are follows.

  i. Animal feed and bioethanol is produced from starch crops by utilizing C6 sugar platform biorefinery.
  ii. Phenols and FT-diesel are produced from straw with the help of syngas platform biorefinery.
  iii. Bioethanol, FT-diesel, and furfural are produced from saw mill residues via C6 and C5 sugar and syngas platform biorefinery.

1.8 Conclusion

Conclusively by adding value to the renewable utilization of biomass, biorefineries can play an advantageous role toward sustainable development. Biorefineries can feed the full bio-based products by producing a spectrum of bioenergy (i.e., fuels, heat, and power) and bio-based products (i.e., chemicals, feed, food, and materials). It should be noticed that the economic situation of market areas, i.e., agriculture, chemical, energy, and forestry is strengthened by reducing feedstocks demands and exploiting biomass conversion effectiveness. According to the international consent availability of biomass is insubstantial that means biomass feedstocks should be utilized as efficiently as possible resulting in development of multi-purpose biorefineries in an organization of deficient feedstocks and energy.

One of the remarkable success factors for biorefineries is bringing together key stakeholders normally operating in different market sectors (e.g., agriculture and forestry, transportation fuels, chemicals, energy, etc.) into multi-disciplinary partnerships to discuss common biorefinery-related topics, to foster necessary R&D direction, and to accelerate the deployment of developed technologies (platform function).

Task 42 can contribute to the growth of biorefineries by identifying the most promising bio-based products, i.e., food, feed, added-value materials, and chemicals (functionalized chemicals and platform chemicals or building blocks) to be co-produced with bioenergy, to optimize overall process economics, and minimize the overall environmental impact. Major initiatives in the immediate future include the preparation of a review and guidance document on approaches for sustainability assessment of biorefineries, and a strategic position paper "Biorefineries: Adding Value to the Sustainable Utilization of Biomass on a Global Scale".

References

1. Wellisch, M., Jungmeier, G., Karbowski, A., Patel, M.K., Rogulska, M., Biorefinery systems–potential contributors to sustainable innovation. *Biofuel. Bioprod. Bior. Innov. Sustain. Econ.*, 4, 3, 275–286, 2010.
2. de Jong, E., van Ree, R., Kwant, I.K., Biorefineries: Adding value to the sustainable utilisation of biomass. *IEA Bioenergy*, 1, 1–16, 2009.
3. Fernando, S., Adhikari, S., Chandrapal, C., Murali, N., Biorefineries: Current status, challenges, and future direction. *Energ. Fuel.*, 20, 4, 1727–1737, 2006.
4. Liu, Z. and Eden, M.R., 22 Biorefinery Principles. *Sustain. Bioenergy Prod.*, 447–474, 2014.
5. E4Tech.com. 2005. Biomass conversion processes. http://www.e-sources.com//biomass_converion.htm (accessed June 30, 2012).
6. Wisconsin Grasslands Bioenergy Network. http://www.wgbn.wisc.edu//biomass-supply-chain (accessed June 30, 2012).
7. Kamm, B. and Kamm, M., Principles of biorefineries. *Appl. Microbiol. Biotechnol.*, 64, 2, 137–145, 2004.
8. Cherubini, F., Jungmeier, G., Wellisch, M., Willke, T., Skiadas, I., Van Ree, R., de Jong, E., Toward a common classification approach for biorefinery systems. *Biofuel. Bioprod. Bior.*, 3, 5, 534–546, 2009.
9. Batsy, D.R., Solvason, C.C., Sammons, N.E., Chambost, V., Bilhartz, D.L., Eden, M.R., Stuart, P.R., Product portfolio selection and process design for the forest biorefinery, in: *Integrated biorefineries:*

Design, analysis, and optimization, P.R. Stuart and M.M. El-Halwagi (Eds.), pp. 837–845, CRC press, Boca Raton, Florida, 2013.
10. McKeown, R., Hopkins, C.A., Rizi, R., Chrystalbridge, M., *Education for sustainable development toolkit*, Energy, Environment and Resources Center, University of Tennessee, Knoxville, 2002.
11. Anastas, P.T. and Warner, J.C., Principles of green chemistry. *Green Chem. Theory Pract.*, 29–56, 1998.

2

Sustainable Biorefinery Concept for Industrial Bioprocessing

Mohd Asyraf Kassim*, Tan Kean Meng, Noor Aziah Serri, Siti Baidurah Yusoff, Nur Artikah Muhammad Shahrin, Khok Yong Seng, Mohamad Hafizi Abu Bakar and Lee Chee Keong

Bioprocess Technology Division, School of Industrial Technology, University Sains Malaysia, Penang, Malaysia

Abstract

In this chapter, the production of various value-added products including biofuels and fine chemicals from renewable resources are reviewed. Initially, the biorefinery concept and its development of different renewable resources are discussed. Then, different processes such as thermochemical, biochemical, and chemical conversion technologies involved in producing various products are reviewed. The processes and steps involved in each conversion approach for wide range chemical production are also discussed. This chapter also reviews the types of biofuels and fine chemicals that can be produced from various renewable resources through biorefinery concept for sustainable industrial processes.

Keywords: Bioethanol, microalgal, lignocellulosic material, renewable, biomass

2.1 Sustainable Industrial Bioprocess

Sustainable development has sparked great interest as an approach to overcome global issues. Nowadays, many countries have put sustainability as a main priority for their policy makers. Sustainable development is a complex approach with the aim to manage natural resources, generate, and improve welfare for future generations. The processes used to achieve sustainability must be reliable where several strategies have been introduced to ensure the success of this approach including waste reduction, pollution reduction, material reuse, and value creation.

Among the technologies that play important role for sustainable development plans, bioprocess technology is identified to contribute especially in the fields of food production, chemical, bioenergy, pollution control, and bioremediation. Modern industrial bioprocess technology is an extension of conventional technique which is important to be transferred to applications in the industry. The industrial bioprocess uses complete living cells or their components including enzymes and metabolites to obtain desired products.

**Corresponding author*: asyrafkassim@usm.my

Arindam Kuila and Mainak Mukhopadhyay (eds.) Biorefinery Production Technologies for Chemicals and Energy, (15–54) © 2020 Scrivener Publishing LLC

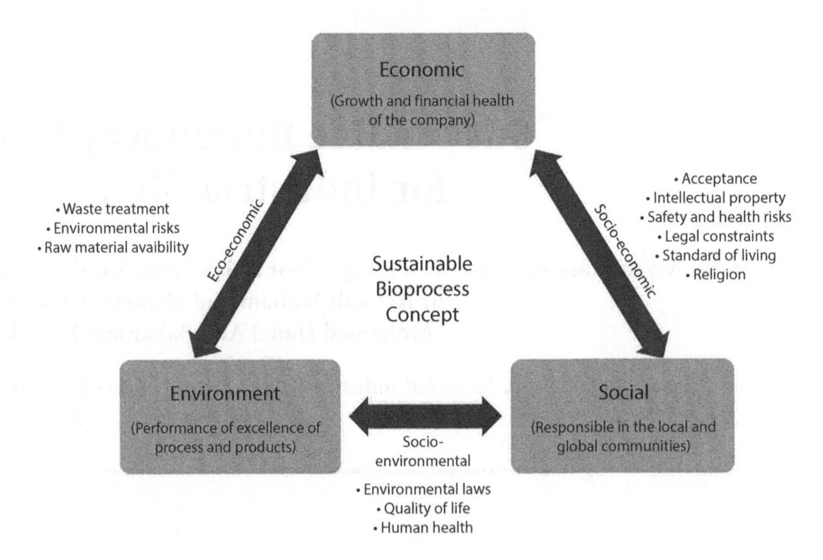

Figure 2.1 Three spheres of sustainable bioprocess.

Sustainable bioprocess has to be developed with the aim to improve social, environment, and socio-economic in the society (Figure 2.1) [1]. These three components are important and not independent of each other [2]. Sustainable bioprocess has more multi-variable situations involving many groups with the aim to minimize environmental impacts and be more economically viable and socially responsible [3]. It has to develop new products and explore new applications of agro-wastes generated from industries. Providing new technologies and industries will create job opportunities and grow special skills for human capital development. Apart from that, this system will diversify economic growth especially in rural areas. Implementation of sustainable bioprocess could also provide carbon neutral technology to assist in reducing global warming. Development of new approach and technology could reduce waste production and pollution especially in foods, chemicals, and agricultural industries. According to Henderson *et al.* [4] in their study in the comparison of 7-ACA's production, they concluded that production using biocatalytic reaction exhibited "greener" technology compared to chemical reaction as this process is safe and less hazardous due to low usage of chemicals.

2.2 Biorefinery

At present, chemicals and energy are among the most important commodities that play important role in daily life. Both are produced globally every year through refining, totally dependent on petroleum-based feedstocks for their production. Overdependence on these has slowly reduced petroleum reserves. It is evident that these activities have led to uncontrolled emission of greenhouse gases (GHG), including carbon dioxide (CO_2), methane, and nitrous oxide (N_2O).

This current scenario has pushed scientific community to explore replacements of renewable carbon sources and development of biorefinery system to produce wide spectrum of products including biofuel, energy, and high-value chemicals from biological-based feedstocks through

Table 2.1 Differences between oil refinery and biorefinery.

	Oil refinery	**Biorefinery**
Feedstock	Petroleum-based Relatively homogeneous	Biological-based Heterogeneous (Cellulose, hemicellulose, lignin, protein, carbohydrate)
Building block/intermediate	Ethylene, propylene, toluene, xylene	Sugar (Glucose, xylose, fructose), fatty acid
Reaction	Chemical process: steam cracking, catalyst reforming	Combination chemical-biotechnology, Involve series of biochemical reaction to obtain building block, Fermentation
Chemical intermediate	Many	Ethanol, methane, furfural, hydroxymethyl, lactic acid, succinic acid, acetic acid, glycerol

a combination of different biomass transformation technologies [5, 6]. A biorefinery is a facility that integrates biomass conversion processes together with the equipment to manufacture products, similar to today's petroleum-based refinery in which chemicals, chemical products, and fuels are produced from crude oil. The products manufactured from the biorefinery process must be able to partially replace chemicals and fuels generated from the oil refinery.

Table 2.1 shows the comparisons between biorefinery and oil refinery processes in chemicals and biofuels production [7]. The major difference between these two refinery concepts is the nature of feedstock or raw materials. Crude oil rich with hydrocarbon and mixture of sulfur with low oxygen contents are typically used as feedstock for oil refinery. While biomass or agro-waste that consists of carbohydrate, protein, cellulose, hemicellulose, and lignin are the common chemical compounds in the biorefinery feedstock. The second aspect that can be observed from these two refinery concepts is the building blocks generated from the reaction of the feedstocks. Building blocks produced are vital as a precursor to the targeted final product formation. In oil refinery process, the main building blocks generated from chemical reaction of petroleum are ethylene, propylene, xylene, toluene, and other isomers. In contrast, organic carbon such as glucose, xylose, and other fatty acids are the main building blocks produced from biochemical reaction of biomass.

Biorefineries can be classified according to the basis of their key characteristics. There are several types of biorefineries such as whole crop biorefinery, lignocellulosic biorefinery, green biorefinery, two-platform biorefinery, and microalgal biorefinery [7, 8]. Most of the biorefinery classifications are based on the generalization on the types of feedstock this concept must fit, even though the system is introduced to achieve similar aim. It is expected that the classification of biorefinery to be more flexible as this concept will deal with various approaches in the biorefinery system [9]. For this chapter, three different biorefineries that support sustainable bioprocess system will be described: starch biorefinery, lignocellulosic biorefinery, and microalgal biorefinery.

2.2.1 Starch Biorefinery

Starch is one of the most abundant storage carbohydrates that can be extracted from wide ranges of agricultural raw materials including seeds, tubers, roots, and fruits. This storage carbohydrate polymer consists of glucose molecules that are linked together by glycosidic bond (alpha 1,4 and alpha 1,6). Most of the starch contain predominantly up to 20%–30% of amylose followed by amylopectin (Figure 2.2). Amylose is a water soluble polysaccharide that forms glucose subunit. While amylopectin is a highly branched polysaccharide consisting of alpha glucose units. This polymer is insoluble in water and it contributes approximately 70% of the starch.

Table 2.2 shows the proximate chemical compound distribution in the various types of starch crops. Genrally, starch-producing crops are a group of plants that carry out photosynthesis and accumulate starch as energy reserves. Examples of starch-producing crops are barley, wheat, potato, bean, and banana.

Starch biorefinery is considered as first generation biorefinery that is based on utilization of agricultural biomass and forestry products. The most common substrates used for the starch biorefinery are cassava and corn. Figure 2.3 shows the common schematic flow process of starch biorefinery. The first step is mechanical separation. At this initial stage, straw is separated from seed. The straw that consists of cellulose, hemicellulose, and lignin can be further processed for production of other valued-added products. While the seed fractions produced at the initial stage may be converted into starch or ground for meal or animal feed. The starch can be processed into other desired products such as sweetener,

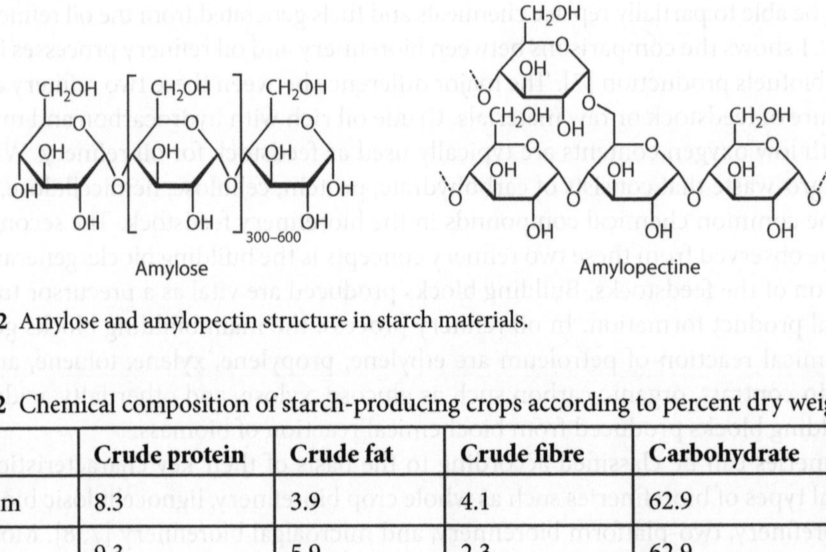

Figure 2.2 Amylose and amylopectin structure in starch materials.

Table 2.2 Chemical composition of starch-producing crops according to percent dry weight.

Cereal	Crude protein	Crude fat	Crude fibre	Carbohydrate	Ash
Sorghum	8.3	3.9	4.1	62.9	2.6
Oats	9.3	5.9	2.3	62.9	2.3
Barley	11.0	3.4	3.7	55.8	1.9
Wheat	10.6	1.9	1.0	69.7	1.4
Potato	2.19	0.25	0.3	24.29	1.2
Cassava	7.81	0.3	2.15	38	0.85

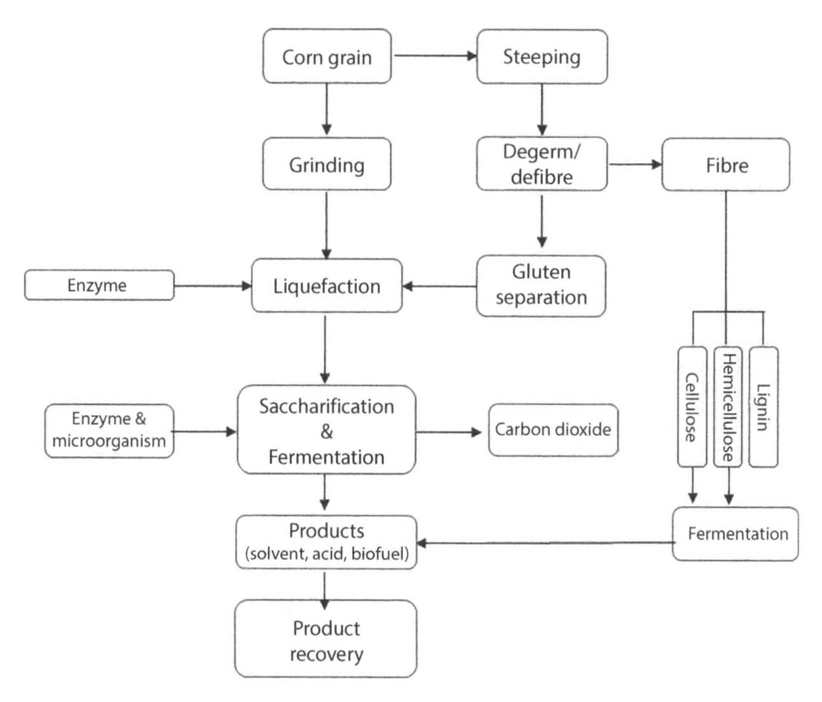

Figure 2.3 General schematic diagram for a starch biorefiney.

acid, alcohol, or biofuel *via* fermentation of other chemical reactions including plasticization, polymerization, and chemical modification.

There are several limitations and issues that have been address associated with the production of biofuel and chemicals in starch biorefinery. Among the most well-known issue is risk of an over consumption of food crops and utilization of land for starch crop production. The limitation of arable land for food crops production has become competative, which also lead to the deteriation of organic quality of the mineral content in soil making fist generation biorefinery not feasible. On the other hand, exhausive utilization of chemical fertilizer for food crop production has significant impact on the environmental sustainability. Hence, exploring new feedstock for biorefinery is crutial to improve the sustainable production of chemicals and fuel from renewable resource is feasible.

2.2.2 Lignocellulosic Biorefinery

Biomass is one of the most important renewable resources on earth. Biomass refers to any organic matters which are renewable or available on recurring basis. It can be food, wood, and crops or waste from plant materials and animals (Table 2.3). Besides, another type of biomass which is always neglected or forgotten is biomass from microorganisms such as bacteria, yeast, and fungi. Lignocellulosic biomass is predominantly derived from cell walls of plants in which cellulose, hemicellulose, and lignin are the three major chemical compositions with minor amount of protein, pectin, and minerals. The quantity of each chemical composition varies depending on the types of plant such as crops, grass, and wood (hardwood, softwood) and the age of the plant tissue [10, 11]. They are cross-linked and bound together *via* different types of bonding to form a very strong and complex structure. A multi-enzyme complex

Table 2.3 Different types of common lignocellulosic biomass.

No	Type of lignocellulosic biomass	Example
1	Woody	Saw dust, forest residues
2	Non-woody	Paddy straw, wheat straw, sugarcane bagasse, palm kernel cake, rice husk Corn stover, wheat bran
3	Microorganisms	Fungi Bacteria Yeast
4	Organic waste	Human sewage sludge Livestock wastes

of cellulase, hemicellulase, and ligninolytic is required to act cooperatively and sometimes synergistically to complete the degradation of lignocellulosic biomass [12–14].

Lignocellulosic biomass has been determined as the most crucial source for the production of wide variety of polymeric materials [11]. This is due to cellulose being one of the most examined biopolymer and has been studied for a long time [15, 16]. It is a major constituent of plant cell wall and is also found in microorganisms (fungi, bacteria, algae) [17]. It is an essential feedstock for many industries such as biofuels production, pulp and paper, wood products, etc. [15, 16]. Biomass conversion process to produce valuable products is designated as the biorefinery process.

Van Dyne *et al.* [18], Kamm and Kamm [19], and Fernando *et al.* [20] have described three types of biorefinery processes known as Phase I, II, and III biorefineries. These three phases are different in terms of number of feedstocks used to produce different number of products *via* different number of processes (Table 2.4) with Phase III being the most advanced/developed type of process. On top of that, lignocellulosic feedstock (LCF) biorefinery is one of the Phase III biorefinery systems actively being conducted in research and development [21].

Lignocellulosic feedstock biorefinery (LFB) system commonly uses "nature-dry" lignocellulosic biomass as feedstock. The feedstock can be bagasse, wood, stover straws, oil palm biomass, etc., which are mainly comprised of six carbon glucose polymers, hemicellulose (five carbon sugar polymers), and lignin (phenol polymer) [20, 21]. These three major chemical constitutes will be fractionated before converted into a variety of products (fuels, chemical products, adhesives, xylites, etc.) [19]. In general, the cleaned feedstock (biomass) will be fractionated into three major fractions *via* chemical (digestion) or biological (enzymatic hydrolysis) process in which they serve as precursors to many other products (Figure 2.4) [19].

Different lignocellulosic biomass requires different process conditions during production and they also need to be optimized for every single feedstock use. Cellulose and hemicellulose can be obtained *via* enzymatic hydrolysis of lignocellulosic biomass using cellulase and hemicellulolytic enzymes complex. These hydrolysis processes will convert cellulose and hemicellulose into their respective sugars and byproducts such as glucose, galactose, mannose and xylose, arabinose, furfural, etc. However, they can also be produced using chemical or heat treatment processes such as autohydrolysis (autoclaving), alkaline (caustic

Table 2.4 A general overview of three types of biorefinery processes.

Phase	Feedstock	Process	Major product
I	Single	Single	Single
	• Rape seed	• Transesterification	• Biodiesel
II	Single	Multiple	Multiple
	• Cereal grains	• Pre-treatment • Enzymatic • Fermentation • Chemical and/or biochemical catalysis	• Polymers • Vitamins • Amino acids • Polyols • Biofuels • Hydrocolloids
III	Multiple	Multiple	Multiple
	• Cereals (straw) • Paper and cellulosic municipal solid waste • Lignocellulosic biomass (reed, reed grass • Forest biomass (wood)	• Pre-treatment • Enzymatic • Fermentation • Grinding • Chemical processes • Biological processes	• Sugars (xylose, glucose) • Fuels (ethanol) • Organic acids (lactic acid) • Solvents (acetone, butanol) • Emulsifiers • Stabilizers • Furfural (furan resins) • Nylon • Natural binder

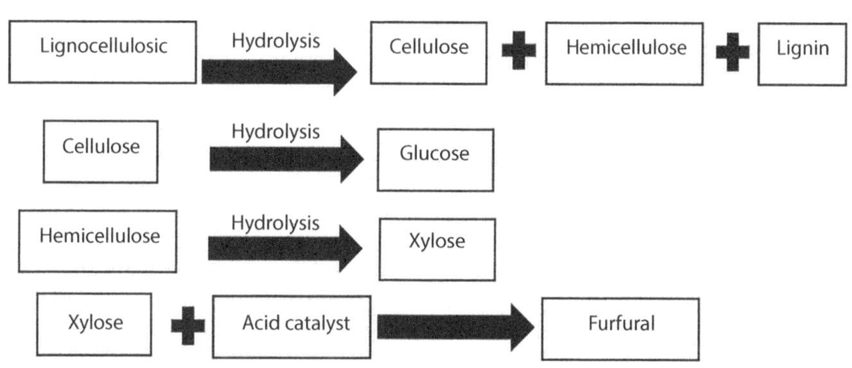

Figure 2.4 A general overview of the conversions that takes place in a LCF biorefinery.

soda), and sulfite (acidic, bisulfite, etc.) [20]. On the other hand, lignin component in lignocellulosic biomass can be broken down using lignin-modifying enzymes which mainly consist of lignin peroxidase, manganese peroxidase, and laccase.

Glucose which is the precursor to a variety of useful products such as ethanol, hydroxymethylfurfural, acetic acid, acetone, and other fermentation products is fractionated from cellulose hydrolysis. On the other hand, xylose which is one of the breakdown products of hemicellulose can be used to produce xylite and furfural (the building blocks for many useful products such as levulinic acid, lubricants, Nylon 6, fuel derivatives, cleaning agents, etc.) [19, 20]. However, lignin has lesser use when compared to cellulose and hemicellulose. So far, it is applied as fuel or binder. Many active researches have been conducted on the use of lignin for formation of more valuable products. Lignin can be pelletized by mixing with biomass to be used as solid fuel for easier transportation [21]. With the increase in lignin application, the overall LCF biorefinery process will be more competitive and value-added in producing useful industrial products such as cellulosic ethanol.

In conclusion, LCF biorefinery process has been successfully demonstrated to produce wide variety of useful industrial products such as biomaterials, chemicals, and energies/fuels. Nevertheless, it has to compete with existing petroleum refinery process which contributes more than 90% of industrial economies [22]. In order to maintain the competitiveness of LCF biorefinery, besides ensuring optimum yield of primary products, establishing value-adds to the byproducts generated are essential to lower the cost of overall process.

2.3 Microalgal Biorefinery

Microalga is a photosynthetic microorganism that is able to undergo photosynthesis by mitigating carbon dioxide (CO_2) from the atmosphere. The biomass produced from photosynthesis is also able to convert to various valuable products such as biofuel, bioethanol, and other wide range of chemicals. Microalga exhibits several interesting properties as a renewable material such as [23]:

i. Easy to grow
ii. Short maturation time
iii. Higher growth rate
iv. No need special attention or culture condition
v. Contains different biochemical components such as carbohydrate, protein, and lipid in biomass

Microalgal biorefinery can be defined as an approach to obtain full valorization of each raw component in the microalgae ranging from species selection, cultivation, harvesting, and lipid extraction [24]. Microalgal biorefinery was initially introduced by Khan, Ahmad [25] in which it was applied in biogas and biofuel industries. This concept has been widespread in various industrial sectors such as food, pharmaceutical, energy, chemical, and feed industries.

Nowadays, microalgal biorefinery has been explored for the synthesis of bio-based products [26]. In order to enhance the efficiency of by-products conversion, this concept has been identified as the most promising way to create biomass-based industry due to its potential in forming multiple products. The design of microalgal biorefineries should maximize outputs and profit from single algae biomass as a raw material source [27]. For example, Burton, Lyons [26] reported that glycerine is the potential co-product of biodiesel production through biorefinery process specifically transesterification of microalgae. The conversion of co-product with higher economic value is a value-add to the microalgal biofuel production. Typically, microalgal biorefinery technology has presented bottlenecks that are mainly associated with upstream processing (USP) and downstream processing (DSP).

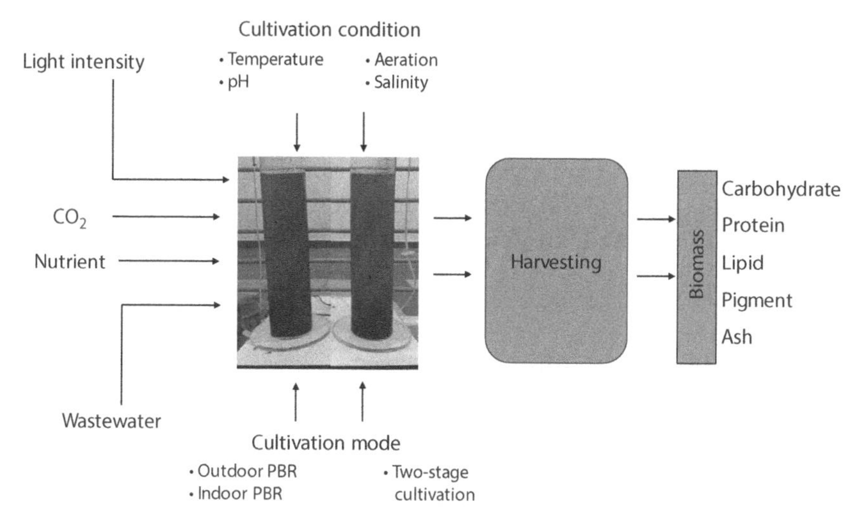

Figure 2.5 The upstream process involved in microalgae cultivation.

2.3.1 Upstream Processing

The efficiencies of the USP are highly dependent on the microalgae strain, illumination, and carbon dioxide (CO_2) supplied and nutrient sources' availability such as nitrogen and phosphorus (Figure 2.5). All these variables were proven by previous studies to be significant to the microalgal growth and biomass production [27]. It was reported by previous studies that different microalgae strains possess different percentage of biochemical components and the data are summarized in Table 2.5.

All of these chemicals are important to ensure the feasibility of DSP. The potential of high value-added chemicals from the microalgal biomass is totally dependent on the chemical components present in it. For example, the microalgal lipid can be converted into biodiesel, while microalgal carbohydrate can be converted into bioethanol or other chemicals *via* fermentation process. Whereas protein is important in food industry [26].

The formation of biochemical compounds in the microalgal biomass could be affected by several factors including mode of cultivation condition and parameters. It was reported that a higher growth rate was observed in *Chlorella* sp. microalgae when illuminated under artificial light condition [31]. This was due to the light source that stimulated photosynthesis

Table 2.5 Differences in biochemical compositions exhibited in different microalgae strain.

	Biochemical compositions			
Microalgae strain	**Carbohydrate**	**Protein**	**Lipid**	**References**
Amphiprora sp.	6.2	10	12	[28]
Chlorella	25.2	53.3	15.7	[29]
Neochloris oleoabundans	37.8	30.1	15.4	[29]
Phaeodactylum tricornutum	46.78	36.67	1.07	[30]

system in microalgae cell and subsequently enhanced its growth rate. Besides that, CO_2 was reported by the previous study to be attributed to microalgae growth rate by directly affecting the environment's pH condition for several species such as *Chlorella* sp., *Zygnema* sp., *Scenedesmus* sp., *Hizikia fusiforme*, *Chaetoceros* spp., *Microcystis aeruginosa*, *Botryococcus braunii*, *Nannochloropsis* sp., *Ulva rigida*, *Chlorococcum* sp., *Spirulina* sp., *Prorocentrum minimum*, and *Mytilus edulis* [32]. Lastly, nutrient sources are vital for microalgae growth production in which they can activate certain genes to enhance microalgae growth and biomass production [33].

2.3.2 Downstream Processing

The microalgal biomass produced in the USP stage is further subjected to DSP before the bio-based products are formed. The DSP process involves all unit processes that are used to extract and purify valuable products from microalgal biomass. Generally, DSP for microalgal biorefinery can be divided into three basic categories: lipid-extracted, thermochemical, and biochemical conversion [34, 35]. The choice of the conversion process is highly dependent on the type and quality of biomass feedstock, economic consideration, and desired end product [27].

2.3.2.1 Lipid-Extracted Microalgae

Lipid-extracted microalgae is a process involving separation of lipid from biomass using non-polar solvent (chloroform, hexane, and diethyl ether) and non-polar solvent (ethanol and methanol) [36, 37]. A non-polar solvent plays an important role by separating the homogeneous into two layers. This layer contains all the lipids, whereas the polar layer contains all the non-lipid components such as carbohydrate, protein, and pigment. The components of interest are then subjected to further processing which includes transesterification, thermochemical, and biochemical conversion.

2.3.2.1.1 Transesterification

Transesterification is a process involving the reaction of triglycerides with alcohol in the presence of catalyst either acid or base to produce fatty acid chain (biodiesel) and glycerol or water [34]. The previous study showed that *Chlorella protothecoides* is able to produce biodiesel of good quality using acidic catalyst under high temperature [38]. This is the most common method to produce biodiesel for commercial applications.

2.3.2.1.2 Thermochemical Conversion

Thermochemical conversion is a process that applies principle of thermal decomposition of organic compounds in biomass to yield fuel products [34]. This conversion process is highly attributed to operating conditions such as temperature, pressure, and duration [39]. Examples of thermochemical conversion process include thermochemical liquefaction, pyrolysis, direct combustion, and gasification (Figure 2.6).

For thermochemical liquefaction, the wet microalgal biomass will decompose into liquid fuel through hot compressed or sub-critical water [40]. Study on the liquefaction of microalgae biomass has been reported by Minowa, Yokoyama [41] who showed that

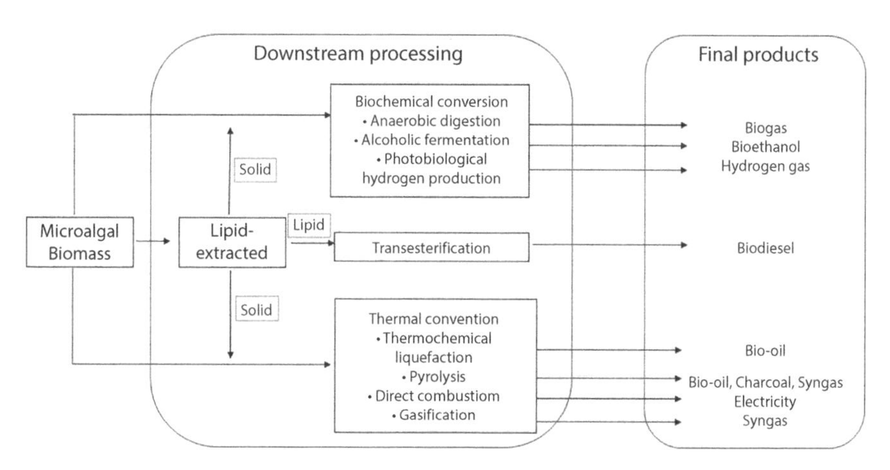

Figure 2.6 The downstream process involved in microalgae and its final products.

Dunaliella tertiolecta biomass with 78.4% moisture content was directly converted into bio-oil by thermochemical liquefaction with the experimental condition of 300°C and 10 MPa. This conversion technology is much simpler by which it can convert entire biomass feedstock within a shorter period with no other chemical additions. On the other hand, pyrolysis depicts the thermal decomposition of microalgal biomass through the cracking of bond into short chain molecules by heat. Then, these particular molecules are cooled into liquid rapidly in the absence of air or oxygen under medium to high temperature (350°C–700°C) [42]. Major products resulting from pyrolysis are bio-oil, charcoal, and syngas. Many studies have been conducted on pyrolysis of microalgal biomass, indicating that temperature, vapor residence time, heating rate, and types of catalysts are highly attributed to the efficiency of pyrolysis process [34]. Demirbaş [43] described the potential of algal biomass to produce high quality bio-oil through pyrolysis conversion, deemed to be applied in large-scale production to replace conventional petroleum fuel under optimum experimental conditions. Direct combustion involves the conversion of stored chemical energy in microalgal biomass into heat, mechanical power, or electricity in the presence of air. This system is only feasible when the moisture content in biomass is less than 50% [44]. Direct combustion normally occurs in furnace, boiler, or steam turbine with higher temperature (>800°C). However, this process is normally associated with pre-treatment steps such as drying, chopping, and grinding which results in additional costs and incurs non-feasibility in the overall process [44]. Later, this technique was improved using a combination of co-combustion of biomass in coal-fired power plants which can produce higher conversion efficiency [27]. Direct combustion of microalgal biomass was well described by Ross, Jones [45] who studied the combustion behavior on different strains of microalgae into bio-fuel *via* TGA analysis. Another technology to convert microalgae biomass into fuel product is gasification. Gasification involves the conversion of biomass (carbonaceous materials) into synthesis gas (syngas) by partial oxidation with a high temperature in the range of 800°C–900°C [44]. Syngas is contributed through a mixture of carbon monoxide, methane, and hydrogen, together with carbon dioxide and nitrogen [46]. According to the Liu *et al.* [47] who suggested that *Chlorella vulgeris* is able to convert into syngas in the presence of iron oxide (Fe2O3) and calcium oxide (CaO) in which they act as oxygen carriers and additives. Syngas can act as a platform to make a variety of fuels and chemicals

Table 2.6 Thermochemical conversion of microalgae and its product formation.

Thermochemical conversion	Microalgae	Temperature	Products	References
Thermochemical liquefaction	*Dunaliella tertiolecta*	300°C	Bio-oil	[41]
Pyrolysis	*Chlorella protothecoides*	502°C	Bio-oil	[43]
Direct combustion	*Fucus vesiculosus, Chorda filum, Laminaria digitata, Fucus serratus, Laminariahyperborea* and *Macrocystis pyrifera*	750°C–800°C	Biofuel	[45]
Gasification	*Chlorella vulgaris*	850°C	Syngas	[47]

intermediates [27]. The thermochemical conversion of microalgae can be summarized in Table 2.6.

2.3.2.1.3 Biochemical Conversion

Biochemical conversion of microalgal biomass is a process involving biological mechanism of microorganisms or enzymes to break down algae into liquid fuels by energy conversion. Examples of biochemical conversion include anaerobic digestion, alcoholic fermentation, and photobiological hydrogen production. In anaerobic digestion, bacteria will convert the microalgal biomass through the breaking down of biomass into biogas with a constituent of primary methane (CH_4), CO_2, and a trace amount of other gases such as hydrogen sulphide. The biogas produced by bacteria under anaerobic conditions has high energy contents up to 20–40% which is beneficial to the industrial applications [48]. The study was done by Dogan *et al.* [49] who showed that using *C. vulgeris* as a feedstock for mixed anaerobic culture has significant potential to produce biogas and biomethane. Besides that, previous studies also suggested that biogas produced from algal biomass is an attractive, alternative method for fuel production to replace the conventional petroleum-based fuel [34, 35]. As for alcoholic fermentation, the organic substrates which contain sugar, starch, or cellulose can be converted into bioethanol with the aid of yeast [48]. Hence, the choice of organic substrate is important to ensure the overall fermentation process becomes feasible. These organic substrates will undergo chemical changes due to the enzyme activities, secreted by the microorganisms. Then, the resulting simple sugar will be converted into bioethanol by yeast [34]. Based on a previous study done by Hirano, Ueda [50] who proved that using microalgae such as *C. vulgeris* for bioethanol production are a good source due to accumulating high starch content (37%) and up to 65% of ethanol-conversion efficiency. Below is the chemical equation for bioethanol production by yeast using algae as a raw material:

$$C_6H_{12}O_6 \Rightarrow 2C_2H_5OH + 2CO_2$$

Demirbaş [46] suggested that a purification process (distillation) is required to remove the unwanted water and impurities in the diluted ethanol product (10%–15% ethanol).

Table 2.7 Biochemical conversions of microalgae and its products formation.

Biochemical conversion	Microalgae	Temperature	Products	References
Anaerobic digestion	*C. vulgeris*	35°C	Biogas, Biomethane	[49]
Alcoholic fermentation	*C. vulgeris*	25°C	Bioethanol	[50]
Photobiological hydrogen production	*C. reinhardtii*	37°C	Hydrogen gas	[54]

Then, the concentrated ethanol (95% volume with single-step distillation process) is drawn off and condensed in liquid form for substitution petrol in cars [51]. Another conversion technology for biochemical conversion is photobiological hydrogen production. The photobiological hydrogen production is the process involved in algae, which involves conversion of water molecules into hydrogen ions (H^+) and oxygen under anaerobic conditions [52]. It can be explained by Ghirardi, Zhang [52] who showed that microalgae possess the necessary genetic, enzymatic, metabolic, and electron-transport machinery to photoproduce H_2 gas. There are two stages for this conversion process in algae. In the first stage, the algae are grown photosynthetically in normal conditions. Then, during the second stage, it involves sulfur deprivation for consistent hydrogen stimulation under anaerobic conditions [53]. A previous study by Hemschemeier and Melis [54] reported of an increase of the H_2 production in *Chlamydomonas reinhardtii* when the second stage of cultivation with the anaerobic conditions was established. This was due to the activation of the hydrogenase gene which controls the hydrogenase enzyme for the H_2 metabolism in microalgae cell [55]. Another study was done by Asada and Miyake [56] who had successfully expressed the hydrogenase gene in recombinant *Synechococcus* PCC7942 to produce hydrogen under anaerobic conditions. Hence, this production system is beneficial as it does not cause adverse environmental impact. Meanwhile, it could add value as a result of mass cultivation of green algae [53]. Table 2.7 summarized the products that can be produced from microalgal biomass *via* biochemical conversions.

2.4 Value Added Products

Development of biorefinery has provided key access to the production of wider range of products including food, feed, chemicals, biopolymer, and fuel using different types of feedstock. Currently, the most common products manufactured from biorefinery process are mainly focused on biofuel (bioethanol, biobutanol, biodiesel), biopolymer, and pharmaceutical products from waste (Figure 2.7).

2.4.1 Biofuel

Production of biofuel using renewable resource as feedstock *via* bioprocess has gained great attention globally. This is due to this feedstock exhibiting advantages such as more sustainable, environmentally friendly, and abundantly available with cheap price. Production of biofuels using renewable feedstock is classified into three different generations as shown

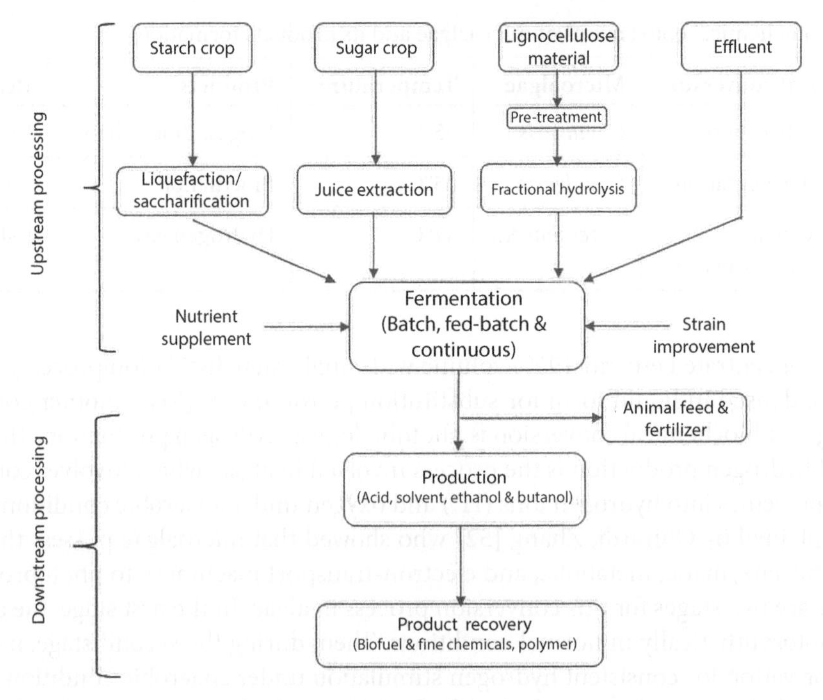

Figure 2.7 Concept of bio-based products from sustainable renewable feedstocks.

Table 2.8 Classification of biofuel generation.

Biofuel	Feedstock	Biofuel
First Generation	Food crops, starch, sugar animal fats, and vegetable oil	Bioalcohol (bioethanol, biobutanol, propanol) and Biodiesel
Second Generation	Non-food biomass, agro-waste, woody crops, lignocellulosic materials	Bioalcohol (bioethanol, biobutanol, propanol), Biogas, Bioether, Biohydrogen
Third Generation	Microalgae, macroalgae	Bioalcohol (bioethanol, biobutanol, propanol), Biogas, Bioether, Biohydrogen

in Table 2.8. The first generation biofuels are produced from agriculture crops such as animal fat, starchy crops, and vegetable oil. The most common biofuels produced using these feedstock are bioethanol and biodiesel. Bioethanol is produced from fermentation of starch materials from feedstock using yeast. While the production of biodiesel is *via* transesterification of fatty acid or lipid extracted from the crops. Several issues and limitations have been reported rising from the production of biofuels using food crops including food versus fuel and utilization of land for biofuel crops.

The second generation biofuels are a new approach that have been introduced to overcome limitations from the previous generation. They are biofuels produced from the conversion

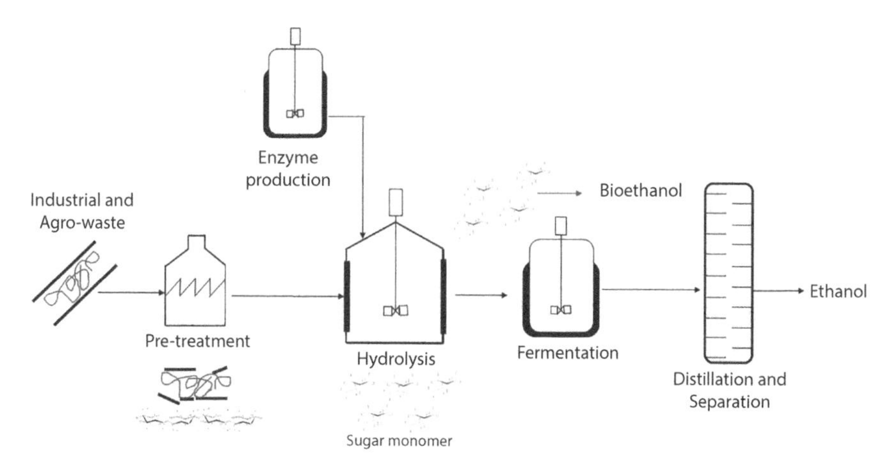

Figure 2.8 Schematic diagram biofuel production from industrial and agro-waste feedstock.

of lignocellulosic or biomass materials obtained from various industrial activities. Apart from that, biofuels produced from by-products or wastes generated from households can be considered in this category. Various types of biofuels can be generated from single feedstock through different technologies including bioethanol, biobutanol, hydrogen, methane, and biogas. To produce these biofuels from biomass and any lignocellulosic material, it has to undergo a series of processes, for example, (1) mechanical preparation, (2) pre-treatment, (3) hydrolysis, and (4) fermentation (Figure 2.8).

Pre-treatment is one of the most crucial steps involved in biofuel production from renewable resources. The main aim of pre-treatment process is to disrupt the biomass structure and provide better access for hydrolysis enzyme to attack the structure in the biomass.

Table 2.9 Common pre-treatment approaches used in biocoversion process.

Physical Methods	**Chemical/Psychialchemical**	**Biological**
Milling: Ball milling, Hammer milling, two-roll milling	Explosion: Steam, ammonia, CO_2 and SO_2	White-rot and Brown-rot fungi (enzyme involve including lignin peroxidase, manganese peroxidase, laccase and cellulase)
Irradiation: Gamma-ray irradiation, Electron beam irradiation, Microwave irradiation	Alkali and Acid: Acid: Sulphuric acid and Phosphoric acid Alkali: KOH and NaOH	Bioalcohol (bioethanol, biobutanol, propanol), Biogas, Bioether, Biohydrogen
Thermal Methods: Hydrothermal, High pressure steaming, Extrusion, Pyrolysis	Ionic liquid/Organosoly	

Selecting the most suitable method to pretreat the biomass is important to obtain maximum reducing sugar and final products. Different types of biomass such as hardwood, softwood, herbaceous plant, aquatic plant, agro-waste, and solid waste consist of different amounts of chemical compositions such as hemicellulose, cellulose, and lignin, which could influence the selection of the right pre-treatment method. Several pre-treatment methods have been introduced for biomass pre-treatment including mechanical, chemical, and biological treatment (Table 2.9).

Enzymatic hydrolysis steps involve biocatalytic reactions for reducing sugar production. Cellulase, hemicellulase, and other accessory enzymes such as peroxidase and laccase are common enzymes used to hydrolyze biomass structure for reducing production. These enzymes degrade cellulose by a synergistic action involving three different major classes of enzyme: (1) endo-glucanase (EGs) (EC3.2.1.4), which randomly cleaves internal bonds at amorphous sites that create new chain ends, and (2) cellobiohydrolases or also known as exoglucanase (CBHs) (EC 3.2.1.91), which cleaves two to four units from the ends of the exposed chains produced by endocellulase, resulting in tetrasaccharides or disaccharides, such as cellobiose (3) beta-glucosidase, also known as beta-glucoside glucohydrolase (EC 3.2.1.21), which hydrolyzes the exocellulase product into individual monosaccharides. The monomer sugar obtained from the hydrolysis process can be further subjected to final biological reaction. The final biocatalytic reaction process for biofuel production is fermentation. The fermentation mode for biofuel production is totally dependent on the final products and microorganisms used in this process. Due to complex biochemical reaction, fermentation using different microorganisms will produce different types of final products.

Current biofuels that have attracted many are third generation biofuels, derived from microalgae and macroalgae feedstocks. These organisms are cultivated and harvested for oil production. Typically, the algae biomass consists of three major chemical fractions which are protein, carbohydrate, and lipid. The oil extracted from the biomass may be converted into fuel such as biodiesel *via* transesterification process.

2.4.1.1 Bioethanol

Bioethanol is a biofuel commonly produced from biological conversion technology of renewable materials to partially replaced petrol. Typical production of bioethanol can be carried out through fermentation of sugar monomer using microorganisms as biocatalyst. Production of bioethanol can be carried out using different types of microorganisms. *Saccharomyces cerevisiae* is the most common microorganism used for bioethanol production. In this fermentation process, one mole of glucose is converted into two moles of ethanol, two moles carbon dioxide (CO_2), and two moles of ATP. Theoretically, 1 ton of sugar (glucose or fructose) yields 511 g of ethanol under anaerobic fermentation condition. Equation (2.1) shows the typical chemical formula for the bioethanol fermentation:

$$C_6H_{12}O_6 \rightarrow 2\ C_2H_5OH + 2\ CO. \tag{2.1}$$

Bioethanol production from renewable resources has gained more attention recently as it can also be produced from wide range of biomass. It was reported that the production of bioethanol from renewable biomass has more advantages which could reduce the accumulation of waste and add more value into the waste generated from various industries.

Numerous studies on bioethanol production from wide range of renewable resources including starch, agricultural residue, forest residue, and post-harvest residue have been reported previously [57, 58].

These lignocellulosic materials consist of three major components, namely, cellulose, hemicellulose, and lignin. To produce bioethanol from this feedstock, it has to involve a series of process to extract sugar prior to fermentation process. Study by Kassim *et al.* [60] on fermentation of empty fruit bunches (EFB) found that almost 90% of sugar obtained from combination of alkali-acid pretreated EFB was converted into bioethanol. Another promising investigation on bioethanol production has also been reported on the fermentation of pretreated rice straw residue [57]. The study reported that approximately 10 to 11 g/L of bioethanol was produced from the rice straw residue that had been treated with the combination of acid-assisted ultrasound treatment method. It was found that the pretreatment method applied prior to fermentation process could influence the bioethanol production. Study by Amani *et al.* [59] also showed that bioethanol can produce 7.23 g/L bioethanol corresponding to 0.12 g ethanol/g sugar from the palm frond hydrolysate.

Apart from agricultural waste, bioethanol production can be produced from food waste [60, 61]. Food wastes generated from household activities consist of complex biomass containing various substrates including carbohydrate, lipid, protein, and cellulose which can be a substrate for bioethanol production by microbial activity [62]. Study by Moon *et al.* [61, 65] reported a very high ethanol yield of 0.23 g/g of dry substrate that was achieved from fermentation of the food waste for 15 h. Similarly, higher ethanol yield of 0.43 g/g total solid was observed for the fermentation of food waste [60]. Further improvement on bioethanol production is currently being carried out, especially for fermentation at high solid concentration. Fermentation at high solid concentration could reduce bioethanol production due to the increase of viscosity in the fermentation system. A study indicated that applying separate liquefaction or saccharification process could reduce viscosity and result in higher bioethanol production [63]. Another study suggested that fermentation *via* fed-batch mode could enhance bioethanol production from the food waste. This is due to the fact that constant feeding of substrates will provide more substrates and maintain the viscosity in the fermentation system [64, 65].

2.4.1.2 Biobutanol

Biobutanol (n-C_4H_9OH) is one of the products formed *via* microbial fermentation using *Clostridium* sp. This gram-positive rod bacteria is an obligate anaerobe and is capable of producing spores in its growth cycle. This bacteria has the capability to degrade a wide range of polysaccharides and produce solvents, acids, and alcohols during fermentation process [66]. Recently, the potential utilization of biobutanol as liquid fuel has gained interest due to its chemical characteristics, which have higher energy contents compared to bioethanol, have lower volatility and can be mixed with gasoline without any engine modification [67].

Production of biobutanol by *Clostridium* sp. involves a very complex biochemical reaction. Figure 2.9 shows the typical biobutanol pathway from the fermentation of monomer sugar by *Clostridium* sp. During fermentation, production of complex products involves biphasic phase starting with acidogenesis, where acetic and butyric acid as primary metabolites are produced at the early stage of the process. The second stage is solventogenesis phase, which is achieved at an early stationary stage. At this stage, the acids produced are re-assimilated into ABE solvents. During the acidogenesis phase, organic acid is produced

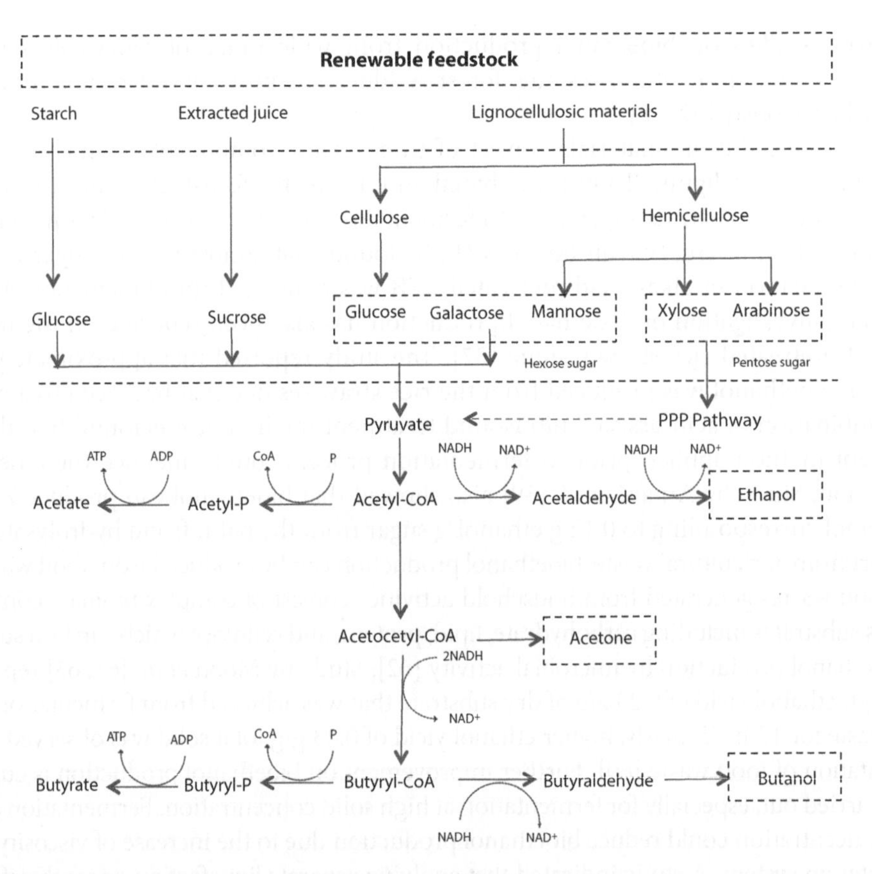

Figure 2.9 Mechanisms of butanol production figure modified from [66].

through an acetyl-CoA and butyryl-CoA pathway, while during solvent production, acetyl-CoA and butyryl-CoA function as the key intermediates for ethanol and butanol production. These pathways produce acetylaldehyde and butyraldehyde, respectively. Ethanol can be produced independently from acetone and butanol by *Clostridium* sp. under certain culture conditions (Figure 2.9) [68]. It is also understood that acetate/acetic acid, butyrate/butyric acid, and ethanol are called primary metabolites, whereas acetone and butanol are called secondary metabolites.

Currently, the production of biobutanol depends on sugar as the substrate and a large-scale production is not feasible due to high feedstock cost and availability. Thus, exploring new strategies for production of biobutanol using renewable resources have been initiated to ensure the bioprocess sustainability. Several cheap renewable resources are agro-waste, lignocellulosic biomass, softwood, and microalgae biomass (Table 2.10).

Development of biobutanol from fermentation using agro-waste is the common focus as an alternative approach to add value on the wastes generated from agricultural industry. Fermentation of wheat straw hydrolysate using *Clostridium beijerinckii* indicated that approximately 28.2 g/L of ABE with productivity and yield of 0.60 g/L and 0.42, respectively can be obtained [70]. Interestingly, their study found that higher butanol titer was achieved from the fermentation of the wheat straw hydrolysate as compared to the control used in the study. Jijosa-Valsero *et al.* [76] reported that approximately 2 g/L acetone, 8 g/L

Table 2.10 Biobutanol production from ABE fermentation of various type of renewable feedstocks.

Strain	Feedstock	Product	Yield	References
Clostridium acetobutylicum NCIM 2877	Orange peel	Butanol 19.5 g/L	-	[69]
Clostridium beijerinkii	Wheat straw	ABE 25 g/L	0.42	[70]
Clostridium acetobutylicum MTCC 481	Rice straw	ABE 15.84 g/L Butanol 12.17 g/L	-	[71]
Clostridium saccharoperbutylacetonicum N1-4	Rice bran	1.5 g/L	0.39	[72]
Clostridium acetobutylicum MTCC 481	Empty fruit bunches	ABE 4.45 g/L Butanol 2.75 g/L	0.18	[73]
Clostridium sp. BOH 3	Cassava starch	ABE 16 g/L Butanol 11 g/L	-	[74]
Clostridium beijerinckii	Pinewood	Butanol 11.6 g/L	-	[75]

butanol, and 1 g/L ethanol were produced from autohydrolysis of substrates. Studies also showed that autohydrolysis is the most effective pre-treatment to obtain maximum sugar production.

Similar observations have also been reported on the fermentation of sugarcane juice, molasses, and sugar from bagasse [77]. Approximately, 0.27-g butanol was obtained from a gram of sugar *via* fermentation of extracted sugarcane juice. Another study on the fermentation of oil palm trunk juice also showed similar findings where higher butanol concentration was obtained compared to the control used in the study [78]. The study found that approximately 12.25 g/L of acetone, 4.56 g/L butanol, and 4.28 g/L of ethanol were obtained from the fermentation of trunk juice without any additional fermentation nutrients. Higher butanol titer obtained from the agro-waste hydrolysate could be attributed to several factors, for instance, the presence of other sugar components and nutrients that could enhance production of solvents which could support the metabolic activity of microorganisms used.

Lignocellulosic biomass is another potential feedstock for biobutanol production which can be obtained abundantly throughout the year. Production of butanol using various lignocellulosic materials including EFB, kernel cake, bagasse, and switchgrass has been reported by various research groups [73, 79]. To produce biobutanol from lignocellulosic material, it has to go through several steps such as pre-treatment, hydrolysis, and fermentation, similar to bioethanol production.

Study by Ibrahim *et al.* [73] on the fermentation of EFB indicated that a total of 2.61 g/L of ABE was produced from the fermentation of untreated EFB. On the other hand, ABE fermentation batch from sugarcane bagasse pretreated with gamma-valerolactone indicated that a total of 14.26 g/L ABE consisted of 4.1 g/L acetone, 9.3 g/L butanol, and 0.86 g/L ethanol were produced. The study also indicated that this process generated low inhibitor content in which it could assist to enhance the solvent production during the fermentation process [80]. This study suggested that selection of suitable pre-treatment is important due to the formation of inhibitor that could affect fermentation performance. The presence of inhibitors such as phenolics, furans, as well as organic acids formed during the fermentation process could limit the production of acetone, butanol, and ethanol from the LCF.

2.4.1.3 Biodiesel

Biodiesel is another alternative fuel that is intended to be used as full replacement or partially blended with petroleum diesel fuel. This biofuel consists of long-chain alkyl (methyl, ethyl, or propyl) esters and can be produced from straight vegetable oil, animal fat or waste cooking oil. The biodiesel is made from conversion fat or lipid which is converted into fatty acid methyl ester (FAME) through transesterification process. The production of biodiesel can be performed *via* three basic routes:

- Base catalyzed transesterification of the oil
- Direct acid catalyzed transesterification of the oil
- Conversion of the fat to fatty acid followed with biodiesel production

Figure 2.10 shows a typical reaction process for production of methyl ester from oil. The transesterification reaction process is reversible in which alcohol should be added in excess to ensure the reaction occurs towards the right complete conversion. In this process, one ester is converted into another ester molecule in the presence of acid or base as catalyst. The formation of biodiesel involves three consequential processes in which monoacylglycerol (MAG) and diacylglycerol (DAG) are intermediates. At the end of the process, the major two products, namely, glycerol and biodiesel are formed.

Biodiesel production *via* chemical reactions has been reported to exhibit many disadvantages such as extensive chemical usage which is not environmentally friendly. Thus, several research groups have changed their direction to explore cleaner and more effective conversion methods to produce biodiesel from renewable resources [81]. Biodiesel production *via* enzymatic reactions using lipase as biocatalyst was introduced to overcome the drawbacks associated with the chemical reaction process. Conversion of lipid using lipase enzyme as biocatalyst was found to be less energy intensive and more robust [82]. Additionally, compared to chemical conversion reaction, enzyme-catalyzed reactions offer more quality products, mild reaction condition, and can be reused several times [83]. Numerous studies have proven that the lipase enzymatic reactions have high catalytic capacity. In this process, lipase enzyme (EC 3.1.1.3), also known as triacylglycerol acylhydrolase, catalyzes the ester bond in the long chain of triacylglycerol (TAGs) and produces free FAME and glycerol as by-product (Figure 2.11).

Study by Li *et al.* [81] reported that maximum conversion yield up to 95% could be obtained from the enzyme-catalyzed transesterification of rapeseed oil performed at

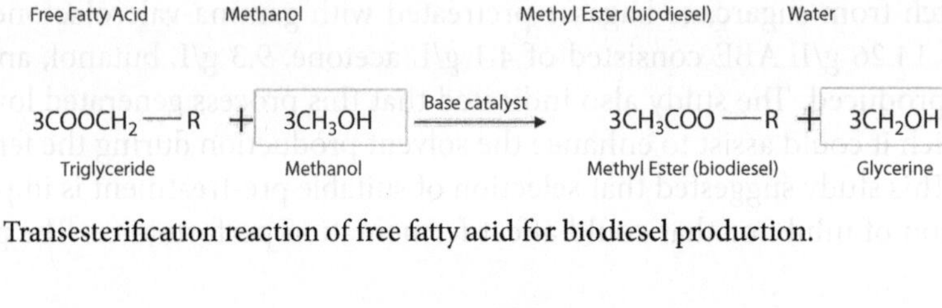

$$\underset{\text{Free Fatty Acid}}{COOH-R_1} + \underset{\text{Methanol}}{CH_3OH} \xleftrightarrow{\text{Acid catalyst}} \underset{\text{Methyl Ester (biodiesel)}}{CH_3COO-R_1} + \underset{\text{Water}}{H_2O}$$

$$\underset{\text{Triglyceride}}{3COOCH_2-R} + \underset{\text{Methanol}}{3CH_3OH} \xrightarrow{\text{Base catalyst}} \underset{\text{Methyl Ester (biodiesel)}}{3CH_3COO-R} + \underset{\text{Glycerine}}{3CH_2OH}$$

Figure 2.10 Transesterification reaction of free fatty acid for biodiesel production.

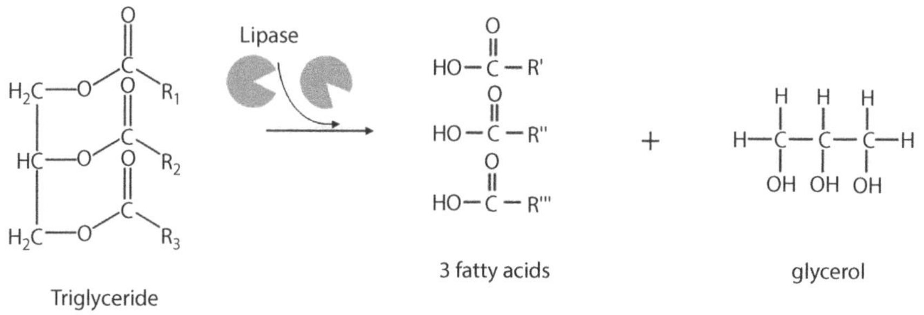

Figure 2.11 Lipase-catalyst reaction for FAME production.

optimum conditions where ratio of solvent to oil is 1:1, 3% lipase at 35°C and 130 rpm for 12 h of reaction. In another study, lipase-catalyzed transesterification of soybean oil showed that 97% of the extracted oil was converted into biodiesel, performed using 2% lipase enzyme at 40°C and 150 rpm, for 15 h of reaction [84].

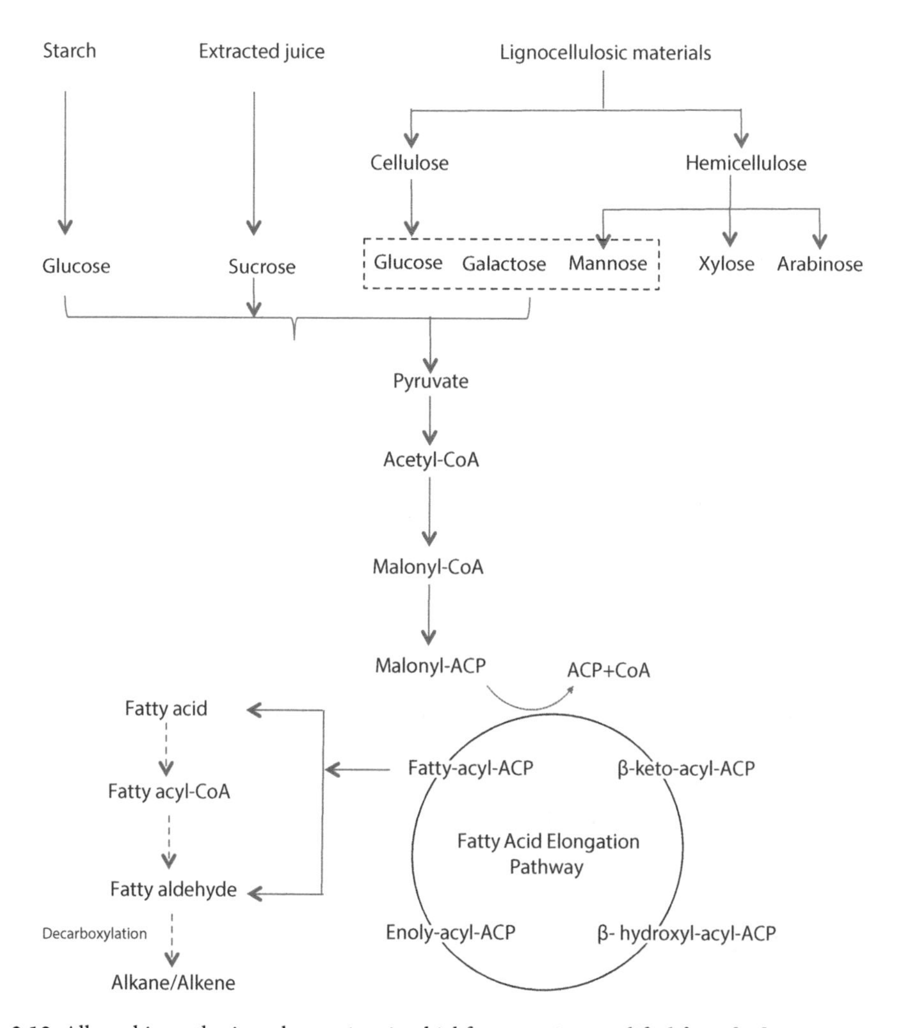

Figure 2.12 Alkane biosynthesis pathway *via* microbial fermentation modified from [85].

2.4.1.4 Short Alkane

Many microorganisms synthesize complex organic molecules particularly hydrocarbon chain in the cells *via de novo* anabolic reaction. Generally, the anabolic process can be categorized into two main stages, which involves biosynthesizing small molecule metabolite as precursor and then assembled to form a polymer as final metabolite during the growth phase (Figure 2.12). The biosynthesis of fatty acid can be composed of wide range polymer with various number of carbon atoms: short chain (<6); medium chain (>14); long chain (>27). Also, the fatty acid produced by microorganisms can be categorized into different degrees including unsaturation (saturated, monounsaturated, and polyunsaturated).

To date, biosynthesis of hydrocarbon such as short alkane, free fatty acid, and fatty alcohol as an alternative fuel has attracted a lot of interest. Hydrocarbons such as alkane and alkene are among the important chemicals used as transportation fuel and plastic. Study by Jimeneze-Diaz *et al.* [86] mentioned that short chain alkane produced from microbial fermentation has a potential to be used as an alternative jet fuel and other chemical applications. It was reported that alkane can be produced from fermentation using various types of microorganisms including *E. coli*, *Sarcina lutea* ATCC 533, cyanobacteria, and yeast.

Typically, these microorganisms produce alkane and alkene naturally in their cells for protection against harsh environmental conditions and threats [85, 87]. Biosynthesis of these hydrocarbon contents in different microorganisms varies over wide ranges. For instance, *Clostridium* sp. produces intracellular hydrocarbon from C_{11} to C_{35} that contains predominantly middle-chain alkane (C_{18}–C_{27}). While yeast has been reported to be able to produce hydrocarbon ranging from C_{10} to C_{34} which contains unsaturated and branched hydrocarbon [88, 89]. On the other hand, a study by Shakeel *et al.* [90] on 50 cyanobacteria strains found that most of the microorganisms are able to accumulate lipid droplets containing alkane predominantly pentadecane and heptadecane in the body cells. The study also found that the highest alkane accumulation was observed in two different cyanobacteria species; freshwater isolate *Oscillatoria* CCC305 and marine water isolate *Oscillatoria formosa* BDU 30603. In another study, Perumuna *et al.* [91] reported that cyanobacteria *Nostocpunctiforme* accumulated lipid droplets containing neutral lipid, and heptadecane as primary metabolite during stationary phase.

Production of short chain *via* fermentation process is still at early research and development stage. Utilization of expensive feedstock is one of the major limitations to ensure feasibility for alkane biosynthesis at commercial stage. Figure 2.13 shows the schematic diagram of production of alkane as bio jet fuel from the fermentation of sugar by microorganisms. In this process, the production of alkane for bio jet fuel can be carried out using cheap materials that can be found abundantly. The main advantages of this process are less chemical requirements during the process, low energy and temperature required to conduct fermentation process, and the conversion can be performed in a single reaction step in a single bioreactor [86, 90].

2.4.2 Polyhydroxyalkanoates (PHA)

Plastics have infiltrated almost every aspect of our modern lifestyle such as clothing, food containers, packaging, and car chassis. Plastics offer the benefits of being lightweight,

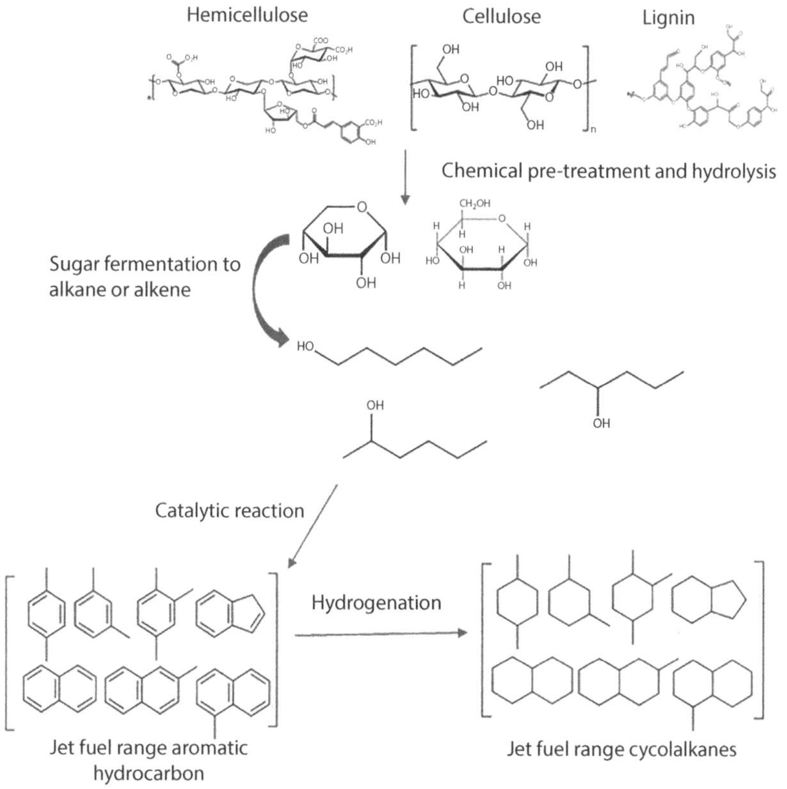

Figure 2.13 Schematic flow on the production of jet fuel from alkane derived from microorganisms' fermentation.

durable, and versatile and come with low manufacturing costs [92]. Currently, majority of plastics are derived from fossil fuel, particularly petroleum. This mineral resource is finite and slowly depleting. It is expected that world oil reserves will deplete in 2050 with current consumption trend. In addition, production and incineration of these synthetic polymers will generate GHG, which will trigger global warming. Oil-based plastic production had generated net heat approximately 0.38×10^{14} kWh from the year of 1939 to 2000 and increased up to 0.49×10^{14} kWh in the year of 2004 [93]. Moreover, improper disposal of these non-biodegradable materials will createan adverse effect on the environment, especially the fragile marine ecosystem. Dumping of plastic waste into water streams will cause flooding during rainy seasons. The toxic residual from decomposition of plastics will seep into soil, accumulate into plants and flow up the food chain. Millions of marine animals die from suffocation per year due to accidental consumption of plastic waste. Combustion of plastic waste will release toxic gas such as dioxin and furan into the atmosphere [94]. Recycling might be a better alternative; however, it is a labor-intensive process. The task of categorizing wide variety of plastics is a time-consuming process. This is made worse by the presence of additives such as pigment, coating, and fillers [95].

Over the past few decades, intensive researches have been conducted on the production of bioplastic from biological origins to reduce reliance on fossil fuel in plastic production. Bioplastic can be defined as a polymer produced from biological resources and can be completely decomposed into water and carbon dioxide by various environmental microorganisms [96, 97]. Typical commercially available bioplastics can be classified into threemain groups based on their origins: (1) chemical synthesis polymers [e.g., poly (lactic acid) (PLA), poly (butylene succinate) (PBS), and polyglycolic acid (PGA)], (2) naturally occurring polymers [e.g., starch, chitin and cellulose acetate], and (3) bacterial synthesis [e.g., Polyhydroxyalkanoates (PHAs), gellan gum, and curdlan] [98].

PHAs are a family of microbial polyester that are entirely produced by biological processes of bacterial fermentation. Polyhydroxybutyrate (PHB) is one of the most common types of PHAs. It had been widely used in biomedical industry due to its biocompatibility and non-toxic properties. Furthermore, copolymerization of 3-hydroxybutyrate monomer with other monomers such as 4-hydroxybutyrate, hydroxyvalerate, and hydroxyhexanoate to form medium chain length polymer with improved physical properties to suit its final application products has been explored.

Figure 2.14 shows the general flow chart of PHAs production. Currently, the main carbon sources used in a large-scale PHAs production are fructose, glucose, and plant oil. However, these carbon sources are expensive which increases the overall production cost of PHAs. The high cost of PHAs limits its economic viability. Thus, it is desirable to utilize low value carbon-rich agriculture and food by-products as alternative carbon sources. Figure 2.15 shows the proposed sustainable PHAs production from various cheap feedstocks as an alternative approach to overcome the limitation associated with current PHAs production.

European Union–funded projects Wheypol and Animpol have successfully produced PHAs at market competitive price utilizing surplus whey and animal fat as main carbon sources for fermentation [99]. The entire process of intracellular PHAs biosynthesis, starting from production of monomers and subsequent polymerization processes, occur mostly in Gram-negative bacterial cells, such as *Cupriavidus necator*, *Bulkhoderia cepacia*, and some Gram-positive bacterial cells, such as *Bacillus megaterium* [100, 101]. After the designated time period of fermentation, the cell biomass is harvested and freeze-dried. The freeze-dried cell biomass is then subjected to solvent extraction to obtain PHAs. Chloroform extraction

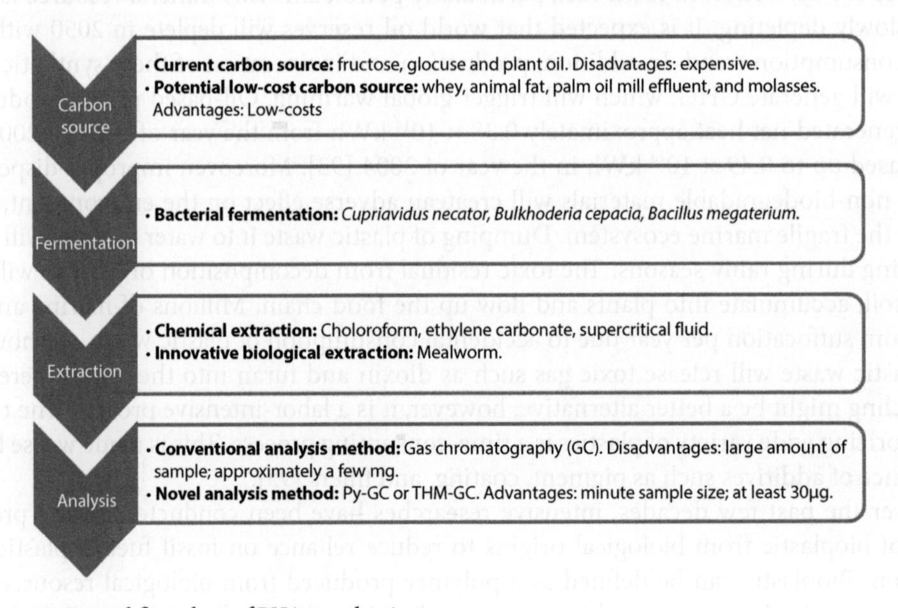

Figure 2.14 General flow chart of PHAs production.

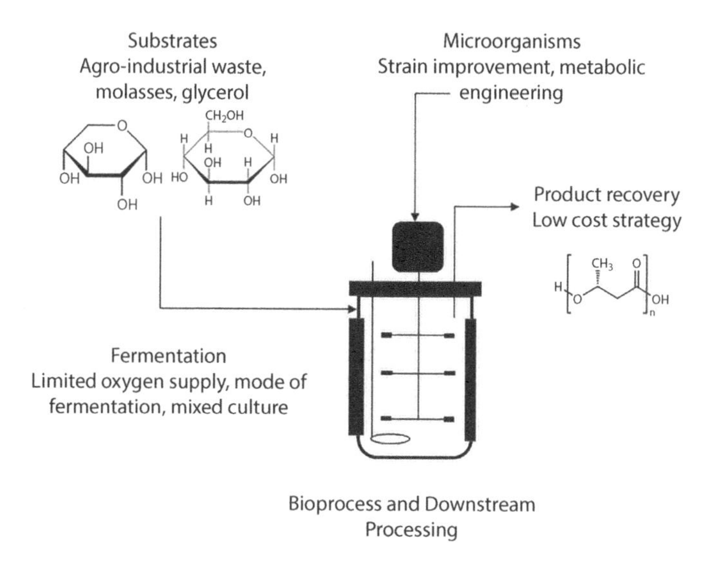

Figure 2.15 Production of bio-degradable polymer as environmentally friendly materials for sustainable bioprocess development.

is one of the most widely used methods in PHAs recovery. Recently, innovative biological extraction of PHAs is proposed to reduce the use of toxic chemicals, by feeding the mealworm, *Tenebrio molitor* with freeze-dried cells and the PHAs granules are excreted in the form of whitish feces [102]. In terms of PHAs determination analysis, gas chromatography [103] is the most extensive and widely used. However, PHAs determination *via* GC analysis involves tedious sample pre-treatment such as transesterification and solvent extraction, where relatively large amount of samples are required, approximately a few milligrams. For these reasons, the GC analysis is not always applied. On the other hand, pyrolysis-gas chromatography (Py-GC) also known as thermally assisted hydrolysis-gas chromatography (THM-GC) in the presence of strong organic alkali is proposed as an alternative rapid and direct analysis method, without using any tedious and/or cumbersome sample pre-treatment [104, 105]. The THM-GC method enables the analysis of PHAs in whole cells directly after fermentation process with sufficient accuracy and precision.

In conclusion, PHAs are being considered as a potential renewable alternative to some petrochemical plastics. This is due to the properties of PHAs resembling the properties of some commercially available plastics. Moreover, PHAs are completely biodegradable in nature. The bio-based and biodegradable nature of PHAs would have the long-term benefits of reducing plastic waste accumulation, global warming, pollution, and dependence on fossil fuels. The availability of cheap and renewable carbon feedstocks, preferably bio-based is expected to encourage PHAs' product prices to be more competitive compared to the conventional petroleum-derived plastics.

2.4.3 Bioactive Compounds From Food Waste Residues

Food wastes are the unwanted by-product derived from various food sectors involving food manufacturing, processing, packaging, and distributing that have not been utilized for any specific use. These waste residues are mainly produced at the entire food life cycle from processing industrial sites and other agro-food processing activities to retail and household levels. This can cause a major havoc at their disposal in municipal landfills due to high rate of biodegradability. Due to current intensification of these industrial activities and rapid grow in human population, it is therefore significant to identify and manage these waste residues in an optimized manner, thus protecting and benefiting nearby communities and environment. Notably, proper utilization of the waste material from food industries may have some positive impacts on food processing chains, thus enhancing economic performance and reducing disposal problems [106].

The search for phytochemicals and recovery of various bioactive compounds derived from various sources including plants and foodstuff has been recorded since ancient times. These bioactive compounds have been utilized for the treatment and prevention of many diseases. Due to diversity of classes and variations in their chemical properties, these compounds can interact with many biological components of cells, tissues, and organs, thus exerting some health promoting effects [107]. In recent times, there is an increasing interest in numerous food industries toward development of functional foods and nutraceutical products. This is due to the increased consumer interest to use food bioactive components as part of their preventive medicines and disease risk reduction. Therefore, the utilization of food wastes from industrial processing of agricultural and household is of great attention to enable effective management of trouble some wastes. This can be achieved through the recovery extraction and isolation of bioactive compounds and various classes of phytochemicals from food wastes [106].

It is well recognized that some areas of pharmaceutical, food, and nutrition have several overlapping interests. Bioactive compounds have been regarded as one of important ingredients in formulation of functional foods, nutraceutical, and food additives [108]. Specifically, fruits, vegetables waste, and their associated by-products are among the most abundant resources that are rich in many bioactive and therapeutic compounds. Some components of these wastes such as peels and seeds are reported to contain several phytochemicals including polyphenols, flavonoids, and carotenoids that have been shown to exhibit antioxidant activity and reduce the risk for developing certain types of cancer [109]. In comparison to the final products, higher levels of total phenolic and flavonoid contents were found in their by-products of seeds and peels. For instance, the peels and seeds of tomatoes contain numerous classes of bioactive phytochemicals including sterols, tocopherols, carotenes, terpenes, and polyphenols compared to that pulp of tomatoes [110]. Besides, higher levels of total flavonoids, phenolic compounds such as lycopene and ascorbic acid were found in tomatoes peels as compared to pulps and seeds [111]. Moreover, high amount of total phenolic compounds was also observed in peels from apples, pomegranates, peaches, pears, yellow, and white flesh nectarines, which was twice the amount of total phenolic compounds in these fruit pulps. The seeds of several fruits, such as avocados, mangos, longans, and jackfruits were reported to contain higher level of total phenolic compounds than that of the edible product [112]. Other classes of phenolic compounds such as mono, oligo, and polymeric pro-anthocyanidins can be found in grape seeds and skins [113]. The bioactive phytochemicals extracted from these waste by-products exhibit significant amount of antioxidant activities. These recovered compounds can be further utilized as the main ingredients in functional foods or can serve as food additives to extend their shelf life by preventing lipid peroxidation and protecting from oxidative damage [110].

Table 2.11 Bioactive phytochemicals in some food waste residues.

Source	Residue	Main bioactive phytochemical	References
Tomatoes	Peels and seeds	Sterols, tocopherols, carotenes, terpenes, polyphenols, and lycopene	[111]
Grape	Seeds and skins	Polyphenols, mono, oligo, and polymeric pro-anthocyanidins	[113]
Citrus fruits	Peels	Phenolics, flavonoids including flavones, flavanones, flavonols, lavones, anthocyanidins, and flavanols	[114]
Mango	Kernel and seeds	Polyphenols, anthocyanins, and carotenoids	[117, 118]
Coffee	Silver skin, spent waste and cherry husk	Polyphenols, tannins, gallic acid, and chlorogenic acid	[120, 121]

Citrus fruit is among the most abundant industrialized crops in the world. The large amount of processed citrus fruits may lead to increased quantities of citrus by-products. Citrus peels, remaining after juice extraction, constitutes more than 50% of the fruit mass. Of note, these generated by-products could be a major source of secondary metabolites including phenolics and flavonoids. Higher level of several classes of flavonoids content including flavones, flavanones, flavonols, lavones, anthocyanidins, and flavanols was found in citrus peels as compared to other parts [114]. These flavonoids were extensively investigated to confer several attributing properties in attenuating various metabolic diseases such as cancer, diabetes, inflammation, and hypertension [115].

Mango wastes contain a considerable number of phenolic compounds, total lipid, and certain amount of crude proteins with the presence of all essential amino acids. It was observed that mango seed extracts are rich in glutamate (13 g/100 g of protein), tannin, and vanillin [116]. Moreover, a significant number of phytochemicals including polyphenols, anthocyanins, and carotenoids were found in acetone extracts of mango peel [117]. In another comparative study, stronger free radical scavenger activities on 1,1-diphenyl-2-picrylhydrazyl and alkyl radicals were noticed in mango peel extracts as compared to edible parts of mango [118], suggesting that these values recovered compound isolated from mango waste could be one of the potential bioresource for extraction and isolation of bioactive compounds.

Coffee has been well recognized as the second most traded commodity after petroleum with the worldwide production of 150 million tons annually. The process of separating the bean from coffee (puling, washing, drying, curing, roasting, and brewing) generates an enormous amount of waste material by-products in the form of pulps, parchment husks, cherry husks, sliver skin, and spent waste. These wastes produced by industrial processing of coffee could be one of the major source for phytochemicals and bioactive ingredients towards development of nutraceutical and functional food industries. Among these coffee by-products, silver skin was found to contain highest phenolic contents, followed by spent waste and cherry husks. Interestingly, further evaluation of the antioxidant capability showed 70% of antioxidant activities in correlation to chlorogenic acid with 1.5 to 2.0 mmol trolox/100 g [119]. Another report on the chemical composition of exhausted coffee waste generated in a soluble coffee industry revealed the presence of polyphenols and tannins in relative amount of 6% and 4%, respectively [120]. Furthermore, the spent ground coffee collected from coffee

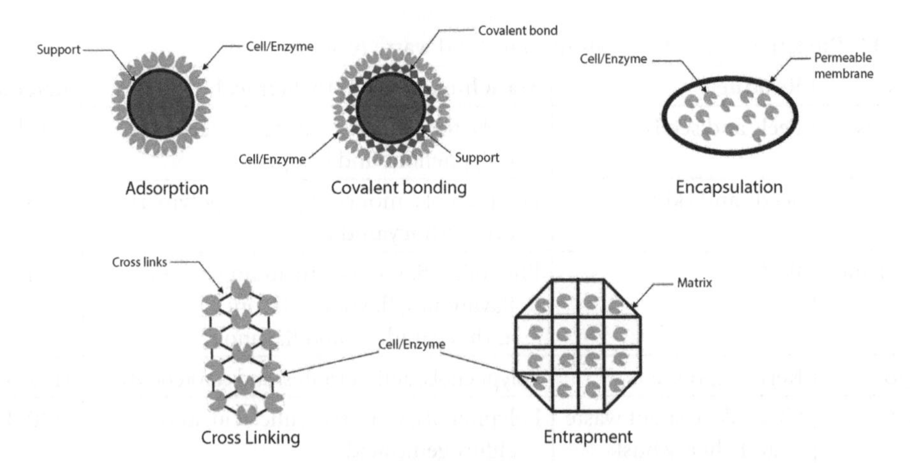

Figure 2.16 Schematic diagram of most common immobilization technique in industrial bioprocess.

bars contains higher level of gallic acid and phenolics compounds with strong antioxidant activities at the value of 0.10 mM Fe(II)/g [121]. Several bioactive phytochemicals found in the some food waste residues are briefly summarized in Table 2.11.

2.5 Novel Immobilize Carrier From Biowaste

The idea of using biowaste has become the prime choice in bioconversion process. Food waste, green waste, and brown waste are among major contributors to the biorefinery platforms. Food and green wastes that come from households, restaurants, food processing plants, and garden or park areas are mostly rich in nitrogen, while biodegradable wastes such as straw, sawdust, and woodchips are predominantly carbon-based, referred to as brown wastes [122–124]. These biowastes are mounting by years and holding huge potential to generate value-added products and resource recovery such as biofuels, energy, bioplastics, chemicals, and enzymes [125–127]. Not only are they beneficial for biorefinery as the main feedstocks, biowastes also attest to have the ability as a carrier in immobilization process. Immobilization approach has been proven by many researchers to attain more advantages over its free form, for instance, improvement in stability towards heat, solvents, or pH; ease of handling and separation; and applicable in continuous fixed-bed reactors. Most importantly, this type of carrier can be recovered and recycled. From existing literature, immobilization methods have been categorized into three groups: support binding (adsorption, ionic, or covalent attachment), entrapment, and cross-linking (carrier-free) (Figure 2.16) [128]. The availability of free sources of nitrogen, carbon, and other elements has made biowastes preferable cost-effective support materials for biorefinery system and indirectly offer an ecological alternative for waste management [129].

2.5.1 Waste Cassava Tuber Fiber

In Thailand, fermentation substrate for ethanol production is having cassava tuber (tapioca root) as a close rival to the molasses. High demand of molasses has caused its shortage, thus urging producers to shift to cassava tuber as an alternative substrate [130]. The waste products generated by using cassava tuber are drained out as insoluble cassava tuber fibers (CTF). These biowastes can be utilized as support materials for yeast cells during fermentation. In such effort, Kunthiphun *et al.* [131] have opted to use *Saccharomyces cerevisiae* to be immobilized on waste CTF *via* physical adsorption. A simple natural adsorption approach has allowed new active yeast cells to be adsorbed by the carrier when dead yeast cells are washed off over time [132, 133]. When comparing carriers among free yeast cell, waste CTF adsorbed, waste CTF adsorbed-calcium alginate entrapped, and calcium alginate entrapped, highest ethanol production was observed when the yeast cells were immobilized on waste CTF with 67.35 g/L ethanol produced. Ethanol produced using waste CTF as carrier was 59.9% higher compared to using only free yeast cells and this might be attributed to the free exogenous nitrogen supply by the waste CTF that eventually favored the increase of ethanol yield [131, 134].

2.5.2 Corn Silk

Corn silk, the female flower stigma of maize that is located on top of corn fruit with yellowish silky strand, is classified as waste by most food manufacturers. Usually, corn silk along with other parts of corn such as cob and husk are discarded as waste for their lack of utilization. Even so, corn silk offers many nutrients and is rich in dietary fiber where some parts of the world still consume it as traditional medicine [103, 135]. For that reason, another attempt to synthesize ethanol was reported using corn silk as a biocarrier for *Zymomonas mobilis*. Corn silk was made into biofilm matrix to hold the bacteria and provide a protective barrier against environmental stresses. This microbial biofilm is seen as a potential biocatalyst for its natural immobilization, stability, and high cell viability even after many cycles [136, 137]. Using rice straw hydrolysate as substrate, two different types of biofilm matrices made from polystyrene and corn silk were compared for their ethanol yields. Todhanakasem *et al.* [138] recorded about 0.51 ± 0.13 g ethanol produced/g glucose consumed when using corn silk as carrier, while microbial biofilm using polystyrene showed slightly lower yield of ethanol, with only 0.40 ± 0.15 g ethanol produced/g glucose consumed [138]. Morphologically fibrous corn silk has high porosity that favors cell immobilization with larger surface area compared to polystyrene. High surface area would facilitate the transmission of substrate and product formation between the support carrier and medium [139].

2.5.3 Sweet Sorghum Bagasse

Sweet sorghum is a sugar-producing cereal crop that makes efficient use of land due to its brief growth cycle and its sugar-containing juice is used as fermentation feedstock for ethanol production [140, 141]. Due to rising worldwide interest of using sweet sorghum as main feedstock for biofuels and bioproducts, the subsequent accumulation of bagasse has found its novel uses such as wood composites, paper pulp, animal feed, and immobilization carrier [142, 143]. A highly fibrous sweet sorghum bagasse has the ability to omit mass transfer limitation for its high porosity [144]. In 2014, Chang *et al.* succeeded in producing acetone-butanol-ethanol (ABE) fermentation by utilizing sweet sorghum bagasse as the immobilization carrier for *Clostridium acetobutylicum*. The yield and productivity of ABE using sweet sorghum bagasse as carrier were found to be 0.41 g/g and 0.37 g/L/h, respectively. These values reflected about 24% yield improvement and 68% higher productivity than those without carrier (free cells) [143].

2.5.4 Coconut Shell Activated Carbon

Once considered as an agricultural waste material by most of coconut millers, coconut shell has found its unique trait by transforming itself into coconut shell activated carbon (CSAC). To create its activated carbon form, the shell will have to undergo steam activation process and during this activation, heaps of pores are formed at the carbon surface that eventually maximize its total surface area. Not only does it have well-developed porosity, but when compared with other commercially available activated carbons, CSAC has become the most promising sorbent material due to its high mechanical strength, good abrasion resistance, and inherent granular structure; hence, it becomes the cheaper alternative [145, 146]. Study by Luthfi *et al.* [147] in 2017 has proven that CSAC as immobilization carrier was able to supersede other carriers (kieselguhr, exclay, vermiculite) in producing succinic acid from hydrolysate oil palm frond bagasse. BET surface area analysis has portrayed CSAC with the highest total surface area and porosity compared to other carriers. Wild-type bacteria cell *Actinobacillus succinogenes* 130Z was immobilized onto each support carrier and subjected to the conversion of the oil palm frond bagasse into succinic acid. *Actinobacillus succinogenes* immobilized on CSAC revealed maximum succinic acid yield with more than 20% improvement over the use of free cells and produced about 40.2 g/L succinic acid in batch-mode system. Even after five cycles of reactions, CSAC was still able to yield high succinic acid, with average of 44.1 g/L over 180 h [148].

2.5.5 Sugar Beet Pulp

After processing, the whole beet to extract sugar for refining as a foodstuff or to be used in fermentation, substantial amount of sugar beet pulp arises as waste. Yet, the resulting fibrous pulp waste is still rich in protein and carbohydrates. Thus, for some regions, it is being utilized as animal feed, while most other places dump the pulp in landfill [33, 34]. There were also works to produce bioethanol and propylene glycol from this sugar beet pulp [34, 35]. A recent study has emerged to prove the ability of sugar beet pulp as biocarrier for *Lactobacillus paracasei* NRRL-B 4564 in lactic acid fermentation using biowaste potato stillage and sugar beet molasses as the main feedstocks. Lactic acid was yielded higher with immobilized cells with average of 0.91 ± 0.03 g/g over three successive fermentation cycles, compared to free cells where only 0.80 ± 0.041 g/g of lactic acid was produced [149]. When compared with other carriers (gluten pellets, cellulosic materials, plastic-composites) used for bacterial immobilization in lactic acid production, beet pulp waste showed stronger adsorption stability due to its high cell viability [147, 150]. This might be explained by the cellular secretion of exopolysaccharides that is able to hold the cell attachment and improve the cell retention on the surface of biowaste [151]. Sugar beet pulp is being utilized as a carrier for *L. paracasei* NRRL-B 4564, assisting biorefinery process in lactic acid and high-quality animal feed production.

2.5.6 Eggshells

Eggshell is largely coming from household biowaste and also a by-product of poultry industry, generated in the million tons per day worldwide. Eggshells are rarely repurposed by public for any value-added products except as fertilizers. However, this biowaste is rich in calcium carbonate ($CaCO_3$) and other organic matters that makes it suitable as support material for heavy metals, enzymes, and microbials [152–154]. Das *et al.* [153] in 2015 had designated eggshell as a biocarrier for *Rhizopus oryzae* 1526 in submerged fermentation to produce fumaric acid and for comparison, a different immobilization carrier, muslin cloth, was also used. Both microbial biofilms formed using eggshell and muslin cloth showed higher fumaric acid production as compared to free cells as mediators. However, muslin cloth as carrier revealed slightly lower yield of fumaric acid, with 43.67 ± 0.32 g/L. While for the immobilized eggshell, higher fumaric acid was yielded at 47.22 ± 0.77 g/L with 56% increment from using free cells. Rich source of $CaCO_3$ in eggshell has indirectly facilitated the fumaric acid synthesis because the acid needs $CaCO_3$ as its neutralizing agent during fermentation [94].

Oftentimes, eco-friendly biorefinery process would involve microbial or enzyme as the catalyst to accelerate the process. Technically, these biocatalysts have their own size control which can be difficult, leading to diffusion limitation in the reaction medium, resulting in the drop in production. Moreover, their functional structures are neither resistant towards environmental stresses (heat, acidity, osmosis, CO_2) nor well-supportive of mass transfer during the reaction [94, 155]. Thus, to overcome these drawbacks, the immobilization technique becomes an alternative approach to enhance the biorefinery productivity.

2.6 Conclusion

Rapid development and increased demand of energy supply for various industrial activities have increased the dependance on fossil fuel resource for enery production. Exhaustive use of this resource has led to the reduction of the fossil fuel reserve and also has given the negative environmental impact due to the excess release of GHG during the process. Sustainable development of industrial activity is recognized as a promising approach to reduce the dependency on the current fossil fuel resources. Production of various chemicals and energy using waste residual generated from the industrial activity for commercial end *via* biorefinery strategy is found to be the attractive approach to solve the problems. In this context, with the use of wide technology tools, utilization of waste residual as renewable feedstock for chemicals and fuel production along with other multiple products such as pharmaceutical compound and carrier for other application can improve the sustainable of the industrial activity. On the other hand, conversion of waste residual using bioprocess technology either using microorganisms or its metabolites could provide environmentally friendly approach and can be performed under mild condition. Even though the concept showed a promising outcome, this sustainable biorefinery in bioprocess are still facing many challenges. To date, most of the technology is still unmatured and not all chemicals can be economically produced *via* bioconversion process. Improvement and further modification of current oil refinery technology availble are essential to ensure the feasibility of the process. Also, further R&D in the conversion technology is needed to be improve the process particularly to reduce reaction time and reduce the utilization of biocatalyst during the process. On the other hand, the bioproduct generated from biological process typically produced in relatively in low concentration. Thus, more research and exploration on the utilization of cell engineering technology, metabolic pathway engineering, and synthetic biology is essential to improve the conversion efficieny reduce the production cost.

References

1. Ok, Y.S., Lee, S.S., Jeon, W.T., Oh, S.E., Usman, A.R., Moon, D.H., Application of eggshell waste for the immobilization of cadmium and lead in a contaminated soil. *Environ. Geochem. Health*, 33, 1, 31–39, 2011.
2. Woinaroschy, A. and Lavric, V., Exploratory investigation of bioprocesses sustainability improvement by multicriteria-multilevel optimization. *Environ. Eng. Manage. J.*, 8, 3, 521–526, 2009.
3. Jiménez-González, C. and Woodley, J.M., Bioprocesses: Modeling needs for process evaluation and sustainability assessment. *Comput. Chem. Eng.*, 34, 7, 1009–1017, 2010.
4. Henderson, R.K., Jiménez-González, C., Preston, C., Constable, D.J., Woodley, J.M., Peer review original research: EHS & LCA assessment for 7-ACA synthesis A case study for comparing biocatalytic & chemical synthesis. *Ind. Biotechnol.*, 4, 2, 180–192, 2008.
5. Cherubini, F., The biorefinery concept: Using biomass instead of oil for producing energy and chemicals. *Energy Convers. Manage.*, 51, 7, 1412–1421, 2010.
6. Ferreira, A.F., Biorefinery concept, in: *Biorefineries: Targeting Energy, High Value Products and Waste Valorisation*, M. Rabaçal (Eds.), pp. 1–20, Springer International Publishing, Cham, Switzerland, 2017.
7. De Jong, E. and Jungmeier, G., Chapter 1 - Biorefinery concepts in comparison to petrochemical refineries, in: *Industrial Biorefineries & White Biotechnology*, A. Pandey (Eds.), pp. 3–33, Elsevier, Amsterdam, 2015.
8. Diep, N.Q., Sakanishi, K., Nakagoshi, N., Fujimoto, S., Minowa, T., Tran, X.D., Biorefinery: Concepts, current status, and development. *Int. J. Biomass Renew.*, 2, 1, 1–8, 2018.
9. Nizami, A., Rehan, M., Waqas, M., Naqvi, M., Ouda, O., Shahzad, K., Miandad, R., Khan, M.Z., Syamsiro, M., Ismail, I.M.I., Waste biorefineries: Enabling circular economies in developing countries. *Bioresour. Technol.*, 241, 1101–1117, 2017.
10. Izydorczyk, M.S. and Biliaderis, C.G., Cereal arabinoxylans: Advances in structure and physicochemical properties. *Carbohydr. Polym.*, 28, 1, 33–48, 1995.
11. Ebringerová, A., Structural diversity and application potential of hemicelluloses. *Macromol. Symp.*, 232, 1, 1–12, 2005.
12. Lee, Y.H., Fan, L.T., Fan, L.S., Kinetics of hydrolysis of insoluble cellulose by cellulase, in: *Advances in Biochemical Engineering*, A. Fiechter (Ed.), pp. 131–168, Berlin, Heidelberg, 1980.
13. Eveleigh, D.E., Cellulase: A perspective. Philosophical transactions of the royal society of London. *Series A Math. Phys. Sci.*, 321, 1561, 435–447, 1987.
14. Shallom, D. and Shoham, Y., Microbial hemicellulases. *Curr. Opin. Microbiol.*, 6, 3, 219–228, 2003.
15. Atalla, R.H., Brady, J.W., Matthews, J.M., Shi-You, D. and Himmel, M.H., Structures of plant cell wall celluloses. In: *Biomass Recalcitrance Deconstructing Plant Cell Wall Bioenergy*, Himmel, M.E. (Eds.), pp. 188–212, Wiley-Blackwell Publishing, Chichester, UK, 2008.
16. Himmel, M.E., Ding, S.Y., Johnson, D.K., Adney, W.S., Nimlos, M.R., Brady, J.W., Foust, T.D., Biomass Recalcitrance: Engineering plants and enzymes for biofuels production. *Science*, 315, 5813, 804–807, 2007.

17. Desvaux, M., *Clostridium cellulolyticum*: Model organism of mesophilic cellulolytic clostridia. *FEMS Microbiol. Rev.*, 29, 4, 741–764, 2005.
18. Van-Dyne, D.L., Blasé, M.G., Clements, L.D., A strategy for returning agriculture and rural america to long-term full employment using biomass refinery, in: *Perspectives on New Crops and New Uses*, J. Janick (Ed.), pp. 114–123, ASHS Press, Alexandria, 1999.
19. Kamm, B. and Kamm, M., Principles of biorefineries. *Appl. Microbiol. Biotechnol.*, 64, 2, 137–145, 2004.
20. Fernando, S., Adhikari, S., Chandrapal, C., Murali, N., Biorefineries: Current status, challenges, and future direction. *Energy Fuels*, 20, 4, 1727–1737, 2006.
21. Clark, J.H. and Deswarte, F.E., The biorefinery concept–An integrated approach, in: *Introduction to Chemicals from Biomass*, C.V. Stevens (Ed.), pp. 1–20, John Wiley & Sons, New York City, 2008.
22. Laird, T., Industrial organic chemicals, in: *Organic Process Research & Development*, H.A. Wittcoff and B.G. Reuben (Eds.), pp. 183–184, Wiley-Interscience, New York, 1996.
23. Mata, T.M., Martins, A.A., Caetano, N.S., Microalgae for biodiesel production and other applications: A review. *Renewable Sustainable Energy Rev.*, 14, 1, 217–232, 2010.
24. Roux, J.M., Lamotte, H., Achard, J.L., An overview of microalgae lipid extraction in a biorefinery framework. *Energy Procedia*, 112, 680–688, 2017.
25. Khan, S., Algal biorefinery: A road towards energy independence and sustainable future. *Int. Rev. Chem. Eng.*, 2, 1, 63–68, 2010.
26. Zhu, L., Biorefinery as a promising approach to promote microalgae industry: An innovative framework. *Renewable Sustainable Energy Rev.*, 41, 1376–1384, 2015.
27. Chew, K.W., Yap, J.Y., Show, P.L., Suan, N.H., Juan, J.C., Ling, T.C., Lee, D.J., Chang, J.S., Microalgae biorefinery: High value products perspectives. *Bioresour. Technol.*, 229, 53–62, 2017.
28. Gonzalez-Delgado, A.D., Martinez, J.B.G., Peralta-Ruiz, Y.Y., Cell disruption and lipid extraction from microalgae *Amphiprora* sp. using acid hydrolysis-solvent extraction route. *Contemp. Eng. Sci.*, 10, 841–849, 2017.
29. Tibbetts, S.M., Milley, J.E., Lall, S.P., Chemical composition and nutritional properties of freshwater and marine microalgal biomass cultured in photobioreactors. *J. Appl. Phycol.*, 27, 3, 1109–1119, 2015.
30. German-Báez, L., Valdez-Flores, M., Félix-Medina, J., Norzagaray-Valenzuela, C., Santos-Ballardo, D., Reyes-Moreno, C., Shelton, L.M., Valdez-Ortiz, A., Chemical composition and physicochemical properties of *Phaeodactylum tricornutum* microalgal residual biomass. *Food Sci. Technol. Int.*, 23, 8, 681–689, 2017.
31. Hsia, S.Y. and Yang, S.K., Enhancing algal growth by stimulation with LED lighting and ultrasound. *J. Nanomater.*, 16, 1, 222, 2015.
32. Singh, S.P. and Singh, P., Effect of CO_2 concentration on algal growth: A review. *Renewable Sustainable Energy Rev.*, 38, 172–179, 2014.
33. Fan, J., Cui, Y., Wan, M., Wang, W., Li, Y., Lipid accumulation and biosynthesis genes response of the oleaginous *Chlorella pyrenoidosa* under three nutrition stressors. *Biotechnol. Biofuels*, 7, 1, 17–17, 2014.
34. Suganya, T., Varman, M., Masjuki, H., Renganathan, S., Macroalgae and microalgae as a potential source for commercial applications along with biofuels production: A biorefinery approach. *Renewable Sustainable Energy Rev.*, 55, 909–941, 2016.
35. Naik, S.N., Goud, V.V., Rout, P.K., Dalai, A.K., Production of first and second generation biofuels: A comprehensive review. *Renewable Sustainable Energy Rev.*, 14, 2, 578–597, 2010.
36. Dong, T., Knoshaug, E.P., Pienkos, P.T., Laurens, L.M., Lipid recovery from wet oleaginous microbial biomass for biofuel production: A critical review. *Appl. Energy*, 177, 879–895, 2016.

37. Bligh, E.G. and Dyer, W.J., A rapid method of total lipid extraction and purification. *Can. J. Biochem. Physiol.*, 37, 8, 911–917, 1959.
38. Miao, X. and Wu, Q., Biodiesel production from heterotrophic microalgal oil. *Bioresour. Technol.*, 97, 6, 841–846, 2006.
39. Chen, W.H., Lin, B.J., Huang, M.Y., Chang, J.S., Thermochemical conversion of microalgal biomass into biofuels: A review. *Bioresour. Technol.*, 184, 314–327, 2015.
40. Barreiro, D.L., Prins, W., Ronsse, F., Brilman, W., Hydrothermal liquefaction (HTL) of microalgae for biofuel production: State of the art review and future prospects. *Biomass Bioenergy*, 53, 113–127, 2013.
41. Minowa, T., Yokoyama, S.Y., Kishimoto, M., Okakura, T., Oil production from algal cells of *Dunaliella tertiolecta* by direct thermochemical liquefaction. *Fuel*, 74, 12, 1735–1738, 1995.
42. Zhang, Q., Chang, J., Wang, T., Xu, Y., Review of biomass pyrolysis oil properties and upgrading research. *Energy Convers. Manage.*, 48, 1, 87–92, 2007.
43. Demirbaş, A., Oily products from mosses and algae *via* pyrolysis. *Energy Sources Part A*, 28, 10, 933–940, 2006.
44. Goyal, H.B., Seal, D., Saxena, R.C., Bio-fuels from thermochemical conversion of renewable resources: A review. *Renewable Sustainable Energy Rev.*, 12, 2, 504–517, 2008.
45. Ross, A., Jones, J., Kubacki, M., Bridgeman, T., Classification of macroalgae as fuel and its thermochemical behaviour. *Bioresour. Technol.*, 99, 14, 6494–6504, 2008.
46. Demirbaş, A., Biomass resource facilities and biomass conversion processing for fuels and chemicals. *Energy Convers. Manage.*, 42, 11, 1357–1378, 2001.
47. Liu, G., Liao, Y., Wu, Y., Ma, X., Synthesis gas production from microalgae gasification in the presence of Fe_2O_3 oxygen carrier and CaO additive. *Appl. Energy*, 212, 955–965, 2018.
48. McKendry, P., Energy production from biomass (part 2): Conversion technologies. *Bioresour. Technol.*, 83, 1, 47–54, 2002.
49. Doğan-Subaşı, E. and Demirer, G.N., Anaerobic digestion of microalgal (*Chlorella vulgaris*) biomass as a source of biogas and biofertilizer. *Environ. Prog. Sustainable Energy*, 35, 4, 936–941, 2016.
50. Hirano, A., Ueda, R., Hirayama, S., Ogushi, Y., CO_2 fixation and ethanol production with microalgal photosynthesis and intracellular anaerobic fermentation. *Energy*, 22, 2, 137–142, 1997.
51. McKendry, P., Energy production from biomass (part 1): Overview of biomass. *Bioresour. Technol.*, 83, 1, 37–46, 2002.
52. Ghirardi, M.L., Zhang, L., Lee, J.W., Flynn, T., Seibert, M., Greenbaum, E., Melis, A., Microalgae: A green source of renewable H_2. *Trends Biotechnol.*, 18, 12, 506–511, 2000.
53. Melis, A. and Happe, T., Hydrogen production. green algae as a source of energy. *Plant Physiol.*, 127, 3, 740, 2001.
54. Hemschemeier, A., Melis, A., Happe, T., Analytical approaches to photobiological hydrogen production in unicellular green algae. *Photosynth. Res.*, 102, 2, 523–540, 2009.
55. Winkler, M., Hemschemeier, A., Gotor, C., Melis, A., Happe, T., [Fe]-hydrogenases in green algae: Photo-fermentation and hydrogen evolution under sulfur deprivation. *Int. J. Hydrogen Energy*, 27, 11, 1431–1439, 2002.
56. Asada, Y. and Miyake, J., Photobiological hydrogen production. *J. Biosci. Bioeng.*, 88, 1, 1–6, 1999.
57. Belal, E.B., Bioethanol production from rice straw residues. *Braz. J. Microbiol.*, 44, 225–234, 2013.
58. Takano, M. and Hoshino, K., Bioethanol production from rice straw by simultaneous saccharification and fermentation with statistical optimized cellulase cocktail and fermenting fungus. *Bioresour. Bioprocess.*, 5, 1, 16, 2018.

59. Amani, A.H.F., Toh, S.M., Tan, J.S., Lee, C.K., The efficiency of using oil palm frond hydrolysate from enzymatic hydrolysis in bioethanol production. *Waste Biomass Valorization*, 9, 4, 539–548, 2018.
60. Kim, J.H., Lee, J.C., Pak, D., Feasibility of producing ethanol from food waste. *Waste Manage.*, 31, 9, 2121–2125, 2011.
61. Moon, H.C., Song, I.S., Kim, J.C., Shirai, Y., Lee, D.H., Kim, J.K., Chung, S.O., Kim, D.H., Oh, K.K., Cho, Y.S., Enzymatic hydrolysis of food waste and ethanol fermentation. *Int. J. Energy Res.*, 33, 2, 164–172, 2009.
62. Alamanou, D.G., Malamis, D., Mamma, D., Kekos, D., Bioethanol from dried household food waste applying non-isothermal simultaneous saccharification and fermentation at high substrate concentration. *Waste Biomass Valorization*, 6, 3, 353–361, 2015.
63. Matsakas, L., Kekos, D., Loizidou, M., Christakopoulos, P., Utilization of household food waste for the production of ethanol at high dry material content. *Biotechnol. Biofuels*, 7, 1, 4, 2014.
64. Chang, Y.H., Chang, K.S., Huang, C.W., Hsu, C.L., Jang, H.D., Comparison of batch and fed-batch fermentations using corncob hydrolysate for bioethanol production. *Fuel*, 97, 166–173, 2012.
65. Laopaiboon, L., Thanonkeo, P., Jaisil, P., Laopaiboon, P., Ethanol production from sweet sorghum juice in batch and fed-batch fermentations by *Saccharomyces cerevisiae. World J. Microbiol. Biotechnol.*, 23, 10, 1497–1501, 2007.
66. Gheshlaghi, R., Scharer, J., Moo-Young, M., Chou, C., Metabolic pathways of clostridia for producing butanol. *Biotechnol. Adv.*, 27, 6, 764–781, 2009.
67. Khamaiseh, E., Hamid, A.A., Yusoff, W.M.W., Kalil, M.S., Effect of some environmental parameters on biobutanol production by *Clostridium acetobutylicum* NCIMB 13357 in date fruit medium. *Pak. J. Biol. Sci.*, 16, 20, 1145–1151, 2013.
68. Kumar, M. and Gayen, K., Developments in biobutanol production: New insights. *Appl. Energy*, 88, 6, 1999–2012, 2011.
69. Joshi, S., Waghmare, J., Sonawane, K., Waghmare, S., Bio-ethanol and bio-butanol production from orange peel waste. *Biofuels*, 6, 1–2, 55–61, 2015.
70. Qureshi, N., Saha, B.C., Cotta, M.A., Butanol production from wheat straw hydrolysate using *Clostridium beijerinckii. Bioprocess Biosyst. Eng.*, 30, 6, 419–427, 2007.
71. Ranjan, A., Mayank, R., Moholkar, V.S., Process optimization for butanol production from developed rice straw hydrolysate using *Clostridium acetobutylicum* MTCC 481 strain. *Biomass Convers. Biorefin.*, 3, 2, 143–155, 2013.
72. Dada, O., Kalil, M.S., Yusoff, W.M.W., Effects of inoculum and substrate concentrations in anaerobic fermentation of treated rice bran to acetone, butanol and ethanol. *Bacteriol. J.*, 2, 4, 79–89, 2012.
73. Ibrahim, M.F., Abd-Aziz, S., Razak, M.N.A., Phang, L.Y., Hassan, M.A., Oil palm empty fruit bunch as alternative substrate for acetone–butanol–ethanol production by *Clostridium butyricum* EB6. *Appl. Biochem. Biotechnol.*, 166, 7, 1615–1625, 2012.
74. Li, T., Yan, Y., He, J., Enhanced direct fermentation of cassava to butanol by *Clostridium* species strain BOH3 in cofactor-mediated medium. *Biotechnol. Biofuels*, 8, 1, 166, 2015.
75. Nanda, S., Dalai, A.K., Kozinski, J.A., Butanol and ethanol production from lignocellulosic feedstock: Biomass pre-treatment and bioconversion. *Energy Sci. Eng.*, 2, 3, 138–148, 2014.
76. Hijosa-Valsero, M., Paniagua-García, A.I., Díez-Antolínez, R., Industrial potato peel as a feedstock for biobutanol production. *New Biotechnol.*, 46, 54–60, 2018.
77. Kim, K. and Day, D.F., Butanol production from sugarcane juice. *Proc. Int. Soc. Sugar Cane Technol.*, 27, 1–7, 2010.
78. Norhazimah, A.H. and Faizal, C.K.M., Kinetic parameters for bioethanol production from oil palm trunk juice. *Int. J. Chem. Mol. Eng.*, 8, 4, 357–359, 2014.

79. Shukor, H., Al-Shorgani, N.K.N., Abdeshahian, P., Hamid, A.A., Anuar, N., Rahman, N.A., Isa, M.H.B.M., Kalil, M.S., Biobutanol production from palm kernel cake (PKC) using *Clostridium saccharoperbutylacetonicum* N1-4 in batch culture fermentation. *BioResources*, 9, 3, 2014.
80. Kong, X., Xu, H., Wu, H., Wang, C., He, A., Ma, J., Ren, X., Jia, H., Wei, C., Jiang, M., Biobutanol production from sugarcane bagasse hydrolysate generated with the assistance of gamma-valerolactone. *Process Biochem.*, 51, 10, 1538–1543, 2016.
81. Li, L., Du, W., Liu, D., Wang, L., Li, Z., Lipase-catalyzed transesterification of rapeseed oils for biodiesel production with a novel organic solvent as the reaction medium. *J. Mol. Catal. B: Enzym.*, 43, 1, 58–62, 2006.
82. Amini, Z., Ilham, Z., Ong, H.C., Mazaheri, H., Chen, W.H., State of the art and prospective of lipase-catalyzed transesterification reaction for biodiesel production. *Energy Convers. Manage.*, 141, 339–353, 2017.
83. Yücel, S., Terzioğlu, P., Özçimen, D., Chapter 8: Lipase applications in biodiesel production, in: *IntechOpen*, Z. Fang (Ed.), pp. 209–250, Shard, London, 2012.
84. Zheng, Y., Quan, J., Ning, X., Zhu, L.M., Jiang, B., He, Z.Y., Lipase-catalyzed transesterification of soybean oil for biodiesel production in tert-amyl alcohol. *World J. Microbiol. Biotechnol.*, 25, 1, 41, 2008.
85. Kang, M.K. and Nielsen, J., Biobased production of alkanes and alkenes through metabolic engineering of microorganisms. *J. Ind. Microbiol. Biotechnol.*, 44, 4, 613–622, 2017.
86. Jiménez-Díaz, L., Caballero, A., Pérez-Hernández, N., Segura, A., Microbial alkane production for jet fuel industry: Motivation, state of the art and perspectives. *Microb. Biotechnol.*, 10, 1, 103–124, 2017.
87. Schirmer, A., Rude, M.A., Li, X., Popova, E., Del Cardayre, S.B., Microbial Biosynthesis of Alkanes. *Science*, 329, 5991, 559, 2010.
88. Zhou, Y.J., Buijs, N.A., Zhu, Z., Qin, J., Siewers, V., Nielsen, J., Production of fatty acid-derived oleochemicals and biofuels by synthetic yeast cell factories. *Nat. Commun.*, 7, 11709, 2016.
89. White, M.J., Hammond, R.C., Rose, A.H., Production of long-chain alcohols by yeasts. *J. Gen. Microbiol.*, 133, 2181–2190, 1987.
90. Shakeel, T., Fatma, Z., Fatma, T., Yazdani, S.S., Heterogeneity of alkane chain length in freshwater and marine cyanobacteria. *Front. Bioeng. Biotechnol.*, 3, 34, 2015.
91. Peramuna, A., Morton, R., Summers, L.M., Enhancing alkane production in cyanobacterial lipid droplets: A model platform for industrially relevant compound production. *Life*, 5, 2, 1111–1126, 2015.
92. Andrady, A.L. and Neal, M.A., Applications and societal benefits of plastics. *Philos. Trans. R. Soc. London. Ser. B*, 364, 1526, 1977–1984, 2009.
93. Gervet, B., The use of crude oil in plastic making contributes to global warming. *Lulea Univ. Technol.*, 1–8, 2007.
94. Das, R.K., Brar, S.K., Verma, M., Chapter 8 - Fumaric acid: Production and application aspects, in platform chemical biorefinery, in: *Future Green Industry*, S. Kaur Brar, S. Jyoti Sarma, K. Pakshirajan (Eds.), pp. 133–157, Elsevier, Amsterdam, 2016.
95. Brandl, H., Gross, R.A., Lenz, R.W., Fuller, R.C., Plastics from bacteria and for bacteria: Poly(β-hydroxyalkanoates) as natural, biocompatible, and biodegradable polyesters, in: *Microbial Bioproducts*, Th. Scheper (Ed.), pp. 77–93, Berlin, Heidelberg, 1990.
96. Snell, K.D. and Peoples, O.P., PHA bioplastic: A value-added coproduct for biomass biorefineries. *Biofuels, Bioprod. Biorefin.*, 3, 4, 456–467, 2009.
97. Keshavarz, T. and Roy, I., Polyhydroxyalkanoates: Bioplastics with a green agenda. *Curr. Opin. Microbiol.*, 13, 3, 321–326, 2010.

98. Flieger, M., Kantorova, M., Prell, A., Řezanka, T., Votruba, J., Biodegradable plastics from renewable sources. *Folia Microbiol.*, 48, 1, 27, 2003.
99. Koller, M. and Braunegg, G., Advanced approaches to produce polyhydroxyalkanoate (PHA) biopolyesters in a sustainable and economic fashion. *EuroBiotech J.*, 2, 2, 89–103, 2018.
100. Sudesh, K., Abe, H., Doi, Y., Synthesis, structure and properties of polyhydroxyalkanoates: Biological polyesters. *Prog. Polym. Sci.*, 25, 10, 1503–1555, 2000.
101. Wong, Y.M., Brigham, C.J., Rha, C., Sinskey, A.J., Sudesh, K., Biosynthesis and characterization of polyhydroxyalkanoate containing high 3-hydroxyhexanoate monomer fraction from crude palm kernel oil by recombinant *Cupriavidus Necator. Bioresour. Technol.*, 121, 320–327, 2012.
102. Murugan, P., Han, L., Gan, C.Y., Maurer, F.H., Sudesh, K., A new biological recovery approach for PHA using mealworm, Tenebrio molitor. *J. Biotechnol.*, 239, 98–105, 2016.
103. Sarepoua, E., Tangwongchai, R., Suriharn, B., Lertrat, K., Influence of variety and harvest maturity on phytochemical content in corn silk. *Food Chem.*, 169, 424–429, 2015.
104. Baidurah, S., Kubo, Y., Kuno, M., Kodera, K., Ishida, Y., Yamane, T., Ohtani, H., Rapid and direct compositional analysis of poly(3-hydroxybutyrate-co-3-hydroxyvalerate) in whole bacterial cells by thermally assisted hydrolysis and methylation-gas chromatography. *Anal. Sci.*, 31, 2, 79–83, 2015.
105. Baidurah, S., Kubo, Y., Ishida, Y., Yamane, T., Direct determination of poly(3-hydroxybutyrate) accumulated in bacteria by thermally assisted hydrolysis and methylation-gas chromatography in the presence of organic alkali. *Pure Appl. Chem.*, 90, 6, 1011–1017, 2018.
106. Schieber, A., Stintzing, F.C., Carle, R., By-products of plant food processing as a source of functional compounds—Recent developments. *Trends Food Sci. Technol.*, 12, 11, 401–413, 2001.
107. Weaver, C.M., Bioactive foods and ingredients for health. *Adv. Nutr.*, 5, 3, 306S–311S, 2014.
108. Joana Gil-Chávez, G., Villa, J.A., Fernando Ayala-Zavala, J., Basilio Heredia, J., Sepulveda, D., Yahia, E.M., González-Aguilar, G.A., Technologies for extraction and production of bioactive compounds to be used as nutraceuticals and food ingredients: An overview. *Compr. Rev. Food Sci. Food Saf.*, 12, 1, 5–23, 2013.
109. Day, L., Seymour, R.B., Pitts, K.F., Konczak, I., Lundin, L., Incorporation of functional ingredients into foods. *Trends Food Sci. Technol.*, 20, 9, 388–395, 2009.
110. Kalogeropoulos, N., Chiou, A., Pyriochou, V., Peristeraki, A., Karathanos, V.T., Bioactive phytochemicals in industrial tomatoes and their processing byproducts. *LWT - Food Sci. Technol.*, 49, 2, 213–216, 2012.
111. George, B., Kaur, C., Khurdiya, D., Kapoor, H., Antioxidants in tomato (*Lycopersium esculentum*) as a function of genotype. *Food Chem.*, 84, 1, 45–51, 2004.
112. Sharma, G., Gupta, A., Ganjewala, D., Gupta, C., Prakash, D., Phytochemical composition, antioxidant and antibacterial potential of underutilized parts of some fruits. *Int. Food Res. J.*, 24, 3, 1167–1173, 2017.
113. Shrikhande, A.J., Wine by-products with health benefits. *Food Res. Int.*, 33, 6, 469–474, 2000.
114. Goulas, V. and Manganaris, G.A., Exploring the phytochemical content and the antioxidant potential of Citrus fruits grown in Cyprus. *Food Chem.*, 131, 1, 39–47, 2012.
115. Alam, M.A., Subhan, N., Rahman, M.M., Uddin, S.J., Reza, H.M., Sarker, S.D., Effect of citrus flavonoids, naringin and naringenin, on metabolic syndrome and their mechanisms of action. *Adv. Nutr.*, 5, 4, 404–417, 2014.
116. Rudra, S.G., Nishad, J., Jakhar, N., Kaur, C., Food industry waste: Mine of nutraceutical. *Int. J. Sci. Environ. Technol.*, 4, 1, 205–229, 2015.
117. Ajila, C., Naidu, K., Bhat, S., Rao, U.P., Bioactive compounds and antioxidant potential of mango peel extract. *Food Chem.*, 105, 3, 982–988, 2007.

118. Kim, H., Moon, J.Y., Kim, H., Lee, D.S., Cho, M., Choi, H.K., Kim, Y.S., Mosaddik, A., Cho, S.K., Antioxidant and antiproliferative activities of mango (*Mangifera indica* L.) flesh and peel. *Food Chem.*, 121, 2, 429–436, 2010.
119. Murthy, P.S. and Naidu, M.M., Recovery of phenolic antioxidants and functional compounds from coffee industry by-products. *Food and Bioprocess Technol.*, 5, 3, 897–903, 2012.
120. Pujol, D., Liu, C., Gominho, J., Olivella, M., Fiol, N., Villaescusa, I., Pereira, H., The chemical composition of exhausted coffee waste. *Ind. Crops Prod.*, 50, 423–429, 2013.
121. Zuorro, A. and Lavecchia, R., Spent coffee grounds as a valuable source of phenolic compounds and bioenergy. *J. Cleaner Prod.*, 34, 49–56, 2012.
122. Kadir, A.A., Ismail, S.N.M., Jamaludin, S.N., Food waste composting study from makanan ringan Mas. *IOP Conf. Ser. Mater. Sci. Eng.*, 136, 1, 012057, 2016.
123. Schwarz, M. and Bonhotal, J., Composting at home: The green and brown alternative. *Cornell Waste Manage. Inst. Department Crop Soil Sci.*, 1–12, 2011.
124. Tong, J., Sun, X., Li, S., Qu, B., Wan, L., Reutilization of green waste as compost for soil improvement in the afforested land of the Beijing plain. *Sustainability*, 10, 7, 1–17, 2018.
125. Tuck, C.O., Pérez, E., Horváth, I.T., Sheldon, R.A., Poliakoff, M., Valorization of biomass: Deriving more value from waste. *Science*, 337, 6095, 695, 2012.
126. Kiran, E.U., Trzcinski, A.P., Ng, W.J., Liu, Y., Bioconversion of food waste to energy: A review. *Fuel*, 134, 389–399, 2014.
127. Kiran, E.U., Trzcinski, A.P., Ng, W.J., Liu, Y., Enzyme production from food wastes using a biorefinery concept. *Waste Biomass Valorization*, 5, 6, 903–917, 2014.
128. Sheldon, R.A., Enzyme immobilization: The quest for optimum performance. *Adv. Synth. Catal.*, 349, 8-9, 1289–1307, 2007.
129. Mandal, S.K. and Das, N., Enhanced biodegradation of high molecular weight pahs using yeast consortia immobilized on modified biowaste material. *J. Microbiol. Biotechnol. Food Sci.*, 7, 6, 594–601, 2018.
130. Sriroth, K., Piyachomkwan, K., Wanlapatit, S., Nivitchanyong, S., The promise of a technology revolution in cassava bioethanol: From Thai practice to the world practice. *Fuel*, 89, 7, 1333–1338, 2010.
131. Kunthiphun, S., Phumikhet, P., Tolieng, V., Tanasupawat, S., Akaracharanya, A., Waste cassava tuber fibers as an immobilization carrier of *Saccharomyces cerevisiae* for ethanol production. *BioResouce*, 12, 1, 157–167, 2017.
132. Bai, F.W., Anderson, W.A., Young, M.M., Ethanol fermentation technologies from sugar and starch feedstocks. *Biotechnol. Adv.*, 26, 1, 89–105, 2008.
133. Genisheva, Z., Teixeira, J.A., Oliveira, J.M., Immobilized cell systems for batch and continuous winemaking. *Trends Food Sci. Technol.*, 40, 1, 33–47, 2014.
134. Albers, E., Larsson, C., Lidén, G., Niklasson, C., Gustafsson, L., Influence of the nitrogen source on *Saccharomyces cerevisiae* anaerobic growth and product formation. *Appl. Environ. Microbiol.*, 62, 9, 3187–3195, 1996.
135. Rahman, N.A. and Wan Rosli, W.I., Nutritional compositions and antioxidative capacity of the silk obtained from immature and mature corn. *J. King Saud Univ. Sci.*, 26, 2, 119–127, 2014.
136. Behera, S., Singh, R., Arora, R., Sharma, N.K., Shukla, M., Kumar, S., Scope of algae as third generation biofuels. *Front. Bioeng. Biotechnol.*, 2, 90, 2014.
137. Cheng, K.C., Demirci, A., Catchmark, J.M., Advances in biofilm reactors for production of value-added products. *Appl. Microbiol. Biotechnol.*, 87, 2, 445–456, 2010.
138. Todhanakasem, T., Tiwari, R., Thanonkeo, P., Development of corn silk as a biocarrier for *Zymomonas mobilis* biofilms in ethanol production from rice straw. *J. Gen. Appl. Microbiol.*, 62, 2, 68–74, 2016.

139. Rattanapan, A., Limtong, S., Phisalaphong, M., Ethanol production by repeated batch and continuous fermentations of blackstrap molasses using immobilized yeast cells on thin-shell silk cocoons. *Appl. Energy*, 88, 12, 4400–4404, 2011.
140. Whitfield, M.B., Chinn, M.S., Veal, M.W., Processing of materials derived from sweet sorghum for biobased products. *Ind. Crops Prod.*, 37, 1, 362–375, 2012.
141. Eggleston, G. and Lima, I., Sustainability issues and opportunities in the sugar and sugar-bioproduct industries. *Sustainability*, 7, 9, 2209–12235, 2015.
142. Ashori, A., Wood–plastic composites as promising green-composites for automotive industries! *Bioresour. Technol.*, 99, 11, 4661–4667, 2008.
143. Chang, Z., Cai, D., Wang, C., Li, L., Han, J., Qin, P., Wang, Z., Sweet sorghum bagasse as an immobilized carrier for ABE fermentation by using *Clostridium acetobutylicum* ABE 1201. *RSC Adv.*, 4, 42, 21819–21825, 2014.
144. Yu, J., Zhang, X., Tan, T., An novel immobilization method of *Saccharomyces cerevisiae to* sorghum bagasse for ethanol production. *J. Biotechnol.*, 129, 3, 415–420, 2007.
145. Babel, S. and Kurniawan, T.A., Cr(VI) removal from synthetic wastewater using coconut shell charcoal and commercial activated carbon modified with oxidizing agents and/or chitosan. *Chemosphere*, 54, 7, 951–967, 2004.
146. Budi, E., Nasbey, H., Yuniarti, B., Nurmayatri, Y., Fahdiana, J., Budi, A., Pore structure of the activated coconut shell charcoal carbon. *AIP*, 1617, 1, 130–133, 2014.
147. Chronopoulos, G., Bekatorou, A., Bezirtzoglou, E., Kaliafas, A., Koutinas, A., Marchant, R., Banat, I., Lactic acid fermentation by *Lactobacillus casei* in free cell form and immobilised on gluten pellets. *Biotechnol. Lett.*, 24, 15, 1233–1236, 2002.
148. Luthfi, A.A.I., Jahim, J.M., Harun, S., Tan, J.P., Mohammad, A.W., Potential use of coconut shell activated carbon as an immobilisation carrier for high conversion of succinic acid from oil palm frond hydrolysate. *RSC Adv.*, 7, 78, 49480–49489, 2017.
149. Mladenović, D., Đukić-Vuković, A., Radosavljević, M., Pejin, J., Kocić-Tanackov, S., Mojović, L., Sugar beet pulp as a carrier for *Lactobacillus paracasei* in lactic acid fermentation of agro-industrial waste. *J. Proc. Energy Agric.*, 21, 1, 41–45, 2017.
150. Kumar, M.N., Gialleli, A.I., Masson, J.B., Kandylis, P., Bekatorou, A., Koutinas, A.A., Kanellaki, M., Lactic acid fermentation by cells immobilised on various porous cellulosic materials and their alginate/poly-lactic acid composites. *Bioresour. Technol.*, 165, 332–335, 2014.
151. Hsu, C., Chu, Y., Argin-Soysal, S., Hahm, T., Lo, Y., Effects of surface characteristics and xanthan polymers on the immobilization of *Xanthomonas campestris* to fibrous matrices. *J. Food Sci.*, 69, 9, E441–E448, 2004.
152. Chatterjee, U., Kumar, A., Sanwal, G.G., Goat liver catalase immobilized on various solid supports. *J. Ferment. Bioeng.*, 70, 6, 429–430, 1990.
153. Das, R.K., Brar, S.K., Verma, M., Valorization of egg shell biowaste and brewery wastewater for the enhanced production of fumaric acid. *Waste Biomass Valorization*, 6, 4, 535–546, 2015.
154. Das, R.K. and Brar, S.K., Verma,M., Enhanced fumaric acid production from brewery wastewater by immobilization technique. *J. Chem. Technol. Biotechnol.*, 90, 8, 1473–1479, 2015.
155. Tesfaw, A. and Assefa, F., Current trends in bioethanol production by *Saccharomyces cerevisiae*: Substrate, inhibitor reduction, growth variables, coculture, and immobilization. *Int. Scholarly Res. Not.*, 11, 1–11, 2014.

3

Biomass Resources for Biorefinery Application

Varsha Upadhayay, Ritika Joshi and Arindam Kuila*

Department of Bioscience and Biotechnology, Banasthali Vidyapith, Vidyapith, Rajasthan, India

Abstract
Biorefinery is a new word to assign two main topics, bioenergy and bioproducts, toward bio-based society; they play very important role. In this chapter, we have focused on modern biorefinery forms as well as its application and future importance on biofuels production. At present, numerous different levels of incorporation in biorefinery are helpful to maintain the feasibility in aspect of environment and economically. Pre-treatment of biomass through enzymatic methods targeted to liberate polysaccharides is a crucial industrial science in the forthcoming biorefinery, and right now, it is governed by speeded up research.

Keywords: Biorefinery, feedstock resources, biorefineries methods, biorefineries application, future approaches

3.1 Introduction

Human development in the product of interruption utilization of natural resources on large scales: renewable energy (e.g., wind, biomass, and sunlight), fossil fuels (e.g., coal oil and gas), water resources, and land resources [1]. Rise in the global population along with (parallel side of) desire of standard life is being the reason for increasing energy consumption at global scale. Nowadays, production of merchandise energy and chemicals is based on crude oil; however, fast diminution of this feedstock is creating frown for especially communication industries like aviation and automobiles, and addition to this statement, visible proof are available regarding release of greenhouse gases, where nitrous oxide (N2O), carbon dioxide (CO2), and methane coming to pass from natural fuel usages and land use change are reason to affect Earth's climatic condition [2]. To an extent, it is necessary to find out economical cost and viable energy which is not only help to lower the greenhouse effect but also take over the fossil fuel dependency. Biomass can be considered most various and willingly present everywhere, which could be responsible for production of bio-product and various non-found polymers [3]. Biofuels generation research is one of the latest work which is adopted by many students in this filed along with biodiesel, bioethanol, and many biomass product which are used in different conversion ways and prominent technologies.

**Corresponding author*: arindammcb@gmail.com

Arindam Kuila and Mainak Mukhopadhyay (eds.) Biorefinery Production Technologies for Chemicals and Energy, (55–66) © 2020 Scrivener Publishing LLC

Biorefinery is a provision which combines biomass conversion techniques and machinery to yield fuel, food, and important chemicals along with energy (heat and power) with the use of biomass. It is a different process from conventional oil refinery process [4].

These days, bioethanol and biodiesel are the main biofuels and are rapidly increasing used additive fuel along with chemical raw material utilization. Presently, in principle, it utilizes 10% gasoline blends with the prospective to reach 85% in supple vehicle fuel cars [5]. Although in first-generation production of biofuels has raised over its sustainability, the lignocelluloses biomass exploitation derived from agro-based or forest-based residues, including sugarcane, can positively contribute to the production biofuels and chemicals without challenging the food chain [6]. Therefore, different sources of energy those have the prospective to provide renewable to explore. To accomplish this, a victorious substitute energy source should have possible criteria [6]:

1. Renewable resources should be easily available
2. Renewable resources should be low-priced
3. Renewable resources should be GHG neutral
4. Renewable resources should not propose a danger to land and food availability

3.2 Concept of Biorefinery

Biorefinery is considered best option as multifunctional method to produce multiple types of material and energy products [2], by that increasing the value of the feedstock material economically practiced during production of waste stream [7]. Biorefineries concept is the fundamental use of biomass to produce complementary chemicals and biomaterial compounds, which has significant impact on chemical industry [8]. Even so, biomass is observed mainly use for the use biofuels production to target the detriment in the fossil fuel use at a complete or partial stage which is majorly responsible for the environmental pollution. In recent years, biomass at basic levels produce important biofuels, fine chemicals,

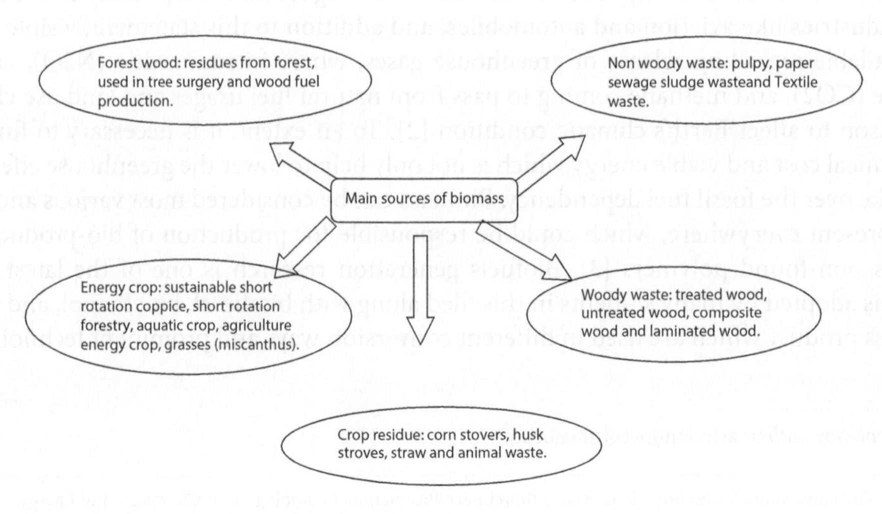

Figure 3.1 Energy production from different sources of biomass.

and biopolymers. This doesn't show new dimensions in the research field only as well as it is a remarkable opportunity for the biofuels industry in environmental and economic aspects [8]. In point of fact, advancement in this area results in a prophecy that nonrenewable oil energy-based economy will be ovecoverd by the upcoming renewable biofuel-based economy [8].

3.3 Biomass Feedstocks

Feedstock selection and the resulting product coming out from it are very important factors for designing of the biorefinery and in addition feedstock initial composition of feedstocks and its possible utilization in various productive streams [9]. Studying through same lane potential biomass feedstock sources has various varieties like starch or sugar containing crops, algae lignocellulosic crops. Conversion processes for the biomass according to the raw material usages are considered as first generation and second generation processes [10]. Nowadays, commercial production of biofuels from the starch, oil seed depending feedstock, and sugar are combined referred as first generation biofuels. Although in comparison to fossil feedstocks, biomass feedstock is low in thermal stability and a high level of functionality so there is requirement for special reaction order [11].

3.3.1 Types of Biomass Feedstocks

3.3.1.1 Biomass of Sugar Industry

Production of saccharose on a large scale uses some crops which are known to store energy in form of sugar like sweet sorghum, sugar beet or sugarcane. Sugar-based biorefineries uses sugar crops such as sugar cane, sugar beet, or sweet sorghum as they have store large

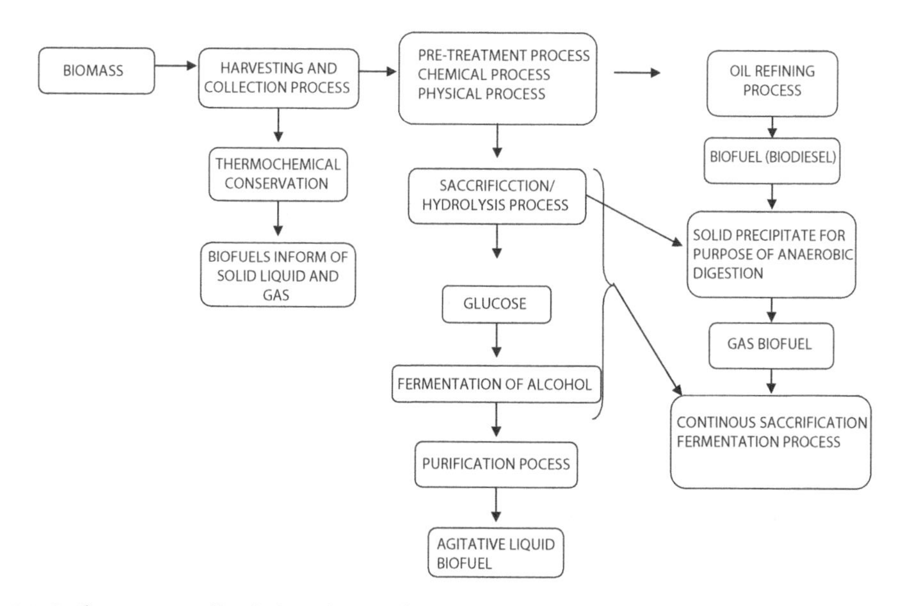

Figure 3.2 Different types of biofuel production from biomass.

amounts of saccharose, and the stored sugar can be extracted from the plant for succeeding fermentation to produce biofuels. Sugarcane is famous in Brazil for bioethanol production at a large scale by using raw materials from past more than 30 years, depending on a biorefinery model where sugar and ethanol are the final product from the sugarcane juice and bagasse that are also utilized by burning to turn them into power and steam energy [12]. Godavari Biorefineries Ltd. which is formerly the Godavari Sugar Mills Ltd. runs two sugar-related biorefinery for the purpose of ethanol, sugar, and electricity production including 20 more products from the renewable sources in India. In some other countries like Colombia, biorefineries which mainly functional with sugarcane to produce biofuels from the molasses, sugarcane juice, and bagasse are used for electricity production and are still moving in a more profitable direction of biorefinery format, which gives a promising future of large-scale sugar production, huge outcome of ethanol for oxygenation program, greenhouse gas release at a sustainable range, less discharge of stillage, and a remarkable social impact with the paid position of generation [30].

3.3.1.2 *Biomass Waste*

Biodiesel production takes almost 70% cost account for the raw materials [31]. So, it is important to note all available options to improve economic value of the final product like biodiesel. Overall availability and the less investment cost are considered as very crucial benchmark for the viable biodiesel feedstock, for example, karanja, jatropha, and castor bean are the main attraction for the biofuels production in brazil and India [13]. Canada and Australia are using tallow and animal fat for biofuels, and in Japan, waste cooking oil is research trend [32]. Second generation crops which are non edible are appraised as optimistic feedstock at a global scale, and they are utilizing marginal lands by saving countless land for edible crops [14]; however, there is lot of economic benefits are associated with the 2nd generation produces biodiesel, but sometimes, it experience with bad impact such as excessive FFA contents and at room temperature accordingly low solidification point [33].

Cassave, wheat, and corn are rich in starch crops which can be enzymatically hydrolyzed to produce sugar solution, later which can ferment and process into chemicals and fuels. Lots of Starch crops processing leads to resulting many important animal feed which are good in energy and proteins as complimentary byproducts. Sweet sorghum is well known inspected to produces many valuable products like wood plastic composite ethanol and butanol [34].

Different types of triacylglycerol depend upon fatty acids of 8 to 24 carbon length chains. [15]. Biofuels production from the oilseed is a unique opportunity and it also provides heavy valued fatty acids which can replace fossil energy source chemical, detergents, and lubricants. Feedstock for the biodiesel production contains plants such as canola, soyabean, and palm fruits [35]. Bouaid, Martinez, and Aracil [16] also have evaluated coconut oil and mixed process for producing important and high and low molecular weight methyl ester fraction; it has utilization as biofuels and recovery of other complimentary important byproducts as bio solvents for selling and bio lubricants so supporting in successful developing a self-assist biorefinery. Rincon, Jaramillo, and Cardona [13] started a united approach to enhance the value added of palm oil mutually producing alcohol and biodiesel from lingonocellulogic precipitate (plant fibers such as palm press and fruit clusters) and methanol or crude glycerol derived from syngas; output is found to numerous products along with low level of energy utilization and raising the maximum utilization of feedstocks.

3.3.1.3 *Sugar and Starch Biomass*

The sugar and starch feedstock cost is very high due to expensive sugar crops and grains. Thus, in lignocelluloses, biomass processing research has been increased; they focused mainly in forestry and agricultural residues which are reasonably renewable, abundant, cheaper, and readily available with no consistency with food industries. Economical method of biorefineries for biomass utilization [17]. Sohel and Jack [18] studied on future advantages of thermodynamic production of mixed geothermal heat into biochemical analysis followed by improved application of both geothermal and biomass resources. Around Pacific Rim, different countries have major geothermal resources where this idea of biomass and collocating geothermal resources for biorefineries can be practical. Still, the energy density linked with transportation in long distance and low economic value is the major limitation associated with uses of agricultural residues [36]. The main component of biomass delivered cost for the fuel recovering overall cost or chemical used in biomass and represents 35%–50% of the total biofuels production cost. In rural or agricultural areas, the site of biorefineries industry abolished in some extent because these areas have large available agricultural residues [19]. In non-agriculture sectors, it will stimulate the formation of job opportunities and increase rural economic growth [37]. Additionally, in biorefineries delivery system, the traffic and transportation crowding can be decreased by raising the bulk density and sharing delivery of agricultural waste in form of wood chips and bales [20].

3.3.1.4 *Algal Biomass*

In modern era, microalgae have obtained great attention due to its strong vitality and quick growth rate. Microalgae have been employed as source of different products including animal feed (i.e., larval bivalves), chemicals (i.e., vitamins and antioxidants), agriculture biomass (i.e., production of biofuels), and oils (i.e., omega-3 fatty acids). *Schizochytrium limacinum*, *Chlorella zofingiensis*, and *Chlorella protothecoids* are well-recognized oil producer and more than 50% oil accumulate the dry body weight. Blue-green algae, seaweeds, and micro-algae are together known as algal biomass can be another potential choice feedstock in biorefineries due to their advanced photosynthetic efficiency [21], oil content, and yield. Besides, algal biomass cannot challenge clean water, food cultures, and cultivable land, and it has chance of being cultivated on a regularly [22].

3.3.1.5 *Lignocelluloses Feedstock*

Lignocelluloses feedstock is the most abundantly present in the biosphere, consider about 50% of the total. It is usually divided into three groups: (1) agro-based residues (i.e., sugarcane and crop residues), (2) forestry-based residues, and (3) woody and herbaceous energy crops [23]. Lignocelluloses feedstock, mainly agro-based and forestry residues, energy-based crops and paper-based waste are becoming latent renewable energy resource [38]. Moreover, in contrast with other biomass, lignocelluloses biomass has a number of benefits as mentioned: (i) reduce competitiveness for water and land for production of food; (ii) enhancement in production of biomass; and (iii) needed lesser growth [24]. Earth is rich in different type of biomasses and lignocelluloses play main role in booming substitute of biofuels. It proves as attractive biomass for the sustainable invention of chemical products and liquid fuels in the biorefineries approach. Different types of lignocelluloses residues, concerning biorefineries feedstock, have been planned and documented in literature such as Switch grass [25], sugarcane wastes [26], and wheat bran [17], etc.

3.3.1.6 Oil Crops for Biodiesel

Biodiesel production is done by adding extract of oil seed and oil- rich nuts with in different with in alcohol during chemical methods called transesterification [27]. In US, Soybean, in Asian countries, palm and coconut oil, and in EU, rapeseeds are the well known oils crops. In rapeseed and soybean, 21% to 35% of oil substance is found, respectively [39]. Palm oil has 40% oil content, and in comparison to other seed, it is highest [17]. On the basis of molecular structure, the major difference between oil feedstock is determine, and in the degree of saturation and unsaturation, diverse group of fatty acid are attach in the triacylglycerol (TAG) [39]. All these quality, factor, expenditure, sequentially, and production affect the processes of biofuels products [39].

Transesterification is the step by step process of alteration of TAG with an alcohol to form esters and glycerol is influence by strong base, procedure mainly used for methanol [27]. Therefore, biofuels production by alkalicatalyzed transesterification process require temperature, low pressure, and economic choice of achieving 98% conversion biofuels production [27]. On the other hand, enzyme-catalyzed process is gaining interest due to low energy costs, high purity of glycerol, and lowering soap structure [28]. Although, exclusive enzyme cost and small reaction rate are the two major troubles are observed in this process. B2 (2% biodiesel and 98% petroleum diesel), B5 (5% biodiesel and 95% petroleum diesel), and B20 (0% biodiesel and 80% petroleum diesel) are very common biofuels blended products.

3.4 Processes

In bio-based products, different processes (Table 3.1) are used and they are divided in four main groups:

1) Physical processes: it is a separation and size reduction process for different components and chemical structure of biomass do not changed in this processes.
2) Biochemical processes: in this process, microorganism or enzyme is added in mild conditions (pressure and temperature).
3) Chemical processes: in substrate chemical, changes occurs
4) Thermo chemical processes: biomass undergoes extreme conditions.

Table 3.1 Different biomass and process involve in biorefinery [3, 26].

S. No.	Conversion technology	Category	Resource type	Product	End use	Examples of resources	Technology status
1.	Gasification	Thermo chemical	Solid biomass	Product gas	Heat electricity transport fuel	Pellets agricultural residues and wood chips	Demonstration/ early commercial
2.	Pyrolysis	Thermo chemical	Solid biomass	Pyrolysis oil, by-products (product gas, char)	Heat electricity	Pellets agricultural residues and wood chips	Demonstration
3.	Combustion	Thermo chemical	Solid biomass	Heat	Heat electricity	Wood logs, chips, and pellets agricultural residues	Commercial
4.	Anaerobic digestion	Biological	Wet biomass	Biogas by-products	Heat electricity transport fuel	Sewage sludge and manure	Commercial
5.	Fermentation	Biological	Solid biomass	Liquid fuels (bioethanol)	Transport fuel	Miscanthus, Sugar beet, wheat, colza, maize and potatoes	Commercial
6.	Transesterification	Chemical	Solid and liquid biomass	Liquid fuels (biodiesel)	Transport fuel, heat and electricity	Fat waste, rape seeds and soybeans oil	Demonstration/ early commercial

3.4.1 Thermo Chemical Processes

The main center of thermo chemical methods has been on production of biofuels and replacements of kerosene for the aviation and road region, even though the gasoline production and other chemicals are also achievable. The most widely used method to convert biomass into energy (power and heat) is combustion and it is documented 90% of energy recovered worldwide from biomass through this method. In biomass combustion system, boiler is the heart of biomass and there are different models and type are present. These are generally classified by the biomass type which is appropriate for use with (e.g., pellet, wet and dry woodchip, bale, etc.), different combustion grate (plane grate, batch-fired, moving grate, and stoker), and thermal output.

They differ from manually feed, normally tiny boilers with some powers, during feed boiler with repeated ignition, controlled system, and remote monitoring. In large-scale application, moving grate systems are more common in 300 kW to 1 MW. Moving grate systems are generally more common in large-scale applications of 300 kW to 1 MW and show extensive tolerance of particle size, type of fuels, and moisture content. In 25 to 300 kW range, plane grate operators are more common and they can use woodchips and pellets both. Stoker systems are used in same ranges and normally they are less sophisticated and cheaper. Lastly, batch-fired systems are cheaper and simple but it requires large amount of biomass and labor power. In combustion technology for power generation grate furnaces are used. In this process, heat is converted into vapor which passes from turbine linked to power generator. In modern plants, typical electric efficiency of 40%–45% equipment technology is used.

In gasification process, first biomass was gasified at 800°C–1,100°C in the presence of different gasification agent, i.e., air, wet vapor, and oxygen then biomass is directly burn. In bio-syngas, biomass is transformed into gas mixture which contains mainly hydrogen and carbon monoxide. After that this, biomass is used for making new and complex molecule, for example, biofuels, material, and chemical (methanol, hydrogen, butanol, and methane are regular ending products of gasification). Overall, in gasification of biomass black liquor, straw and wood, etc., are required, but in different range of feedstocks with cheap MSW, agro-based residues and oil pyrolysis is used. In physical or chemical pre-treatment processes, it is compulsory to improve the lignocelluloses biomass conversion, and before the synthesis process, conditioning and cleaning of product gas is required. In most cases, gasification appears more appropriate for power and bio-based chemicals generation than for heat generation.

The thermal degradation method which occur anerobically is known as pyrolysis. Tars, fuel gas, and pyrolysis oil are various products which are produced during this process. Gasification (co-combustion of diesel engine and plants), hydrogenation, and fractionation are secondary products of this process. Most of the pyrolysis methods are only appropriate for dry feedstock with restricted changeability, for example, cheap lignocelluloses stuff and wastes. Pyrolysis technologies are not so far entirely commercially established. At present, pilot plants are utilize in worldwide; however, the technology is prepared for advance up scaling. In Combined Heat and Power (CHP), pyrolysis method can be economically profitable or in a biorefinery theory where both energy and other products of higher value are produced.

Under low temperature and high pressure, liquefaction changes solid biomass into liquid in the presence of catalyst and hydrogen. This method needs high reactor difficulty and very expensive due to which it is not used for pyrolysis. In mild thermal treatment, torrefaction occurs when biomass is treated under low-oxygen state due to which homogeneous fuel in good quality, less impurities, and higher energy density. Through palletization, torrified biomass can be densified ensuing energy-dense product. Numerous torrefaction methods are developed and reached commercial production. Torrified biomass has number of prospective applications; combustion and co-firing are most common. In torrified biomass combustion, chips and wood pellets are also used in boilers in various applications and it has high market value. In conclusion, hydrogenation process animal and vegetable oils are converted into high quality product which can be used as fossil fuel substitute, for example, Hydrotreated Vegetable Oils (HVO) and Hydrotreated Renewable Diesel (HDRD), etc.

3.4.2 Biochemical Processes

Biochemical processes mainly focus on increasing alcohols production, for example, bio-based chemicals, butanol, and ethanol for biofuels and bioplastics. Since World War I, fermentation process is used for industrial production for chemicals and energy production. The main rule of fermentation is to change biomass into sugars specific product with the help of microorganism. In recent years, biotechnology development has advanced the opportunity to design microorganism production for selected chemicals which unable to produce well by microorganisms present in the environment. In the bio-based production, point of view with the help fermentation many biological chemicals can be produced in large scale, for example, citric acid, ethanol, amino acids, and lactate, etc., are latent in production of large scale of platform chemicals, by this way, starch and sugar biomass can be simply fermented in conventional methods. Lignocelluloses biomass and dedicated crops are mainly grown for advanced fermentation systems. The fermentation way for bioethanol production is developing and more effective. Biosurfactants method is used second generation fermentation. In this biosurfactants process, microorganisms are capable to grow on water miscible and oil substrate, and here, microorganisms remain bind in surface of microbial cell in the medium. The different types of microorganisms with different molecular structures, for example, fungi, bacteria, and yeasts, are able to produce biosurfactants.

Decomposition of organic matter in the airproof reactor tanks in absence of air is known as anaerobic digestion; it is the microbiological process. The three main steps of this process are as follows:

1) hydrolysis
2) acidogenesis
3) methanogenesis

The microbial activities of different microorganisms are characterized in these processes, and they are reliant upon each other. Digestate and biogas are the end-products of these processes. The digestate are used in the form of fertilizer and very rich in nutrients. The biogas properties and composition are obtained which depend upon feedstock input, operational factors, and type of digester. Digesters can work in batch flow or continuous and in multiple or single steps. Batch flow digesters have low working costs and multiple digesters allow higher organization but investment is very high. Up flow sludge blanket, plug flow, covered lagoon, and complete mix are some basic digester. The biomass are needed to check regularly and pre-treatments are required (either mixing or separation) to make sure that suitable composition is entering inside the digester. Co-digestion of many types of biomass is a well-known observation for achieving the best stability between methods stability and biogas yield, mainly in medium- to large-scale AD plants. The natural anaerobic digester are landfills; they produce biogas. The revival of biogas from landfill is partial, mainly in developing countries, and the appropriate revival techniques vary extensively as a function of landfill and waste characteristics.

3.4.3 Biobased Products and the Biorefinery Concept

In a biorefinery, feedstock is processed to be improving to one or more expensive products for example bio-based chemicals, biodiesel, and energy resources. The biorefinery idea suggests the well-organized improvement of feedstock and the generation of costly products in an eco-friendly sustainable, social, and economic way [29]. The word does not used in detailed process industry or system but indeed covers a wide range of technical systems. A biorefinery can be anything from one single machine for alteration of feedstock up to a complex polygeneration plant integrated with other industries and energy systems, refining different kinds of biomass feedstock into one or many products using chemical, biochemical, and thermochemical transformations.

IEA Bio-energy Task 42 has classified biorefineries based on their key characteristics (listed in order of importance):

- Raised area
- Manufactured goods
- Biomass
- Methods

With the grouping of these four characteristics, different biorefinery patterns can be described in a dependable approach. The stages are transitional from which ending products are resulting in an idea which is worn in petrochemical industry and crude oil is fractionated into various intermediate methods and from product. The area may be reached using various conversion methods useful to various biomasses and may lead to a range of marketable foodstuffs using multiple methods. The number of area involved in a biorefineries is a needle of system convolution.

3.5 Conclusions

Biomass has unbelievable prospective to complete chemical and future energy demands. Biochemicals and biofuels are measured to have inborn advantages on environmental protection, value adding, and energy safety. A biorefinery is a provision that combine different biomass conversion methods and apparatus to produce biomaterials, biofuels, power, biochemical, and heat. Biorefinery must be planned and managed systematically to decrease the production cost and environmental issues. Yet, the analysis and design of biorefinery systems is always demanding because of the difficulty passed by the processer under oberservation, the optimality preferred by the choice maker, and the doubt related to the statistics and information. Numerous methodologies and devices have been made newly to handle those provocations and assist the task of plan and investigation of biorefinery.

References

1. Jiang, L. and Zhang, J., Biodegradable Polymers and Polymer Blends, in: *Handbook of Biopolymers and Biodegradable Plastics*, Elsevier, USA, 2013.
2. Cherubini, F., Strømman, H.A., Ulgiati, S., Influence of allocation methods on the environmental performance of biorefinery products—A case study. *Resourc. Conserv. Recycl.*, 55, 11, 1070–1077, 2011.
3. Cherubini, F., The biorefinery concept: Using biomass instead of oil for producing energy and chemicals. *Energy Convers. Manage.*, 51, 1412–, 1421, 2010.
4. Amidon, T.E. and Liu, S., Water-based woody biorefinery. *Biotechnol. Adv.*, 27, 542–550, 2009.
5. Michailos, S.E. and Webb, C., Biorefinery Approach for Ethanol Production From Bagasse. *Bioethanol. Prod. From Food Crops*, 319, 341, 2019.
6. Saini, J.K., Saini, R., Tewari, L., Lignocellulosic agriculture wastes as biomass feedstocks for second-generation bioethanol production: Concepts and recent developments. *Biotech*, 5, 337–353, 2015.
7. Thomsen, H.M., Complex media from processing of agricultural crops for microbial fermentation. *Appl. Microbiol. Biotechnol.*, 68, 598–606, 2005.
8. Cherubini, F. and Ulgiati, S., Crop Residues as Raw Materials for Biorefinery Systems–A LCA. *Case Study Appl. Energy*, 87, 47–57, 2010.
9. Mabee, W.E., Gregg, D.J., Saddler, J.N., Assessing the emerging biorefinery sector in Canada. *App. Biochem. Biotech.*, 5, 0273–2289, 2005.
10. Lyko, H., Deerberg, G., Weidner, E., Coupled production in biorefineries—Combined use of biomass as a source of energy. *Fuels Mater. J. Biotech.*, 142, 78–86, 2009.
11. Huber, J.W. and Dumesic, J.A., An overview of aqueous-phase catalytic processes for production of hydrogen and alkanes in a biorefinery. *Catal. Today*, 111, 119–132, 2006.
12. Mariano, A.P., Dias, M.O.S., Junqueira, T.L., Cunha, M.P., Bonomi, A., Filho, R.M., Utilization of pentoses from sugarcane biomass: Techno-economics of biogas vs. butanol production. *Bioresour. Tech.*, 142, 390–399, 2013.
13. Rincon, L.E., Jaramillo, J.J., Cardona, C.A., Comparison of feedstocks and technologies for biodiesel production: An environmental and techno-economic evaluation. *Renew. Energy*, 69, 479–48, 2014.
14. Barnwal, B.K. and Sharma, M.P., Prospects of biodiesel production from vegetable oils in India. *Renew. Sustain. Energy Rev.*, 9, 363–378, 2005.
15. Daniel, T. and Octave, S., Biorefinery: Toward an industrial metabolism. *Biochimie*, 91, 659–664, 2009.
16. Bouaid, A., Martínez, M., Aracil, J., Biorefinery approach for coconut oil valorisation: A statistical study. *Bioresour. Technol.*, 101, 4006–4012, 2010.
17. Kongjan, P., Min, B., Angelidaki, I., Biohydrogen production from xylose at extreme thermophilic temperatures (70 °C) by mixed culture fermentation. *Water Res.*, 43, 1414–1424, 2009.

18. Sohel, M.I. and Jack, M., Efficiency improvements by geothermal heat integration in a lignocellulosic biorefinery. *Biores. Technol.*, 101, 9342–9347, 2010.
19. Lopolito, A., Nardone, G., Prosperi, M., Sisto R Stasi, A., Modeling the bio-refinery industry in rural areas: A participatory approach for policy options comparison. *Ecolog. Econom.*, 72, 18–27, 2011.
20. Sultana, A. and Kumar, A., Development of energy and emission parameters for densified form of lignocellulosic biomass. *Energy*, 36, 2716–, 2732, 2011.
21. Singh, A., Poonam, S., Nigam, Murphy, J.D., Renewable fuels from algae: An answer to debatable land based fuels. *Biores. Technol.*, 35, 9, 3865–3876, 2011.
22. Rosenberg, J.N., Mathias, A., Karen Korth, K., Betenbaugh, M.J., Oyler, G.A., Microalgal biomass production and carbon dioxide sequestration from an integrated ethanol biorefinery in Iowa: A technical appraisal and economic feasibility evaluation. *Biomass Bioenergy*, 35, 3865–3876, 2011.
23. Carriquiry, M.A., Du, X., Timilsina, G.R., Second generation biofuels: Economics and policies. *Energy Pol.*, 39, 4222–423, 2011.
24. Perrin, R., Vogel, K., Schmer, M., Mitchell, R., Farm-Scale Production Cost of Switchgrass for Biomass. *Bioenerg. Res.*, 1, 91–97, 2008.
25. Cherubini, F. and Jungmeier, G., LCA of a biorefinery concept producing bioethanol, bioenergy, and chemicals from switchgrass. *Int. J. Life Cycle Assess.*, 15, 53–66, 2010.
26. Rabelo, S.C. and Costa, A.C., Production of bioethanol, methane and heat from sugarcane bagasse in a biorefinery concept. *Biores. Technol.*, 102, 7887–7895, 2011.
27. Balat, M. and Balat, H., Progress in biodiesel processing. *Appl. Energy*, 87, 6, 1815–1835, 2010.
28. Christopher, L.P., Yao, B., Ji, Y., Lignin biodegradation with laccase-mediator systems. *Energy Res. Rev. Article*, 2, 1–13, 2014.
29. Zhang, E.D., Wang, B., Wang, Q.H., Zhang, S.B., Zhao, B.D., Ammonia-nitrogen and orthophosphate removal by immobilized Scenedesmus sp isolated from municipal wastewater for potential use in tertiary treatment. *Bioresour. Technol.*, 99, 3787–3793, 2008.
30. Moncada, J., El-Halwagi, M.M., Cardona, C.A., Techno-economic analysis for a sugarcane biorefinery: Colombian case. *Biores. Technol.*, 135, 533–543, 2013.
31. Math, M.C., Kumar, S.P., Chetty, S.V., Technologies for biodiesel production from used cooking oil—A review. *Energ. Sust. Dev.*, 14, 339–345, 2010.
32. Bhattacharyya, S.C., The economics of renewable energy supply, in: *Energy Economics: Concepts, Issues, Markets and Governance*, pp. 249–271, Springer-Verlag, London, 2011.
33. Leung, D.Y.C., Wu, X., Leung, M.K.H., A review on biodiesel production using catalyzed transesterification. *Appl. Energ.*, 87, 1083–1095, 2010.
34. Yu, J., Zhang, T., Zhong, J., Zhang, X., Tan, T., Biorefinery of sweet sorghum stem. *Biotechnol. Adv.*, 30, 811–816, 2012.
35. Demirbas, A., Biofuels sources, biofuel policy, biofuel economy and global biofuel projections. *Energ. Convers. Manage.*, 49, 2106–2116, 2007.
36. Mayfield, C.A., Foster, C.D., Smith, C.T., Gan, J.B., Fox, S., Opportunities, barriers, and strategies for forest bioenergy and bio-based product development in the Southern United States. *Biomass Bioenerg.*, 31, 631–637, 2007.
37. Bailey, C. and Dyer, J.F., Assessing the rural development potential of lignocellulosic biofuels in Alabama. *Biomass Bioenerg.*, 35, 1408–1417, 2011.
38. Zhang, X., Ruoshui, W., Molin, H., Martinot, E., A study of the role played by renewable energies in China's sustainable energy supply. *Energy*, 35, 4392–4399, 2010.
39. Ramos, M.J., Fernández, C.M., Casas, A., Rodríguez, L., Pérez, A., Influence of fatty acid composition of raw materials on biodiesel properties. *Bioresource Technol.*, 100, 261–268, 2009.

4

Evaluation of the Refinery Efficiency and Indicators for Sustainability and Economic Performance

Rituparna Saha[1,2] and Mainak Mukhopadhyay[1]*

[1]Department of Biotechnology, JIS University, Agarpara, West Bengal, India
[2]Department of Biochemistry, University of Calcutta, Ballygunge, West Bengal, India

Abstract

The biofuel industry and its biorefinery system have evolved into an alternative energy industry that can replace the non-renewable fossil fuel sources and its role in the transportation sector. An integrated biorefinery system can utilize the whole biomass and use it for the production of sugars, ethanol, and several value-added products. Sustainability development for biorefinery systems is measured through the indicators—efficiency, socio-economic viability, and effect on the environment. Various methodologies and strategies are utilized to assess and develop these indicators, but there have been limitations involved due to several policies which make it difficult to implement and assess these indicators on an industrial scale. A detailed investigation is required for fulfilling and overcome the challenges concerned with a sustainable biorefinery system.

This chapter will provide an outlook into the indicators involved with the sustainable development and assessment of biorefinery systems involved with the production of biofuels using various biomasses. The different types of biorefinery systems used for measurement of sustainability and economic performance will also be discussed.

Keywords: Biorefinery, bioethanol, refinery efficiency, sustainability indicators, economic feasibility

4.1 Introduction

Presently, the whole population and industry sector depend on fossil fuels as the primary source of energy, which contributes about 80% with 58% going to the transport sector. The major sources of fossil fuels are non-renewable and are getting depleted very swiftly, to meet the global demand [1]. Until recently, the use of fossil fuels has been determined to be the source of major greenhouse gases leading to biodiversity loss, melting of glaciers, rise in sea level, ultimately playing a major role in global warming and climate change. This has led the industrialization sector to look for a clean and green alternative source of energy, which can meet the global energy demands without being

**Corresponding author*: m.mukhopadhyay85@gmail.com

Arindam Kuila and Mainak Mukhopadhyay (eds.) Biorefinery Production Technologies for Chemicals and Energy, (67–76) © 2020 Scrivener Publishing LLC

harmful to the environment [2]. Over the years, researchers have identified quite a few alternative energy sources, out of which biofuels produced from sustainable biomass has proved to be one of the most efficient renewable alternatives to replace non-renewable fossil fuel sources [3].

Biofuels are mainly extracted from biomass and are of three types: solid, liquid, and gaseous biofuels. Based on the chemical and complex nature of the biomass from which biofuels have been extracted, they are divided into first, second, and third-generation biofuels [4]. For example, biodiesel and vegetable oils produced from crop plants are first-generation biofuels, biohydrogen, and bioethanol derived from agricultural and forestry by-products, lignocellulosic biomass is considered to be second-generation biofuels, whereas third-generation biofuels are mostly biogas and biobutanol, which are produced from marine resources like seaweeds, cyanobacteria, and algae because of their high biomass content and no land requirement for their growth [4, 5].

Predominantly, biomass is considered to be one of the most natural and inexpensive sources of energy that can be utilized in any form and at any time. Currently, the availability of biomass especially in tropical countries is estimated to be in million metric tons, which constitutes agricultural, forestry, and lignocellulosic biomass. Besides getting consumed in the form of food and meeting other reliable needs in the form of their by-products, these biomasses generate a huge amount of waste which are mostly thrown away in landfills where they get decayed or burnt [6]. The residues when discarded either cause air pollution or when released as sewage effluents get in the water bodies and get leached onto the groundwater and soil, wreaking havoc to the environment. Thus, biomass, if utilized effectively to produce biofuels, can help to reduce greenhouse emissions, environmental pollution and also aid in uplifting the rural economy [7].

In the past decade, there has been a boost in producing energy through agricultural development which has been directly related to economic growth. And with the rapid innovation in hydrocarbon technology, more improvement has been made towards biofuels and its biorefineries [8]. The biofuel economy has the means to take over the industry and transport sector in the future because of its eco-friendly nature, and advantages like extraction, high combustibility, and sustainability [9].

Biofuel sustainability is multi-faceted and inter-connected with the environment, economy, and socio-economic impact, which varies widely and depends upon the type of biofuels, its source, and production, considering the alternative sources and outcomes are regularly compared and evaluated [10]. Overall, studies and data have indicated the presence of both positive and negative impacts related to biofuels as a measure of sustainability. The positive impact of the production and utility of biofuels provides energy security and economic advancement in rural areas, improves the quality of urban life, and decreases air pollution by reducing global greenhouse gas emissions; whereas the problems related to the negative impact of biomass production related to land-use in agriculture, food crop competition, degrading water quality, water shortage, increased rate of deforestation, all contribute to the disadvantages of the biofuel economy [11, 12].

Thus, it is highly important to ensure a balance among the various aspects related to the sustainability of biofuels, in terms of both short-term and long-term benefits. And, it is also essential to innovate and develop more advanced technology to make the whole process of biofuels, from source to production, more efficient and help in its market growth.

4.2 Biofuels and Biorefineries: Sustainability Development and Economic Performance

The biorefinery system is an industrial technology developed for the upstream and downstream processing of different kinds of biomass which generates from forestry, agriculture containing both crop and lignocellulosic residues, as well as residues that arise from the various industries. A biorefinery mostly produces a whole range of intermediate and final products, which includes feedstocks, food materials, and chemicals, and also produces energy in the form of fuels, power, and heat. Biorefineries have become a contributor to the commercial and financial progress, with its immense benefits to the economy of several countries. It is touted as an emerging new market providing new job opportunities and competences, in a socially and environment-friendly manner [13, 14].

The main objective behind the development, innovation, and implementation of biorefineries for the production of biofuels, arises from the driving need to find alternatives to fossil fuels and gaining energy independence and security, climate change mitigation through reduced emissions, and improvement in the rural economy [15]. Despite the progress and expansion of the biofuel market, its development has slowed down over the years due to major conflicts in policy and challenges over the structure of biorefineries. This has led to the rise of sustainability issues with regard to food versus fuel competition, accounting the emission of greenhouse gases through the whole source to the production process, and the percentage of land committed to the cause [16]. By assessing and working out these sustainability and policy issues, the alternative fuel industry has the potential to become a key contributor to the agriculture and transport domain in the future.

One way to achieve sustainability is by following through respective design methodologies that are proposed according to the biorefinery concept in the industry, such as design, calibration, and optimization before it reaches to industrial scale. Over the last decade, improvement in integrated biorefinery systems has led to allowing assessment of features like raw materials, end products, processing routes, and technologies required along with energy analysis [17]. Other indicators which require to be taken into consideration before building and transitioning to a biofuel based economy are the economic outcome and environmental impact by the biorefinery systems and should be evaluated and examined by the decision-makers and the policymakers in the government or the business world to compare the technology and sustainability from a dual point of view [18].

Energy forms a viable and one of the most important indicators related to sustainable development. It is important in societal reforms like removing poverty, improving living standards, and helping in human welfare. The energy resources currently available are unsustainable and don't even reach millions of people especially in developing countries. To remedy this, governments and policymakers have added sustainable energy development as their forefront mission for the transition to a bio-based economy, giving special attention to providing an uplifting rural economy in developing countries [19]. However, models like the computable general equilibrium (CGE) made by the Scottish economy put forth the idea that increase in energy efficiency causes a rebound in the economy and ultimately backfires by decreasing the GDP to CO_2 emissions ratio, suggesting that the policies concerned with the enhancement of energy efficiency are not the only dependent factor related to environmental improvement [20].

The idea of biorefineries is similar to that of petroleum refineries in structure and function, irrespective of the different types of raw materials that are used in the two systems. To make the biorefineries more sustainable by reducing the costs of production and making the process more competitive, it's better to understand the impact and implications of the separation technologies that are involved in the biorefineries, with key challenges to certain separation processes that are not applicable and are unsustainable in the long run [21]. Other priorities in the biorefinery system include the type of biomass, their supply, and processing the feedstock according to their composition. The total agricultural input and emissions related to it are also taken into consideration, along with the preference of feedstock—which is much more sustainable if biomass rich in cellulose and hemicellulose is used in the integrated biorefinery system [22].

The study of a biorefinery complex in Thailand for sustainability assessment revealed the relationships involved between the maximal use of biomass to the reduction of greenhouse gas emissions, improved living conditions for the rural population, as well as a profitable economy based on the input and the output of the biorefinery [23]. Another type of biorefinery enterprise includes the production of multiple end-products and is dependent on the supply chains of the biomass. When an input-output model was integrated to assess the supply-chains of a biorefinery of thistle oil and its residues, in Sardinia (Italy), the findings showed that the economic viability and environmental sustainability is driven by the transportation distance of the raw material, which helps to decide the price level between the thistle producers and the owners of the biorefinery [24].

The use of the whole biomass like sugarcane has helped to develop a bio-based circular economy, where biorefineries are used to produce materials like biopolymers, biochemicals, biofuels, and bioenergy. Techno-economic assessment of various technologies that could be used successfully for this type of economy could help raise investment opportunities for government and stakeholders [25].

Research into cellulosic bioethanol biorefineries has led to the innovation of two alternative systems: centralized and distributed systems, whose comparison has led to the assessment of sustainability in terms of economic stability, suitability price of ethanol, and production efficiency concerning eco-friendliness. The centralized system has better economic stability in small-scale biorefineries, whereas the distributed system in large-scale biorefineries; the distributed system has a better environmental impact as compared to the centralized system, and also the distributed system proves to highly eco-efficient in large scale as to centralized systems being eco-efficient in small-scale systems [26]. Each type has its characteristics and can be attributed to the population as per the requirement.

Over the years, first-generation biofuels have garnered wide-scale criticism due to their competition with food crops, increased energy inputs, generating low yields; as compared to lignocellulosic feedstocks which are produced in abundance throughout the year and are easily available, and under biorefinery systems could produce a variety of by-products along with biofuels. Thus, life-cycle assessment as a property to measure the sustainability of lignocellulosic biomass biorefineries in terms of its environmental, economic, and social impacts could help to develop and increase the usage of biorefineries where lignocellulosic biomass is used [27]. Though lignocellulosic biorefineries have economic and environmental advantages, it hasn't seen any vibrant improvement in its systems because of the expenditures involved with its operation, seasonal biomass supply, unavailability of standardized procedures, and difficulty related with scaling-up of the system [28]. To counter the problems, it is essential to develop an integrated biorefinery system that can maximize the utilization of the whole of lignocellulosic biomass and increases the efficiency in synthesis of the value-added products [29].

Forestry residues are another important bioeconomy sector where the amalgamation of bioeconomy and circular economy has been used consistently to emphasize on the renewable sources as well as ameliorate the weaknesses involved with sustainability development. Analysis of a group of 22 forest biorefinery systems set in Finland and Sweden offered an insight into the sustainability of the combined system with increased benefits to the economy and environment [30].

Microalgae is also a valuable feedstock and has received a lot of interest as an important biomass for the generation of biofuels. But, its utilization for biofuel production only has made this resource economically unprofitable due to the overwhelming investment required for the whole production process. This has led to the design and innovation of special integrated microalgal biorefineries for the synthesis and production of multiple high-value products and different biofuels as well [31]. One such possibility of an integrated biorefinery system was by combining microalgae into sugarcane biorefinery. The same scheme of sugarcane biorefinery system was incorporated with microalgae to make the system economically and environmentally sustainable. Two simulations were performed—in the first case, streams rich in CO_2 were utilized for the cultivation and harvesting of microalgal biomass, after which the microalgal oil is further transformed to biodiesel, thus, jointly producing sugar, biofuel, and electricity. In the second case, *Chlorella* sp. is used in the for cultivation and harvesting of the microalgae in the CO_2 containing streams, from where the algal oil is used for the production of biofuels, sugar, electricity, as well as other value-added products, making the process both economically and environmentally viable [32].

Other commercialization techniques concerning microalgae have also proved effective in improving the microalgal biorefinery systems, like the de-oiled microalgal biomass, which is produced as a surplus by-product through the biorefinery system has shown a range of applications, for further use. This biomass is rich in protein, carbohydrates, and minerals that have been used as a feedstock, fertilizer, as well as substrates for the production of bioethanol and/or bio-methane. Also, further conversion of the biomass has produced industrially important chemicals and fuels along with novel materials like nanoparticles and carbon-dot, which has proved to be of importance in medical science. The de-oiled biomass has also found application as an adsorbent of toxic substances, heavy metals, and dyes that get released in the industrial effluents, which has been known to cause environmental harm [33]. Thus, microalgae have always proved to be a promising bioresource for biofuels production and other industrially important compounds, but challenges still exist for commercialization of microalgal biorefinery. One of the major challenges is the development of a cost-effective model of microalgal biorefinery which can be used for the production of algal biomass on an industrial scale [34].

The government and stakeholders should provide enough funds to promote as well as boost the biofuel industry by spending more on research to develop models and simulation for a sustainable and economically developed biorefinery for the industry.

4.3 Future Developments Required for Building a Sustainable Biorefinery System

The existing model of biorefineries that are used to assess its sustainability and economic efficiency are not sufficient and are mostly incomplete. The strategy used to design and simulate needs to consider the alternatives depending upon the composition of the biomass, whereas the optimization limit should be finalized keeping the industrial scale in mind [35]. If the optimization and alternative techniques work on the laboratory scale, it doesn't necessarily mean it's going to be successful in the industrial scale.

Biorefinery design is equally important to maintain the sustainability of the process with regard to efficiency, economic, social, and environmental standards. A biorefinery system should be able to maintain viable alternatives for the specific biomass it uses; otherwise, the biorefinery would not be successful in the long run. This is the reason academicians should closely work with stakeholders in the biofuel industries to understand the limitations and the requirements for the implementation of eco-efficient biorefineries on a massive scale [36].

The main challenges that are needed to overcome the limitations of an integrated biorefinery system include the development of a more efficient and sustainable biorefinery, innovation of structures and models that work in an industrial scale, and also maintains the socio-economic and environmental policies put forth by the government [37]. Progress is also required in areas of biomass supply chains that are to be maintained throughout the year, focusing on lignocellulosic biomass and also to address the food-fuel competition in terms of land requirement and CO_2 emissions [38].

Other improvements are also needed to be made in small-scale biorefineries for complete utilization of any biomass and along with the production of wholesale biofuels, also aid and develop the production of several value-added products [39]. A more detailed understanding and an investigation are required into small-scale biorefineries to understand the implications and its economic sustainability, to develop its efficiency and further innovate to study its outcome and limitations [40].

4.4 Conclusion

The alternative biofuel industry has developed over the last few decades to become a rising force against the fossil fuel industry. Biofuels produced from renewable natural sources can provide energy, feedstocks, and value-added products, which are economically and environmentally advantageous, as compared to the rising carbon emissions from the fossil fuel industry. To make biofuel production more sustainable, biorefinery systems modeled on the petroleum refineries have been developed to make the whole process more efficient, technologically advanced and environmentally friendly.

Recently, studies have revealed the refinery efficiency and other indicators assessed for measuring the sustainability of a biorefinery system needs to further investigated and improved and innovated to increase the implementation and reduce the limitations of the system. Research into the integrated biorefinery systems for utilization of whole biomass for effective biofuel production concerning alternative systems is one area where new methodologies and technologies could be modeled and simulated for further innovation.

References

1. Gaurav, N., Sivasankari, S., Kiran, G.S., Ninawe, A., Selvin, J., Utilization of bioresources for sustainable biofuels: A review. *Renew. Sust. Energ.*, 73, 205–214, 2017.
2. Chang, W., Hwang, J., Wu, W., Environmental impact and sustainability study on biofuels for transportation applications. *Renew. Sust. Energ. Rev.*, 67, 227–288, 2017.
3. Demirbas, A., Biofuels sources, biofuel policy, biofuel economy and global biofuel projections. *Energy Convers. Manage.*, 49, 8, 2106–2116, 2008.
4. Nigam, P.S. and Singh, A., Production of liquid biofuels from renewable resources. *Prog. Energy Combust. Sci.*, 37, 1, 52–68, 2011.
5. Bhowmick, G.D., Sarmah, A.K., Sen, R., Lignocellulosic biorefinery as a model for sustainable development of biofuels and value added products. *Bioresour. Technol.*, 247, 1144–1154, 2018.
6. Hamelinck, C.N. and Faaij, A.P.C., Outlook for advanced biofuels. *Energy Policy*, 34, 17, 3268–3283, 2006.
7. Escobar, J.C., Lora, E.S., Venturini, O.J., Yanez, E.E., Castillo, Almazan, O., Biofuels: Environment, technology and food security. *Renew. Sust. Energ. Rev.*, 13, 6–7, 1275–1287, 2009.
8. de Man, R. and German, L., Certifying the sustainability of biofuels: Promise and reality. *Energy Policy*, 109, 871–883, 2017.
9. Rajak, U., Nashine, P., Singh, T.S., Verma, T.N., Numerical investigation of performance, combustion and emission characteristics of various biofuels. *Energy Convers. Manage.*, 156, 235–252, 2018.
10. Robertson, G.P., Dale, V.H., Doering, O.C., Hamburg, S.P., Melillo, J.M., Wander, M.M., Wilhelm, W.W. *et al.*, Agriculture: Sustainable Biofuels Redux. *Science*, 322, 5898, 49–50, 2008.
11. Ghosh, P., Westhoff, P., Debnath, D., Biofuels, food security, and sustainability. *Biofuels, Bioenergy Food Sec.*, 11, 211–229, 2019.
12. Solomon, B.D., Biofuels and sustainability. *Ann. N. Y. Acad. Sci.*, 1185, 119–134, 2010.
13. de Jong, E. and Jungmeier, G., Biorefinery concepts in comparison to petrochemical refineries. *Ind. Biorefineries White Biotechnol.*, 3–33, 2015.
14. Luo, L., van der Voet, E., Huppes, G., Biorefining of lignocellulosic feedstock – Technical, economic and environmental considerations. *Bioresour. Technol.*, 101, 13, 5023–5032, 2010.
15. Luque, R., Herrero-Davila, L., Campelo, J.M., Clark, J.H., Hidalgo, J.M., Luna, D., Marinas, J.M., Romero, A.A., Biofuels: A technological perspective. *Energy Environ. Sci.*, 1, 5, 542, 2008.
16. Araújo, K., Mahajan, D., Kerr, R., da Silva, M., Global biofuels at crossroads: An overview of technical, policy, and investment complexities in the sustainability of biofuel development. *Agriculture*, 7, 4, 32, 2017.
17. Moncada, J., Aristizábal, M.V., Cardona A., C.A., Design strategies for sustainable biorefineries. *Biochem. Eng. J.*, 116, 122–134, 2016.
18. Gavrilescu, M., Biorefinery systems: An overview. *Bioenergy Res. Adv. Appl.*, 219–241, 2014.

19. Vera, I. and Langlois, L., Energy indicators for sustainable development. *Energy*, 32, 6, 875–882, 2007.
20. Hanley, N., McGregor, P.G., Swales, J.K., Turner, K., Do increases in energy efficiency improve environmental quality and sustainability? *Ecol. Econ.*, 68, 3, 692–709, 2009.
21. Kiss, A.A., Lange, J., Schurr, B., Brilman, D.W.F., van der Ham, A.G.J., Kersten, S.R.A., Separation technology - Making a difference in biorefineries. *Biomass Bioenerg.*, 95, 296–309, 2016.
22. Parajuli, R., Dalgaard, T., Jørgensen, U., Adamsen, A.P.S., Knudsen, M.T., Birkved, M., Gylling, M., Schjørring, J.K., Biorefining in the prevailing energy and materials crisis: A review of sustainable pathways for biorefinery value chains and sustainability assessment technologies. *Renew. Sust. Energ. Rev.*, 43, 244–263, 2015.
23. Gheewala, S.H., Bonnet, S., Prueksakorn, K., Nilsalab, P., Sustainability assessment of a biorefinery complex in Thailand. *Sustainability*, 3, 3, 518–530, 2011.
24. Yazan, D.M., Mandras, G., Garau, G., Environmental and economic sustainability of integrated production in bio-refineries: The thistle case in Sardinia. *Renew. Energ.*, 102, 349–360, 2017.
25. Mandegari, M.A., Farzad, S., Görgens, J.F., Recent trends on techno-economic assessment (TEA) of sugarcane biorefineries. *Biofuel Res. J.*, 15, 704–712, 2017.
26. Kim, S. and Dale, B.E., Comparing alternative cellulosic biomass biorefining systems: Centralized versus distributed processing systems. *Biomass Bioenerg.*, 74, 135–147, 2015.
27. Nanda., S., Azargohar, R., Dalai, A.K., Kozinski, J.A., An assessment on the sustainability of lignocellulosic biomass for biorefining. *Renew. Sust. Energ. Rev.*, 50, 925–941, 2015.
28. Chandel, A.K., Garlapati, V.K., Singh, A.K., Antunes, F.A.F., da Silva, S.S., The path forward for lignocellulosic biorefineries: Bottlenecks, solutions, and perspective on commercialization. *Bioresour. Technol.*, 264, 370–381, 2018.
29. Saini, J.K., Gupta, R., Hemansi, Verma, A., Gaur, P., Saini, R., Shukla, R., Kuhad, R.C., Integrated lignocellulosic biorefinery for sustainable bio-based economy. *Sustainable Approaches Biofuels Prod. Technol.*, 7, 25–46, 2018.
30. Temmes, A. and Peck, P., Do forest biorefineries fit with working principles of a circular bioeconomy? A case of Finnish and Swedish initiatives. *For. Policy Econ.*, 110, 101896, 2019.
31. Zhu, L., Biorefinery as a promising approach to promote microalgae industry: An innovative framework. *Renew. Sust. Energ. Rev.*, 41, 1376–1384, 2015.
32. Moncada, J., Tamayo, J.A., Cardona, C.A., Integrating first, second, and third generation biorefineries: Incorporating microalgae into the sugarcane biorefinery. *Chem. Eng. Sci.*, 118, 126–140, 2014.
33. Maurya, R., Paliwal, C., Ghosh, T., Pancha, I., Chokshi, K., Mitra, M., Ghosh, A., Mishra, S., Applications od de-oiled microalgal biomass towards development of sustainable biorefinery. *Bioresour. Technol.*, 214, 787–796, 2016.
34. Chia, S.R., Chew, K.W., Show, P.L., Yap, Y.J., Ong, H.C., Ling, T.C., Chang, J., Analysis of economic and environmental aspects of microalgae biorefinery for biofuels production: A review. *Biotechnol. J.*, 13, 1700618, 2018.
35. Acheampong, M., Ertem, F.C., Kappler, B., Neubauer, P., In pursuit of Sustainable Development Goal (SDG) number 7: Will biofuels be reliable? *Renew. Sust. Energ. Rev.*, 75, 927–937, 2017.
36. Hingsamer, M. and Jungmeier, G., Biorefineries. The role of Bioenergy in the Bioeconomy, 179–222, 2019.
37. Aristizábal-Marulanda, V. and Cardona Alzate, C.A., Methods for designing and assessing biorefineries: Review. *Biofuels, Bioprod. Biorefin.*, 13, 789–808, 2018.
38. Schaidle, J.A., Moline, C.J., Savage, P.E., Biorefinery sustainability assessment. *Environ. Prog. Sust. Energ.*, 30, 4, 743–753, 2010.

39. Loaiza, S.S., Aroca, G., Cardona, C.A., Small-scale biorefineries: Future and perspectives. *Biorefineries: Concepts Adv. Res.*, 39–72, 2017.
40. Wellisch, M., Jungmeier, G., Karbowski, A., Patel, M.K., Rogulska, M., Biorefinery systems – potential contributors to sustainable innovation. *Biofuels, Bioprod. Biorefin.*, 4, 3, 275–286, 2010.

5

Biorefinery: A Future Key of Potential Energy

Anirudha Paul[1], Sampad Ghosh[2], Saptarshi Konar[3] and Anirban Ray[4]*

[1] India Dairy Products Limited, Dankuni, West Bengal, India
[2]Department of Chemistry, Nalanda College of Engineering, Nalanda, Bihar, India
[3] Molecular Biology and Biotechnology Department, University of Kalyani, Kalyani, West Bengal, Kalyani, India
[4]Department of Microbiology, R.G. Kar Medical College, Kolkata, West Bengal, India

Abstract
For a long time, fossil fuels provide endless source of energy to civilization and development of human races globally. However, abrupt hike in fossil fuel price, depletion of underground reserves, localization or distribution pattern of those reserves, discharge of toxic byproducts and fatal climatic changes, and processes like biorefinery have drawn attentions of scientists, which is more bio-based, eco-friendly, and cost effective energy source that may fulfill future endless energy demand worldwide. Besides being eco-friendly and cost effective, this process could take a hand over sustainable waste management, even may reduce greenhouse gas emission. Biorefinery facility provides equipment for conversion of biomass into energy (heat, fuel, gas) and production or separation of multiple value added chemical by-products. However, being on starting footsteps, a variety of further research contribution is needed to stabilize and establish biorefinery industries. Both governmental and non-governmental approaches are required to fulfill the future goal.

Keywords: Biorefinery, fossil fuel, waste management, biomass, biodegradable, biomass processing, types of biorefinery, green technology

5.1 Introduction

Life means energy. Without energy, there is no life and no progression of life. In every single cell, ATP is the primary source of energy or fuel that helps a cell for proper functioning. Therefore, fuel is the most essential source for development. Apart from survival, progression and civilization are also fuel dependent. For a long time, fossil fuels play a key role as being major source of energy. But fossil fuel is not renewable and will be exhausted in near future [1]. Furthermore, the reservoirs of fossil fuel are also localized to some limited places [2]. Apart from that, from the environmental safety point of view, processing and refining of such fuel causes release of green house gases such as CO_2 in great amount and the waste materials are also toxic and non-biodegradable. It can also be predicted that due to high consumption rate and public demand, the price of fossil fuel will become high in next decades and so on.

**Corresponding author*: anirbanrayiitkgp@gmail.com

Arindam Kuila and Mainak Mukhopadhyay (eds.) Biorefinery Production Technologies for Chemicals and Energy, (77–88) © 2020 Scrivener Publishing LLC

To overcome such dilemma and to find a new alternative source of energy, a great amounts of researches are now oriented towards biorefinery concepts [3]. Further study indicates that such biofuels derived from biomass have great impact on environmental safety. Processing of such biofuels involve no production of sulfur-containing byproducts and non-biodegradable cyclic aromatic hydrocarbon [4]. Biorefinery concept is quite similar to petrochemical refinery but it uses bio-based raw materials to generate fuel, energy, and heat, although the working principle and processes are quite some different [5]. It is a multifunctional process which can maximize the economic value of the feedstock along with waste management in a sustainable way [6].

5.2 Biorefinery: Definitions and Descriptions

Over the years, different organizations and scientists had described biorefinery by different way to define its activity and fruitfulness. Here, some definitions of biorefinery are mentioned. The first ever definition was first introduced in 1997 to clarify green biorefinery. According to this, the Green biorefineries are complex systems based on ecological technology for comprehensive (holistic), material, and energy utilization of renewable resources and natural materials using green and waste biomass and focalizing on sustainable regional land utilization [7]. According to American Department of Energy (DOE), a biorefinery is an overall concept of a processing plant where biomass feedstock are converted and extracted into a spectrum of valuable products. Its operation is similar to that of petrochemical refineries [8, 9].

According to the American National Renewable Energy Laboratory (NREL), a biorefinery is a facility that integrates biomass conversion processes and equipment to produce fuels, power, and chemicals from biomass. The biorefinery concept is analogous to today's petroleum refineries, which produce multiple fuels and products from petroleum. Industrial biorefineries have been identified as the most promising route to the creation of a new domestic bio-based industry. Whereas, International Energy Agency (IEA) considers the biorefineries as the sustainable processing of biomass into a spectrum of marketable products (food, feed, materials, chemicals) and energy (fuels, power, heat) [10]. Some other definitions are as follows: Biorefining intensifies the uses of biomass for building platform molecules [11]. A forest biorefinery is a multi-product factory that integrates biomass conversion processes and equipment to produce fuels and chemicals from wood-based biomass [12]. The biorefinery is an integrated system of bio-based firms, able to produce a wide range of goods from biomass raw materials (chemicals, bio-fuels, food and feed ingredients, biomaterials, including fibers and power) using a variety of technologies, maximizing the value of the biomass [13]. Biorefining is the transfer of the efficiency and logic of fossil-based chemistry and substantial converting industry as well as energy production onto the biomass industry [14]. Biorefinery systems are systems in which biomass can be utilized entirely by conversion through multiple processes into a number of valuable products [15]; therefore, a biorefinery can utilize a variety of raw materials including green land, marine, and forestry materials; the process also helps in the crop waste management as the land field wastes can be used in some biorefinery industry. Other than this, a biorefinery can able to produce different value added products along with beneficial chemicals. A biorefinery may apply a variety of processes including, thermal, mechanical, chemical, and biological processes simultaneously. On the other hand, it should encourage the sustainability in different various aspects from environmental, social, and economical value point. It should also be capable of manufacturing a variety of beneficial products that are market competitive from economical point of view. High price will not able to encourage population demands. Apart from this, a biorefinery is also able to produce bio-fuels from different raw materials which in turn lowers the emission the green house gases as in case of petro chemical refining.

5.3 Modus Operandi of Different Biorefineries

Depending on the concept and fruitful results, mainly four types of processes are carried out in different biorefinery industries. These include thermo-chemical, mechanical, biochemical, and chemical processing. These processes vary depending on the feedstocks and on the desirable end products and on intermediates.

5.3.1 Thermochemical Processing

In this category, feedstocks are subjects to either gasification or pyrolysis or direct combustion to produce energy, heat, and fuel. In gasification, biomass is treated at high temperature (~700°C) and in low oxygen concentration. In such condition, biomass is converted to syngas (CO, CH_4, H_2, and CO_2). Further downstream processing yields a variety of chemicals and fuel can be used as a source of heat/electricity. The pyrolysis process is maintained at low temperature (300°C–600°C) compared to gasification and at anaerobic environment. The resultant will be pyrolysis oil, charcoal, and combination of some gases. Such pyrolysis oil and charcoal can be used as fuel [16]. In direct combustion process, feedstocks are subjected to oxidation for heat and electricity generation.

5.3.2 Mechanical Processing

In this technique, raw materials are subjected to cutting for achieving particular/desirable size and shape. Mechanical processing is also applied in some other processes such are those including filtration and separation of intermediate and final products [17].

5.3.3 Biochemical Processing

Fermentation and digestion are the two main processes under this group. In fermentation process, either microbes or different enzymes convert raw materials into products such as alcohol or organic acids. Special classes of microbes or microbial enzymes are applied depending on the choice of raw materials and desirable end products [18]. In another process, industrial effluent, sewage, and land field wastes are subjected to anaerobic digestion. Different combination of biogas is formed in this process including CH_4 and CO_2 including some artefacts. Sometimes, depending on future application, biogas is converted to bio methane and can be used in transportation and power grid [19].

5.3.4 Chemical Processing

Depending on the choice of final products, a variety of chemical processes are applied on raw materials. Such in case of polysaccharides, proteins and fats, generally, hydrolysis and trans-esterification, are mainly applied. These processes convert the polymers into simple constituents for further treatment. Some other processes are like conversion of syngas to ethanol [20] or conversion of syngas to CH_4 [21].

5.4 Types of Biorefineries

Before 2008, there was no definite classification system on the basis of which biorefineries can be classified. However, biorefinery systems were categorized on different aspects [22, 23] on that time, which includes

1. Types of conversion process (thermochemical biorefineries, biochemical biorefineries)
2. Types of raw materials (feedstock biorefineries, oleochemical biorefineries, green biorefineries, etc.)
3. Types of intermediates (sugar platform biorefineries, syngas platform biorefineries)
4. Types of technology implied (first generation, second generation, third generation biorefineries, conventional, and advanced biorefineries).

In the year of 2008, a more realistic and suitable classification system was developed by IEA Bioenergy Task 42 [24–26]. According to this system, every step, *viz.*, from starting biomass to beneficial end product production, can be represented schematically. This new classification system can improve the development of future biorefinery industries leading to the production of cost effective and efficient production of biofuels along with economically beneficial bio-based by-products. This biorefinery classification system was developed depending on four main attributes such as feedstock, platforms, products, and conversion process.

5.4.1 Lignocellulose Feedstock Biorefinery

This type of biorefinery systems utilizes lignocellulose biomass such as wood to yield bioenergy *via* intermediates such as lignin, cellulose, and hemicellulose. This cellulose and hemicellulose are mainly consisting of polymers of five or six carbon sugars, which is in tight association with lignin. In this type of biorefinery, lignocellulosic biomass is being converted to cellulose and hemicellulose by acid or alkali treatment. Furthermore, these intermediates are subjected to enzymatic degradation to produce five and six carbon sugars [27]. This lignocellulose-based biorefinery is a good source of concurrent production of pulp and energy [28, 29], whereas the pulp-dependent biorefineries provide economically beneficial products such as lignosulfonates, vanillin, rosin, and tall oil [30]. These C5 and C6 sugars are also being used as feedstock in different fermentation processes to yield biofuels or sometimes for catalytic conversion or to produce value added end products [27, 31]. This type of biorefinery draws high level of attention as it can produce fuels and value added chemicals, thus minimizes petroleum dependency, which in turn reduces green house gas emission and improve air quality.

5.4.2 Syngas Platform Biorefinery

The main working principle behind the syngas or synthesis gas biorefinery is that either by extreme heat or by application of autotrophic acetogenic bacteria, a wide range of carbonaceous feedstocks is subjected to degradation. Syngas is mainly consist of CO and H_2, which can be further subjected to a value added chemicals and biofuels [32]. Additionally, methanol, ethanol, and ammonia can also be produced by syngas fermentation [33]. The main obstacles in this biorefinery system are the complexity of technology and production cost. However, suitable biochemical or thermochemical techniques can add values to this system.

5.4.3 Marine Biorefinery

Earth consists of near about 71% of water covering its surface, among which 97% is sea water. Sea water possesses a variety of minerals and nutrients, which confers it as a suitable media for bio-production or fermentation [34]. Beside this, as it is accessible to many countries worldwide and renewable property sea water can be used as excellent media for bio-fermentation [35]. In marine biorefinery, mainly microalgae and seaweeds are used as starting material or feedstock. The main advantages in this process are there is no competition for production land with crops or foods and high bio growth rate. Besides, microalgae and seaweeds could provide high amounts of lipid, protein, and carbohydrate [36]. They can also be used as bio-fuel, bioethanol, and bio-gas production [37]. Brown algae contain fucoidans and laminaran having potential health beneficial activities [38, 39]. Several different industries such as agri food, biopolymer, and cosmetics industries use seaweeds as feedstock [40, 41]. In addition, this type of biorefinery can add value to other application such as aqua culture, wastewater treatment, and biogas installation. Due to worldwide accessibility of sea water, this process could minimize the transportation cost of oil/fuel.

5.4.4 Oleochemical Biorefinery

Chemical compounds such as plant and animal fats are collectively termed as oleochemicals. These olechemicals include fatty acids, fatty esters, glycerol, methyl esters, and fatty amines [42, 43]. Such bio-based feedstocks are available worldwide. In other hand, such chemicals can be cultivated easily with very low production cost. Having such advantages, oleochemicals draw attention as sustainable and cost effective energy supply [44, 45]. Generally, crops are subjected high temperature to produce triglycerides. Now, these triglycerides are further converted to basic components such as glycerol and fatty acids esters by trans-esterification. Biodiesel is safer than petro diesel as it is less toxic and less flammable and biodegradable [46]. It can also improve lubricating property. Besides biodiesel, surfactants can also be produced from plant oils [23]. Glycerol, produced by the transesterification reaction, also can be used as fermentation and anaerobic digestion.

5.4.5 Green Biorefinery

Green biorefinery is generally referred to combine use of green/grassland biomass for the production of energy and chemicals. In this process, silage or grass is preferred as raw material. Green biomass is primarily subjected to chemical, physical, or biological treatment to separate liquid fraction (press juice) and solid fraction (press cake). The solid fraction generally contains fibrous materials of plant biomass (cellulose, hemicelluloses, lignin), whereas liquid fraction contains proteins, amino acids, organic acids, minerals, and sugars. However, the composition of liquid fraction may vary depending on plant species and harvesting time [47]. Now, separated fresh juice is being used for bioethanol production [48] or feedstock for other value added chemicals such as plant biostimulant [49], lactic acids [50], and proteins [51]. The waste materials can be used as feedstock for renewable energy source such as biogas in anaerobic digester. The fiber cake produced during separation also can add value. It can be further used as feedstock for cellulose production [52] and bioethanol production [53].

5.4.6 Whole Crop Biorefinery

As the name indicates, generally, different cereals including wheat, rye, maize, and triticale are generally used as the raw materials in whole crop biorefinery. In this technique, mechanical process is utilized to separate corn and straw. The processed straw can be further utilized in lignocellulose biorefinery or can be used as the raw material for the syngas production. The corn portion can be used as meal or can be subjected chemical modification or biochemical modification to produce starch [54].

5.5 Some Biorefinery Industries

From the developmental and functioning point of view, a biorefinery is considered to be an industrial consortium where different units/factories work together as symbiotic relationship by providing each other intermediate products and thus enhance its production and acquisition from market demand point of view. Many biorefinery companies or entities have developed over time, considering these key features. Here are some examples of those.

5.5.1 European Biorefinery Companies

British Sugar

British sugar, a United Kingdom–based company, generally involves in the production of sugar, bioethanol, and animal feed. Primarily sugar is extracted from sugar beet and the residual pulp is served as animal feed. Both dried and moist pulps are processed here. Following biological fermentation sugar is converted to bioethanol. (http://www.britishsugar.co.uk/Bioethanol.aspx).

Grassa

It is a Dutch biorefinery company based on green biorefinery. This company mainly utilizes ryegrass as raw materials to produce proteins for animal feed and waste being involved in pulp and paper manufacturing (http://www.grassanederland.nl).

5.5.2 Biorefinery Companies in USA

Enerkem

This is a Canada-based company that generally involved in the production of biofuels and green chemicals from non-recyclable wastes. Such biofuel is renewable, non-toxic, clean,

water soluble, and highly biodegradable in nature. Apart from ethanol, other chemicals such as methanol, n-propanol, acrylic acid, olefins, and acetic anhydride are also produced here (http://www.energy-xprt.com/companies/enerkem-35896).
Spero Energy
This company focuses on the production of renewable fuels and value added chemicals from non-food biomass. This company utilizes a special extraction mechanism to isolate ferulic acid, which is used in the production of vanilla. Other than biofuel, some biodegradable chelates are processed here, which are used in agriculture, cosmetics and food (http://www.speroenergy.com).
Arbiom
Arbiom generally involves in the production of beneficial protein feed from non-food feedstocks such as wood. Such high value added protein rich products are generally used as animal feed and in aquaculture (http://www.arbiom.com).

5.5.3 Biorefinery Companies in Asia

Sinopec
Sinopec is a China-based company that involves in the production of bio fuels from waste oil. A variety of other products such as lubricants and catalysts are also processed here (http://www.sinopecgroup.com/group/en/).
Wilmar
Asia's largest and leading agribusiness group is now involved in this tradition. It is mainly involved in the sugar milling and refining, oil cultivation, oil seed crushing, oleochemicals, biodiesel, and fertilizer production (http://www.wilmar-international.com).

5.6 Conclusion and Future of Biorefinery

From the earlier discussion, we have come to know that biorefinery system has remarkable potential to be the most effective renewable source of energy in near future. Other than being renewable energy source, there are some other advantages such as

1. The system is eco-friendly and bio-based.
2. Decreases the emission rate of green house gases comparing to petrochemical refining.
3. Can provide sustainable waste management, in case where waste materials can be used as raw materials.
4. Single type of biomass can be converted into a variety of desirable beneficial intermediates and end products.
5. Raw material availability is not always localized as in case of fossil fuel.

However, to make the process more effective in near future, more research is needed on product evaluation, process integration, and on economical sustainability point of view. Biorefinery can provide a variety of beneficial eco friendly biodegradable products and intermediates. But most of the research and production is now limited to biofuel or biodiesel production. Further research should be done in the production of other value added chemicals to minimize or compensate production cost. Large amount of other products with low market price or low amount beneficial products with high market demand can be equally important to cope up with production cost. From the technological point of view, care should be taken to minimize the process energy requirement. A better and improve technological research facilities are also require in case of intermediate product separation and purification. Research up gradation also required to fuse different processes for accurate functioning. Apart from construction and development, market analysis is another key point, considering the future position of biorefinery companies. Market is not any existing characteristics. It has to be built by analyzing public demand, seasonal factors, rate of product supply, and transportation facilities. Markets are generally formed where there is better transport communication and public localization. From economic point of view better government interference or good market policies are also needed to strengthen market stability. Sellers and buyers should have equal proper information along with awareness. A special economical research and market survey will guide future biorefinery companies to have a strong hold over the existing market. Although the biofuel market is surprisingly successful and has a better economical growth, but the public demand towards other biorefinery chemicals is quite stagnant. Some time choice of market place or production unit is another key factor. Generally, a biorefinery having production capacity of animal feeds and biofertilizer along with biofuel should have better impact in rural areas rather than in township. Rural awareness will increase in such case, as most of the rural civilization is dependent on farming and production of crop. Such process integration will minimize the transportation cost. In some cases, transportation of raw material or feedstock is a point of concern. Feedstocks having high water content are the main entity in such case. Several months might require transporting raw material from harvesting area to production unit. On the other hand, harvesting is mainly limited to shorter period of time over a year; therefore, special storage facilities are required to preserve the raw materials. However, being a renewable energy source, much better knowledge is also necessary on other existing renewable energy (air, water) sources for implementation and development of biorefinery. Attention should be taken in the choice of raw material. Raw materials should not be in competition with substances that are the primary source of food.

References

1. Woldeyohannes, A.D., Woldemichael, D.E., Baheta, A.T., Sustainable renewable energy resources utilization in rural areas. *Renew. Sust. Energy Rev.*, 66, 1–9, 2016.
2. Daza, L.V., Solarte, J.C., Serna, S., Chacón, Y., Cardona, C.A., Agricultural waste management through energy producing biorefineries: The Colombian case. *Waste Biomass Valorization*, 7, 4, 789–798, 2016.
3. Chew, K.W., Yap, J.Y., Show, P.L., Suan, N.H., Juan, J.C., Ling, T.C., Lee, D.J., Chang, J.S., Microalgae biorefinery: High value products perspectives. *Bioresour. Technol.*, 229, 53–62, 2017.
4. Sarbolouki, M.N. and Moacanin, J., Chemicals from biomass. The U.S. prospects for the turn of the century. *Sol. Energy*, 25, 303–315, 1980.
5. van Haveren, J., Scott, E.L., Sanders, J., Bulk chemicals from biomass. *Biofuels Bioprod. Bioref.*, 2, 1, 41–57, 2008.

6. Thomsen, M., Complex media from processing of agricultural crops for microbial fermentation. *Appl. Microbiol. Biotechnol.*, 68, 598–606, 2005.
7. Kamm, B., Kamm, M., Soyez, K. (Eds.) Die Gr€une Bioraffinerie/The Green Biorefinery, in: *Technologiekonzept, Proceedings of 1st international symposium green biorefinery/Gru¨ne Bioraffinerie*, Berlin, Neuruppin, Oct 1997, 1998.
8. Kamm, B., Kamm, M., Gruber, P., Biorefinery systems – an overview, in: *Biorefineries – Industrial processes and products. Status Quo and future directions*, vol. 1, B. Kamm, M. Kamm, P. Gruber (Eds.), pp. 3–40, Wiley-VCH, Weinheim, 2006.
9. Kamm, B., Gruber, P., Kamm, M., *Biorefineries – industrial processes and products. Ullmann's encyclopedia of industrial chemistry Bio-based Chemicals – Value Added Products from Biorefineries*, Ullmann's Encyclopedia of Industrial Chemistry, 2016, https://doi.org/10.1002/14356007.l04_l01.pub2
10. de Jong, E., Higson, A., Walsh, P., Wellisch, M., Product developments in the bio-based chemicals arena. *Biofuels Bioprod. Bioref.*, 6, 606–24, 2012.
11. Debref, R., The Paradoxes of Environmental Innovations: The Case of Green Chemistry. *J. Innov. Econ.*, 9, 83, 2012.
12. Hämäläinen, S., Näyhä, A., Pesonen, H.-L., Forest biorefineries – A business opportunity for the Finnish forest cluster. *J. Clean Prod.*, 19, 1884–1891, 2011.
13. Lopolito, A., Morone, P., Sisto, R., Innovation niches and socio-technical transition: A case study of bio-refinery production. *Futures*, 43, 27–38, 2011.
14. Kamm, B., Gruber, P.R., Kamm, M., *Ullmann's Encycl Ind Chem*, pp. 659–688, Wiley, USA, 2012.
15. Menrad, K., Klein, A., Kurka, S., Interest of industrial actors in biorefinery concepts in Europe. *Biofuels Bioprod. Bioref.*, 3, 384–394, 2009.
16. Cherubini, F., *Life cycle assessment of biorefinery systems based on lignocellulosic raw materials – concept development, classification and environmental evaluation (PhD thesis)*, University of Technology at Graz, Austria, 2009.
17. Huang, H.-J., Ramaswamy, S., Tschirner, U.W., Ramarao, B.V., A review of separation technologies in current and future biorefineries. *Sep. Purif. Technol.*, 62, 1, 1–21, 2008.
18. Kaparaju, P., Serrano, M., Thomsen, A.B., Kongjan, P., Angelidaki, I., Bioethanol, biohydrogen and biogas production from wheat straw in a biorefinery concept. *Bioresour. Technol.*, 100, 9, 2562–2568, 2009.
19. Bruijstens, A.J., Beuman, W.P.H., Molen, M.v.d., Rijke, J.d., Cloudt, R.P.M., Kadijk, G., Camp, O.o.d., Bleuanus, S., *Biogas Composition and Engine Performance, Including Database and Biogas Property Model*, BIOGASMAX, 2008, January. http://www.biogasmax.eu/media/r3_report_on_biogas_composition_and_engine_performance__092122100_1411_21072009.pdf.
20. Naik, S.N., Production of first and second generation biofuels: A comprehensive review. *Renew. Sust. Energy Rev.*, 14, 578–597, 2010.
21. Gassner, M. and Maréchal, F., Thermo-economic process model for thermochemical production of Synthetic Natural Gas (SNG) from lignocellulosic biomass. *Biomass Bioener.*, 33, 11, 1587–1604, 2009.
22. Kamm, B. and Kamm, M., Biorefinery – systems. *Chem. Biochem. Eng. Q.*, 18, 1, 1–6, 2004.
23. Kamm, B. and Kamm, M., Principles of biorefineries. *Appl. Microbiol. Biotechnol.*, 64, 137–45, 2004.
24. Cherubini, F., The biorefinery concept: Using biomass instead of oil for producing energy and chemicals. *Energy Convers. Manage.*, 51, 7, 1412–1421, 2010.
25. Cherubini, F., Jungmeier, G., Wellisch, M., Willke, T., Skiadas, I., van Ree, R. *et al.*, Toward a common classification approach for biorefinery systems. *Biofuels Bioprod. Bioref.*, 3, 5, 534–46, 2009.

26. Jungmeier, G., Cherubini, F., Dohy, M., de Jong, E., Jørgensen, H., Mandl, M. *et al.*, Definition and classification of biorefinery systems? The approach in IEA Bioenergy Task 42 biorefineries. *Presentation held at the biorefinery course adding value to the sustainable utilisation of biomass.* Ghent, Belgium, June 12, 2009.
27. de Jong, E. and Gosselink, R.J.A., Lignocellulose-based chemical products, in: *Bioenergy research: Advances and applications*, V.K. Gupta, Kubicek CP, J. Saddler, F. Xu, Tuohy MG (Eds.), pp. 277–313, Elsevier;, Amsterdam (The Netherlands), 2014.
28. Chambost, V. and Stuart, P.R., Selecting the most appropriate products for the forest biorefinery. *Ind. Biotechnol.*, 3, 2, 112–9, 2007.
29. Chambost, V., Eamer, B., Stuart, P.R., Forest biorefinery: Getting on with the job. *Pulp. Pap. Can.*, 108, 2, 18–22, 2007.
30. Mleziva, M.M. and Wang, J.H., In: Matyjaszewski K, Möller M, editors. *Polymer science: a comprehensive reference*, in: *Polymers for a sustainable environment and green energy*, vol. 10, J.E. McGrath, M.A. Hickner, R. Höfer (Eds.), pp. 397–410, Elsevier, Amsterdam; Waltham, Oxford, 2012.
31. de Jong, W.E., Higson, A., Walsh, P., Wellisch, M. (Eds.). *Bio-based Chemicals - Value Added Products from Biorefineries*, IEA Bioenergy - Task 42 Biorefinery, Wiley-VCH, (2012) 2007.
32. Huber, G.W., *Breaking the chemical and engineering barriers to lignocellulosic biofuels: Next generation hydrocarbon biorefineries*, National Science Foundation. Chemical, Bioengineering, Environmental, and Transport Systems Division, 2008.
33. Ineosbio, [Online] [Cited: 17 January 2011.] www.ineosbio.com/57-Welcome_to_INEOS_Bio.html.
34. Lin, C.S.K., Luque, R., Clark, J.H., Webb, C., Du, C., A seawater-based biorefining strategy for fermentative production and chemical transformations of succinic acid. *Energy Environ. Sci.*, 4, 1471–1479, 2011.
35. Zaky, A.S., Tucker, G.A., Daw, Z.Y., Du, C., Marine yeast isolation and industrial application. *FEMS Yeast Res.*, 14, 813–825, 2014.
36. Roesijadi, G., Jones, S.B., Zhu, Y., Macroalgae as a Biomass Feedstock: A Preliminary Analysis. *Analysis*, 1–50, 2010.
37. Laurens, L.M.L., McMillan, J.D., Baxter, D., Cowie, A.L., Saddler, J.N., Barbosa, M. *et al.*, *State of Technology Review - Algae Bioenergy*, 2017. https://www.ieabioenergy.com/wp-content/uploads/2016/01/Laurens-Algae-Bioenergy-Report-IEA-webinar-170124-final-rev1.pdf.
38. Maehre, H.K., Malde, M.K., Eilertsen, K.-E., Elvevoll, E.O., Characterization of protein, lipid and mineral contents in common Norwegian seaweeds and evaluation of their potential as food and feed. *J. Sci. Food Agric.*, 94, 3281–90, 2014.
39. Rioux, L.E., Turgeon, S.L., Beaulieu, M., Structural characterization of laminaran and galactofucan extracted from the brown seaweed Saccharina longicruris. *Phytochemistry*, 71, 1586–95, 2010.
40. Abenavoli, L.M., Cuzzupoli, F., Chiaravalloti, V., Proto, A.R., Traceability system of olive oil: A case study based on the performance of a new software cloud. *Agron. Res.*, 14, 4, 1247–1256, 2016.
41. Davis, T.A., Volesky, B., Mucci, A., A review of the biochemistry of heavy metal biosorption by brown algae. *Water Res.*, 37, 4311–30, 2003.
42. Noor Armylisas, A.H., Siti Hazirah, M.F., Yeong, S.K., Hazimah, A.H., Modification of olefinic double bonds of unsaturated fatty acids and other vegetable oil derivatives *via* epoxidation: A review. *Grasas Aceites*, 68, 1, 2017. http://grasasyaceites.revistas.csic.es/index.php/grasasyaceites/rt/captureCite/1640/2040/ApaCitationPlugin.
43. Rupilius, W. and Ahmad, S., Palm oil and palm kernel oil as raw materials for basic oleochemicals and biodiesel. *Eur. J. Lipid Sci, Technol.*, 109, 4, 433–439, 2007.

44. Fiorentino, G., Ripa, M., Ulgiati, S., Chemicals from biomass: Technological versus environmental feasibility. A review. *Biofuels Bioprod. Bioref.*, 11, 1, 195–214, 2017.
45. Biermann, U., Bornscheuer, U., Meier, M.A.R., Metzger, J.O., Schäfer, H.J., Oils and fats as renewable raw materials in chemistry. *Angew. Chem.*, 50, 3854–3871, 2011.
46. Tan, S.G. and Chow, W.S., Biobased epoxidized vegetable oils and its greener epoxy blends: A review. *Polymer-Plastics Technol. Eng.*, 49, 15, 1581–1590, 2010.
47. Andersen, M. and Kiel, P., *Method for treating organic waste materials*, European Patent Application WO 00/56912, 1999.
48. Martel, C.M., Warrilow, A.G.S., Jackson, C.J., Mullins, J.G.L., Togawa, R.C., Parker, J.E., Morris, M.S., Donnison, I.S., Kelly, D.E., Kelly, S.L., Expression, purification and use of the soluble domain of Lactobacillus paracasei b fructosidase to optimise production of bioethanol from grass fructans. *Bioresour. Technol.*, 101, 4395–4402, 2010.
49. Mora, V., Bacaicoa, E., Zamarreno, E.-M., Aguirre, E., Garnica, M., Fuentes, M., Garcia-Mina, J.M., Action of humic acid on promotion of cucumber shoot growth involves nitrate related changes associated with the root-to-shoot distribution of cytokinins, polyamines and mineral nutrients. *J. Plant Physiol.*, 167, 633–642, 2010.
50. Thomsen, M.H., Bech, D., Kiel, P., Manufacturing of stabilized brown juice for L-lysine production—From university lab scale over pilot scale to industrial production. *Chem. Biochem. Eng. Q.*, 18, 37–46, 2004.
51. Hulst, A.C., Ketelaars, J.J., Sanders, J.P., Separating and recovering components from plants, 2000. Patent WO 00/40788; PCT/NL99/00805.
52. Sharma, H.S.S., Carmichael, E., Muhamad, M., McCall, D., Andrews, F., Lyons, G., McRoberts, C., Hornsby, P.R., Biorefining of perennial ryegrass for the production of nanofibrillated cellulose. *RSC Adv.*, 2, 6424–6437, 2012.
53. Neureiter, M., Danner, H., Fruhauf, S., Kromus, S., Thomasser, C., Braun, R., Narodoslawsky, M., Dilute acid hydrolysis of press cakes from silage and grass to recover hemicellulose-derived sugars. *Bioresour. Technol.*, 92, 21–29, 2004.
54. Nonato, R.V., Mantellato, P.E., Rossel, C.E.V., Integrated production of biodegradable plastic, sugar and ethanol. *Appl. Microbiol. Biotechnol.*, 57, 1–5, 2001.

Part 2

BIOREFINERY FOR PRODUCTION OF CHEMICALS

6

Biorefinery for Innovative Production of Bioactive Compounds from Vegetable Biomass

Massimo Lucarini[1*], Alessandra Durazzo[1], Ginevra Lombardi-Boccia[1], Annalisa Romani[2], Gianni Sagratini[3], Noemi Bevilacqua[4], Francesca Ieri[2], Pamela Vignolini[2], Margherita Campo[2] and Francesca Cecchini[4]

[1]CREA-Research Centre for Food and Nutrition, Via Ardeatina, Rome, Italy
[2]PHYTOLAB, University of Florence, Sesto Fiorentino, Firenze, Italy
[3]School of Pharmacy, University of Camerino, Via Sant'Agostino 1, Camerino, Italy
[4]CREA - Research Centre for Vitinculture and Enology, Velletri, Roma, Italy

Abstract
The "Universal Recovery Strategy" for the commercial recovery of valuable compounds from food wastes represents new goal of the circular bioeconomy and the biorefinery concept: food waste is recycled inside the whole food value chain from field to fork and represents a sustainable alternative source of biologically active compounds in order to formulate functional foods and nutraceuticals. This chapter is dedicated to give an update shot on: *-Waste from grape and during vinification, -Waste from olive and during oil production* and *-Bioactive compounds in legume residues*. Each paragraph consists in a description of the bioactive compounds that characterize foodstuffs with a focus on related wastes coming from food processing industry or agricultural production. The overview also take into account the most used extraction technologies for natural value-added compounds, with the aim moving towards sustainable products for application in different industrial sectors.

Keywords: Biorefinery, circular economy, food waste, olive oil, grapes, legumes

6.1 Introduction

The "Universal Recovery Strategy" for the commercial recapture of valuable compounds from food wastes represents new goal of the circular bioeconomy and the biorefinery concept: food waste is recycled inside food chain from field to fork and represents a sustainable alternative source of biologically active compounds [1–6] in order to formulate functional foods and nutraceuticals [7]. Several authors reported the advantages of the transformation of biomass waste to bulk chemicals respect to that of animal feed and fuel [8, 9]. Several researches of [10] showed how the agro-industrial field generate the large quantities of waste and by-products; in particular by-products derived from processing of fruit and vegetables are source of active compounds, with potential favorable technological or biological properties;

**Corresponding author*: massimo.lucarini@crea.gov.it

Arindam Kuila and Mainak Mukhopadhyay (eds.) Biorefinery Production Technologies for Chemicals and Energy, (91–128) © 2020 Scrivener Publishing LLC

moreover, the same authors marked how the use of new technologies is, currently, utilized to reinforce and increase the "Green Economy" in agriculture and agro-industry. The recent review of Santana-Méridas *et al.* [11] identified and described the agricultural residues as a source of bioactive natural products, gives an overview of the potential of agricultural residues as raw materials for the production of bioactive products considering their availability, processing, and their chemical and biological properties. The extraction of new biomolecules from agri-food by-products to be employed as active principles is addressed toward different industrial fields, such as food, feed, cosmetics, biomedical, and agronomic applications. In this context, it is worth mentioning the work of Baiano *et al.* [12], that describes and summarizes the type and amounts of food wastes, their legislation, and conventional and emerging techniques for the extraction of bioactive compounds; also the future trends in nutraceutical, cosmetic, and pharmaceutical sectors were discovered [12].

6.2 Waste From Grape and During Vinification: Bioactive Compounds and Innovative Production

6.2.1 Grape

The grape is a fruiting berry of the deciduous woody vines of the botanical genus *Vitis*. Grapes can be eaten raw or they can be used for wine making, jam, juice, jelly, raisins, vinegar, and grape seed oil. Grapes are a non-climacteric type of fruit, generally occurring in clusters. *Vitis vinifera*, L. (common grape vine) is a species of *Vitis*, whereas the term subsp. *Vitis vinifera* is restricted to cultivated forms.

Grapes are a type of fruit that grow in clusters berries. From a structural point of view, the berry is composed of three different types of tissue: skin, pulp, and seeds [13]. The berry can be black, dark blue, yellow, green, and pink. "White" grapes are usually green-yellow in color and are evolutionarily derived from the purple grape. Figure 6.1 give a picture of different varieties cultivated in CREA - Research Centre for Viticulture and Enology.

The anthocyanins, which are responsible for the color of the skin of purple grapes, coincide with expression of the gene encoding the final step of anthocyanins biosynthesis, UDP-glucose: flavonoid 3-O-glucosyl transferase (UFGT) [14]. In white grapes, the gene encoding UFGT is not expressed [15]. Anthocyanins and other pigment chemicals of the larger family of polyphenols in red grapes are responsible for the varying hue of red color in wines [16]. Grapes are typically a rounded or spherical ellipsoid shape.

6.2.2 Polyphenols

Vitis vinifera fruit (grape) contains various phenolic compounds, flavonoids, and stilbenes [17]. In detail, structurally, phenolic compounds contain an aromatic ring bonded to hydroxyl group (–OH). They have a weak acidity due to the hydroxyl group bonded to an unsaturated ring. These molecules are bioactive secondary metabolites widely distributed in all higher plants and are mainly synthesized by the shikimic acid pathway. Polyphenols are generally categorized into four major groups that include flavonoids, phenolic acids, lignans, and stilbenes (Figure 6.2), and they have a wide therapeutic potential as these can contribute toward the antioxidant activities [18].

Figure 6.1 A picture of different varieties cultivated in CREA - Research Centre for Viticulture and Enology. (a) *Angelica* variety, (b) *Bombino Nero* variety, (c) *Bellone* variety, and (d) *Bombino bianco* variety.

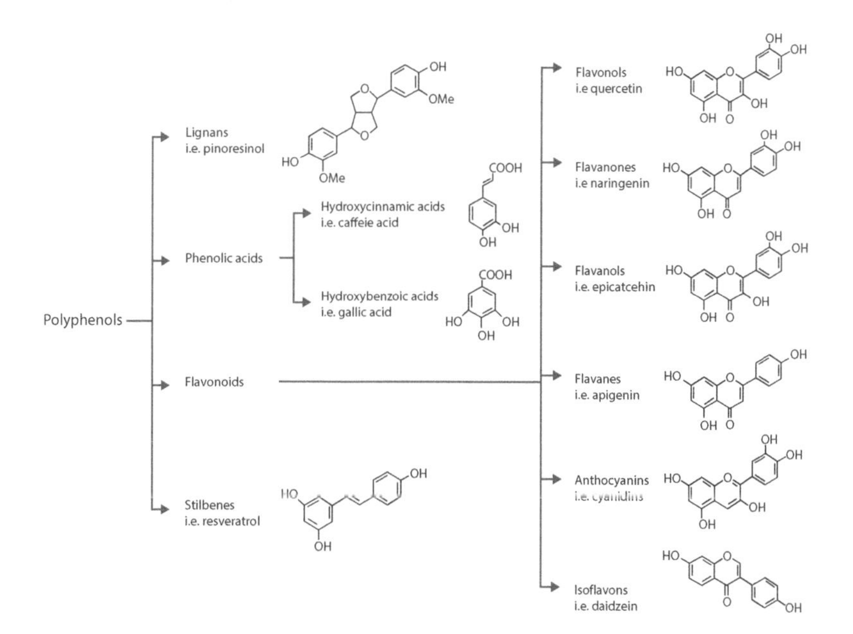

Figure 6.2 Polyphenols subfamilies organization and classification.

The most common hydroxybenzoic acids in grape berry include p-hydroxybenzoic acid, vanillic acid, gallic acid, gentisic acid, and salicylic acid. The hydroxycinnamic acids include p-cumaric acids, caffeic, and ferulic acids. These three hydroxycinnamic acids are present primarily as trans isomers.

They differ by type and number of substituents on the aromatic ring. When these hydroxycinnamic acids are esterified with tartaric acid, they are named coutaric acid (trans-p-coumaroyl-tartaric acids), caftaric acid (trans-caffeoyl-tartaric acid), and fertaric acid (trans-feruloly-tartaric acid).

The flavonoids are formed by diphenylpropane (C6C3C6) skeleton. The basic structure of the fused A and C rings with the phenyl B ring attached through its 1′- position to the 2-position of the C ring (numbered from the pyran oxygen).

These flavonoids contain several phenolic hydroxyl functions attached to ring structures, designated as A, B, and C. Structural variations within the rings subdivide the flavonoids into several families: monomeric flavanols, flavanones, anthocyanidins, flavones, flavonols, and isoflavones. The flavonoid variants are all related by a common biosynthetic pathway, incorporating precursors from both the shikimate and the acetate-malonate pathway.

Another class of polyphenols is the stilbenes, phytoalexins that occur in edible plants, especially in grapes. The main stilbene found in grape, in wine and winemaking pomace is resveratrol. The most important stilbenes are resveratrol-3-O-β-d-glucopiranoside, cis and trans resveratrol, picetanol and resveratrol dimers, being some glycosylated and isomeric form also found [19]. One of main compounds of grapes is resveratrol [20]. Resveratrol (Figure 6.3) belongs to family of stilbenes, based on structure backbone of two benzene rings linked by a double bond.

Several authors summarize the beneficial properties of resveratrol [21–23].

The tannins can be divided in in two main groups, hydrolyzable, and condensed tannins. Hydrolyzable are water soluble and further divided in gallotannins that release glucose and gallic acid by hydrolysis at low ambient pH and ellagitannins, made from ellagic acid glycosides. Condensed tannins insoluble in water formed by more highly polymerized procyanidins.

6.2.3 Antioxidant Activity and Health Properties of Grape

The grape is a polyphenols rich-fruit and is one of the earliest domesticated fruit crops. It has been widely cultivated all over the world. According to the Organisation Internationale de la Vigne et du Vin [24], world vineyards reached a total area surface of 7,534 mha

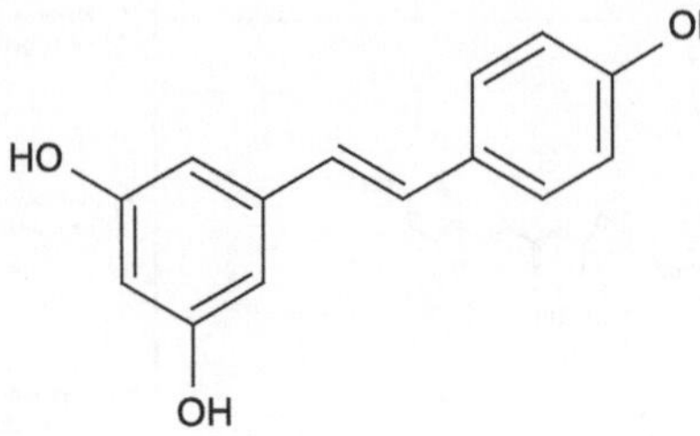

OH
HO
OH

Figure 6.3 Chemical structure of resveratrol.

(millions of hectares) in 2017. There are more than 10,000 different wine grape varieties in the world. Global grape production in 2017 was 73 mt (millions of tons) [24]. Approximately, 71% of world grape production is used for wine, 27% as fresh fruit, and 2% as dried fruit. Grapes have been also shown to be a rich source of high-added-value bioactive compounds, named phytochemicals. Phenolic compounds are the most important phytochemicals in grapes because they possess many biological activities and health promoting benefits [25–27]. The phenols represent the third most abundant constituent in grapes, after carbohydrates and fruit acids. The phenolic compounds are broadly distributed inside grapes. In grape berries, phenolics are mainly accumulated in skin and seeds. Grape skin is proved to be a rich source of anthocyanins, hydroxycinnamic acids, flavanols, and flavanol glycosides, whereas flavanols are mainly accumulated in the seeds [28–30]. The total extractable phenolics in grapes are present at about 28%–35% in the skin, 10% or less in the pulp, 60%–70% in the seeds [31]. The different antioxidant activity in skin, pulp, seeds, and in red and white grapes depends on their different phenolic composition [28, 32]. As regards different grape parts, the seeds are known for their remarkable antioxidant properties. The seeds contain lipids, proteins, carbohydrates, and 5%–8% polyphenols depending on the cultivar [31]. Phenolic compounds in grape seeds are mainly flavon-3-ols including gallic acid, the monomeric flavan-3-ols (+)-catechin, (−)-epicatechin, gallocatechin and epicathechin 3-O-gallate, procyanidin dimers, trimers, and more highly polymerized procyanidins.

These compounds possess strong antioxidant properties that enable them to scavenge free radicals, donate hydrogen, chelate metals, break radical chain reactions, and quench singlet oxygen *in vitro* and *in vivo* [33–35]. Flavonols are well-suited to modulate auxin transport and signaling, given their ability to affect the activities of a wide range of proteins [36–39], as well as to scavenge ROS [40, 41]. It has been demonstrated that procyanidins reduce risk of cardiovascular diseases (CVDs) [42], blood clotting [43], and cancer [44–49]. All these properties enable them to act in the prevention of oxidative-stress–related diseases [50]. Grape seed extract (GSE) is a complex mixture of polyphenols including gallic acid (GA), catechin (Cat), epicatechin (Epi), and procyanidins-oligomers of Cat and Epi, some of which are esterified with GA. Agarwall *et al.* [51] in studies to identify the GSE components cytotoxic to human prostate carcinoma (DU145) cells demonstrated that GA and several crude chromatographic fractions containing procyanidin dimers and trimers were biologically active. They showed that the most active procyanidin identified by mass spectrometry and enzymatic hydrolysis was the 3,3′-di-O-gallate ester of procyanidin dimer B2 (Epi–Epi). B2-digallate exhibited dose-dependent effects on DU145 cells over the range 25–100 μM, whereas GA exhibited comparable activity at lower doses but was highly lethal at 100 μM.

Simonetti *et al.* [52] for the first time demonstrated the positive effect of GSE extracted from *Vitis vinifera* cultivar, on the mucosal infection as vaginal candidiasis in mice. They demonstrated that a high antifungal activity was correlated at a high content of polymeric flavan-3-ols, with a polymerization degree ≥4.

The different antioxidant activity of the skin, pulp, and seeds in red and white grapes depend on their different phenolic composition. The phenol composition and antioxidant activity of the grape is clearly affected by many biotic and abiotic factors [53–55] such as grape variety, climate, temperature, light, maturation degree, ripeness, pathogen attacked, defoliation, and various agricultural techniques [28, 53–55].

Wine production is one of the most important agricultural activities throughout the world. Global wine production was 250 mhl in 2017 and it is estimated for 2018 in 279 mhl [24]. In Europe, Italy (42.5 mhl) confirm its position as the leading word wine producer, followed by France (36.7 mhl) and Spain (32.1 mhl).

Some of the by-products formed during winemaking can be used for different purposes. Thus, grape pomace and seeds can be used to obtain anthocyanin colorants or oils, phenol monomers, catechin, procyanidins procyanidin dimers, trimers, and more highly polymerized procyanidins. The lees (basically remains of dead yeasts) have been considered for use as a supplement in animal feed, but the yeasts in lees (recovered by wine after filtration) have an exceedingly poor nutrient value, because that they contained the β-glucans that represent the soluble fiber.

There are two principal type of winemaking techniques: white vinification which consists in fermentation of the must without the presence of the solid part (consisting of skin and seeds) of the grape; red vinification which consists in fermentation of the must with the presence of the solid part of the grape.

6.2.4 Winemaking Technologies

The main phases of white vinification the grapes are crushing, destemming, dripping, pomace pressing, sulfiting, and clarification from which the clarified must obtained is inoculated with selected yeast strain (usually *Saccharomyces cerevisiae*) and sent to fermentation. In the red vinification, the grapes are crushing, destemming, the stalk removed, the obtained must is sulfited, inoculated with selected yeast strain and added with solid part. The fermentation occurs together with maceration of the solid part of the grape, where the phenolic compounds extracted pass of the skin of the grape to the must-wine. During the fermentation, the CO_2, going up to the top of the tank, brings the pomace with it to form a cap. The main operation consists in pumping the must from the bottom over the cap. The immersion of the floating cap is necessary to diffuse the yeast from the cup to the must and facilitate the aeration of the must.

6.2.5 Winemaking By-Products

Vinification process produces a considerable amount of by-products, such as pomace, stalk, seeds, and lees. In recent years, wine industry has invested in the reuse wine by-products and sludge valorization. Skins and seeds of grapes are produced in large quantities by the winemaking industry. The pomace representing about 20%–30% of the original grape weight [56] depends on the grape cultivar and pressing, as well as on the fermentation steps. Grape seeds represent about 38%–52% of solid wastes produced by wine industries [56]. These by-products have become valuable raw materials due to their high content of polyphenols and to their antioxidant activity. The composition of grape seeds is about (w/w) 40% fiber, 16% essential oil (oleic and linoleic), 11% protein, 7% complex phenolic compounds like tannins, and also sugars and minerals [57]. Standardized GSEs contain 74%–78% oligomeric proacyanidins and less than approximately 6% of free flavanol monomers on a dry weight basis. These can esterified with gallic acid to form gallate esters and ultimately glycosides [58–60]. During fermentation only the 30%–40% of polyphenols are extracted that depend on varying factors such as grape variety, fermentation technology adopted (destemming, crushing, maceration, pressing, etc.) [61]. Winemaking is a biochemical process of transformation of the grape must in wine, carried out by yeasts. The main yeast used in vinification is *Saccaromyces cerevisiae*, whose cell wall consists of three layers: an inner layer of alkali-insoluble β-glucan (30%–35%); a middle layer of alkali-soluble β-glucan (20%–22%); an outer layer of glycoprotein (30%) in which the carbohydrate is composed of phosphorylated mannan. The cell wall *Saccaromyces cerevisiae* contains two types of β-glucans following the mode of glucose linkages. Long chains of β-(1→3)-glucan account about 85 %, whereas a short chain of β-(1→6)-glucan represents about 15% of the total yeast cell wall β-glucans [62]. The amount of 3,6-branched glucose residues varies with the yeast species and it was about 7% in *Saccaromyces cerevisiae* [62]. These extensively researched substances are the soluble fraction of dietary fiber and help the body to regulate the blood glucose and cholesterol levels [63–67]. Many works have showed that β-glucans enhance innate immune system and include both adaptative and innate responses [68–70].

The main difference in chemical structure between soluble and insoluble glucan is the number of β-(1→6)-linked glucose residues which are present in the long sequences of β-(1→3)-linked glucose chains. At the end of winemaking process, from clarification and decanting process, the yeast lees are isolated.

Several research reported the extraction of β-glucans from yeast cell wall of *Saccharomyces cerevisiae* [71]. In recent work, Cecchini *et al.* [72] show that the β-glucans content extract from of yeast less, compared to the same yeast used for fermentation, still contains the 38% of β-glucans.

6.2.6 Extraction Technologies

The main extraction technologies for polyphenols for grape pomace are: solid-liquid extraction using a mixture of water and organic solvent (mainly ethanol or methanol); microwave-assisted extraction, supercritical fluid extraction (SFE), and ultrasound-assisted extraction. In the solid-liquid extraction, the extraction efficiency can be improved mainly by changes in the solvent type, particle size, temperature, extraction time, and presence of interfering substances in the matrix [19]. Microwave-assisted extraction, SFE, and ultrasound-assisted extraction are novel extraction technologies that reduced extraction time and low use of organic solvent.

Extraction of β-glucans from yeast lees consists of two major steps separation of cell wall and extraction. There main extraction methods are Chemical (N_aOH, HCl, acetic acid, citric acid) [73]; physical (sonication, high pressure) [74].

The recyclable by-products can have a relevant role in the reduction of carbon dioxide emission, considering the huge amount of land dedicated to this culture. The Italian National Agri-food Waste Assessment [75] has estimated the production of around to 18 and 6 kg for pomace and lees, respectively, for each hectoliter of wine produced. Considering that the global wine production in 2017 was about 250 mhl [24], we can assess that were produced of about 4,500 mt of pomace and 1,500 mt of yeast lees. The presence of β-glucans in lees can represent an important factor for the oenological sector. The extraction of polyphenols from pomace and of β-glucans from the lees can transform these by-products in nutrients source and to transform a cost due to disposal, into an economical and environmental benefits. In conclusion, the recovery of phitochemicals from wine industry by-products is of great importance, not only because of their health properties but also because it could exploit a large amount of the wine industry wastes (about 6 million tons per year). In the perspective of circular economy, these wastes could be a secondary raw material to use in cosmetic, pharmaceutical, and food industry and, consequently, to reduce their environmental impact.

Winemaking produces a series of by-products, easily exploitable for new productions and in various supply chains. In this regard, several authors have shown how the quality and quantity of these by-products depend on a set of cofactors, i.e., the type of vinification [76, 77]. Grape seeds have great biological potential that could be exploited by extracting bioactive compounds with high added value [58, 78–81] before using biomass for energy purposes, to obtain extracts and semifinished products useful for agronomic, cosmetics, feed, food, nutraceutical, and pharmaceutical purposes [82].

Grape and wine by-products have been extensively studied for the recovery of phenolic compounds with antioxidant activity and a variety of biological actions. As instance, Soto *et al.* [83] have investigated the recovery and concentration of antioxidants from winery wastes. The recent work of Jara-Palacios *et al.* [84] proposed a cyclic voltammetry to evaluate the antioxidant potential in winemaking by-products. Extraction techniques, innovative technologies, and chemometrics are being developed and applied in this sector [85–90].

A recent review of Yammine *et al.* [91] explored the different recovery stages used by both conventional and alternative techniques and processes for extraction and purification of high added value compounds from by-products of the winemaking. Yammine *et al.* [91] described and summarized alternative pre-treatments techniques such as ultrasounds, pulsed electric fields and high voltage discharges, nonconventional solvent extraction under high pressure, specifically, SFE, and subcritical water extraction, alternative purification technologies, for example, membrane processing.

As instance, Merchante *et al.* [92] studied the effect of the oven-drying treatment on the phenolic composition in winemaking by-products: oven-drying at 45°C caused a significant decrease on the total phenolic content, which implied a reduction of the antioxidant capacity of the extracts and it produced a decrease in total and individual flavan-3-ols, stilbenes, and flavonols, being greater in those extracts from stems. Moreover, the authors showed how respect to anthocyanins, which were only identified in grape pomace extracts, oven-drying caused an important decrease, being the peonidin-3-O-glucoside the more thermosensitive compound. Finally, the authors concluded how natural extracts from fresh or oven-dried winemaking by-products could be used in other food industries as a valuable source of phenolic compounds with antioxidant properties, and further studies on other drying methods are required for addressing the preservation of phenolic compounds from winery by-products successfully [92].

Another work of Carmona-Jiménez *et al.* [93] studied the effect of drying on the phenolic content and antioxidant activity of red grape pomace and proposed a new methodology for drying grape pomace in a climatic chamber: drying process is both feasible and beneficial because significant increases in the extractability of phenolic compounds and antioxidant activity were achieved.

Spectroscopic techniques coupled to chemometrics could represent a valid green alternative to conventional methods for determination of bioactive compounds in winery waste [94–96].

Also a picture of the Italian experience applied to the by-products of the wine industry is presented by Lucarini *et al.* [82]. As described by several authors [82, 97], for wood or hard matrices such as grape seeds, yields of aqueous extraction can also be raised by working at high temperatures, compatibly with the stability of the active principles that have to be preserved: in subcritical water extraction processes at high temperatures (150°C) of grape seeds, there have been observed high yields on procyanidins extraction, but also hydrolysis processes of galloylated compounds, and consequently increased concentrations of free gallic acid [82, 97].

As instance, the recent work of Rodriguez *et al.* [98] presents studies about the gasification of the lignocellulosic winery wastes in fluidized bed to obtain energy: the thermal decomposition of the studied biomass wastes at three heating rate, 5, 10, and 15°C/min under steam/air mixture atmosphere show that the gasification takes place in three visible stage: water vaporization, pyrolysis, and the last step associated with the reaction of the char by CO_2.

Another recent research of Da Ros *et al.* [99] has assessed the potential phytotoxicity of digestate from winery wastes using macrophytes and evaluating the potential contribution of ammonium and copper: The effluent did not meet the entire amendment quality standard defined by Italian law (Decree 75/2010 germination index > 60% with solution of 30% v/v of digestate), but bio-stimulation was observed at low doses (3.15%–6.25% v/v) for *S. alba* and *S. saccharatum*. Moreover, the authors showed how the beneficial concentration agreed with Nitrate Directive dose and suggested that limited addition of digestate could have several positive effects on soil characteristics and on crop growth. Also, the authors reported that specific test using ammonium and copper solutions showed that these pollutants were not directly correlated to observed phytotoxicity.

An efficient alternative for the management and valorization of the huge amount of seasonal wastes generated by winery industries is shown by Guardia *et al.* [100]: sustainable porous carbons were successfully prepared by one-pot activation of grape bagasse, stalks, and oil free-seeds. The authors demonstrated how they are excellent sorbents for CO_2 capture under post- and pre-combustion conditions (16 and 69 wt%, respectively) and achieve superior electrochemical capacitance of nearly 300 F/g in aqueous electrolyte and 180 F/g in ionic liquid based-medium.

6.3 Waste from Olive and During Oil Production: Bioactive Compounds and Innovative Process

The Extra Virgin Olive Oil (EVOO) industry is a significant productive sector in the European Union with 75% of world production concentrated in Europe. Nevertheless, EVOO production, under the pressure of international competition, the decline in profitability of agricultural activities, and the uncertainty related products and consumers' safeguard, must face enormous challenges. Moreover, the EVOO extraction also poses a significant environmental issue, in view of the more than 30 million tons of waste generated each year in Europe with related operating cost of several billion and the creation of an uncontrolled pollution of the water (rivers and sea) [101].

According to the study "EU EVOO farms report" [102] published by the EU DG Agriculture and Rural Development, the economic situation of EVOO farms has been deteriorating significantly over recent years. In the member states, the trend in income of EVOO farms has been worse than the national average: 25% of farms earned a year less than 5.000 EUR per family work unit (FWU), with 30% in Italy, and 37% in Greece. On the other hand, 11% of Spanish EVOO farms earned more than 30.000 EUR/ WU, with 10%

in Italy and 3% in Greece. COLDIRETTI, the main Italian Agriculture Association, has established in 760 euro/hectare the annual income of an intensive system. Therefore, such low profit margin threatens the survival of the EU EVOO production industry [102]. Around 30 million tons of olive mill waste is produced annually in the Mediterranean area [102]. This waste cannot be sent to ordinary treatment systems; thus, safe disposal is of serious environmental concern. Moreover, olive processing waste (OPW) is not easily biodegradable and needs to be detoxified before it can properly be used in agricultural and other industrial processes. This is too expensive for small and medium producers. It is estimated that around 58% of the mills pour their waste into water courses, 11.5% in the sea and 19.5% in the ground affecting the soil and, through leaching, may potentially be contaminating groundwater [102].

6.3.1 Olive Oil Quality, its Components, and Beneficial Properties

Quality is a key factor in increasing consumer confidence and building consumption in the market. The strengths of European EVOO sector are primarily product of quality and positive image, thus related to EVOO health benefits also stated by EFSA for its health benefits in an approved health claim: "EVOO polyphenols contribute to the protection of blood lipids from oxidative stress" [103]. EVOO is normally sold at a higher price than other vegetable oils and fraudulent activities are tempting. To preserve the image of EVOO, it is necessary to guarantee its quality and authenticity. EVOO characteristics are regulated at EU level by Regulation (EEC) N° 2568/91 [104] which establishes a list of physical, chemical, and organoleptic characteristics.

Quality and its control on the antioxidants level are extremely important to preserve and promote the image of the European EVOO and to better protect/inform the consumers. On the health side, clinical trials have proven that the intake is also decreasing the numbers of various types of cancers, increasing the antioxidants activity in the body and positively affecting blood circulation system. So, a high quality EVOO will also have very positive consequences on the Public Health budget.

The quality of virgin olive oil is largely dependent on the amount and the composition of phenolic compounds of fruits at harvesting. These compounds determine oil taste and stability against oxidation, as well as the oil nutritional and therapeutic characteristics. More hydrophilic compounds, such as verbascoside, that cannot be found in the oil but are present in olive oil industry wastewaters, are still of interest from a pharmacological point of view because of their peculiar antioxidant properties. Hence, there is much interest in developing agronomic techniques that maximize particular classes of olive phenolics. Several factors affect the phenolic composition and quantity in olive fruits, such as cultivar, growing site, climatic conditions, alternate bearing, and ripening stage.

Because of this plurality of factors, finding a general objective criterion to establish the best harvest time related to the highest quality of olive oil is a difficult task.

A widely accepted practice is to use the maturation index suggested two decades ago by the International Olive Oil Council (IOC) derived from the visual classification of the skin color of olives. Indeed, fruit ripening is accompanied by a change in the skin color due to a modification of pigment concentration in the superficial tissues. Usually, color development from green to purple is modulated by an accumulation of anthocyanins (Anths) together with the degradation of chlorophylls (Chls) and carotenoids. This feature suggests the use of optical methods to follow the maturation process, as applied to other kinds of fruits. Spectroscopic techniques can assess rapidly and nondestructively both photosynthetic and accessory pigments in fruits, more advantageously than standard, time-consuming extraction procedures. The drawback of noninvasive optical methods is that only superficial layers can be analyzed due to the limited penetration of radiation inside the sample. However, spectral signatures of fruit peculiarities localized inside the pulp, which are correlated to a surface-related compound, can still be obtained.

Anths are suitable for *in situ* spectroscopic detection since they accumulate in the outer layers of olives and present a characteristic absorption band in the green region around 530 nm. Because Anths are synthesized through the phenylpropanoid pathway, it is expected that their increase in concentration during ripening be correlated to a change in the content of other phenolic compounds. Such a relationship was proven for Anths and flavonols during the ripening of bilberries. In olives, studies have reported a general decrease of total phenolics as maturation proceeds, while specific compounds may have different trends during ripening. Metabolic links between oleuropein, verbascoside, and caffeic acid during ripening have been suggested; however, a direct quantitative relationship between Anths and other phenolic compounds is still lacking. Although several reports on phenolic compounds in olives have been published, only few data about Anths in *O. europaea* fruits are available.

Recently, methods using chlorophyll fluorescence (ChlF) to assess *in situ* the UV-absorbing phenolics in the leaf epidermis have been developed. Agati *et al.* [105] applied the ChlF excitation method to nondestructively assess the Anth content in olives with different skin pigmentation, corresponding to different maturation levels. The method is based on the comparison of the ChlF excitation spectra from olives with different pigmentation from green to green-red, reddish-purple, and purple. The logarithm of the ratio between the fluorescence excitation spectra (logFER) from two different colored zones gave the difference in the absorption spectrum between them. The absorbance spectrum derived from the logFER between a red olive and the same olive devoid of the skin showed the typical Anth green band (at 550 nm). It matched that recorded by microspectrophotometry on a single pulp cell and the *in vitro* absorbance spectrum of the olive skin extract. As expected, the *in vivo* Anths absorption maximum increased in intensity going from less to more mature olives and was higher in the sun-exposed olive side with respect to the sun-shaded side.

They found that the logFER approach is suitable to detect the *in vivo* Anth absorption spectrum. Furthermore, the deduced Anth absorbances were nicely correlated, going from green to purple skin colors, to those measured spectrophotometrically *in vitro* after ethanol/water extraction of the very same olive area used for the ChlF measurements. In addition, they analyzed the phenolic composition of single olive samples in order to evaluate a possible correlation between specific phenolic compounds and Anths during fruit ripening. The main result was a net increase of verbascoside with increasing Anths content.

According to these results, new portable optoelectronic devices for the nondestructive monitoring of olive development and ripening can be developed. They can provide automatic maturation indexes directly related to pigment concentrations and, therefore, are expected to be more reliable than those merely based on the subjective visual classification through olive skin colors. A LogFER-based ripening index should also be better than colorimetric procedures, which use broad bandwidth reflectance signals. It can solve the problem of carefully defining the olive maturation stages that is fundamental for comparing the ripening-related features observed in different studies [105].

Among bioactive compounds of olive oil, there is growing interest in the identification of lignans in EVOO as these metabolites can contribute to the healthy effects of EVOOs. Publications in recent years have shown that lignans from sesame, from several vegetal matrices, and from olive oil are *in vitro* antioxidants and in some tests they show high antioxidant capacities, similar to that of hydroxytyrosol and some oleuropein derivatives put in evidence a high antioxidant activity in a liposome model system probably due to the chelating ability of lignans. Enterolactone, produced by intestinal microorganisms from lignan precursors in diet, has been associated with protection against LDL oxidation, and recently, an analogous activity was revealed for lignans from olive oil. Similarly, studies carried out on another important source of lignans, flaxseed have shown that these compounds have the capacity to inhibit cancerous cell growth in skin, breast, prostate, colon, and lung tissue.

In general, current milling processes use the whole fruit in the production of olive oil. However, in recent years, due to the application of new technology able to efficiently remove the stone without loss of pulp, commercial extra virgin oils produced from olives without stones are available on the market. The aim of such milling process is to produce virgin olive oils of higher quality and the possibility to valorize the stones as a discarded matrix.

Moreover, the removal of pits, through the elimination of most of the lipooxigenase and peroxidase enzymes from the olive mash, could reduce oxidative processes in the oil.

Only over the last 6 years, lignan compounds, such as pinoresinol (Pr) and acetoxypinoresinol (Ac–Pr), have been shown to be widely present among the minor polar compound (MPC) fractions from EVOOs from Arbequina, Taggiasca, Frantoio, and Picual cultivars (Figure 6.4).

Oliveras Lopez *et al.* [106], investigate the distribution of lignans, such as Pr and Ac–Pr, in the stone and pulp of selected Italian and Spanish varieties, known to produce EVOOs containing these metabolites. To the best of the authors' knowledge, a systematic approach

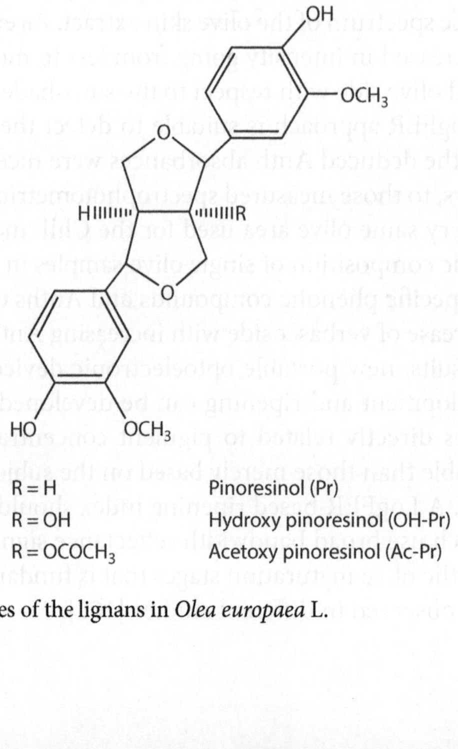

R = H Pinoresinol (Pr)
R = OH Hydroxy pinoresinol (OH-Pr)
R = $OCOCH_3$ Acetoxy pinoresinol (Ac-Pr)

Figure 6.4 Chemical structures of the lignans in *Olea europaea* L.

focused on evaluating the lignan origin in *Olea europaea* L. fruit has not yet been developed. Through the use of HPLC/ESI/MS technique, a quantitative estimation of their content in the stones was also performed.

Analyses were carried out by HPLC/DAD/MS on four cultivars: Frantoio and Taggiasca from Italy and Arbequina and Picual from Spain. The main results can be summarized as follows: (a) lignans were demonstrated to be present in the stone; (b) acetoxypinoresinol, pinoresinol, previously detected in the respective extra virgin olive oils, and OH-pinoresinol were found in the samples; (c) the total lignan content ranged between 0.1 and 0.29 mg/g of dried stone weight. These values agree with the average lignan content reported in literature for the respective extra virgin olive oils.

It should be emphasized that (i) the recovery of these metabolite from the fruits is not obtained applying the common extraction procedures by the use of hydroalcoholic mixtures at room temperature; (ii) the main source of the lignans of the oil is indubitably the stone and not the pulp.

This preliminary work could be extended in order to analyze the possibility of including the lignan fraction in the parameters for the phytochemical characterization of cultivars or to control the quality of the EVOOs from destoned fruit.

In light of the interesting biological activities attributed to this group of lignans, including antitumoral and antioxidant action, the estimation of their content as a by-product of the milling process could be of particular interest [106].

Phenolic compound content is an important parameter for the evaluation of EVOO quality as phenols largely contribute to oil flavor and taste as long as it is protected from autooxidation. Virgin olive oil contains a large number of phenolic compounds including phenyl alcohols, such as tyrosol (Tyr) and hydroxytyrosol (OH–Tyr), phenolic acids, flavonoids, like luteolin and apigenin, as well as other more complex secoiridoid derivatives from oleuropein and ligstroside. Since 2000, the presence of lignans in the phenolic fraction, with (+)-pinoresinol and (+)-1-cetoxypinoresinol as major components, has been described in some olive oils.

The application of techniques such as HPLC–DAD and HPLC–MS has made it possible to recognize novel bioactive molecules from extra virgin olive oil and to better compare the phenolic profiles in order to explain biological activities of olive oil. Among phenols from olive fruit and virgin olive oil, a special significance has been revealed for those with the o-diphenolic group, mainly OH–Tyr and oleuropein derivatives, which have shown antiatherogenic effects and an antioxidant capacity higher than that of other known antioxidants such as Vitamins E and C.

Although the antioxidant content of virgin olive oil is actually widely researched, it is necessary to compare in detail the phenolic profile of diverse monocultivar oils to better correlate their differences among *in vitro* and *in vivo* effects.

Oliveras-Lopez *et al.* [107] compare the phenolic content in selected monocultivar extra virgin olive oils. Analyses were carried out by HPLC/DAD/MS on Picual, Picuda, Arbequina, and Hojiblanca oils from Spain and Seggianese and Taggiasca oils from Italy. Oils from cultivar Picual showed similar characteristics to those of Seggianese oils, with total amounts of secoiridoids of 498.7 and 619.2 mg/L, respectively. The phenolic composition of Arbequina oils is close to that of the Taggiasca variety with lignans among the main compounds. The total amounts of OH–Tyr and Tyr, including their bound forms, were evaluated after a quantitative chemical hydrolysis.

The determination of free and linked OH–Tyr, by way of an acid hydrolysis, represents a rapid and suitable method, especially when standards are not available, to determine antioxidant potentialities in terms of MPC, particularly for fresh extra virgin olive oils rich in secoiridoidic derivatives [107].

Currently, the two most commonly used methods to evaluate polyphenolic content of olive oil are the well-known Folin-Ciocalteu (FC) colorimetric assay and high-performance liquid chromatography (HPLC). FC is a simple and highly efficient procedure, but it is limited by a low specificity. HPLC is sensitive and specific, but it is very time-consuming and requires special expertise and a laboratory. The IOC itself adopts the HPLC method but gives also a provisional approval to a method for the colorimetric determination of o-diphenols. Alessandri *et al.* [108] evaluate, in EVOO analysis, the correlation of FC's results with the results by HPLC-DAD-MS focusing also on the relationship between the single molecules or classes of MPCs quantified by HPLC, and the corresponding FC results, and aiming to a consistent interpretation of FC results, for olive oils characterized by very different phenolic profiles. The Pearson correlation coefficients were calculated between HPLC and FC results. The highest, positive R were related with deacetoxyoleuropein aglycone (R = 0.93) and oleuropein aglycone (R = 0.93) as single compounds and with the sum of orthodiphenols (R = 0.94) and the sum of all compounds (R = 0.95), showing that both estimations of total phenols content are reliably correlated, regardless for the absolute contents and are independent of the relative composition of the phenolic fraction. On the other hand, the HPLC quantifications of apigenin and lignans showed no significant correlation with FC. These results, supported also by principal component analysis, may suggest caution about the interpretation of FC results to compare olive oils with very different phenolic profiles [108].

EVOO, the primary source of fat of Mediterranean diet, differs significantly in composition from dietary lipids that are consumed by other populations. The several minor constituents of virgin olive oil, there are vitamins such as alpha- and gamma-tocopherols (around 200 ppm) and beta-carotene, phytosterols, pigments, terpenic acids, flavonoids, squalene, and a number of phenolic compounds, usually grouped under the rubric "polyphenols". The formulation of an antioxidant/atherosclerosis hypothesis stimulated experimental and epidemiological studies on the possible role of antioxidants, including olive oil phenolics, in the protection from CHD observed in the Mediterranean area [109].

The influence of sex and gender is particularly relevant in CVDs as well as in several aspects of drug pharmacodynamics and pharmacokinetics. Anatomical and physiological differences between the sexes may influence the activity of many drugs, including the possibility of their interaction with other drugs, bioactive compounds, foods, and beverages. Phenolic compounds could interact with our organism at organ, cellular, and molecular levels triggering a preventive action against chronic diseases, including CVD. Campesi *et al.* [110] review the role of sex on the activity of these bioactive molecules, considering the existence of sex differences in oxidative stress. It describes the pharmacokinetics of phenolic compounds, their effects on vessels, on cardiovascular system, and during development, including the role of nuclear receptors and microbiota. Although there is a large gap between the knowledge of the sex differences in the phenolic compounds' activity and safety, and the urgent need for more research, available data underlie the possibility that plant-derived phenolic compounds could differently influence the health of male and female subjects [110]. For example, it has been reported that the phenolic compounds of extra-virgin olive oil, in a sex-specific manner, increase glutathione reductase activity and GPx and GSH levels by activating the Nrf2 pathway in rat VSMCs [111].

Epidemiological studies suggest that Mediterranean diets are associated with a reduced risk of CVDs, the lower incidence of CVD being associated with greater adherence to the Mediterranean diet. Recently, it has been shown in more than 3,000 individuals, without clinical evidence of CVD, that total plasma antioxidant capacity and low oxidized low-density lipoproteins (LDLs) are associated with greater adherence to the Mediterranean diet.

Compared with a saturated fat diet, the Mediterranean diet, rich in oleic acid, has also been associated with lower LDL and total triglycerides with maintenance of high-c lipoproteins and lower blood pressure. It was reported that the antioxidant activity of monovarietal extra virgin olive oils was increased in those oils extracted from destoned fruits and that this effect was variety dependent.

It is known that some single MPC increase the resistance of LDL against oxidation *in vitro*, but the single MPC approach fails to account for the interactions among MPC and does not take into consideration that some MPC are correlated between them. In this respect, it is important to note that the mixture of phenols may exert different activity in comparison with the single phenols, because they may cooperate, thereby modifying biological activity. Interactions among phenols seem also to depend on the relative amount of single polyphenols. Thus, individual olive oils, which differ qualitatively and quantitatively, could have different biological activities. At present, little is known about the antioxidant effect of the total olive oil extract. In addition, the relationship between antioxidant activity of whole virgin olive oil and its global MPC content is not yet known. The foregoing observations may account for the conflicting results obtained in clinical trials designed to determine the effect of olive oil. Interestingly, in two of these studies, the beneficial effects of virgin olive oil were related to MPC amount. Olive oil is a basic component of the Mediterranean diet; thus, it is important to identify and quantify the antioxidant compounds in individual virgin oils so as to determine which have the healthiest effects.

Whole virgin olive extracts were studied by Franconi *et al.* [112] to determine whether they maintain the antioxidant activity and whether this last is linked to MPC composition of a single virgin oil. Using HPLC-DAD the MPC content in Taggiasca and Seggianese virgin olive oils was measured. Taggiasca oil was less rich in total MPC (208.5 mg/L) than Seggianese oil (441.9 mg/L). In addition, the major compounds of Taggiasca oil were lignan derivatives, whereas the major compounds in Seggianese oils were secoiridoid derivatives. Moreover, Taggiasca oil was practically free of 5-hydroxytyrosol and 5-hydroxytyrosol derivatives, deacetoxy-oleuropein aglycone and oleuropein aglycone. The antioxidant activity of the oils on human LDL was evaluated by measuring malondialdehyde and conjugate diene generation induced by copper ions. In both tests, the oil extracts dose-dependently reduced malondialdehyde and conjugate diene generation. Moreover, antioxidant potency correlated with total MPC; thus, Seggianese extract was more active. The two oils differed quantitatively and qualitatively, and these differences influenced their biological activities; thus, clinical trials focused on studying the effects of olive oils should specify the oils used.

In a small clinical study carried out with olive oils containing different amounts of MPC, the short-term inhibition of LDL oxidation is greater with oils rich in MPC. On the basis of this finding and data from Franconi *et al.* [112], authors suggest that clinical trials designed to evaluate the healthy properties of virgin olive oils should specify the type of virgin oil being tested and possibly its chemical composition, or at least the amount of MPC contained in the oil, so that results from different studies can be compared. It seems that the presence of 5-hydroxytyrosol and its derivatives and high levels of secoiridoids enhance antioxidant activities, suggesting that virgin olive oil rich in these compounds could have more potent health-protecting properties because the antioxidant activity of oils could have beneficial effect according to the oxidative hypothesis of atherosclerosis [112].

EVOO quality depends on many factors, such as condition of the fruits (which must be undamaged and rapidly processed), harvesting time, crushing, and storage methods. EVOO quality is also strictly correlated with the properties of the cultivar used for its production. In the past few years, interest has focused on monocultivar extra virgin olive oils, especially from autochthonous cultivars, as their unique features are yet to be fully investigated and exploited.

A unique characteristic of EVOO is the equilibrium between stability and autooxidation phenomena, often evaluated by determining the peroxide content and mainly dependent on the concentration of antioxidant molecules. In fact, as MPCs are predominantly responsible for the olive oil resistance to oxidations mainly owing to molecules containing an orthodiphenolic group (e.g., 5-hydroxytyrosol), a method of assessing oil stability consists of the evaluation of MPC over time. Yet, their time-wise pattern of decrease and the specific antioxidant role of single MPC are still to be fully understood. As mentioned, interest in the pharmacological activities of MPC, in particular 5-hydroxytyrosol and oleuropein, is constantly growing, as data that demonstrate the cardioprotective (and possibly chemopreventive) potential of such compounds are accumulating. Indeed, antioxidant molecules such as MPC might provide health benefits, but primarily discriminate the quality of the oil and maintain it over time. By investigating, the time-wise MPC concentration profile in bottled oil, nutritional guidelines might be fine-tuned.

Both simple and complex phenolic antioxidants are detectable in the polar fraction of extra virgin olive oil: examples are 5-hydroxytyrosol and tyrosol derivatives, secoiridoids, and lignans. MPC concentration in extra virgin olive oil depends on several factors such as olive cultivar, and agronomic and technological aspects of production. Moreover, different storage conditions might decrease phenols concentration in the oil, adding to other factors such as temperature and exposure to light. Because of their antioxidant properties, phenolic compounds are of paramount importance to the shelf life of extra virgin olive oils.

Oil consumption usually occurs within 1 year from its production, but, according to the current legislation [113], extra virgin olive oil can be consumed within 18 months from bottling. Consequently, we took an 18-month period of storage as approximation of an extended shelf life and investigated chemical and organoleptic changes of bottled oil during this period.

Romani *et al.* [114] characterize "Olivastra Seggianese" extra virgin olive oil and to evaluate its chemical and sensory characteristics and antioxidant and antiradical activity during storage under novel conditions. Furthermore, studied the correlation of MPC concentration with their antioxidant capacity and calculated how this could be predictive of olive oil stability.

Two oils (A and B) were analyzed for the commodity characteristics at blending (t0) and after 9, 12, and 18 months; panel tests were performed and MPCs content was assessed at blending (t0) and after 6, 9, 12, and 18 months. Antioxidant and antiradical activities *in vitro* were evaluated at t0 and after 12 months, by human LDL and 1,1-diphenyl-2-picrylhydrazil radical (DPPH) tests. Oil A, which had an initially higher MPC content, possessed "harder" organoleptic characteristics than oil B, which had a lower MPC content and was endowed with a "smoother" taste profile. Statistical analyses showed that secoiridoids, particularly deacetoxy-oleuropein aglycone, should be quantified to evaluate EVOO stability during storage. The antioxidant activity toward human LDL was linked to MPC content and to storage time. The tests on the stable free radical DPPH confirmed the results on human LDL and authors proposed this as an additional parameter to evaluate olive oil quality and stability over time.

In conclusion, this paper reports on the study of two EVOOs from the Seggianese cultivar, with different organoleptic features and MPC content, bottled into 100 ml bottles made of special UV-filtering, capped with nitrogen, and sealed with synthetic stoppers. Under these packaging conditions, the MPC content, the biological properties, and the organoleptic characteristics of the two EVOOs were rather stable for 12 months after bottling. Statistical analyses indicated that secoiridoids, particularly DacOLagl, should be selected to evaluate EVOO stability during storage; the antioxidant activity toward human LDL was linked to MPC content and to storage time.

In synthesis, the long-term preservation of the potentially beneficial properties of olive oil and of its organoleptic characteristics, so much appreciated by consumers, might be improved by optimizing all of the production steps, with special attention paid to bottling [114].

Diets rich in plant-derived polyphenols such as olive oil and/or catechins such as epigallocatechin 3-gallate (EGCG) have been shown to reduce the incidence of CVDs, potentially by improving endothelial function, an important surrogate for atherosclerosis. The possible augmentation of endothelial function with the combined efforts of EVOO and EGCG is intriguing, yet unknown.

In a study by Widmer *et al.* [115], 82 patients with early atherosclerosis (presence of endothelial dysfunction) were enrolled in this double-blind, randomized trial. It was the aim to compare the effect of a daily intake of 30-ml simple EVOO, with 30 ml of EGCG supplemented EVOO, on endothelial function as well as on inflammation and oxidative stress after a period of four months. Endothelial function was assessed non-invasively via peripheral arterial tonometry (Endo-PATR).

After 4 months, EVOO and EGCG supplemented EVOO significantly improved endothelial function (RHI, 1.59 {0.25 to 1.75} 0.45, $p < 0.05$). However, there were no significant differences in results between the two olive oil groups. Interestingly, with EVOO supplementation there was a significant reduction in inflammatory parameters (sICAM from 196 ng/ml to 183 ng/ml, $p = <0.001$; white blood cells (WBCs) (6.0×109/L to 5.8×109/L, $p < 0.05$); monocytes from 0.48×109/L to 0.44×109/L, $p = 0.05$ and lymphocytes from 1.85×109/L to 1.6×109/L, $p = 0.01$).

Improvement in endothelial dysfunction in patients with early atherosclerosis in association with significant reduction in leukocytes may suggest an important role of early cellular inflammatory mediators on endothelial function. The study supports one potential mechanism for the role of olive oil, independent of EGCG, modestly supplemented to a healthy cardiovascular diet [115].

Both *in vivo* and *in vitro* studies have shown that EVOO components have positive effects on metabolic parameters, such as plasma lipoproteins, oxidative damage, inflammatory markers, platelet function, and antimicrobial activity. Pampaloni *et al.* [116] have investigated the possible interactions between two extracts of extra virgin olive oil and estrogen receptor b (ERb) in an *in vitro* model of colon cancer. The qualification and quantification of the components of the two samples tested showed that phenolic compounds—hydroxytyrosol, secoiridoids, and lignans—are the major represented compounds. EVOO extracts were tested on a colon cancer cell line engineered to overexpress ERb (HCT8-b8). By using custom made Oligo microarray, gene expression profiles of colon cancer cells challenged with EVOO-T extracts when compared with those of cells exposed to 17b-estradiol (17b-E2).

This study demonstrated that the EVOO extracts tested showed an antiproliferative effect on colon cancer cells through the interaction with estrogen-dependent signals involved in tumor cell growth. Specifically, the ability of EVOO extracts to inhibit cell proliferation was superimposable to the activation of the ERb receptor, similar to what was observed after 17b-E2 challenge.

The study evaluated the effect of the extract of two different varieties of EVOO (M and T) derived from two different Italian regions on cell proliferation in an *in vitro* model of human colon cancer engineered to overexpress ERb. The action of these extracts was further characterized evaluating the response in terms of expression of genes selected for their impact in the estrogen response pathway.

Through transactivation, EVOO-M and -T extracts showed an interaction with ERb similar to 17b-E2. However, the EVOO-T extract appears to influence gene expression in a colon cancer cell line overexpressing ERb in a different manner when compared to estrogen, suggesting tumorigenesis inhibition [116].

6.3.2 Olive Oil By-Products

As presented before in the Mediterranean area, the olive mill waste that is produced annually is 30 million tons. This waste is not easily biodegradable and needs to be detoxified but the detoxification solutions necessary are too expensive. The final result is that around 58% of the mills pour their waste into watercourses, 11.5% in the sea, and 19.5% in the ground affecting the soil and, through leaching, may potentially contaminating groundwater.

The adoption of green technologies for the extraction of biomolecules from the solid mill waste, coupled with the use of residual biomass for the energy production in an eco-efficient process, integrate innovative conditioning and production technologies, which results in a total cutoff the energy consumption, raw materials inputs, and water usage. This approach improves resource efficiency through a zero-waste process.

The process of olive oil production creates a large amount of waste, including olive mill wastewaters, olive pulp, and leaves.

Olive mill wastewaters are the main waste produced from three-phase olive processing while olive pulp is the main waste deriving from two-stage olive oil processing. This last technique, promoted by the European Community, was recently applied in Spain and Italy to eliminate the production of olive mill wastewaters characterized by the high level of toxicity and disposal costs. Olive leaves can be considered as a waste because they derive from both olive processing and pruning practices.

These raw materials are a precious source of bioactive compounds including low-molecular weight phenols. In olive leaves, the main constituent is oleuropein (Figure 6.5), a phenolic secoiridoid glycoside, which by enzymatic or chemical hydrolysis produces hydroxytyrosol (HTyr), elenolic acid, and glucose. Among these, HTyr is an outstanding

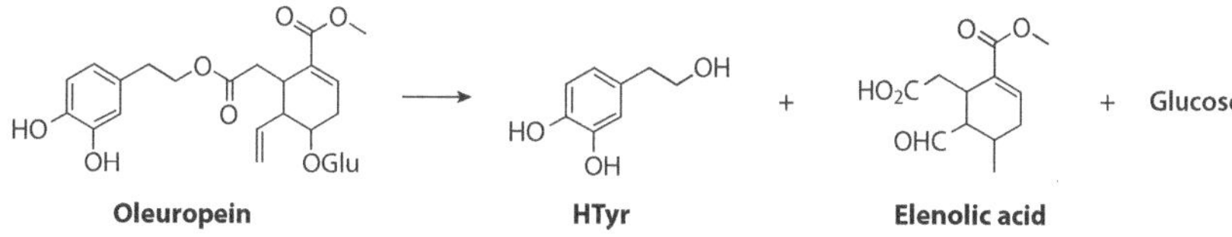

Figure 6.5 Products of the hydrolysis of oleuropein.

biological compound due to its properties, in particular its strong antioxidant activity. These peculiar properties and the absence of genotoxicity make HTyr a good candidate for use as a preservative, thus potentially replacing synthetic food and cosmetic additives such as butylated hydroxytoluene (BHT) and butylated hydroxyanisole (BHA). It is worth mentioning that these additives have recently raised concerns about their possible mutagenic and carcinogenic effects. At the same time, HTyr plays an important role in pharmaceutical applications for its beneficial health properties such as anticancer activity.

In recent years, international scientific research has proven several biological effects of the polyphenols present in olive leaves and by-products.

Several extraction procedures have been optimized to recover polyphenols from olive waste or from olive leaves and fruit [117]. Among them, membrane separation techniques have been recently developed to fractionate olive mill wastewaters. This technology offers several advantages over traditional techniques mainly in terms of low energy consumption, no additive requirements, and no phase change [118]. Unfortunately, in almost cases, these technologies are applied on a laboratory scale or are not economically feasible.

The case-study described by Romani *et al.* [119] represents an original example of a circular economy process applied to the agricultural system of olive oil processing on an industrial scale. It concerns the valorization of *Olea europaea* L. leaves and pitted olive pulp as a source of bioactive polyphenols to produce standardized extracts to be used in the food, nutraceutical, pharmaceutical, feed, and agronomic fields.

Polyphenolic extracts were obtained from Olea europaea matrices (leaves, olives, oil, and by-products) using a sustainable patented technology [119] (PCT/IT/2009/09425529) consisting of the following steps: a) water extraction; b) physicochemical pre-treatment of the raw materials with enzymes and acidification; c) fractionation by membranes technology (microfiltration, ultrafiltration, nanofiltration, and reverse osmosis. HTyr-enriched fractions were obtained after concentration by evaporation at low temperature or spray-dried technique of the aqueous extracts.

The technology used to obtain these extracts is both environmentally and economically sustainable and already applied at the industrial level to produce extracts for stabilized and functionalized foods, pharmaceuticals, and cosmetic conventional and innovative formulations [119].

The pilot plant allows for a membranes fractionation process that follows the aqueous extraction, and also, a spray drying step may be performed to simulate the whole kilo-scale process.

A sustainable kilo-scale extractive technology has been applied followed by membrane separation methods, in particular by-products of the processing of olive oil (and artichokes), in order to obtain different extracts and fractions. The entire treatment, described by Romani *et al.* [120], consists of water extraction of the vegetal material followed by selective fractionation in five steps: i) physicochemical pre-treatment with enzymes and acidifying substances (pre-filtration); ii) microfiltration (MF); iii) ultrafiltration (UF); iv) nanofiltration (NF); and v) reverse osmosis (RO). The fractions were characterized and quantified by HPLC/DAD-ESI/MS [120, 121].

The kilo-scale green-extraction of *Olea* and *Cynara* was performed in a Rapid Extractor Timatic series (from Tecnolab S.r.l., Perugia, Italy) using a solid-liquid extraction technology. The extraction was performed with water, in a stainless-steel basket at a temperature of 60°C. The working cycle is fully automatic and alternates between a dynamic phase, obtained with a set pressure (7–9 Bar), and a static phase necessary for transferring the substance into the extraction solvent. Forced percolation is generated during the stationary phase, which, thanks to the programmable recirculation, ensures a continuous flow of solvent to the interior of the plant matrix. This avoids over-saturation and the formation of preferential channels, thus ensuring total extraction of the active principles from the vegetal matrix [121].

This innovative separation process performed with physical technologies [122] is defined as BAT (Best Available Technology) and recognized by the EPA (Environmental Protection Agency). The technology studied consists of an integrated system of all the filtration stages: Micro (MF), Ultra (UF), Nano (NF), and Reverse Osmosis (RO). The different filtration stages are characterized by different molecular weights with cut-off and filtration degrees. During the manufacturing process, the MF stage is carried out with tubular ceramic membranes in titanium oxide, and the UF, NF, and RO stages are conducted with spiral wound module membranes in polyethersulfone (PES). This design maximizes the surface area in a minimum amount of space. Less expensive but more sensitive to pollution, this ecofriendly system consists of consecutive layers of large membranes and supporting material in an envelope-type design rolled up around a perforated steel tube The diagram of the sustainable industrial plant is shown in Figure 6.6.

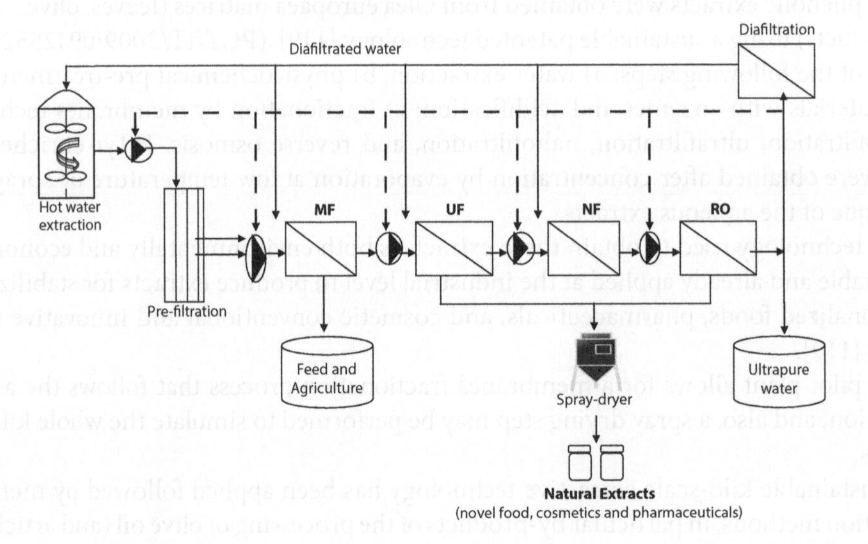

Figure 6.6 Industrial plant scheme of the sustainable process, for the recovery of phenolic fractions and pure water from Olea europaea L. and Cynara scolymus L. matrices [121].

By way of example, the chromatographic profile registered at 280 and 240 nm of the fraction OLEA CRO GL is reported in Figure 6.7, with the list of each identified compound. The quantitative data of each polyphenol in the raw fractions of the industrial plant is reported in Table 6.1.

The GL Olea CMF fraction is more diluted than the others (CNF and CRO) because of the higher cut-off of the membrane, allowing for the passage of the main part of the molecules, and then concentrated in the following nanofiltration and reverse osmosis steps. The concentrations of CNF and CRO from GL are very similar, with a titer in total polyphenols of 3.9% p/V. The main phenolic compounds in all the fractions, the secoiridoids, show the lowest quantity in the CRO fraction obtained from dried leaves due to the decomposition of oleuropein, one of the main secoiridoid compounds, during the drying process itself. The table also illustrates the concentrated fractions obtained by using a heat pump evaporator, as the initial extracting materials are not only GL and DL, but also pitted olive pulp. This last fraction was called Olea OH-Tyr since the main compounds are hydroxytyrosol (OH-Tyr) and its derivatives. The low concentration of secoiridoids is still evident in the concentrated fractions or soft extracts from olive leaves after their drying process. The titer in total polyphenols of the soft extracts obtained from green leaves and pitted olive pulp are very similar, 24.4%–28.9% p/p, whereas that coming from the dried leaves is 5.7% p/p. To partially avoid the secoiridoid loss and, particularly, in order to increase the percentage of secoiridoids in the DL fraction, a spray drying process can be applied to obtain a powder with a final titer of 5.7% p/p (44.2% of secoiridoids; 27.8% of hydroxytyrosol derivatives; 16.2% of elenolic acid derivatives; 5.3% of flavonoids; 3.9% of verbascoside; and 2.5% of hydroxycinnamic acid derivatives). Table 7.1 also shows the spray-dried composition of GL extract, with a final titer of 12.8% p/p (60.8% of secoiridoids; 18.3% of hydroxytyrosol derivatives; 13.2% of elenolic acid derivatives; 3.4% of flavonoids; 3.2% of verbascoside; 1% of hydroxycinnamic acid derivatives).

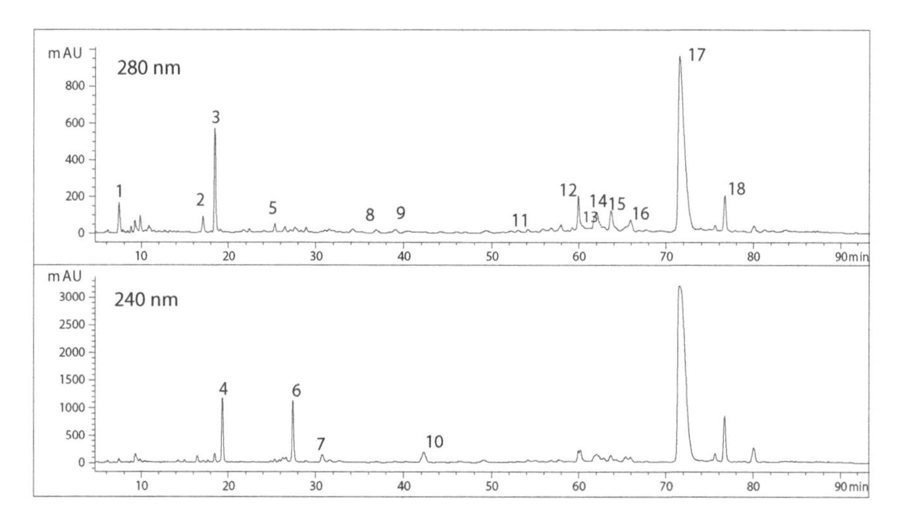

Figure 6.7 Chromatogram of Olea CRO GL. Peaks: 1. Hydroxytyrosol derivative; 2. Hydroxytyrosol; 3. Hydroxytyrosol glucoside; 4. Oleoside; 5. Esculin; 6. Demetyl elenolic acid diglucoside; 7. Elenolic acid glucoside; 8. Olivile; 9. Hydroxycinnamic derivative; 10. Elenolic acid glucoside derivative; 11. β-OH-verbascoside; 12. Verbascoside; 13. Luteolin 7-O-glucoside; 14. Pinoresinol; 15. Verbascoside isomer; 16. Acetoxypinoresinol; 17. Oleuropein; 18. Oleuropein isomer [121].

Table 6.1 HPLC/DAD quantitative analyses of Olea plant fractions concentrate fractions and powders (spray dried). Data are mean values of triplicate analyses (±SD) [121].

	Plant fractions (g/L)				**Concentrated fractions (mg/g)**			**Spray dried (mg/g)**	
	GL Olea CMF	**GL Olea CNF**	**GL Olea CRO**	**DL Olea CRO**	**Soft extract Olea OH-Tyr**	**Soft extract Olea GL**	**Soft extract Olea DL**	**Olea GL**	**Olea DL**
Hydroxytyrosol derivatives	0.29 ± 0.10	4.69 ± 0.67	6.18 ± 0.58	3.63 ± 0.64	279.89 ± 18.24	24.69 ± 3.47	25.21 ± 1.56	23.55 ± 0.03	15.98 ± 0.96
Secoiridoid der.	2.74 ± 1.75	25.13 ± 8.88	26.62 ± 8.14	2.44 ± 1.74	nd	164.19 ± 1.47	11.09 ± 0.45	78.18 ± 16.70	25.41 ± 11.20
Elenolic acid der.	0.82 ± 0.28	4.05 ± 1.33	4.15 ± 0.45	1.05 ± 0.37	0.51 ± 0.04	28.34 ± 0.43	7.54 ± 0.40	16.98 ± 1.17	9.30 ± 4.46
Hydroxycinnamic derivatives	0.03 ± 0.02	0.24 ± 0.13	0.30 ± 0.67	0.21 ± 0.12	7.83 ± 0.25	1.42 ± 0.06	4.30 ± 0.31	1.26 ± 0.88	1.49 ± 0.61
Flavonoids	0.15 ± 0.09	0.56 ± 0.18	0.83 ± 0.13	0.29 ± 0.21	nd	1.27 ± 0.01	1.00 ± 0.41	4.38 ± 1.63	3.08 ± 1.05
Verbascoside	0.09 ± 0.03	0.99 ± 0.31	0.83 ± 0.23	0.71 ± 0.49	1.69 ± 0.17	6.76 ± 0.10	5.85 ± 1.05	4.13 ± 0.44	2.27 ± 0.37
Lignans	nd	3.18 ± 1.16	nd	nd	nd	17.48 ± 0.01	2.65 ± 0.23	nd	nd
Total polyphenols	**4.12 ± 2.12**	**38.84 ± 10.31**	**38.91 ± 8.24**	**8.33 ± 2.51**	**289.93 ± 18.70**	**244.15 ± 5.54**	**57.63 ± 4.42**	**128.48 ± 20.84**	**57.53 ± 18.66**

CMF = concentrate of microfiltration; CNF = concentrate of nanofiltration; CRO = concentrate of reverse osmosis; GL = green leaves; DL = dried leaves; nd = not detected.

The main fractions have been produced by evaporation of CNF and CRO to obtain the previously described soft extracts with a different polyphenolic composition, depending on the initial extracting material (olive leaves and pitted olive pulp). It is worth noting that the soft extract Olea OH-Tyr is composed almost exclusively by hydroxytyrosol and derivatives, 96.5% of the total polyphenols. Another commercial fraction can be obtained from appropriate mixing of the concentrated fractions of CNF and CRO of pitted olive pulp, in order to obtain a standardized titer in polyphenolic compounds, generally ranging from 2.3 to 4.5% p/p, depending on how intense the production process is (data not shown).

6.3.3 Olive Oil, Tradition, Biodiversity, Territory, and Sustainability

Olive oil manufacture is one of the most typical and oldest Italian productions. A great variety of typologies, over 400 "cultivar", are produced in this country, 68 of which only in Tuscany [123]. Territory preservation and biodiversity are key concepts on which both the agricultural and food production chain are based. Appropriate production techniques and respect of natural resources are fundamental aspects of a correct management system for companies involved in this field. Only recent studies have applied the LCA analysis on extra-virgin olive oil production [124].

Olivieri *et al.* [125] underlined the necessity to preserve typical agricultural and gastronomic traditions of this region, trying to demonstrate how these are still based on sustainable patterns. The analysis, following a preliminary calculation where the environmentally most significant activities were identified, wants to present a more detailed study that illustrates the importance of a co-generative use of waste materials. This evaluation shows how the recoveries from the phases of olive transformation, olive pulp allocation, wastewater reuse, and leaves and limbs reprocessing, represent an important gain in the total environmental assessment. The project involves OTA (Olivicoltori Toscani Associati, Florence, Italy), LCA-*lab* spin off ENEA and the University of Florence with the collaboration of the Industrial Ecology master at Chalmers Institute of Technology.

For the study, the EPS 2000 method and the Sima Pro 5.0 code has been used for Life Cycle Impact Assessment (LCIA) phase, and the database Standard, Ivamlca3, Idemat, Eth-Esu, Data Archive, Ecoinvent for Life Cycle Inventory Analysis phase [126, 127].

In EPS 2000 [125] method (Environmental Priority Strategies in product design), the impact categories are identified from five safe guard subjects: human health, ecosystem production capacity, abiotic stock resource, biodiversity, and cultural and recreational values. The Human Health indicators are: life expectancy expressed in years of life lost (person/year), severe morbidity and suffering including starvation (person/year), morbidity like cold or flue (person/year), severe nuisance (person/year), and nuisance which causes irritation, but not any direct reaction (person/year). The impact categories of ecosystems production capacity are: crop production capacity (kg of harvest), wood production capacity (dry kg), fish and meat production capacity (entire weight of species in kg), base cations capacity (H^+ mole equivalents), production capacity of water (kg) with respect to persistent toxic substances, and production capacity of drinking water (kg) fulfilling WHO's criteria on drinking water. Abiotic stock resource indicators are: depletion of elemental or mineral reserves and depletion of fossil reserves. The weight factors, which are another distinctive element of the EPS 2000 method, stand for the willingness to pay (WTP) to avoid any change that may cause damages to the environment and to human health. The WTP is an economic concept and defines a method of evaluation meant to establish the maximal amount of money that a subject is willing to pay for a certain benefit.

The goal of this LCA study is to determine the environmental damage due to the olive oil production in the conventional Tuscan company (OTA) [125]; the object of the analysis is the virgin olive oil production; the functional unit is 0.75l of 1 olive oil bottle; the system boundaries includes the olives transformation phases until bottling phase, including solid olive residue allocation, olive oil wastewater reuse, and leaves and limbs reprocessing. In Figure 6.8 and Table 6.2, the principal results of weigthing analysis.

The damages incurring in the olive oil production phases are a result of a large use of electricity. The −2% gain is consequent of olive oil wastewater reuse for fertigation, which results in avoiding the use of fertilizing products. The −0.13% gain is due to the reuse for fertigation of leaves and limbs in the defoliation phase.

LCA analysis can represent a valid tool to numerically quantify the environmental impact of a productive cycle in the agricultural food industry with the use of international indicators and specific software. This analysis' objective is to guarantee hygiene and to preserve

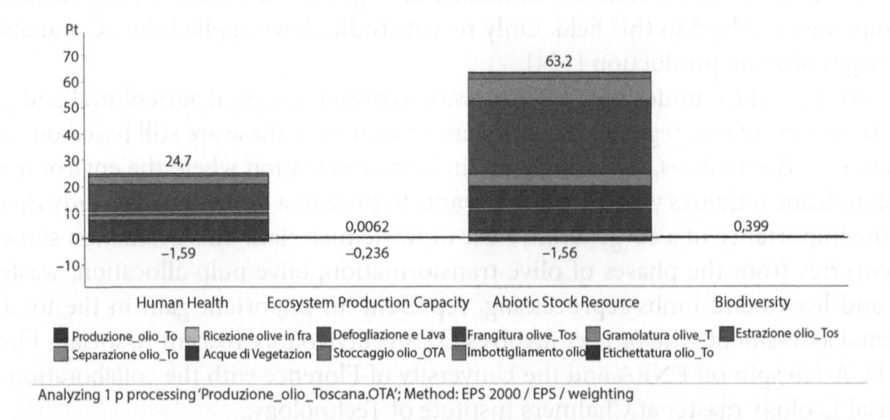

Figure 6.8 Diagram of weighting analysis of one bottle of Olive oil production [125].

Table 6.2 The results (%) of weighting analysis of one bottle of Olive oil production [125].

The processes that produces the most damage	% point of damage
Extraction	47
Crushing	27
Separation	13
Kneading	7
Bottling	4
The processes that produces the gain	**% point of damage**
Wastewater reuse	−2
Defoliation	−0.13

quality through an accurate evaluation of the chemical, physical, and biological risks of the entire olive oil weaving factory (from cultivation to oil retail).

The environmental impact analysis divided into productive phases wants to clarify the criticalities of each productive step and creates a manual listing of the "Best Production Practices for Environmental Sustainability" for the sector [128].

Olivieri *et al.* [129] applied a preliminary Life Cycle Assessment to the new integrated technology for Olive Oil Wastewater (OOWW) treatment and polyphenols recovery from biphasic olive mill. Treatment and disposal of OOWW are serious environmental problems for the agricultural and olive oil sector due to the high pollution load of the organic compounds. The OOWW treatment process consists in the OOWW selective fractionation in five steps: the physico-chemical pre-treatment with enzymes and acidifying substances; the microfiltration; the ultrafiltration; the nanofiltration; and the reverse osmosis. Once removed toxic potential pollutant of OOWW components, the concentrated organic substances obtained from the tangential streams in each filtration step are of high economic value for "novel food" [130], fitoterapic, or cosmetic industries. The benefits of this procedure are the following: it treats a sewage that otherwise it would be a waste (containing pollutants such as COD and BOD_5) and, at the same time, produces natural products with a potential economic value. In the present study, a preliminary way, the sensitivity analysis is carried out to compare the OOWW treatment with tangential flow membrane technology (Best Available Techniques) with a traditional wastewater treatment for removal COD pollutant.

The improvement of innovative physic-chemical processes, with high performances and low environmental impact, is important for a sustainable development. In this perspective, membrane technology can offer important new opportunities in the design, rationalization and optimization of processes, products, and wastewater treatments. In particular, the main advantages of the membrane technology system studied are that: the discharge level is almost equal to zero, that treats wastewater and, at the same time, produces the "novel products", and finally, the environmental assessment of novel products and processes is important for "novel food" producers, since they have introduced sustainability as a core company goal.

The preliminary result of impact assessment showed that the energy consumptions are the most significant impacts; therefore, next aim of the study could be to improve the system to reduce the use of energy from nonrenewable resources. Hence, the categories where impacts are more significant are Climate Change and Depletion of Resources. A sensitivity analysis was undertaken to compare the new technology with the traditional wastewater treatment for to remove the COD pollutant. The result of the preliminary sensitivity analysis show that the LCA of new OOWW process is less impacting for an overall percentage of 57% respect to traditional process [129].

6.4 Bioactive Compounds in Legume Residues

Legumes are dried seeds that have long been one of the most important traditional dietary components worldwide, supplying proteins, dietary fibers, minerals, and vitamins [131]. To highlight their important nutritional properties and the low environmental impact of pulse production worldwide, the Food and Agriculture Organization of the United Nations (FAO) declared 2016 the International Year of Legumes [132]. Pulses for human consumption include lentils, chickpeas, beans, grass peas, green peas, broad bean, and soy beans [133]. India is the largest producer (around 25 %) of pulses in the world and the Indian subcontinent covers the leading area in terms of consumption of pulses.

Moreover nutrients, legumes are very rich in bioactive compounds typically occurring in small quantities (when compared with macronutrients); pulses contain plenty of bioactives that exert metabolic effects on human body upon consumption [134]. Some of these substances play important role in defense mechanisms of the plants against environmental conditions or predators, while others act as reserve compounds that are accumulated (mainly for germination) in seeds as energy stores. These bioactive components include dietary fibers, resistant starches, and bioactive phytochemicals (mainly polyphenols and phytosterols) which make pulses suitable for application in wide range of food products.

6.4.1 Polyphenols

Phenolic acids and their derivatives, flavanols, flavan-3-ols, anthocyanins, condensed tannins, are the main polyphenol categories present in legume seeds. Dehulling of pulses results in removal of large amounts of polyphenols as these are concentrated in the seed coat portions. The presence of polyphenols as chlorogenic acid, catechin, delphinidin 3,5-diglucoside, kaempferol 3-glucoside, and ferulic acid is often related to the color of the pulse and to its antioxidant activity. Caprioli *et al.* [135] reported that the highest polyphenol levels were found in beans, especially black beans (458 mg kg^{-1}), ruviotto beans (189 mg kg^{-1}), violet pavoni beans (162 mg kg^{-1}), borlotti beans (158 mg kg^{-1}), and pavoni beans (150 mg kg^{-1}). Lentils also displayed high levels of polyphenols, particularly black lentils (137 mg kg^{-1}), quality gold lentils (132 mg kg^{-1}), and eston lentils (118 mg kg^{-1}).

6.4.2 Phytosterols and Squalene

Phytosterols are the plant sterols having a structure similar to cholesterol. These compounds are extensively studied as functional food ingredients and well known for their wide variety of health benefits such as lowering of blood cholesterol levels and reducing the intestinal absorption of dietary and endogenous cholesterol. Phytosterol concentration in kidney beans and chickpeas was reported to be 127 and 35 mg/100 g, respectively [136]. Ryan *et al.* [137] studied the total phytosterol content in butter beans, chickpeas, kidney beans, lentils, and peas and reported it in the range of 134 mg/100 g DW (kidney beans) to 242 mg/ 100 g DW (peas). In addition, chickpeas and lentils had a phytosterol content of 205 and 158 mg/100 g DW, respectively. They further measured the beta-sitosterol, campesterol, and stigmasterol content and reported that the beta-sitosterol was the dominant phytosterol in all the studied legumes. Pulses have a high percentage of glycosylated sterols, and Nystrom *et al.* [138] reported that most of the beans and lentils had glycosylated sterols ranging between 100 and 150 mg/g DW. Squalene is a triterpene primarily known as an intermediate in the biosynthesis of sterols in plant and animal world. During the last decades research has indicated that squalene exhibits antioxidant properties and it contributes to the inhibition of several forms of cancer. Kalogeropoulos *et al.* reported that lupines contained the higher amounts of squalene −1.74 mg/100 g, followed by beans varieties which contained 0.45–0.94 mg/100 g, and the rest of legumes which contained 0.12–0.32 mg/100 g [139].

6.4.3 Dietary Fiber and Resistant Starch

Pulses are good source of dietary fiber and contain both soluble (SDF) and insoluble (IDF), the later being the major portion of pulse dietary fibers. Pulse SDF constitutes gums, pectins, fructans, inulins, and some hemicelluloses, whereas IDF included cellulose, some hemicelluloses, lignins, and arabinoxylan [140]. Resistant starch (RS) is a kind of starch that is not digested in the small intestine by endogenous enzymes consequently pass to large intestine where it functions physiologically similar to dietary fibers as it increases faecal bulk, reduces colonic pH, lowers serum cholesterol and triglycerides levels [141], and reduces postprandial glycemic responses. Similar to dietary fibers, RS also functions as prebiotics and promotes the growth of probiotics in human gut. The RS content of different pulses (beans, chickpeas, lentils, and peas) ranged from 3.95% to 5.09 % against 1.42% to 2.85 %, respectively, for cereals like wheat, rice, and barley.

6.4.4 Soyasaponins

Legumes are a primary dietary source of food saponins, which are bioactive compounds that have been demonstrated to possess multiple health-promoting properties, such as reduction of cholesterol levels, anticarcinogenic and antihepatotoxic properties, and antireplicative effects against HIV. Soyasaponins are triterpenoidal glycosides, structurally divided into two groups, called A (bidesmosidic) and B (monodesmosidic). Lentils mainly contain soyasaponin I (soyasaponin Bb) and soyasaponin βg (also called soyasaponin VI) (Figure 6.9), both of which belong to the B group of soyasaponins.

Lentils are a good source of soyasaponins, with a content of soyasaponin I that ranged from 636 to 735 mg kg^{-1} and of soyasaponin βg from 672 to 1807 mg kg^{-1}. The cooking

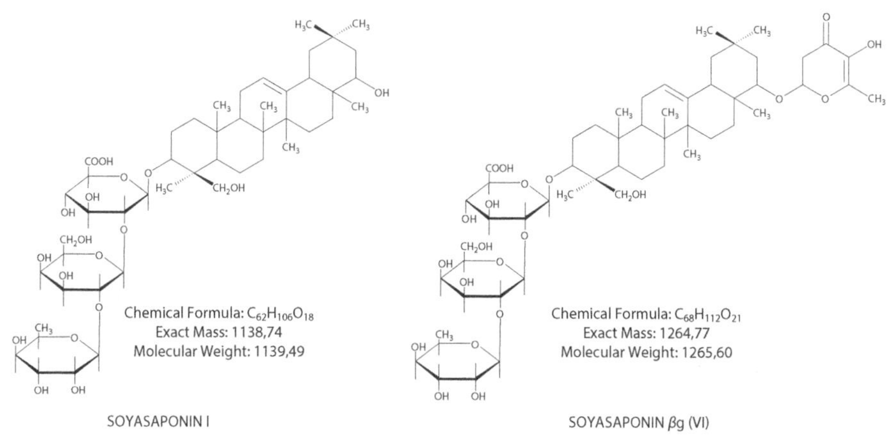

Figure 6.9 Chemical structures of soyasaponins I and βg (VI).

process produced a small loss of soyasaponins in water, that is, 4.8%–8.7%, and partially converted soyasaponin βg into soyasaponin I. Peas show a content of soyasaponin I that ranged from 702 to 907 mg kg^{-1} and of soyasaponin βg from 1070 to 1411 mg kg^{-1}, where chickpeas showed a content of soyasaponin I that ranged from 688 to 761 mg kg^{-1} and of soyasaponin βg from 866 to 1,412 mg kg^{-1}. Soybean is another good source of soyasaponins with a content of 2492 mg kg^{-1} [142].

Tocopherols and ascorbic acid. Pulses contain all four tocopherol isomers (alpha, beta, gamma, and delta) although no tocotrienols have been detected. However, the levels of these isomers vary quite markedly, with gamma-tocopherol being the predominant one. The greatest gamma-tocopherol levels were reported in lentils and peas in a concentration value around 100 mg kg^{-1} [143]. Leguminous seeds are not a rich source of ascorbic acid (vitamin C), but this vitamin's contribution to the antioxidant capacity of seeds, especially for the part generated by hydrophilic compounds, is high. In foods of plant origin, vitamin C is determined as ascorbic and dehydroascorbic acid. The content of ascorbic acid in peas ranged from 0.40 to 1.48 mmol/g.

6.4.5 Bioactive Peptides

Pulses are a rich source of protein (20%–30%, dry weight) and in addition to their role in modulating plasma lipids, they appear to lower CVD risk by displacing dietary saturated fats found in animal-derived protein with plant protein [144]. Further, bioactive peptides, which are short sequences of amino acids that are released from parent proteins through proteolytic processes occurring during digestion and/or food processing, are receiving attention for their hypotensive activity by angiotensin converting enzyme (ACE) inhibition and antioxidant activity through free radical scavenging. Significant peptide sequences capable of ACE inhibition, dipeptidyl peptidase IV (DPP-IV) inhibition, and antioxidant activity were identified in 87.5% of sequenced peptides in bean protein isolates made from 15 cultivars of common bean from Mexico and Brazil [145], and hypocholesterolemic and antioxidant properties have been demonstrated in chickpea protein hydrosylate.

In recognition of their multiple benefits on environment, food security, and human health, nowadays, valorization of legumes co-products and by-products pulses are attracting attention for nutraceutical scopes and for package application and energy production.

Herb and root parts of legumes are not commonly used in human consumption, since they are traditionally considered waste material. Although inedible, these plant parts have a complex chemical composition, with potential biological activities. They are especially rich in polyphenols, such as flavonoids, isoflavonoids, and phenolic acids. Šibul *et al.* [146] have studied the phenolic profile, antioxidant and anti-inflammatory potential of herb and root extracts of seven selected legumes - [soybean (*Glycine max* L.), common bean (*Phaseolus vulgaris* L.), pea (*Pisum sativum* L.), broad bean (*Vicia faba* L.), chickpea (*Cicer arietinum* L.), white lupin (*Lupinus albus* L.), and grass pea (*Lathyrus sativus* L.)]: high levels of isoflavones genistein and daidzein were found in *G. max*, *P. vulgaris*, and *L. albus* root extracts (1.02–1.53, 0.309–0.648, and 2.81 mg/g, respectively), whereas C. arietinum, L. albus, and P. vulgaris herb extracts were found to be the richest sources of flavonoid compounds (13.1–19.4, 5.98, and 2.36–3.64 mg/g, respectively). Moreover, the same authors reported that all samples demonstrated the ability to scavenge both DPPH· and NO· radical, and exhibited a good

FRAP reducing power and, in particular, *V. faba* exhibited the highest antioxidant activity (DPPH· IC50 11.7–20.7 μg/ml, NO·IC30 0.080–0.34mg/ml, FRAP 32.6–66.9 AAE/g dw).

Another work of Khetarpaul and Chand [147], has investigated the antioxidant activity in leaves of chickpea (*Cicer arietinum* L.) varieties desi and kabuli (on dry matter basis) in India: Diphenyl-I-Picrylhydrazyl (DPPH) and flavonoids activity also increased in leaves of both the desi and kabuli chickpea varieties as the growth period of leaves increased.

Arya *et al.* [148] investigated the degradation of anthropogenic pollutant and organic dyes by biosynthesized silver nano-chetatalyst from cicer arietinum leaves.

Gutöhrlein *et al.* [149] described the isolation and quantification of galacturonic acid from pea hulls.

Previous studies reported the high content of cellulose of approximately 65% of pea hulls [150] as well as 9 to 15% of uronic acids [151, 152].

Belghith-Fendri *et al.* [153] described the identification and extraction procedures from pea pod by-products and their properties: the main monosaccharides in polysaccharide from pea pod were galactose, xylose, and arabinose, whereas the major functional groups identified from FT-IR spectrum included CO, OH, and CH; moreover, the same authors have reported for PPP significant antioxidant activity and good antibacterial activity, by marking how it can potentially be used as additive in food, pharmaceutical, and cosmetic preparations.

Hidalgo-Cuadrado *et al.* [154] investigate the potential use of post harvest agricultural wastes as alternative sources of peroxidases, with focus on lentil (*Lens culinaris* L.) stubble: lentil stubble crude extract was seen to provide one of the highest specific peroxidase activities, catalyzing the oxidation of guaiacol in the presence of hydrogen peroxide to tetraguaiacol, and was used for further studies.

It is worth to mentioning the recent review of Zhong *et al.* [155] on characterization, processing, and applications of seed coats of pulses biochemical and physicochemical functionalities of seed coats of six globally important pulses: chickpea, field pea, faba/broad bean, lentil, and mung bean with a special emphasis on the emerging food pulse lupin as a food ingredient. As described by the authors, the seed coat is a major by-product of pulse processing, and it has been used as ruminant feed and very limited use in high fiber foods. Recently, recent researches have suggested how this underutilized by-product can represent a functional food ingredient for its dietary fiber content. The same authors summarized the food process modification and recent human food applications of the seed coats and discussed bio-availability of the seed coat compounds, and phomopsins contaminated lupin seed coats as a typical example of safety issue. The authors concluded that high levels of dietary fiber, minerals, and potential health-promoting phytochemicals in the seed coats indicate their great potential to be used as a natural "nutritious dietary fiber". However, further in-depth studies are required to improve their desirable nutritional, physiological, and techno-functional properties while minimizing any undesirable ones.

Cheng *et al.* [156] described the pea soluble polysaccharides obtained from two enzyme-assisted extraction methods, spray-drying and ethanol precipitation oven drying, and their application as acidified milk drinks stabilizers.

Belghith-Fendri *et al.* [157] focused their studies on pea and broad bean pods as natural source of dietary fiber; in particular, the authors investigated the impact on texture and sensory properties of cake after addition of substituted of 5%, 10%, 15%, 20%, 25%, and 30% of pea pod (PP) and broad bean pod (BBP) flours: Cakes hardness increased, whereas L* and a* color values decreased. The overall acceptability rate showed that a maximum of 15% of PP and BBP flours can be added to prepare acceptable quality cakes.

References

1. Banach, M., Mikhailidis, D.P., Serban, M.C., Sahebkar, A., Editorial: Natural products as the integral part of the therapy? *Curr. Pharm. Des.*, 23, 2411–2413, 2017.
2. Kumar, H., Yadav, A.N., Kumar, V., Vyas, P., Dhaliwal, H.S., Food waste: A potential bioresource for extraction of nutraceuticals and bioactive compounds. *Bioresour. Bioprocess.*, 4, 18, 2017.
3. Galanakis, C.M., Cvejic, J., Verardo, V., Segura Carretero, A., Food use for social innovation by optimizing food waste recovery strategies, in: *Innovation Strategies in the Food Industry. Tools for Implementation*, C.M. Galanakis (Ed.), pp. 211–236, Academic Press, London, UK, 2016.
4. Thompson, P.B., *From field to fork, food ethics for everyone*, p. 346, Oxford University Press, New York, NY, USA, 2015.
5. Lin, C.S.K., Koutinas, A.A., Stamatelatou, K., Mubofu, E.B., Matharu, A.S., Kopsahelis, N., Pfaltzgraff, L.A., Clark, J.H., Papanikolaou, S., Kwan, T.H. *et al.*, Current and future trends in food waste valorization for the production of chemicals, materials and fuels: A global perspective. *Biofuels Bioprod. Bioref.*, 8, 686–715, 2014.
6. Galanakis, C.M., Recovery of high added-value components from food wastes: Conventional, emerging technologies and commercialized applications. *Trends Food Sci. Technol.*, 6, 68–87, 2012.
7. Varzakas, T., Zakynthinos, G., Verpoort, F., Plant food residues as a source of nutraceuticals and functional foods. *Foods*, 5, 88, 2016.
8. Tuck, C.O., Pérez, E., Horváth, I.T., Sheldon, R.A., Poliakoff, M., Valorization of biomass: Deriving more value from waste. *Science*, 337, 6095, 695–699, 2012.
9. Pfaltzgraff, L.A., De bruyn, M., Cooper, E.C., Budarin, V., Clark, J.H., Food waste biomass: A resource for high-value Chemicals. *Green Chem.*, 15, 307–314, 2013.
10. Romani, A., Ciani Scarnicci, M., Scardigli, A., Paiano, A., Circular economy as a new Model for the exploitation of the agroindustrial biomass, in: *Proceedings of the 20th IGWT Symposium, Commodity Science in a Changing World*, pp. 12–16, University of Economics, Varna, Bulgaria, September 2016.
11. Santana-Méridas, O., González-Coloma, A., Sánchez-Vioque, R., Agricultural residues as a source of bioactive natural products. *Phytochem. Rev.*, 11, 447, 2012.
12. Baiano, A., Recovery of biomolecules from food wastes—A review. *Molecules*, 19, 14821–14842, 2014.
13. Fregoni, M., Viticultura di qualità, in: *Informatore Agrario*, Ed. L'Informatore Agrario, 2th ed., pp. 25–48, 1999.
14. Boss, P.K., Davies, C., Robinson, S.P., Expression of anthocyanin biosynthesis pathway genes in red and white grapes. *Plant. Mol. Biol.*, 32, 652–663, 1996.
15. Boss, P.K., Davies, C., Robinson, S.P., Analysis of the expression of anthocyanin pathway genes in developing Vitis vinifera L. cv. Shiraz grape berries and the implications for pathway regulation. *Plant Physiol.*, 111, 1059–1066, 1996.
16. Waterhouse, A.L., Wine phenolics. *Ann. N. Y. Acad. Sci.*, 957, 21–36, 2002.
17. Nassiri-Asl, M. and Hosseinzadeh, H., Review of the pharmacological effects of *Vitis vinifera* (Grape) and its bioactive constituents: An update. *Phyt. Res.*, 30, 1392–1403, 2016.

18. Singh, N., Functional and physicochemical properties of pulse starch, in: *Pulse foods: processing, quality and nutraceutical applications*, pp. 91–119, 2011.
19. Beres, C., Costa, G.N.S., Cabezudo, I., da Silva-James, N.K., Teles, A.S.C., Cruz, A.P.G. *et al.*, Towards integral utilization of grape pomace from winemaking process. A review. *Waste Manag.*, 68, 581–594, 2017.
20. Tabeshpour, J., Mehri, S., Shaebani Behbahani, F., Hosseinzadeh, H., Protective effects of Vitis vinifera (grapes) and one of its biologically active constituents, resveratrol, against natural and chemical toxicities: A comprehensive review. *Phytother. Res.*, 32, 11, 2164–2190, 2018.
21. Annunziata, G., Tenore, G.C., Novellino, E., Resveratrol-based Nutraceuticals for the Management of Diabetes and Obesity: Real Therapeutic Potential or a mere Palliative? *Arch. Diab. Obes.*, 1, 2, 34–35, 2018.
22. Koushki, M., Amiri-Dashatan, N., Ahmadi, N., Abbaszadeh, H.A., Rezaei-Tavirani, M., Resveratrol: A miraculous natural compound for diseases treatment. *Food. Sci. Nutr.*, 6, 8, 2473–2490, 2018.
23. Ramírez-Garza, S.L., Laveriano-Santos, E.P., Marhuenda-Muñoz, M., Storniolo, C.E., Tresserra-Rimbau, A., Vallverdú-Queralt, A., Lamuela-Raventós, R.M., Health Effects of Resveratrol: Results from Human Intervention Trials. *Nutrients*, 10, 1892, 1–18, 2018.
24. OIV (Organization International de la Vigne et du Vin). Statistical report on world vitiviniculture 2018. International organization of Vine and Wine, 1–27, 2008. http://www.oiv.int/public/medias/6371/oiv-statistical-report-on-world-vitiviniculture-2018.pdf.
25. Croft, K.D., Dietary polyphenols: Antioxidants or not? *Arch. Biochem. Biophys*, 595, 120–124, 2016.
26. Teixeira, A., Baenas, N., Dominguez-Perles, R., Barros, A., Rosa, E., Moreno, D.A., Garcia-Viguerra, C., Natural bioactive compounds from winery by-products as health promoters: A review. *Int. J. Mol. Sci.*, 15, 9, 15638–15678, 2014.
27. Shrikhande, A.J., Wine by-products with health benefits. *Food Res. Intern*, 33, 469– 474, 2000.
28. Cecchini, F., Factors affecting antioxidant activity of grape tissues, in: *Grape production phenolic composition and potential biomedical effects*, J S. Câmara (Ed.), pp. 107–133, Nova Science Publisher, New York, 2014.
29. Guerrero, R.F., Liazid, A., Palma, M., Puertas, B., Gonzàlez-Barrio, R., Gil-Izquierdo, A. *et al.*, Phenolic characterisation of redgrapes autochthonous to Andalusia. *Food Chem.*, 112, 949–955, 2009.
30. Cortell, J.M. and Kennedy, J.A., Effect of shading on accumulation of flavonoid compounds in (Vitis vinifera L.) Pinot Noir fruit and extraction in a model system. *J. Agric. Food Chem.*, 54, 8510–8520, 2006.
31. Shi, J., Yu, J.J., Pohorly, E., Kakuda, Y., Polyphenolics in grape seeds -biochemistry and functionality polyphenols in grape seeds-Biochemistry and functionality. *J. Med. Food.*, 6, 291–299, 2003.
32. Ylmaz, Y., Göksel, Z., Erdoğan, S.S., Öztürk, A., Atak, A., Özer, C., Antioxidant activity and phenolic content of seed, skin and pulp parts of 22 grape (*Vitis vinifera* L.) cultivars (4 common and 18 registered or candidate for registration). *J. Food Process. Preserv.*, 39, 1682–1691, 2015.
33. Fischer, B.B., Krieger-Liszkay, A., Hideg, E., Snyrychová, I., Wisendanger, M., Egger, R.I.L., Role of singlet oxygen in chloroplast to nucleus retrograde signaling in Chlamydomonas reinhardtii. *FEBS Lett.*, 581, 5555–5560, 2007.
34. Havsteen, B.H., The biochemistry and medical significance of the flavonoids. *Pharmacol. Therap*, 96, 2–3, 67–202, 2002.
35. Rice-Evans, C.A., Miller, N.J., Paganga, G., Reviews Antioxidant properties of phenolic compounds. *Trend Plant. Sci.*, 2, 152–159, 1997.

36. Peer, W.A., Blakeslee, J.J., Hanga, H., Murphy, A.S., Seven things we think we know about auxin transport. *Mol. Plant.*, 4, 487–504, 2011.
37. Lewis, D.R., Ramirez, M.V., Miller, N.D., Vallabhaneni, P., Ray, W.K., Helm, R.F. *et al.*, Auxin and ethylene induce flavonol accumulation through distinct transcriptional networks. *Plant Physiol.*, 156, 144–164, 2011.
38. Santelia, D., Henrich, S., Vincenzetti, V., Sauer, M., Biglewer, L., Klein, M. *et al.*, Flavonoids redirect PIN-mediated polar auxin fluxes during root gravitropic responses. *J. Biol. Chem.*, 283, 31218, 2008.
39. Peer, W.A. and Murphy, A.S., Flavonoids and auxin transport: Modulators or regulators? *Trends Plant Sci.*, 12, 556–563, 2007.
40. Peer, W.A., Cheng, Y., Murphy, A.S., Evidence of oxidative attenuation of auxin signalling. *J. Exp. Bot.*, 64, 2629–2639, 2013.
41. Agati, G. and Tattini, M., Multiple functional roles of flavonoids in photoprotection. *New Phytol.*, 186, 786–793, 2010.
42. Rasmussen, S.E., Frederiksen, H., Struntze Krogholm, K., Poulsen, L., Dietary proanthocyanidins: Occurence, dietary intake, bioavailability, and protection against cardiovascular disease. *Mol Nutr Food Res.*, 49, 159–174, 2005.
43. Murphy, K.J., Chronopoulos, A.K., Singh, I., Francis, M.A., Moriarty, H., Pike, M.J. *et al.*, Dietary flavanols and procyanidin oligomers from cocoa (*Theobroma cacao*) inhibit platelet function. *Amer. J. Clin. Nutr.*, 77, 1466–1473, 2003.
44. Santosh, K.K., Harish, C.P., Ram, P., Review Dietary proanthocyanidins prevent ultraviolet radiation-induced non-melanoma skin cancer through enhanced repair of damaged DNA-dependent activation of immune sensitivity. *Semin. Canc. Biol.*, 46, 138–145, 2017.
45. Cecchini, F., Moretti, S., Mulinacci, N., Innocenti, M., Pasqua, G., Ciriolo, M.R. *et al.*, Anti-proliferative effects in cervical cancer cells (HeLa) of grape seed extracts from Vitis viniferacultivars of central Italy, in: *Proceedings of XXVI International Conferences on Polyphenols*, vol. 2, pp. 471–472, 2012.
46. Dinicola, S., Cucina, A., Pasqualato, A., D'Anselmi, F., Proietti, S., Lisi, E. *et al.*, Antiproliferative and apoptotic effects triggered by grape seed extract (GSE) versus epigallocatechin and procyanidins on colon cancer cell lines. *Int. J. Mol. Sci.*, 13, 651–664, 2012.
47. Dinicola, S., Cucina, A., Pasqualato, A., Proietti, S., Danselmi, F., Pasqua, G. *et al.*, Apoptosis-inducing factor ad caspase dependent apoptotic pathways triggered by different grape seed extracts on human colon cancer cell line Caco2. *Brit. J. Nut*, 104, 824–832, 2010.
48. Shen-Chien, C., Kaur, M., Thompson, J.A., Agarwal, R., Agarwal, C., Influence of gallate esterification on the activity of procyanidin B2 in androgen-dependent human prostate carcinoma LNCaP cells. *Pharmaceut. Res.*, 27, 619–927, 2010.
49. Kaur, M., Agarwal, C., Agarwal, R., Anticancer and Cancer Chemopreventive potential of grape seed extract and other grape-based products. *J. Nutr.*, 139, 9, 1806–1812, 2009.
50. Brunetti, C., Fini, A., Sebastiani, F., Gori, A., Tattini, M., Modulation of Phytohormone Signaling: A Primary Function of Flavonoids in Plant–Environment Interactions. *Front. Plant Sci.*, 9, 1042, 1–8, 2018.
51. Agarwal, C., Veluri, R., Kaur, M., Chou, S.C., Thompson, J.A., Agarwal, R., Fractionation of high molecular weight tannins in grape seed extract and identification of procyanidin B2-3,3′-di-O-gallate as a major active constituent causing growth inhibition and apoptotic death of DU145 human prostate carcinoma cells. *Carcinogenesis*, 28, 7, 1478–1484, 2007.
52. Simonetti, G., Santamaria, A.R., D'Auria, F.D., Mulinacci, N., Innocenti, M., Cecchini, F., Pericolini, E., Gabrielli, E., Panella, S., Antonacci, D., Palamara, A.T., Vecchiarelli, A., Pasqua, G., Evaluation of anti-candida activity of *Vitis vinifera* L. seed extract obtaine from wine and table cultivars. *Biomed. Res. Int.*, 2014, 127021, 2014.

53. Cecchini, F., Giannini, B., Morassut, M., Moretti, S., Effect of the basic terroir on the nutraceutical proprieties in *Vitis vinifera* L. cv Dolcetto red grape variety. *Le Progrès agricole et Viticole*, 5, 88–92, 2012.
54. Cecchini, F. and Giannini, B., Comparison of the free radical-scavenging activity in Oidium and sound Dolcetto grape cultivars grown in a terroir of central Italy, in: *Proceedings VIII International Terroir Congress*, Soave Italy, 14-18 June 2010, pp. 41–46, 2010.
55. Downey, M.O., Dokoozlian, N.K., Krstic, M.P., Cultural practice and environmental impacts on the flavonoids composition of grapes and wine: a review of recent research. *Am. J. Enol. Vitic.*, 57, 3, 257–267, 2006.
56. Dwyer, K., Hosseinian, F., Rod, M., The market potential of grape waste alternatives. *J. Food. Res.*, 3, 91–96, 2014.
57. Campos, L.M.A.S., Leimann, F.V., Pedrosa, R.C., Ferreira., S.R.S., Free radical scavenging of grape pomace extracts from Cabernet Sauvignon (*Vitis vinifera*). *Bioresour. Technol.*, 99, 8413–8420, 2008.
58. Giannini, B., Mulinacci, N., Pasqua, G., Innocenti, M., Valletta, A., Cecchini, F., Phenolics and antioxidant activity in different cultivars/clones of *Vitis vinifera* L. seeds over two years. *J. Plant Biosyst.*, 150, 1408–1416, 2016.
59. Weber, H.A., Hodges, A.E., Guthrie, J.R., O'Brien, B.M., Robaugh, D., Clark, A.P., Harris, R.K., Algaier, J.W., Smith, C.S., Comparison of proanthocyanidins in commercial antioxidants: Grape seed and pine bark extracts. *J. Agric. Food Chem*, 55, 148–156, 2007.
60. Negro, Tommasi, L., Miceli, A., Phenolic compounds and antioxidant activity from red grape marc extracts. *Bioresour. Technol.*, 87, 41–44, 2003.
61. Ky, I., Lorrain, B., Kolbs, N., Crozier, A., Teissedre, P.L., Wine by-products: Phenolic Characterization and antioxidant activity evaluation of grape and grape pomaces from six different French grape varieties. *Molecules*, 19, 482–506, 2014.
62. Klis, F., Mol, P., Hellingwerf, K., Brul, S., Dynam1c of cell wall structure in Saccharomyces cerevisiae. *FEMS Mycrob. Rev.*, 26, 239–256, 2002.
63. Langella, C., Naviglio, D., Marino, M., Gallo, M., Study of the effects of a diet supplemented with active components on lipid and glycemic profiles. *Nutrition*, 31, 1, 180–186, 2015.
64. Francelino Andrade, E., Vieira Lobato, R., Vasques Araújo, T., Gilberto Zangerônimo, M., Vicente Sousa, R., José Pereira, L., Effect of betaglucans in the control of blood glucose levels of diabetic patients: A systematic review. *Nutr Hosp.*, 1, 170–177, 2014.
65. Moghadasian, M.H. and Jones, P.J., Cholesterol-lowering effects of oat β-glucan. *Nutr. Rev*, 69, 6, 299–309, 2011.
66. Rop, O., Mlcek, J., Jurikova, T., Beta-glucans in higher fungi and their health effects. *Nutr. Rev.*, 67, 11, 624–631, 2016.
67. Jenkins, A.L., Jenkins, D.J.A., Zdravkovic, U. *et al.*, Depression of the glycemic index by high levels of ß-glucan fibers in two functional foods tested in type 2 diabetes. *Eur. J. Clin. Nutr.*, 56, 622–628, 2002.
68. Brown, G.D. and Gordon, S., Immune recognition of fungal β-glucans. *Cell. Microbiol.*, 7, 4, 471–479, 2005.
69. Tsiapali, E., Whaley, S., Kalbfleisch, J., Ensley, H., Browder, I.W., Williams, D.L., Glucans exhibit weak antioxidant activity, but stimulate macrophage free radical activety. *Free Rad. Biol. Med.*, 30, 393–402, 2001.
70. Müller, A., Raptis, J., Rice, P., Kalbfleisch, J.H., Stout, R., Ensley, H. *et al.*, The influence of glucan polymer structure and solution conformation on binding to (1→3)-beta-D-glucan receptors in human monocyte-like cell line. *Glycobiology*, 10, 339–346, 2000.
71. Pengkumsri., N., Sivamaruthi, B.S., Sirilun, S., Peerajan, S., Keiska, P., Chaiyasut, K., Chaiyasut, C., Extraction of β-glucan from *Saccharomyces cerevisiae*: Comparasion of different extraction

methods and *in vivo* assessment of immunomodulatory effect in mice. *Food Sci. Technol.*, 37, 1, 124–130, 2017.

72. Cecchini, F., Bevilacqua, N., Morassut, M., Bioactive compounds of winery by-products: The sustainable development of oenological field, in: *Proceedings The 6th International virtual Conferences on advanced Sientific results*, pp. 236–240, 2018.
73. Asif, A., Faqir, M.A., Tahr, Z., Haq Nawaz, Z.A., Extraction and characterization of β-D-glucan from oat for industrial utilization. *Int. J. Biol. Macromol.*, 46, 304–9, 2010.
74. Shokri, H., Asadi, F., Khosravi, A.R., Isolation of β-glucan from the cell wall of Saccharomyces cerevisiae. *Nat. Product Res.*, 22, 5, 414–421, 2008.
75. ANPA Agenzia Nazionale per la Protezione dell'Ambiente, *I rifiuti del comparto agroalimentare", Rapporti 11/2001*, ANPA - Unità Normativa Tecnica, Roma, 2001.
76. Novello, V., Filiera vitivinicola: Valorizzare residui e sottoprodotti. *Informatore Agrario*, 33, 61–63, 2015.
77. Ruggieri, L., Cadena, E., Martínez-Blanco, J., Gasol, C.M., Rieradevall, J., Gabarrell, X., Gea, T., Sort, X., Sánchez, A., Recovery of organic wastes in the Spanish wine industry. Technical, economic and environmental analyses of the composting process. *J. Clean. Prod.*, 17, 830–838, 2009.
78. Durante, M., Montefusco, A., Marrese, P.P., Soccio, M., Pastore, D., Piro, G., Mita, G., Lenucci, M.S, Seeds of pomegranate, tomato and grapes: An underestimated source of natural bioactive molecules and antioxidants from agri-food by-products. *J. Agric. Food Chem.*, 63, 65–72, 2017.
79. Shinagawa, F.B., de Santana, F.C., Araujo, E., Purgatto, E., Mancini-Filho, J., Chemical composition of cold pressed Brazilian grape seed oil. *Food Sci. Technol.*, 38, 164–171, 2017.
80. Garavaglia, J., Markoski, M.M., Oliveira, A., Marcadenti, A., Grape seed oil compounds: Biological and chemical actions for health. *Nutr. Metab. Insights*, 9, 59–64, 2016.
81. Xu, C., Zhang, Y., Zhang, Y., Jun, W., Lu, J., Extraction, distribution and characterisation of phenolic compounds and oil in grapeseeds. *Food Chem.*, 122, 688–694, 2010.
82. Lucarini, M., Durazzo, A., Romani, A., Campo, M., Lombardi-Boccia, G., Cecchini, F., Bio-based compounds from grape seeds: a biorefinery approach. *Molecules*, 23, 1888, 2018.
83. Soto, M.L., Conde, E., González-López, N., Conde, M.J., Moure, A., Sineiro, J., Falqué, E., Domínguez, H., Núñez, M.J., Parajó, J.C., Recovery and concentration of antioxidants from winery wastes. *Molecules*, 17, 3008–3024, 2012.
84. Jara-Palacios, J.M., Escudero-Gilete, M.L., Hernández-Hierro, J.M., Heredia, F.J., Hernanz, D., Cyclic voltammetry to evaluate the antioxidant potential in winemaking by-products. *Talanta*, 165, 211–215, 2017.
85. Martinez, G.A., Rebecchi, S., Decorti, D., Domingos, J.M., Natolino, A., Del Rio, D., Bertin, L., Da Porto, C., Fava, F., Towards multi-purpose biorefinery platforms for the valorisation of red grape pomace: Production of polyphenols, volatile fatty acids, polyhydroxyalkanoates and biogas. *Green Chem.*, 18, 261–270, 2016.
86. Chamorro, S., Viveros, A., Vega, E., Brenes, A., Changes in polyphenol and polysaccharide content of grape seed extract and grape pomase after enzymatic treatment. *Food Chem.*, 133, 308–314, 2012.
87. Ghafoor, K., Al-Juhaimi, F.Y., Choi, Y.H., Supercritical fluid extraction of phenolic compounds and antioxidants from grape (*Vitis labrusca* B.) seeds. *Plant Foods Hum. Nutr.*, 67, 407–414, 2012.
88. Li, Y., Skouroumounis, G.K., Elsey, G.M., Taylor, D., Microwave-assistance provides very rapid and efficient extraction of grape seed polyphenols. *Food Chem.*, 129, 570–576, 2011.
89. Casazza, A.A., Aliakbarian, B., Mantegna, S., Cravotto, G., Perego, P., Extraction of phenolics from Vitis vinifera wastes using non-conventional techniques. *J. Food Eng.*, 100, 50–55, 2010.

90. Vilkhu, K., Mawsoa, R., Simons, L., Bates, D., Applications and opportunities for ultrasound assisted extraction in the food industry—A review. *Innov. Food Sci. Emerg. Technol.*, 9, 161–169, 2008.
91. Yammine, S., Brianceau, S., Manteau, S., Turk, M., Ghidossi, R., Vorobiev, E., Mietton-Peuchot, M., Extraction and purification of high added value compounds from by-products of the winemakingchain using alternative/nonconventional processes/technologies. *Crit. Rev. Food Sci. Nutr.*, 58, 8, 1375–1390, 2018.
92. Marchante, L., Gómez Alonso, S., Alañón, M.E., Pérez-Coello, M.S., Díaz-Maroto, M.C., Natural extracts from fresh and oven-dried winemaking by-products as valuable source of antioxidant compounds. *Food Sci Nutr.*, 6, 6, 1564–1574, 2018.
93. Carmona-Jiménez, Y., García-Moreno, M.V., García-Barroso, C., Effect of drying on the phenolic content and antioxidant activity of red grape pomace. *Plant Foods Hum. Nutr.*, 73, 1, 74–81, 2018.
94. Nogales-Bueno, J., Baca-Bocanegra, B., Rooney, A., Hernández-Hierro, J.M., Heredia, F.J., Byrne, H.J., Linking ATR-FTIR and Raman features to phenolic extractability and other attributes in grape skin. *Talanta*, 167, 44–50, 2017.
95. Nogales-Bueno, J., Baca-Bocanegra, B., Rooney, A., Hernández-Hierro, J.M., Byrne, H.J., Heredia, F.J., Study of phenolic extractability in grape seeds by means of ATR-FTIR and Raman spectroscopy. *Food Chem.*, 232, 602–609, 2017.
96. Nogales-Bueno, J., Baca-Bocanegra, B., Rodríguez-Pulido, F.J., Heredia, F.J., Hernández-Hierro, J.M., Use of near infrared hyperspectral tools for the screening of extractable polyphenols in red grape skins. *Food Chem.*, 172, 559–564, 2015.
97. Campo, M., Pinelli, P., Romani, A., Hydrolyzable Tannins from Sweet Chestnut Fractions Obtained by a Sustainable and Eco-friendly Industrial Process. *Nat. Prod. Comm.*, 11, 3, 409–415, 2016.
98. Rodriguez, R., Mazza, G., Fernandez, A., Saffe, A., Echegaray, M., Prediction of the lignocellulosic winery wastes behavior during gasification process in fluidized bed: Experimental and theoretical study. *J. Environm.Chem. Engin.*, 6, 4, 5570–5579, 2018.
99. Da Ros, C., Libralato, G., Ghirardini, A.V., Radaelli, M., Cavinato, C., Assessing the potential phytotoxicity of digestate from winery wastes. *Ecotox. Environm. Safety*, 150, 26–33, 2018.
100. Guardia, L., Suárez, L., Querejeta, N., Pevida, C., Centeno, T.A., Winery wastes as precursors of sustainable porous carbons for environmental applications. *J. Clean. Prod.*, 193, 614–624, 2018.
101. Boz, O., Ogut, D., Kir, K., Dogan, N., Olive processing waste as a method of weed control for okra, fava bean, and onion. *Weed Technol.*, 23, 569–573, 2009.
102. European Commission: Market situation in the olive oil and table olives sectors. Committee for the Common Organisation of the Agricultural Markets- Arable crops and olive oil -. Production, Consumption & Ending Stocks. 28 May 2020, 1–27, 2020. available at: https://ec.europa.eu/info/sites/info/files/food-farming-fisheries/plants_and_plant_products/documents/market-situation-olive-oil-table-olives_en.pdf
103. Commission Regulation (EU), N° 432/2012 of 16 May 2012 establishing a list of permitted health claims made on foods, other than those referring to the reduction of disease risk and to children's development and health. Official Journal of the European Union, 2012. available at: https://eur-lex.europa.eu/LexUriServ/LexUriServ.do?uri=OJ:L:2012:136:0001:0040:EN:PDF
104. Commission Regulation (EU), N° 2568/91 of 11 July 1991 on the characteristics of olive oil and olive-residue oil and on the relevant methods of analysis. Official Journal of the European Union, 1991 available at: https://eur-lex.europa.eu/legal-content/EN/TXT/PDF/?uri=CELEX:01991R2568-20151016&from=EN.
105. Agati, G., Pinelli, P., Cortes Ebner, S., Romani, A., Cartelat, A., Cerovic, Z.G., Nondestructive evaluation of anthocyanins in olive (*Olea europaea*) fruits by *in situ* chlorophyll fluorescence spectroscopy. *J. Agric. Food Chem.*, 53, 1354–1363, 2005.

106. Oliveras-Lopez, M.J., Innocenti, M., Ieri, F., Giaccherini, C., Romani, A., Mulinacci, N., HPLC/DAD/ESI/MS detection of lignans from Spanish and Italian Olea europaea L. fruits. *J. Food Compos Anal.*, 21, 62–70, 2008.
107. Oliveras-Lopez, M.J., Innocenti, M., Giaccherini, C., Ieri, F., Romani, A., Mulinacci, N., Study of the phenolic composition of spanish and italian monocultivar extra virgin olive oils: Distribution of lignans, secoiridoidic, simple phenols and flavonoids. *Talanta*, 73, 4, 726–732, 2007.
108. Alessandri, S., Ieri, F., Romani, A., Minor polar compounds in extra virgin olive oil: Correlation between HPLC-DAD-MS and the Folin-Ciocalteu spectrophotometric method. *J. Agric. Food Chem.*, 62, 826–835, 2014.
109. Visioli, F., Ieri, F., Mulinacci, N., Vincier, i F. F., Romani, A., Olive-oil phenolics and health: Potential biologically properties. *Nat. Prod. Comm*, 3, 2085–2088, 2008.
110. Campesi, I., Marino, M., Cipolletti, M., Romani, A., Franconi, F., Put "gender glasses" on the effects of phenolic compounds on cardiovascular function and diseases. *Eur. J. Nutr.*, 57, 8, 2677–2691, 2018.
111. Ambrosini, G., Romani, A., Scazzocchio, B., Varì, R., Vona, R., Campesi, I., Banelli, L., Straface, E., Masella, R., Malorni, W., Franconi, F., Sex-gender influences the expression of antioxidant/detoxifying enzyme induced by extra rirgin olive oil extracts in rat vascular smooth muscle cells, in: *XXVIth international conference on polyphenols*, Florence Italy, 22–26, July, pp. 437–438, 2012.
112. Franconi, F., Coinu, R., Carta, S., Urgeghe, P.P., Ieri, F., Mulinacci, N., Romani, A., Antioxidant effect of two virgin olive oils depends on the concentration and composition of minor polar compounds. *J. Agric. Food Chem.*, 54, 8, 3121–3125, 2006.
113. RIS-2/78-IV/98 COI/T.15/NC n 2/Rev. 8, 25/11/1998.
114. Romani, A., Lapucci, C., Cantini, C., Ieri, F., Mulinacci, N., Visioli, F., Evolution of minor polar compounds and antioxidant capacity during storage of bottle extra virgin olive oil. *J. Agric. Food Chem.*, 55, 1315–1320, 2007.
115. Widmer, R.J., Freund, M.A., Flammer, A.J., Sexton, J., Lennon, R., Romani, A., Mulinacci, N., Vinceri, F.F., Lerman, L.O., Lerman, A., Beneficial effects of polyphenol-rich olive oil in patients with early atherosclerosis. *Eur. J. Nutr.*, 52, 3, 1223–1231, 2013.
116. Pampaloni, B., Mavilia, C., Fabbri, S., Romani, A., Ieri, F., Tanini, A., Tonelli, F., Brandi, M.L., In *vitro* effects of extracts of extra virgin olive oil on human colon cancer cells. *Nutr Canc.*, 66, 7, 1228-1236, 2014.
117. Ciriminna, R., Meneguzzo, F., Fidalgo, A., Ilharco, L.M., Pagliaro, M., Extraction, benefits and valorization of olive polyphenols. *Eur. J. Lipid Sci. Technol.*, 118, 503–511, 2016.
118. Cassano, A., Conidi, C., Giorno, L., Drioli, E., Fractionation of olive mill wastewaters by membrane separation techniques. *J. Haz. Mat.*, 248–249, 185-193, 2013.
119. Romani, A., Pinelli, P., Ieri, F., Bernini, R., Innovation, and green chemistry in the production and valorization of phenolic extracts from *Olea europaea* L. *Sustainability*, 8, 1002, 2016.
120. Romani, A., Vita, C., Campo, M., Scardigli, A., Recupero di sottoprodotti agroindustriali: un modello di economia circolare. *Tecnologie Alimentari Sistemi per il Produttore*, Anno XXVII, 5, 50-55, 2016.
121. Romani, A., Scardigli, A., Pinelli., P., An environmentally friendly process for the production of extracts rich in phenolic antioxidants from *Olea europaea* L. and *Cynara scolymus* L. matrices. *Eur. Food Res. Technol.*, 10, 12, 1959, 2017.
122. Pizzichini, D., Russo, C., Vitagliano, M., Pizzichini, M., Romani, A., Ieri, F., Pinelli, P., Vignolini, P., Phenofarm, S.R.L., Process for producing concentrated and refined actives from tissues and byproducts of Olea europaea with membrane technologies, 2011. Number European Patent: EP 2338500 (A1).
123. Cimato, A., Cantini, C., Sani, G., Romani, A., Autino, A., Cresti, M., Il germoplasma autoctono dell'olivo in toscana LCD s.r.l., Report ARSIA. 1–55, 2004.

124. Notarnicola, B., Tassielli, G., Nicoletti, G.M., LCA and LCC of extra-virgin olive oil: Organic vs. conventional, in: *Proceedings of the 4th International Conference: Life Cycle Assessment in the Agri-food sector*, Horsens, 6-8 October, pp. 289–292, 2003.
125. Olivieri, G., Falconi, F., Pergreffi, R., Neri, P., Romani, A., Life Cycle Assessment for environmental integrated system in the olive oil tuscan company, in: *Proceeding of the 12th SETAC Europe LCA Case Studies Synposium, 2004/2005 European Meeting of the International Society for Industrial Ecology*, pp. 127–130, 10-11 January 2005, Bologna, Italy, printed by Setac.
126. CPM., A systematic approach to environmental priority strategies in product development (EPS). CPM report, Chalmers University of Technology, Version 2000.
127. The Eco-indicator 99, Methodology Report- Annex, Manual for designers, 2000.
128. Neri, P., L'anailisi ambientale dei prodotti agroalimentari con il metodo del Life Cycle Assessment, Report ARPA SICILIA, 1–124, 2010. Available at: http://www.arpa.sicilia.it/wp-content/uploads/2014/06/Analisi_col_metodo_LCA.pdf
129. Olivieri, G., Romani, A., Falconi, F., The preliminary LCA of membrane processes for olive oil wastewater polyphenols recovery, *Paper presented at VII International Conference on Life Scycle Assessment in the Agri Food Sector*, Bari, Italy, 2010.
130. Regulation (EC) No 258/97 of the European Parliament and of the Council of 27 January 1997 concerning novel foods and novel food ingredients. Available at:https://eurlex.europa.eu/LexUriServ/LexUriServ.do?uri=CONSLEG:1997R0258:20090120:EN: PDF.
131. FAO/INFOODS Global Food Composition Database for Pulses Version 1.0 - Pulses 1.0. Rome, FAO, 2017.
132. Calles, T., del Castello, R., Baratelli, M., Xipsiti, M., Navarro, D.K., The International Year of Pulses – Final report. Rome. FAO. 40 pp. Licence: CC BY-NC-SA 3.0 IGO, 2019. Available at: http://www.fao.org/3/CA2853EN/ca2853en.pdf.
133. Giusti, F., Caprioli, G., Ricciutelli, M., Vittori, S., Sagratini, G., Determination of fourteen polyphenols in pulses by high performance liquid chromatography-diode array detection (HPLC-DAD) and correlation study with antioxidant activity and colour. *Food Chem.*, 221, 689–697, 2017.
134. Rochfort, S. and Panozzo, J., Phytochemicals for health, the role of pulses. *J. Agric. Food Chem.*, 55, 7981–7994, 2007.
135. Caprioli, G., Kamgang Nzekoue, F., Giusti, F., Vittori, S., Sagratini, G., Optimization of an extraction method for the simultaneous quantification of sixteen polyphenols in thirty-one pulse samples by using HPLC-MS/MS dynamic-MRM triple quadrupole. *Food Chem.*, 266, 490–497, 2018.
136. Weihrauch, J.L. and Gardner, J.M., Sterol content of foods of plant origin. *J. Am. Diet. Assoc.*, 73, 39–44, 1978.
137. Ryan, E., Galvin, K., O'Connor, T.P., Maguire, A.R., O'Brien, N.M., Phytosterol, squalene, tocopherol content and fatty acid profile of selected seeds, grains, and legumes. *Plant Food Hum. Nutr.*, 62, 85–91, 2007.
138. Nystrom, L., Schar, A., Lampi, A.M., Steryl glycosides and acylated steryl glycosides in plant foods reflect unique sterol patterns. *Eur. J. Lipid Sci. Tech.*, 114, 656–669, 2012.
139. Kalogeropoulos, N., Chiou, A., Ioannou, M., Karathanos, V.T., Hassapidou, M., Andrikopoulos, N.K., Nutritional evaluation and bioactive microconstituents (phytosterols, tocopherols, polyphenols, triterpenic acids) in cooked dry legumes usually consumed in the Mediterranean countries. *Food Chem.*, 121, 682–690, 2010.
140. Rebello, C.J., Greenway, F.L., Finley, J.W., A review of the nutritional value of legumes and their effects on obesity and its related co-morbidities. *Obes. Rev.*, 15, 392–407, 2014.
141. Fuentes-Zaragoza, E., Riquelme-Navarrete, M.J., Sanchez-Zapata, E., Perez-Alvarez, J.A., Resistant starch as functional ingredient: A review. *Food Res. Int.*, 43, 931–942, 2010.

142. Sagratini, G., Caprioli, G., Maggi, F., Font, G., Giardinà, D., Mañes, J., Meca, G., Ricciutelli, M., Sirocchi, V., Torregiani, E., Vittori, S., Determination of soyasaponins I and βg in raw and cooked legumes by Solid Phase Extraction (SPE) coupled to Liquid Chromatography (LC)–Mass Spectrometry (MS) and assessment of their bioaccessibility by an *in vitro* digestion model. *J. Agric. Food Chem.*, 61, 1702–1709, 2013.
143. Amarowicz, R. and Pegg, R.B., Legumes as a source of natural antioxidants. *Europ. J. Lipid Sci. Technol.*, 110, 10, 865– 878, 2008.
144. Padhi, E.M.T. and Ramdath, D.D., A review of the relationship between pulse consumption and reduction of cardiovascular disease risk factors. *J. Funct. Foods*, 38, 635–643, 2017.
145. Mojica, L., Chen, K., González de Mejía, E., Impact of commercial precooking of common bean (*Phaseolus vulgaris*) on the generation of peptides, after pepsin–pancreatin hydrolysis, capable to inhibit dipeptidyl peptidase-IV. *Food Chem.*, 80, 1, H188–H198, 2015.
146. Šibul, F., Orčić, D., Vasić, M., Anačkov, G., Nađpal, J., Savić, A., Mimica-Dukića, N., Phenolic profile, antioxidant and anti-inflammatory potential of herb and root extracts of seven selected legumes. Indust.Crops Prod., 83, 641–643, 2016.
147. Khetarpaul, S.N. and Chand, G., Minerals profile and antioxidants properties of chickpea Leave of Desi and Kabuli varieties at different stages of maturity. *Int. J. Curr. Microbiol. App. Sci.*, 7, 3, 3171–3177, 2018.
148. Arya, G., Sharma, N., Ahmed, J., Gupta, N., Kumar, A., Chandra, R., Nimesh, S., Degradation of anthropogenic pollutant and organic dyes by biosynthesized silver nano-catalyst from Cicer arietinum leaves. *J. Photochem. Photobiol. B.*, 174, 90–96, 2017.
149. Gutöhrlein, F., Drusch, S., Schalow, S., Towards by-product utilisation of pea hulls: Isolation and quantification of galacturonic acid. *Foods*, 7, 12, pii: E203, 2018.
150. Le Goff, A., Renard, C.M.G.C., Bonnin, E., Thibault, J.-F., Extraction, purification and chemical characterisation of xylogalacturonans from pea hulls. *Carbohyd. Polym*, 45, 325–334, 2001.
151. Leterme, P., Théwis, A., Leeuwen, P.V., Monmart, T., Huisman, J., Chemical composition of pea fibre isolates and their effect on the endogenous amino acid flow at the ileum of the pig. *J. Sci. Food Agric.*, 72, 127–134, 1996.
152. Weightman, R., Renard, C.M.G.C., Thibault, J.-F., Structure and properties of the polysaccharides from pea hulls. Part 1: Chemical extraction and fractionation of the polysaccharides. *Carbohyd. Polym*, 24, 139–148, 1994.
153. Belghith-Fendri, L., Chaari, F., Kallel, F., Zouari-Ellouzi, S., Ghorbel, R., Besbes, S., Ellouz-Chaabouni, S., Ghribi-Aydi, D.J., Pea and broad bean pods as a natural source of dietary fiber: The impact on texture and sensory properties of cake. *Food Sci.*, 81, 10, C2360–C2366, 2016.
154. Hidalgo-Cuadrado, N., Pérez-Galende, P., Manzano, T., De Maria, C.G., Shnyrov, V.L., Roig, M.G.J., Screening of postharvest agricultural wastes as alternative sources of peroxidases: Characterization and kinetics of a novel peroxidase from lentil (*Lens culinaris* L.) stubble. *J. Agric. Food Chem.*, 60, 19, 4765–72, 2012.
155. Zhong, L., Fang, Z., Wahlqvist, M.L., Wu, G, Hodgson, J.M., Johnson, S.K., Seed coats of pulses as a food ingredient: Characterization, processing, and applications. *Trend. Food Sci. Technol.*, 80, 35–42, 2018.
156. Cheng, M., Qi, J.R., Feng, J.L., Cao, J., Wang, J.M., Yang, X.Q., Pea soluble polysaccharides obtained from two enzyme-assisted extraction methods and their application as acidified milk drinks stabilizers. *Food Res Int.*, 109, 544–551, 2018.
157. Belghith-Fendri, L., Chaari, F., Jeddou, K.B., Kallel, F., Bouaziz, F., Helbert, C.B., Abdelkefi-Mesrati, L., Ellouz-Chaabouni, S., Ghribi-Aydi, D., Identification of polysaccharides extracted from pea pod by-products and evaluation of their biological and functional properties. *Int. J. Biol Macromol.*, 116, 947–954, 2018.

7

Prospects of Bacterial Tannase Catalyzed Biotransformation of Agro and Industrial Tannin Waste to High Value Gallic Acid

Sunny Dhiman and Gunjan Mukherjee*

University Institute of Biotechnology, Chandigarh University, Gharuan, Mohali, Punjab, India

Abstract

Globally, around five billion metric tons of agro waste is produced every year. In India alone, solid wastes exceeding 960 million tons is produced annually as by-products of municipal, agricultural, mining, industrial activities, etc. Several industries utilizing plant constituents as raw and processing materials generate extensively high volumes of wastewater having plentiful tannins. Toxic tannery effluent wastes from tannery industries also possess superabundant levels of tannins as tannic acid. Thus, there is a continuously growing apprehension about these stockpiling wastes which are creating pollution and health hazards. A worthy and sustainable approach to deal with these accumulating wastes can be to utilize them as alternates of high cost raw materials for economic production of high value products like gallic acid. The biocatalytic potential of bacterial tannase in hydrolyzing natural tannins and tannic acid may be exploited for biotransformation of tannins into gallic acid.

Keywords: Biotransformation, tannins, tannase, gallic acid

7.1 Introduction

Gallic acid (3, 4, 5-trihydroxybenzoic acid) is a type of phenolic acid occurring naturally in a broad range of fruits, vegetables, herbs, and several beverages, etc. Gallic acid is a planar molecule comprising of an aromatic ring which has a carboxylic acid group and three phenolic hydroxyl groups bonded to it in an ortho position with respect to each other (Figure 7.1). Gallic acid has been named so after oak galls which were historically used in preparation of tannic acid. Gallic acid is broadly distributed in plant kingdom and majorly occurs either in free form or as its derivatives in different food stuffs like grapes, blueberries, walnut, pomegranate, nuts, apple, tea, wine, etc. Tannin acyl hydrolase (EC 3.1.1.20), commonly known as tannase, is widely used for bioconversion of hydrolyzable tannins to simple phenolics such as gallic acid. Various sources of hydrolyzable tannins principally utilized for gallic acid production are tara fruit pod, sumac leaves, gall nuts, chestnut

**Corresponding author:* gunjanmukherjee@gmail.com

Arindam Kuila and Mainak Mukhopadhyay (eds.) Biorefinery Production Technologies for Chemicals and Energy, (129–144) © 2020 Scrivener Publishing LLC

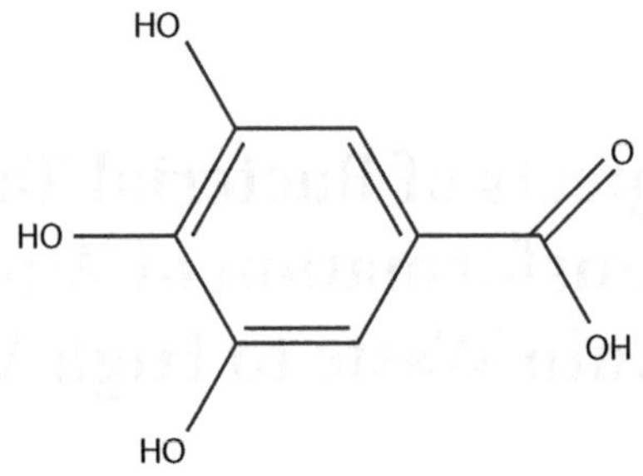

Figure 7.1 Structure of gallic acid.

(*Castanea sativa*), teri pod cover, *Myrobalan*, etc. Although the production of tannases by various microorganisms, along with their characterization, immobilization, and role in biotransformation of tannins to gallic acid have been extensively studied. Despite this, the commercial sources of tannase are limited to very few only [1]. Biotransformation process offers several advantages over chemical technologies that include better yields, credibility, environment friendly, reproducibility of the process, ease in operation, etc. Tannin acyl hydrolase, universally known as tannase (EC 3.1.1.20), plays a pivotal role in biotransformation of hydrolyzable tannins to produce gallic acid, a versatile molecule having broad spectrum of pharmacological and industrial applications. Gallic acid is the major hydrolytic product of tannic acid. Tannic acid is basically the commercial form of gallotannins. Over the past two decades, gallic acid has been found to have tremendous therapeutic value as antimicrobial, antiviral, antitumor, a potential drug, a major component of antimalarial drug Trimethoprim and radioprotective agent [2–6]. In addition, its utilization in agriculture as a crop protection agent renders Gallic acid, a molecule of enormous commercial value [7]. The production of gallic acid at industrial level is being done either through extraction or by hydrolyzing the tannins through acidic/alkaline treatment. The extraction of valuable chemical compounds from plants is difficult and uneconomical due to their complex nature and interference by other compounds and impurities. On the other hand, hydrolyzing the tannins through acidic/alkaline treatment generates toxic effluents thus posing health hazards. The drawbacks of these methods clearly suggest the need for developing inexpensive, high-yielding, and environment-friendly processes of manufacturing gallic acid. Nowadays, tannase has been one of the most talked enzymes due to its tremendous biocatalytic potential and commercial importance. Tannase is a hydrolase that belongs to esterase family. The hydrolysis of tannins using microbial tannases has gained pace in recent years [8]. However, this is a well-documented fact that bacteria have the ability for efficient degradation of naturally occurring hydrolyzable tannins and tannic acid [9]. Also, bacteria may prove as a source of thermo stable tannase which might be able to work over a broad range of temperature in terms of enzyme activity and stability [10, 11]. However, there are very few research investigations involving tannase and gallic acid production on fermenter level [12]. The enzymatic hydrolysis of hydrolyzable tannins especially utilizing bacterial tannase can overcome the shortcomings of the chemical methods of gallic acid production because of better process control and optimization, ability to genetically alter the microbes, safety, reproducibility, as well as credibility.

7.2 Bacterial Tannase Producers

Microbes are utilized as the major sources of industrial enzymes on account of their biochemical diversity, ease in culturing, amenability to genetic alteration, reproducibility, credibility, and safety [13]. In addition, microbes can produce higher titers of enzymes in a consistent manner [8]. Even though tannase can be produced by a wide range of microorganisms, still it remains a challenging task to find out a strain with higher productivity and yields [14]. A vast majority of tannase producing microorganisms are fungi especially those native to *Aspergillus*, *Penicillium*, and *Trichoderma* genera. Industrially, the problem with fungi resides in their slow tannins degradation and difficult to manipulate genetically [11] documented that fungal degradation of tannins is relatively slow. In addition, it is quite toilsome to genetically alter them. This makes them less suitable for industrial level degradation of tannins. Bacteria, on the other hand, have received a minor or very little attention for production of tannase.

However, the interest in bacterial tannases has gained momentum in last 25 years owing to their broader applicability, amenability to undergo genetic modifications along with potential to live under extreme temperature conditions. Thus, in this context, various tannase-producing bacteria have been explored until now. Majority of these bacterial strains pertain to genera such as *Lactobacillus* [15], *Serratia* [16], *Bacillus* [12, 17], *Staphylococcus* [18], *Pseudomonas* [19], *Enterobacter* [20], *Azobacter* [21], *Klebsiella* [22], *Citrobacter* [26], *Pantonea* [24], and *Lonepinella* [25]. Some of the tannase-producing bacteria reported in previous years have been listed in Table 7.1.

7.3 Bacterial Tannase Production

SMF has been most preferred approach for bacterial tannase production with higher enzyme titers [45]. Obtaining high titers of bacterial tannase depends mainly on the culture medium composition, the bacterial strain, and the process optimization of culture conditions. In recent years, interest in bacterial tannases has increased because of their wide applications; hence, several new strains have been isolated. Investigations have been centered on the search for bacterial strains capable of synthesizing high titers of tannase. Das *et al.* [46] utilized tannins from eight unalike plant extracts for production of tannase with *Bacillus licheniformis* KBR6 in SMF and recorded a higher activity with the tannin of *Anacardium occidentale*. Selwal *et al.* [19] studied the production of tannase enzyme using *Pseudomonas aeruginosa* IIIB 8914 under submerged fermentation with the leaves of *Phylanthus emblica* (amla), *Acacia nilotica* (keekar), *Eugenia cuspidate* (Jamoa), and *Syzygium cumini* (Jamun) as substrates and reported a maximum tannase yield. Kannan *et al.* [37] reported maximal enzyme activity (5.22 U/ml) of tannase from *Lactobacillus plantarum* MTCC 1407 under submerged fermentation. A great majority of lab-scale tannase production from bacteria is done within 250-ml Erlenmeyer flask [46]. However, some research investigations involving bacterial tannase production have been carried out on fermenter scale [12]. Research investigations have been conducted with major emphasis on exploring the bacterial strains with the potential of producing higher enzyme titers. Maximal tannase production (16.54 U/ml) was recorded by Raghuwanshi *et al.* [12] from *Bacillus sphaericus* with a 30-L fermenter. SSF approach has been utilized to a relatively lesser extent for production of tannase as compared to submerged fermentation [39]. The research investigations conducted in recent years have

Table 7.1 Bacteria used for the production of tannase.

Bacteria	Reference
Achomobacter sp.	[26]
Bacillus polymyxa	[9]
Klebsiella planticola	[9]
Paenibacillus polymyxa	[9]
Corynebacterium sp.	[9]
Pseudomonas solanacearum	[27]
Lonepinella koalarum	[28]
Citrobacter freundii	[29]
Lactobacillus paraplantarum	[30]
Bacillus cereus KBR9	[31]
Lactobacillus plantarum	[32]
Citrobacter freundii	[33]
Enterococcus faecalis	[34]
Lactobacillus sp.	[35]
Staphylococcus lugdunensis	[18]
Lactobacillus plantarum	[36]
Pediococcus pentosaceus	[15]
Serratia ficaria	[16]
Pseudomonas aeruginosa	[19]
Lactobacillus plantarum	[37]
Bacillus sphaericus	[12]
Gluconacetobacter hansenii	[38]
Bacillus subtilis	[39]
Lactobacillus plantarum CIR1	[40]
Erwinia carotovora	[41]
Bacillus gotthelii M2S2	[42]
Escherichia coli	[43]
Streptomyces sp.	[44]

claimed enhanced tannase production and better stability in accordance with pH and temperature deviations. However, the majority of literature suggests suitability of SSF for fungal tannase production by utilizing natural tannin containing agro residues as they imitate the natural conditions indispensable for fungal growth. It is widely accepted that the optimal conditions for tannase production might vary considerably depending on the microorganism, culture conditions, type of fermentation, and experimental process. Therefore, several methodologies for the optimization process of bacterial tannase production have been developed. Conventional statistical methods are used commonly for the optimization of tannase production. However, these methodologies are time consuming, are expensive, and do not take into account the interactions of the factors [10, 19, 47]. For that reason, orthogonal arrays such as response surface methodology and Taguchi methodology have been employed for the optimization of bacterial tannase production. The former allows one to determine the influence of factors over the response and to optimize these variables to achieve the maximum yield under the best possible economic conditions [12, 48, 49]. The Taguchi methodology identifies the influence of individual factors and establishes the relationship between the variables and the operative conditions; also, experimental data from analysis of variance gives a statistical relationship of system production [50].

7.4 Hydrolyzable Tannins: A Substrate for Gallic Acid Production

Tannins are polyphenolic compounds present in plants where they play an important role to prevent the attack of viruses, bacteria, and fungi. Tannins are high molecular weight (500 to 3,000 kDa) polyphenolic compounds that exist abundantly in different parts of plants such as fruits, leaves, and bark [51, 52]. Tannins have the ability to precipitate macromolecules (such as proteins, cellulose, starch, etc.) and minerals by forming strong complexes. Tannins are the second most important group of natural phenolic compounds after lignin [51]. Tannins are widely distributed in different vascular plant structures. They are considered as secondary metabolic products due to them do not participate directly in biosynthesis, biodegradation, or any transformation of energy process. Tannins have been classified into four major groups: Gallotannins, Ellagitannins, Condensed tannins, and Complex tannins [53]. Gallotannins are the simplest form of hydrolyzable tannins. They are formed by galloyl or di-galloylunites esterified with a core of glucose or a polyvalentalcohol (such as glucitol, shikimic acid, quinicacid, etc.). Tannic acid or pentagalloyl-glucose is an example of such compounds. A characteristic of gallotannins is their easy hydrolysis by heat, acid oralkali conditions, and for tannase. Tannic acid is the commercial form of gallotannins. It mainly consists of glucose esters of gallic acid. Chinese gallotannin (*Rhus semilata*) is the principally utilized natural hydrolyzable tannin for gallic acid production.

7.5 Tannins as Waste

7.5.1 Agro-Waste

Tannins are present in several naturally occurring agricultural wastes such as red gram husk, green gram husk, black gram husk, tamarind seed powder, tea dust, rice bran, and groundnut shell which can be utilized in industrial bioprocess for the production of value added products such as gallic acid through fermentation. With the advancements in biotechnology, efforts are being progressively made to replace the costlier raw materials with the cheaper and easily available agro waste (e.g., pomegranate rind, tamarind seed, mango kerenel, grape seed, red gram husk, tea dust, cashew waste, etc.) for the production of valuable products of immense commercial importance like gallic acid. The principally utilized commercial sources of hydrolyzable tannins include tannins from Chinese gall (*Rhus semialata*), Keekar (*Acacianilotica*) leaves, red gram husk and cashew waste testa (*Anacardium occidentales*), and *Myrobalan* nuts (*Terminalia chebula*). A number of research investigations utilizing a concoction of agro-industrial wastes like paddy husk, wheat bran, palm kernel cake, cashew waste, apple baggasse, and rice bran for enhancing the tannase and gallic acid production have reported [52, 54–57].

7.5.2 Industrial Waste

Tannery wastewater is reckoned as one of the highest ranked environmental pollutant among all forms of wastewater generated from various industries [58, 59]. Emergence of tannery wastewater as an absolute pollutant in countries such as China has eventually posed catastrophic threat to mankind and aquatic life. Tannins are used in tanneries for processing of leather in the form of tannic acid. During the process of leather tanning in tanneries, huge volumes of water and tannins (tannic acid) are used along with several other chemicals for processing of raw hides and skins. The entire process generates an approximate 30–35 m^3 volume of wastewater per ton of raw hide/skins processed [60]. In addition, the industries utilizing plant constituents as raw and processing materials eventually produce enormously high levels of wastewater rich in tannins. The currently available conventional chemical and biological wastewater treatment methods are not good enough to remove these pollutants especially tannins because of their recalcitrant nature and low biodegradability. This necessitates the development and utilization of appropriate and effective treatment methods for bioprocessing of such enormously high volumes of tannery wastewater. Bioremediation of these tannin rich wastewaters using suitable enzymes may result in their effective biodegradation. The ability of bacterial tannase to efficiently degrade natural tannins and tannic acid can be utilized for bioremediation of tannery wastewater high in tannins and eventually for gallic acid production.

7.6 Bacterial Biotransformation of Tannins

Bacterial tannase-catalyzed biotransformation of tannins can be efficiently utilized for producing significant levels of gallic acid. Enzymatic biodegradation of tannins is most eminent approach for effectively biotransforming larger tannin molecules into relatively smaller molecules of higher market value. The enzymatic approach of making gallic acid principally involves cleavage of ester and depside bonds in hydrolyzable tannins and gallic acid esters using tannase enzyme. The tannase catalyzed hydrolysis of tannins particularly the gallotannins results in liberation of gallic acid and glucose molecules.

Gallic acid production and tannase production are interconnected with each other since tannase catalyzes the depolymerization of hydrolyzable tannins thus releasing gallic acid. A great majority of tannin hydrolyzing organisms reported to date pertain to bacteria, fungi, and yeast. Tannase from filamentous fungi pertaining to genera *Aspergillus* and *Penicillium* have been documented as the most widely utilized tannase worldwide for tannin hydrolysis [33, 61]. However, this is a well-established fact that bacteria can effectively disintegrate natural tannins as well as tannic acid [9]. In recent years, the reliance on microbial tannases for hydrolyzing the tannins has accelerated tremendously [8]. Tannin degradation potential significantly varies among different microbes like bacteria, fungi, and yeast. Yeast can effectively degrade gallotannins but loses its effectiveness in degrading elagitannins. Bacteria have stupendous potential to efficiently degrade gallotannins as well as ellagitannins [9]. Some bacterial strains pertaining to genera such as *Bacillus* [12], *Pseudomonas* [19], *Staphylococcus* [18], *Klebsiella* [22], *Lactobacillus* [15], *Citrobacter* [23], *Serratia* [16], *Pantonea* [24], *Azobacter* [21], and *Enterobacter* [20] have been documented with the ability to degrade tannins. The enzymes involved in depolymerization of tannins include tannase and gallic acid decarboxylase. However, as a matter of fact, tannase has been the most widely investigated and utilized enzyme for tannin degradation. Microbial tannase holds utmost importance in tannin degradation as compared to tannase from plant and animal sources [51]. Tannase breaks the ester and depside bonds in different types of tannins. However, their effectiveness in degrading condensed tannins is limited by their inability to affect C-C bonds [62]. Gallic acid decarboxylase mediates the decarboxylation to gallic acid to pyrogallol; however, the enzyme is extremely unstable due to its relatively higher sensitivity to oxygen which renders its isolation and purification quite ardous [63]. There are certain bacteria like *Selenomonas gallolyticus* and *E.coli* that catalyzes decarboxylation of gallic acid to pyrogallol. Further transformation of this compound does not take place possibly because of it having lesser toxicity or its production being thermodynamically more viable [64]. The degradation pathway of gallotannins is schematically represented in Figure 7.2. Various sources of hydrolyzable tannins principally utilized for gallic acid production

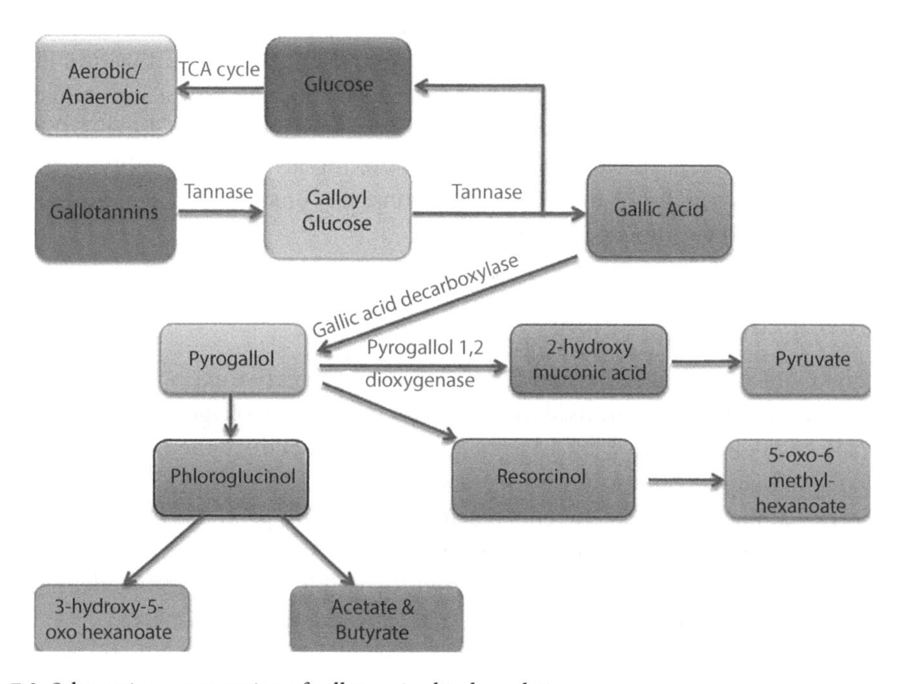

Figure 7.2 Schematic representation of gallotannins biodegradation.

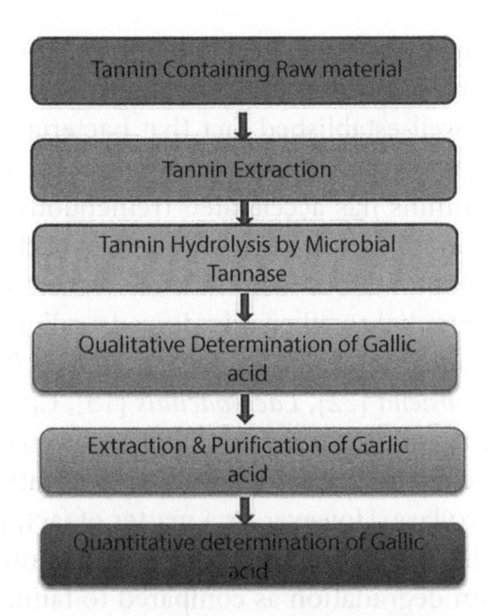

Figure 7.3 General methodology of tannase catalyzed production of gallic acid from tannins.

are tara fruit pod [65], sumac leaves [66], gall nuts [67], teri pod cover [68], *Myrobalan* [69], *Cassia siamea* [70], *Larrea tridentate* [71], *Quercus infectoria* [72], *Acer ginnala* [73], Mango seed kernel [74], Cashew testa [75], and Pine needles [76]. The general methodology of tannase catalyzed production of gallic acid from tannins has been illustrated in Figure 7.3.

7.7 Applications of Gallic Acid

Gallic acid has always been a molecule of immense interest to research communities worldwide due to its broad spectrum of pharmacological and industrial applications. Over the last two decades, it has witnessed a continuously expanding rise in its utilization in numerous applications ranging from healthcare and food sector to agriculture and industrial sectors. The worldwide demand of gallic acid is 8,000 tons per annum. China is the leading manufacture of gallic acid in world (Table 7.2). Looking at the wide utility of gallic acid in both pharmacological and industrial applications; its consumption around the globe is expected to rise significantly in the coming years.

7.7.1 Therapeutic Applications

Antioxidant: Gallic acid has been reported to exhibit a protective action toward DNA, thus preventing its degradation from oxidative damage by reactive oxygen species (ROS) [77]. Further, gallic acid along with EDTA has been reported to reduce the motor and oxidative damages arising because of lead poisoning [78].

Antiulcer: Gallic acid has been reported to enhance the production of mucus rich in glycoproteins that help in protecting the gastric lining [79].

Antiangiogenic: Gallic acid has been reported to exhibit antiangiogenic effect in growth inhibition and apoptotic death of human prostate carcinoma DU145 cells [4].

Anti-Inflammatory: Gallic acid has been reported to play antagonistic role against inflammatory cytokines like IL-10, IL-1β, and TNF-α [80].

Table 7.2 Gallic acid manufacturers from China.

Name of the company	Product grade
Xian Plant Bio Engineering Co., Ltd.	Food grade
Xi'an Salus Nutra Bio-Tech Inc.	Food grade
Shaanxi Undersun Biomedtech Co., Ltd.	Food grade
Dalian Sinobio Chemistry Co., Ltd.	Industry grade
Xi'an Lyphar Biotech Co., Ltd.	Medicine grade, food grade
Xi'an Ceres Biotech Co., Ltd.	Food and pharmaceutical grade
Shaanxi Sangherb Bio-Tech Inc.	AAAAA, Food & Medical grade
Hugestone Enterprise Co., Ltd. (Nanjing)	Food, Food & Medical grade
Xi'an Sgonek Biological Technology Co., Ltd.	AAAAA
Xi'an Sonwu Biotech Co., Ltd.	Food grade

Antibacterial: Gallic acid has been reported to exhibit antibacterial activity against a number of pathogens such as *Staphylococcus aureus*, *Pseudomonas aeruginosa*, *Escherichia coli*, *Klebsiella pneumonia*, etc. [81]. Gallic acid is also used as a precursor for a broad spectrum antimalarial drug Trimethoprim [82, 83].

Neuroprotective: Gallic acid has been reported to exhibit neuroprotective action against neurodegenerative diseases like Parkinson's disease, Dementia, Epilepsy, and Alzheimer's disease [84–86].

Hepatoprotective: Gallic acid has been reported to exhibit hepatoprotective action to hepatocytes against oxidative stress generated due to hydrogen peroxide and carbon tetrachloride [87, 88].

7.7.2 Industrial Applications

As a preservative: It prevents the rancidity and spoilage of fats and oils due to its antioxidant nature facilitating its application as food additives in various eatable materials like baked goods, candy, and chewing gums.

In cosmetic industry: It protects the cells from UV-B or ionizing irradiation and thus can be used in cosmetics. A study on B16F10 melanocyte cells showed that gallic acid inhibited melanogenesis, which could make it an additive in cosmetics to overcome pigmentation [89, 90]. Gallic acid has also been utilized in the development of a brown color hair dye [91].

In paper industry: Gallic acid in association with laccase enzyme has been reported to strengthen the tensile strength of paper by improving the quality of kraft pulp [92].

In leather industry: Gallic acid has been reported to enhance chromium uptake during tanning process in leather industry [93].

Radioprotective agent: Gallic acid has been reported to provide protection against harmful gamma radiations [94].

As an analytical reagent: Gallic acid is used an analytical reagent in pharmaceutical industry and other analytical studies for estimation of the phenolic content of analytes under analysis. The resulting phenolic content is expressed in terms of gallic acid equivalents (GAE) [95, 96].

Crop protection: Gallic acid serves as an antifungal agent and thus has been reported for preservation of grains such as wheat, corn, and nuts by inhibiting aflatoxin synthesis [7].

Biodiesel stabilizer: Gallic acid has been reported to act as a stabilizer by providing protection to biodiesel oxidation [97].

Stabilizer of collagen in cosmetic fillers: Gallic acid esters epigallocatechin gallate, and epicatechin gallate have been reported to inhibit the degradation of collagen based cosmetic fillers by collagenase [98].

7.8 Conclusions

The agro waste and industrial wastes are continuously piling up because of various agricultural and industrial practices worldwide. It is a common practice to burn these accumulating wastes in open or to dispose them in nearby natural water resources, thus posing threat to the aquatic organisms as well as to the mankind. With the advancements in tool and techniques in biotechnology, efforts are being continuously made to utilize cheaper and easily available substrates like agro waste and industrial wastes rich in tannins as substituents for costlier raw materials. Even though, tannase of fungal origin have been immensely studied and utilized for gallic acid production. Bacteria, on the other hand, have been least utilized for gallic acid production. However, the scientific studies reveal that bacteria possess enormous potential for biodegradation of natural tannins and tannic acid. In addition, bacteria may also prove as a source of thermo stable tannase which might be able to withstand higher tannic acid concentration which otherwise proves lethal to microorganisms utilizing tannic acid as a carbon source. Gallic acid is one of the most valuable biomolecules in past two decades. The currently available chemical methodologies of making gallic acid are uneconomical and also generate effluents which might prove detrimental to mankind. Thus, the biodegradation potential of bacterial tannase may prove fruitful in degrading the natural tannins as well as tannic acid to manufacture gallic acid to meet its current global demand and utility in numerous therapeutic and industrial applications.

References

1. Van de Lagemaat, J. and Pyle, D.L., Modelling the uptake and growth kinetics of *Penicillium glabrum* in tannic acid-containing solid state fermentation for tannase production. *Process Biochem.*, 40, 5, 1773–1782, 2005.
2. Mukherjee, G. and Banerjee, R., Production of gallic acid. Biotechnological routes (part 1). *Chim Oggi.*, 21, 59–62, 2003.
3. Gandhi, N.M. and Nair, G.K.K., Protection of DNA and membrane from gamma radiation induced damage by gallic acid. *Mol. Cell. Biochem.*, 278, 111–117, 2005.
4. Veluri, R., Singh, R.P., Liu, Z., Thompson, J.A., Agarwal, R., Agarwal, C., Fractionation of grape seed extract and identification of gallic acid as one of the major active constituents causing growth inhibition and apoptotic death of DU145 human prostate carcinoma cells. *Carcinogenesis.*, 27, 1445–1453, 2006.

5. Hsu, C.L., Lo, W.H., Yen, G.C., Gallic acid induces apoptosis in3T3-L1 pre-adipocytes fas and mitochondrial-mediated pathway. *J. Agri. Food Chem.*, 55, 7359–7365, 2007.
6. Aithal, M. and Belur, P.D., Enhancement of propyl gallate yield in nonaqueous medium using novel cell-associated tannase of *Bacillus massiliensis*. *Prep. Biochem. Biotechnol.*, 43, 445–455, 2013.
7. Mahoney, N. and Molyneux, R.J., Phytochemical inhibition of aflatoxigenicity in *Aspergillus flavus* by constituents of walnut (*Juglans regia*). *J. Agric. Food Chem.*, 52, 1882–1889, 2004.
8. Dhiman, S., Mukherjee, G., Singh, A.K., Recent trends and advancements in microbial tannase-catalyzed biotransformation of tannins: a review. *Int. Microbiol.*, 21, 175–195, 2018.
9. Deschamps, A.M., Otuk, G., Lebault, J.M., Production of tannase and degradation of chestnut tannins by bacteria. *J. Ferment. Technol.*, 61, 55–59, 1983.
10. Beniwal, V., Chhokar, V., Singh, N., Sharma, J., Optimization of process parameters for the production of tannase and gallic acid by *Enterobacter cloacae* MTCC 9125. *J. Am. Sci.*, 6, 8, 389–397, 2010.
11. Beniwal, V., Kumar, A., Goel, G., Chhokar, V., A novel low molecular weight acido-thermophilic tannase from *Enterobacter cloacae* MTCC 9125. *Biocatal. Agric. Biotechnol.*, 2, 132–137, 2013.
12. Raghuwanshi, S., Dutt, K., Gupta, P., Misra, S., Saxena, R.K., *Bacillus sphaericus*: The highest bacterial tannase producer with potential for gallic acid synthesis. *J. Biosci. Bioeng.*, 111, 635–640, 2011.
13. Trevino, L., Contreras-Esquivel, J., Rodriguez-Herrera, R., Aguilar, C., Effects of polyurethane matrices on fungal tannase and gallic acid production under solid state culture. *J. Zhejiang Univ Sci. B.*, 8, 771–776, 2007.
14. Jana, A., Halder, S.K., Banerjee, A., Paul, T., Pati, B.R., Mondal, K.C., Mohapatra, P.K.D., Biosynthesis, structural architecture and biotechnological potential of bacterial tannase: A molecular advancement. *Biores. Technol.*, 157, 327–340, 2014.
15. Guzman-Lopez, O., Loera, O., Parada, J.L., Castillo-Morales, A., Martinez-Ramirez, C., Augur, C., Gaime-Perraud, I., Saucedo-Castaneda, G., Microcultures of lactic acid bacteria: Characterization and selection of strains, optimization of nutrients and Gallic acid concentration. *J Ind. Microbiol. Biotechnol.*, 36, 11–20, 2009.
16. Belur, P.D., Mugeraya, G., Kuppalu, N.R., Temperature and pH stability of a novel cell-associated tannase of *Serratia ficaria* DTC. *Int. J. Biotechnol. Biochem.*, 6, 667–674, 2010.
17. Muhammad, N.A., Hamid, M., Ikram-Ul-Haq, Production and characterization of tannase from a newly isolated *Bacillus Subtilis*. *Pak. J. Bot.*, 48, 3, 1263–1271, 2016.
18. Noguchi, N., Ohashi, T., Shiratori, T., Association of tannase producing *Staphylococcus lugdunensis* with colon cancer and characterization of a novel tannase gene. *J. Gastroenterol.*, 42, 346–351, 2007.
19. Selwal, M.K., Yadav., A., Selwal, K.K., Aggarwal, N.K., Gupta, R., Gautam, S.K., Optimization of cultural conditions for tannase production by *Pseudomonas aeruginosa* IIIB8914 under submerged fermentation. *World J. Microbiol. Biotechnol.*, 26, 599–605, 2010.
20. Mandal, S. and Ghosh, K., Isolation of tannase-producing microbiota from the gastrointestinal tracts of some freshwater fish. *J. Appl. Ichthyol.*, 29, 145–153, 2013.
21. Gauri, S.S., Mandal, S.M., Atta, S., Dey, S., Pati, B.R., Novel route of tannic acid biotransformation and their effect on major biopolymer synthesis in *Azotobacter* sp. SSB81. *J. Appl. Microbiol.*, 114, 84–95, 2012.
22. Sivashanmugam, K. and Jayaraman, G., Production and partial purification of extracellular tannase by *Klebsiella pneumonia* MTCC 7162 isolated from tannery effluent. *Afr. J. Biotechnol.*, 10, 1364–1374, 2013.
23. Wilson, P.A., Rojan, P.J., Kumar, P., Sabu, T., Tannin acyl hydrolase production by *Citrobacter* sp. isolated from tannin rich environment, using *Tamarindus indica* seed powder. *J. Appl. Sci. Environ. Manage.*, 13, 95–97, 2009.

24. Pepi, M., Lampariello, L.R., Altieri, R., Esposito, A., Perra, G., Renzi, M., Lobianco, A., Feola, A., Gasperini, S., Focardi, S.E., Tannic acid degradation by bacterial strains *Serratia* spp. and *Pantoea* sp isolated from olive mill waste mixtures. *Int. Biodeterior. Biodegrad.*, 64, 73– 80, 2010.
25. Goel, G., Puniya, A.K., Singh, K., Phenotypic characterization of tannin-protein complex degrading bacteria from faeces of goat. *Small Rumin. Res.*, 69, 217–220, 2007.
26. Lewis, J.A. and Starkey, R.L., Decomposition of plant tannins by some soil microorganism. *Soil Sci.*, 107, 4, 235–240, 1969.
27. Deschamps, A.M. and Lebeault, J.M., Production of gallic acid from tara (*Caesalpinia spinosa*) tannin by bacterial strains. *Biotechnol. Lett.*, 6, 237–242, 1984.
28. Osawa, R., Fujisawa, T., Sly, L.I., *Streptococcus gallolyticus* sp. nov., gallate degrading organisms formerly assigned to *Streptococcus bovis*. *Syst. Appl. Microbiol.*, 18, 74–78, 1995.
29. Kumar, R.A., Gunasekaran, P., Lakshmanan, M., Biodegradation of tannic acid by *Citrobacter freundii* isolated from a tannery effluent. *J. Basic Microbiol.*, 39, 161–168, 1999.
30. Osawa, R., Kuroiso, K., Goto, S., Shimizu, A., Isolation of tannin degrading *Lactobacilli* from humans and fermented foods. *Appl. Environ. Microbiol.*, 66, 3093–3097, 2000.
31. Mondal, K.C., Banerjee, D., Banerjee, R., Pati, B.R., Production and characterization of tannase from *Bacillus cereus* KBR 9. *J. Gen. Appl. Microbiol.*, 47, 263–267, 2001.
32. Ayed, L. and Hamdi, M., Culture conditions of tannase production by *Lactobacillus plantarum*. *Biotechnol. Lett.*, 24, 1763–1765, 2002.
33. Belmares, R., Contreras-Esquivel, J.C., Rodriguez-Herrera, R., Coronel, A.R., Aguilar, C.N., Microbial production of tannase: An enzyme with potential use in food industry. *LWT Food Sci. Technol.*, 37, 857–864, 2004.
34. Goel, G., Puniya, A.K., Aguilar, C.N., Singh, mK., Interaction of gut microflora with tannins in feeds. *Naturewissenchaften.*, 92, 947–503, 2005.
35. Sabu, A., Augur, C., Swati, C., Pandey, A., Tannase production by *Lactobacillus* sp ASR-S1 under solid-state fermentation. *Process Biochem.*, 41, 575–580, 2006.
36. Iwamoto, K., Tsurutab, H., Nishitainia, Y., Osawa, R., Identification and cloning of a gene encoding tannase (tannin acyl hydrolase) from *Lactobacillus plantarum* ATCC 14917. *Syst. Appl. Microbiol.*, 31, 269–277, 2008.
37. Kannan, N., Aravindan, R., Viruthagiri, T., Effect of culture conditions and kinetic studies on extracellular tannase production by *Lactobacillus plantarum* MTCC 1407. *Ind. J. Biotechnol.*, 10, 321–328, 2011.
38. Rani, U.M. and Appaiah, A.K.A., *Gluconacetobacter hansenii* UAC09 mediated transformation of polyphenols and pectin of coffee cherry husk extract. *Food Chem.*, 130, 243–247, 2012.
39. Jana, A., Maity, C., Halder, S.K., Das, A., Pati, B.R., Mondal, K.C., Das, M.P.K., Structural characterization of thermostable, solvent tolerant, cytosafe tannase from *Bacillus subtilis* PAB2. *Biochem. Eng. J.*, 77, 161–170, 2013.
40. Aguilar-Zarate, P., Cruz-Hernandez, M.A., Montanez, J.C., Belmares-Cerda, R.E., Aguilar, C.N., Enhancement of tannase production by *Lactobacillus plantarum* CIR1: Validation in gas-lift bioreactor. *Bioproc. Biosyst. Eng.*, 37, 2305–2316, 2014.
41. Muslim, S.N., Mahammed, A.N., Musafer, H.K. *et al.*, Detection of the optimal conditions for tannase productivity and activity by *Erwinia Carotovora*. *J. Medical Bioeng.*, 4, 3, 198–205, 2015.
42. Subbulaxmi, S. and Murty, V.R., Process optimization for tannase production by *Bacillus gottheilii* M2S2 on inert polyurethane foam support. *Biocatal. Agric. Biotechnol.*, 7, 48–55, 2016.
43. Thakur, N. and Nath, A.K., Isolation of tannase producing bacteria from sheep excreta. *Indian J. Small Rumi.*, 23, 2, 264–266, 2017.
44. Roy, S., Parvin, R., Ghosh, S., Bhattacharya, S., Maity, S., Banerjee, D., Occurrence of a novel tannase (tan B LP) in endophytic *Streptomyces* sp. AL1L from the leaf of *Ailanthus excelsa* Roxb. *3 Biotech.*, 8, 1, 33, 2018.

45. Belur, P.D. and Mugeraya, G., Microbial production of tannase. *Res. J. Microbiol.*, 6, 25–40, 2011.
46. Das, M.P.K., Mondal, K.C., Pati, B.R., Production of tannase through submerged fermentation of tannin-containing plant extracts by *Bacillus licheniformis* KBR6. *Pol. J. Microbiol.*, 55, 297–301, 2006.
47. Rao, R.S., Kumar, C.G., Prakasham, R.S., Hobbs, P.J., The Taguchi methodology as a statistical tool for biotechnological applications: A critical appraisal. *Biotechnol.J.*, 3, 4, 510–523, 2008.
48. Naidu, R.B., Saisubramanian, N., Sivasubramanian, S., Selvakumar, D., Janardhan, S., Puvanakrishnan, R., Optimization of tannase production from *Aspergillus foetidus* using statistical design methods. *Cur. Tre. Biotechnol. Pharm.*, 2, 523–530, 2008.
49. Das, M.P.K., Maity, C., Rao, R.S., Pati, B.R., Mondal, K.C., Tannase production by *Bacillus licheniformis* KBR6: Optimization of submerged culture conditions by Taguchi DOE methodology. *Food Res. Int.*, 42, 430–435, 2009.
50. Taguchi, G., *Introduction to quality engineering*, UNIPUB/Kraus International Publications, White Plains, NY, USA, 1986.
51. Aguilar, C.N., Rodriguez, R., Gutierrez-Sanchez, G., Augur, C., Favela-Torres, E., Prado-Barragan, L.A., Ramirez-Coronel, A., Contreras-Esquivel, J.C., Microbial Tannases: Advances and perspectives. *Appl. Microbiol. Biotechnol.*, 76, 47–59, 2007.
52. Rodriguez, H., de las Rivas, B., Gomez-Cordoves, C., Munoz, R., Characterization of tannase activity in cell-free extracts of *Lactobacillus plantarum* CECT 748. *Int. J. Food Microbiol.*, 121, 92–98, 2008.
53. Khanbabee, K. and Van Ree, T., Tannins: Classification and definition. *Nat. Prod. Rep.*, 18, 641–649, 2001.
54. Lekha, P.K. and Lonsane, B.K., Comparative titres, location and properties of tannin acyl hydrolase produced by *Aspergillus niger* PKL104 in solid-state, liquid surface and submerged fermentations. *Process Biochem.*, 29, 497–503, 1994.
55. Sabu, A., Pandey, A., Jaafar Daud, M., Szakacs, G., Tamarind seed powder and palm kernel cake: Two novel agro residues for the production of tannase under solid state fermentation by *Aspergillus niger* ATCC 16620. *Bioresour Technol.*, 96, 1223–1228, 2005.
56. Battestin, V. and Macedo, G.A., Effects of temperature, pH and additives on the activity of tannase produced by *Paecilomyces variotii*. *Electro. J. Biotechnol.*, 10, 191–199, 2007.
57. Paranthaman, R., Vidyalakshmi, R., Murugesh, S., Singaravadivel, K., Manipulation of fermentation conditions on production of tannase from agricultural by-products with *Aspergillus oryzae*. *Afr. J. Microbiol. Res*, 4, 13, 1440–1445, 2010.
58. Verma, T., Ramteke, P.W., Garg, S.K., Quality assessment of treated tannery wastewater with special emphasis on pathogenic *E. coli* detection through serotyping. *Environ. Monit. Assess.*, 145, 1–3, 243–249, 2008.
59. Gupta, K., Gaumat, S., Mishra, K., Studies on phyto-genotoxic assessment of tannery effluent and chromium on *Allium cepa*. *J. Environ. Biol.*, 33, 3, 557–563, 2012.
60. Islam, B.I., Musa, A.E., Ibrahim, E.H., Sharafa, S.A.A., Elfaki, B.M., Evaluation and characterization of tannery wastewater. *J. For Prod. Ind.*, 3, 141–150, 2014.
61. Macedo, G.A., Matsuda, L.K., Battestin, V., Seleçao de fungos produtores de tanase em resıduos vegetais ricos em taninos. *Cienc. Agrotec.*, 29, 833–838, 2005.
62. Haslam, E. and Stangroom, J.E., The esterase and depsidase activities of tannase. *Biochem. J.*, 99, 1, 28–31, 1966.
63. Zeida, M., Wieser, M., Yoshida, T., Sugio, T., Nagasawa, T., Purification and characterization of gallic acid decarboxylase from *Pantoea agglomerans* T71. *Appl. Env. Microbiol.*, 64, 4743–4747, 1998.
64. Mingshu, L.I., Kai, Y., Qiang, H., Dongying, J., Biodegradation of gallotannins and ellagitannins. *J. Basic Microbiol.*, 46, 1, 68–84, 2006.

65. Pourrat, H., Regerat, F., Pourrat, A., Daniel, J., Production of Gallic acid from tara by a strain of *Aspergillus niger*. *J. Ferment. Technol.*, 63, 401–403, 1985.
66. Pourrat, H., Regerat, F., Morvan, P., Pourrat, A., Microbiological production of gallic acid. *Biotechnol Lett.*, 9, 731–734, 1987.
67. Regerat, F., Pourrat, H., Pourrat, A., Hydrolysis by fermentation of tannins from gall nuts. *Jalca.*, 84, 323–328, 1989.
68. Kar, B., Banerjee, R., Bhattacharyya, B.C., Microbial production of gallic acid by modified solid state fermentation. *J. Ind. Microbiol. Biotechnol.*, 23, 173–177, 1999.
69. Mukherjee, G. and Banerjee, R., Biosynthesis of tannase and gallic acid from tannin rich substrates by *Rhizopus oryzae* and *Aspergillus foetidus*. *J. Basic Microbiol.*, 44, 42–48, 2004.
70. Banerjee, D., Mondal, K.C., Pati, B.R., Tannase production by *Aspergillus aculeatus* DBF9 through solid-state fermentation. *Acta Microbiol. Immuno. Hung.*, 54, 2, 159–166, 2007.
71. Trevino-Cueto, B., Luis, M., Contreras-Esquivel, J.C., Rodriguez, R., Aguilera, A., Aguilar, C.N., Gallic acid and tannase accumulation during fungal solid state culture of a tannin-rich desert plant (*Larrea tridentata* Cov.). *Biores. Technol.*, 98, 721–724, 2007.
72. Sarıozlu, N.Y. and Kıvanc, M., Isolation of gallic acid-producing microorganisms and their use in the production of gallic acid from gall nuts and sumac. *Afr. J. Biotechnol.*, 8, 6, 1110–1115, 2009.
73. Qi, F.H., Jing, T.Z., Wang, Z.X., Zhan, Y.G., Fungal endophytes from Acer ginnala maxim: Isolation, identification and their yield of gallic acid. *Lett. Appl. Microbiol.*, 49, 98–104, 2009.
74. El-Fouly, M.Z., El Awamry, Z., Shahin Azza, A.M., El-Bialy, H.A., Narem, E., El Saeed, G.E., Gallic acid formation from gallotaninns rich agricultural wastes using *Aspergillus niger* UMC4301 or its tannase enzyme. *Ar. J. Nucl. Sci. Appl.*, 45, 2, 489–496, 2012.
75. Lokeshwari, N., Utilization of natural tannins from Anacardium occidentales testa for producing the industrially important Gallic acid through submerged fermentation. *W. J. Pharma. Res.*, 5, 8, 861–864, 2016.
76. Thakur, N. and Nath, A.K., Detection and production of Gallic acid from novel fungal strain-*Penicillium crustosum* AN3 KJ820682. *Curr. Trends Biotechnol. Pharm.*, 11, 1, 60–66, 2017.
77. Ferk, F., Chakraborty, A., Jager, W., Kundi, M., Bichler, J., Misík, M., Wagner, K., Grasl-Kraupp, B., Sagmeister, S., Haidinger, G., Hoelzl, C., Nersesyan, A., Dusinska, M., Simic, T., Knasmüller, S., Potent protection of gallic acid against DNA oxidation: Results of human and animal experiments. *Mutat. Res.*, 715, 61–71, 2011.
78. Reckziegel, P., Dias, V.T., Benvegnú, D., Boufleur, N., Barcelos, R.C.S., Segat, H.J., Pase, C.S., Santos, C.M.M., Flores, E.M.M., Burger, M.E., Locomotor damage and brain oxidative stress induced by lead exposure are attenuated by gallic acid treatment. *Tox. Let.*, 203, 74–81, 2011.
79. Sen, S., Asokkumar, K., Umamaheswari, M., Sivashanmugam, A.T., Subhadradevi, V., Antiulcerogenic effect of gallic acid in rats and its effect on oxidant and antioxidant parameters in stomach tissue. *Indian J. Pharm. Sci.*, 75, 2, 149–155, 2013.
80. Chen, H., Ma, C.Y., Chen, X., Geng, Y., Yang, C., Jiang, H.Z., Wang, X., The effect of rhein and gallic acid on the content of IL10, IL-1β and TNF-α in serum of rats with endotoxemia. *J. Chem. Pharm. Res.*, 6, 10, 296–299, 2014.
81. Vaquero, M.J., Alberto, M.R., de Nadra, M.C., Antibacterial effect of phenolic compounds from different wines. *Food Control.*, 18, 2, 93–101, 2007.
82. Anderson, R., Grabow, G., Oosthuizen, R., Theron, A., Van Rensburg, A.J., Effects of sulfamethoxazole and trimethoprim on human neutrophil and lymphocyte functions *in vitro*: *In vivo* effects of co-trimoxazole. *Antimicro. Agents Chemother.*, 17, 322–326, 1980.
83. Ow, Y.Y. and Stupans, I., Gallic acid and gallic acid derivatives: Effects on drug metabolizing enzymes. *Curr. Drug Metab.*, 4, 241–248, 2003.

84. Cho, Y.S., Kim, S.K., Ahn, C.B., Je, J.Y., Inhibition of acetylcholine esterase by Gallic acid-grafted-chitosans. *Carboh. Polym.*, 84, 690–693, 2011.
85. Huang, H.L., Lin, C.C., Jeng, K.C.G., Yao, P.W., Chuang, L.T., Kuo, S.L., Hou, C.W., Fresh green tea and gallic acid ameliorate oxidative stress in kainic acid-induced status epilepticus. *J. Agric. Food Chem.*, 60, 2328–2336, 2012.
86. Mansouri, M.T., Farbood, Y., Sameri, M.J., Sarkaki, A., Naghizadeh, B., Rafeirad, M., Neuroprotective effects of oral gallic acid against oxidative stress induced by 6-hydroxydopamine in rats. *Food Chem.*, 138, 2–3, 1028–1033, 2013.
87. Tung, Y.T., Wu, J.H., Huang, C.C., Peng, H.C., Chen, Y.L., Yang, S.C., Chang, S.T., Protective effect of *Acacia confusa* bark extract and its active compound gallic acid against carbon tetrachloride induced chronic liver injury in rats. *Food Chem. Tox.*, 47, 1385–1392, 2009.
88. Li, T., Zhang, X., Zhao, X., Powerful protective effects of gallic acid and tea polyphenols on human hepatocytes injury induced by hydrogen peroxide or carbon tetrachloride *in vitro*. *J. Med. Plant. Res.*, 4, 3, 247–254, 2010.
89. Sawa, T., Nakao, M., Akaike, T., Ono, K., Maeda, H., Alkyl peroxyl radical scavenging activity of various flavonoids and other phenolic compounds: Implications for the anti-tumor promoter effect of vegetables. *J. Agr. Food Chem.*, 47, 2, 397–402, 1999.
90. Su, T.R., Lin, J.J., Tsai, C.C. *et al.*, Inhibition of melanogenesis by gallic acid: Possible involvement of the I3K/Akt, MEK/ERK and Wnt/β-catenin signaling pathways in B16F10 cells. *Int. J. Mol. Sci.*, 14, 20443–20458, 2013.
91. Jeon, J., Kim, E., Murugesan, K., Park, H., Kim, Y., Kwon, J., Kim, W., Lee, J., Chang, Y., Laccase-catalysed polymeric dye synthesis from plant-derived phenols for potential application in hair dyeing: Enzymatic colourations driven by homo- or hetero-polymer synthesis. *Microbial. Biotechnol.*, 3, 3, 324–335, 2010.
92. Chandra, R.P., Lehtonen, L.K., Ragauskas, A.J., Modification of high lignin content kraft pulps with laccase to improve paper strength properties, laccase treatment in the presence of gallic acid. *Biotechnol. Prog.*, 20, 255–261, 2004.
93. Ramamurthy, G., Krishnamoorthy, G., Sastry, T.P., Mandal, A.B., Rationalized method to enhance the chromium uptake in tanning process: Role of Gallic acid. *Clean Technol. Enviro. Pol.*, 16, 3, 647–654, 2014.
94. Nair, G.G. and Nair, C.K., Radioprotective Effects of Gallic Acid in Mice. *BioMed Res. Int.*, 2013, 13, 2013. Article ID 953079 https://doi.org/10.1155/2013/953079.
95. Damiani, E., Bacchetti, T., Padella, L., Tiano, L., Carloni, P., Antioxidant activity of different white teas: Comparison of hot and cold tea infusions. *J. Food Composition Anal.*, 33, 59–66, 2014.
96. Roshanak, S., Rahimmalek, M., Goli, S.A., Evaluation of seven different drying treatments in respect to total flavonoid, phenolic, vitamin C content, chlorophyll, antioxidant activity and color of green tea (*Camellia sinensis* or *C. assamica*) leaves. *J. Food Sci. Technol.*, 53, 721–729, 2016.
97. Chen, Y. and Luo, Y., Oxidation stability of biodiesel derived from free fatty acids associated with kinetics of antioxidants. *Fuel Pro. Technol.*, 92, 1387–1393, 2011.
98. Jackson, J.K., Zhao, J., Wong, W., Burt, H.M., The inhibition of collagenase induced degradation of collagen by the galloyl-containing polyphenols tannic acid, epigallocatechin gallate and epicatechin gallate. *J Mat. Sci: Mat. Med.*, 21, 1435–1443, 2010.

8

Biorefinery Approach for Production of Industrially Important C4, C5, and C6 Chemicals

Shritoma Sengupta[1,2] and Aparna Sen[1]*

[1]*Department of Microbiology, Lady Brabourne College, University of Calcutta, Kolkata, West Bengal, India*
[2]*Department of Biotechnology, JIS University, Agarpara, Kolkata, West Bengal, India*

Abstract

The production of bio-based chemicals is not new and is trending for bio-based chemicals, which include non-food starch, cellulose fibers and cellulose derivatives, tall oils, fatty acids, and fermentation products such as ethanol, malic acid, citric acid, fumaric acid, butanol, and others. Biorefinery is the sustainable processing of biomass into a range of marketable products and energy, which refers it to be a facility, a process, a plant, or even a cluster of facilities for the transformation of biomass to different industrial viable products. A biorefinery serves an integral part of upstream, midstream, and downstream processing of biomass into a wide range of products. A biorefinery can utilize various forms of biomass for production of bio-based chemicals obtained from agriculture, forestry, or aquaculture, and some from industrial residues or households' residues which includes wood, agricultural crops, or organic residues. Recent advancement in biotechnology allows the industry to target fermentation products such as chemicals of 4 carbon, 5 carbon, and 6 carbon carbohydrates which may be obtained mainly from hemicellulosic substances, such as starch, sucrose, or from cellulose, lignin, plant or algal-based oils, organic solutions from grasses or pyrolytic liquids. In this chapter, the study describes the various biorefinery approaches taken for production C4, C5, and C6 chemicals of industrial importance, while maintaining its high yields of production.

Keywords: Biorefinery, saccharification, fermentation, malic acid, glutamic acid, citric acid, biochemical platform, biomass sugars

8.1 Introduction

In the 21st century, conservation and organization of the diverse natural resources are vital tasks to substitute sustainable development. Safe and sustainable resources for industrial production are the main requirement for sustainable economic growth that have a long-term and assured ecological safety, sustainable life, and work perspectives for the public benefit [1]. "Biorefinery" is a term that draws from GreenTech, Industrial Biotech, and Synthetic Biology. The concept of the biorefinery evolved during the late 1990's. The US

**Corresponding author*: aparnasen_mac@yahoo.co.in

Arindam Kuila and Mainak Mukhopadhyay (eds.) Biorefinery Production Technologies for Chemicals and Energy, (145–162) © 2020 Scrivener Publishing LLC

Department of Energy defines a biorefinery as an overall concept of a processing plant where biomass feedstocks are converted and/or extracted into a spectrum of valuable products [2]. The core concept behind the term "biorefinery" is relatively simple as these plants mimic the traditional petroleumrefinery, but using biomass instead. The biorefinery can help us produce energy and co-products while reducing pollution and greenhouse gas emissions from hydrocarbon-based production. A biorefinery, which utilizes a renewable biomass as a feedstock resource, may offer a more sustainable solution for the conversion of harvested and waste biomass into platform chemicals [3]. A biorefinery could also bring about sustainable growth, along with environmental advantages in the reduction of overall greenhouse gas emissions (NREL) and air toxics [4]. Just as petroleum refineries transform crude oil into fuels and chemical building blocks that become a part of many different end products, a biorefinery can use all kinds of biomass from forestry, agriculture, aquaculture, and residues from industry and households including wood, agricultural crops, organic residues (both plant and animal derived), forestresidues, and aquatic biomass (algae and seaweeds) to make bioproducts. A biorefinery is not a new concept. Many of the traditional biomass converting technologies such as the sugar, starch, and pulp and paper industry can be (partly) considered as biorefineries. The production of bio-based chemicals is neither novel nor a historic object. Current global bio-based chemical and polymer production (excluding biofuels) is estimated to be around 50 million tonnes. Remarkably, some instances of bio-based chemicals include non-food starch, cellulose fibers and cellulose derivatives, tall oils, fatty acids, and fermentation products such as ethanol and citric acid.

The key factors that should be addressed for the bio-based production of chemicals include the development of natural or engineered microorganisms for efficient production, compatibility with existing industrial infrastructures, optimization of downstream processes, and access to cost-competitive and sustainable resources [5, 6]. The development of efficient microorganisms capable of producing target compoundsat a sufficiently high titers, yields, and industrial production levels remains a significant challenge and a major limiting step. The optimization of downstream processes is interrelated with the development of host microorganisms, and both will play important roles in reducing production costs andallowing bio-based processes to compete against the current petrochemical processes [7]. Green biorefineries are integrated technologies and technology systems for production of materials and energy processing of green plants and parts of greenplants. Above all, green biorefinery technologies are based on traditional technologies of green forage preservation, leaf-protein extraction, chlorophyll production, and modern biotechnological and chemical conversion methods.

- C4 chemicals (e.g., butanol, fumaric acid, malic acid), from substrates like glucose and xylose.
- C5 chemicals (e.g., such as xylose, arabinose, xylitol, itaconic acid) produced from hydrolysis of hemicellulose and food materials.
- C6 chemicals (e.g., glucose, fructose, galactose, citric acid), from hydrolysis of sucrose, starch, cellulose, and hemicellulose.

A biorefinery is not a completely new concept. It is a fundamental upstream, midstream, and downstream processing of biomass into a range of products. The classification system has differentiated between mechanical pre-treatments (extraction, fractionation, separation), thermochemical conversions, chemical conversions, enzymatic conversions, and microbial (fermentation both aerobic, anaerobic) conversions. Many of the traditional biomass converting technologies such as the sugar, starch, and pulp and paper industry can be considered as an integral part of biorefineries. A biorefinery can use several types of biomass for different chemical productionsuch as from forestry, agriculture, aquaculture, and residues from different industry and households which may include wood, agricultural crops, and others. It also includes forest residues, aquatic biomass such as algae and seaweeds, and organic residues which may be plant or animal derived. Several economic and environmental factors such as energy conservation, global warming, and agricultural policies have also engaged to those industries which utilizes biomass to further advance their operations in a biorefinery manner. This should result in improved integration and optimization aspects of all the biorefinery subsystems.

These biorefineries are expected to contribute to an increased competitiveness and wealth of the countries by responding to the need for supplying a wide range of bio-based products and energy in an economically, socially, and environmentally sustainable manner. Biorefineries show potentials for both types of countries: industrialized and developing. New competencies, new job opportunities, and new markets are also expected to intricate the biorefinery processes. The biorefinery systems are classified based on the facts that are dependent on platforms, products, feedstocks, and the processes for the production. Some examples of classifications are the following:

- C5 and C6 sugars, electricity and heat, lignin biorefinery using wood chips for bioethanol, electricity, heat, and phenols.
- C6 sugar platform biorefinery for bioethanol and animal feed from starch crops.
- Oil biorefinery using oilseed crops for biodiesel, glycerin, and feed
- Syngas platform biorefinery for Fischer-Tropsch (FT) diesel and phenols from straw.

8.2 Role of Biorefinery in Industrially Important Chemical Production

Biorefinery is the sustainable processing of biomass into a spectrum of marketable products and energy. Around the world, small but noticeable steps are being taken to move from today's fossil-based economy to a more sustainable economy based on greater use of renewable resources. As a multifunctional process, biorefinery applies a varied range of raw materials to sustainably and concurrently generate a range of diverse intermediates and products including food and feed. The biorefinery chain consists of the pre-treatment and separation of biomass components (primary biorefinery) and the subsequent conversion (secondary biorefinery) [8]. Using green and sustainable processes, new generations of biorefinery techniques combine various types of biomass waste resources from different origins, following chemical extraction and finally synthesis of biomaterials, biofuels, or bioenergy. Utilization of biomass as a sustainable renewable resource is the only way to replace carbon from fossil sources for the production of carbon-based products such as chemicals, materials, and liquid fuels with a remarkable reduction of CO_2 releases into the atmosphere.

A biorefinery is not a completely new concept. The foundation of biorefineries is for the benefit toward environment and is also anaspect of sustainability. Biorefinery involves a process which includes integral upstream, midstream, and downstream processing of biomass into a wide range of products. A biorefinery system can use various kinds of biomass such as from agriculture, forestry, aquaculture, and residues from industry and households which may include agricultural crops, wood, organic residues derived from both plant and animal system. It also includes forest residues, and aquatic biomass such as algae and seaweeds. Many of the traditional biomass converting technologies are partially considered as biorefineries which includes sugar, starch, and pulp and paper industry. This biorefinery system can be integrated with food or feed production, as seen in the instance with first generation ethanol production. Essential elements of a biorefinery are:

- multiple feedstock capability and a tolerance of wide variation in those feedstocks;
- feedstock-processing by enzymes to fermentable sugars (and by-productstreams);
- biocatalyst, which converts sugars to desired product(s); and
- co-products which are used in the process, recycled through the process, or sold.

Bio-based products such as chemicals and different materials can be produced in single product processes with the help of biorefinery system; where in the production processes, it produces both bio-based products and secondary energy carriers such as fuels, power, and heat in equivalence with oil refineries. In a future bio-based economy, it is probably a more resourceful approach to have a sustainable valorization process for biomass resources. Biomass can be transformed into a wide range of chemicals in two main ways exploiting the latest developments in biotechnology particularly by:

- fermentation
- thermochemical processing routes

Chemo-catalytic process or bio-forming is the conversion of simple carbohydrates obtained from biomass into important intermediates at mild temperatures in aqueous solution can be considered as a third and evolving route that is gaining much interest. These main routes represent processes that can be used in a biorefinery and provides the structure for the discussion in this unit. The important chemicals currently being produced from biomass are discussed in the following sections of this chapter, including ethene and propene and the polymers like polyethene and polypropene, aromatic hydrocarbons, liquid fuels (naphtha, kerosine, and diesel), various organic acids like citric acid, adipic acid different alcohols like xylitol, butanol, and others.

Manufacture of chemicals from biomass by fermentation: Chemicals such as ethanol, butanol, and several organic acids are currently produced by fermentation processes that are discussed in detail in the later part of this chapter. By fermentation process, a lot of chemicals are produced, predominantly C-4, C-5, and C-6 chemicals with the help of several microorganisms utilizing various biomass for the production.

Manufacture of chemicals from biomass by thermochemical processes: Chemicals are also produced by different thermochemical processes utilizing the biomass for production. There are two main routes using thermochemical processing:

- One involves heating biomass to a high temperature and under pressure in a controlled amount of oxygen which leads to the production of synthesis gas known as syngas (a mixture of carbon monoxide and hydrogen). This process is known as gasification. Synthesis gas can be converted into several important industrial compounds.
- The second method again involves heating the biomass to a high temperature, but in this case in the absence of air. This process is known as pyrolysis. In order to form useful products, the reaction time must be very short; otherwise, the major product will be carbon (char). This process is thus called fast pyrolysis and the major product is an oil known as bio-oil.

Biorefineries are classified into three different types [1, 9] as mentioned below:

Phase I Biorefinery: This type of biorefinery utilizes only one feedstock material, has fixed processing capability, and produces a single primary product, which includes examples such as biodiesel from vegetable oil, pulp, and paper mills [10].

Phase II Biorefinery: This type of biorefinery uses only one feedstock but is capable of producing various products which includes the production of various chemicals from starch and the production of multiple carbohydrate derivatives and bioethanol from cereal grains [4].

Phase III Biorefinery: These are advanced biorefineries and can utilize various types of feedstocks, processing technologies, and produce multiple types of products. There are four classes of Phase III Biorefinery systems [9]. These are identified as: 1. Whole-crop biorefinery, 2. Green biorefinery, 3. Lignocellulosic biorefinery, 4. Two-platform concept biorefinery [4].

The process of biochemical fermentation followed by biorefineries for several chemical productions is comprised of three steps as follows: Pre-treatment, Saccharification, and Fermentation [11].

Pre-treatment: This is an important aspect in the processing of biomass for the production of ethanol. The pre-treatment process helps in increased susceptibility of the biomass to hydrolysis and results in generating higher yields of monomer sugars [4].

Saccharification: In this second step, complex carbohydrates are converted into simple monomers by hydrolysis using hydrolase group of enzymes such as cellulase and hemicellulase [10, 12]. The hydrolysis of the cellulose leads to the formation of the initial material, i.e., glucose. But on the hydrolysis of hemicellulose, it produces numerous isomers of pentoses and hexoses.

Fermentation: The biomass sugars obtained upon saccharification are then used for fermentation by various microorganisms. An idyllic microorganism for the commercial production of alcohol should be proficient of utilizing various types of sugars which can be used to produce their respective alcohols in high yields [12].

8.3 Production of C4 Chemicals

In this section, the bio-based production of representative C4 platform chemicals, including carboxylic acid (butyric acid), dicarboxylic acids (succinic, malic, andfumaric acids), alcohols (butanol and iso-butanol), diols (2,3-butanediol) will be discussed.

Butanol: Butanol is a 4-carbon alcohol with the molecular formula $C_4H_{10}O$. There are four isomeric structures for butanol, they aren-butanol, iso-butanol, 2-butanol, and tert-butanol. All of them are currently produced mainly *via* petrochemical routes and used as chemical intermediates and industrial solvents. Compared to ethanol, n-butanol and isobutanol are superior liquid fuels in many respects.

n-Butanol: Oxo process which includes hydroformylation of propylene to butyraldehyde, followed with hydrogenation. n-butanol is naturally produced by a number of clostridia. Typically, clostridia are rod-shaped, sporeforming, and gram-positive strict anaerobes. These naturally solventogenic clostridia allcan produce butanol with a high yield; however, their substrate utilization ability is very different from each other, as well as their optimal pH, temperature, and product profiles. Most of these species can fermentpentose and hexose sugars, as well as starch, while somestrains also possess the ability to utilize syngas and glycerolas the carbon source [13].

Solventogenic clostridia utilizes substrates such as glucose, xylose, arabinose, cellobiose, starch, sucrose, cellobiose, sucrose, molasses, mannose, galactose, fructose, and even syngas for production of C-4 chemicals such as butanol or butyrate [14–18]. Several fermentation modes such as batch, fed-batch, and continuous have been used and recovery techniques such as gas stripping, pervaporation, liquid-liquid extraction, adsorption, and others have been used *in situ* recovery of butanol and higher alcohols [15, 19–22, 85, 86]. A different approach for n-butanol production was seen by engineering *E. coli* by reversing its β-oxidation cycle in combination with endogenous dehydrogenases and thioesterases for its synthesis, attaining a high titer value and a high yield by utilizing glucose as substrate [23].

Isobutanol: Isobutanol is produced by fermentation with genetically engineered microbes. It is co-produced with n-butanolin the Oxo process; it also helps in carbonylation of propylene.

2-Butanol: 2-Butanol can be produced from acetolactate and acetoin *via* 3-amino-2-butanol and 2-butanone by some *Lactobacilli* sp. strains, but the production level is low due to its toxicity to cells. The other forms of butanol obtained are tert-Butanol which is produced by a co-product of propylene oxide production from isobutane.

2,3-Butanediol: 2,3-Butanediol has a chemical formula of $(CH_3CHOH)_2$. It is a 4-carbon diol with hydroxylgroups at 2 and 3 carbon positions and isalso known as 2,3-butylene glycol, 2,3-dihydroxybutane, or dimethylethylene glycol [24]. It has a high boiling point of 177°C and a low freezing point of −60°C, hence used as an antifreezing agent [25]. Due to the shortage of fossil fuel supplies and rising oil prices, the biotechnology for the production of 2,3-butanediol has received much attention recently [26–29]. Furthermore, the derivatives of 2,3-butanediol, such as methyl ethyl ketone and 1,3-butadiene, have attracted much attention recently because those chemicals, which are now produced by a petroleum refineryprocess, can be produced from biologically produced 2,3-butanediol. Methyl ethyl ketone can be produced from 2,3-butanediol by dehydration, and it can be used as an effective fuel additive because of a higher heat of combustion than ethanol [30, 31]. Biological production of 2,3-butanediol has been investigated since Harden and Walpole [32] reported 2,3-butanediol production from glucose and mannitol. That work was followed by Donker [33] using *Bacillus polymyxa*. Industrial production was first proposed by Fulmer *et al.* [34], and pilot scale fermentation had already been performed during World War II with *K. oxytoca* and *B. polymyxa* [35, 36]. 2,3-Butanediol being an industrially useful chemical has numerous applications in the production of printing inks, synthetic perfumes, softening, moistening agents, and others. A solvent for lacquers and resins can be used as a carrier for drugs and pharmaceuticals importance [30, 37].

Succinic Acid: Succinic acid is a C-4 chemical having a molecular formula $C_4H_6O_4$. It is currently a chemical produced by catalytic hydrogenation of petrochemical maleic acid or a hydride having high-volume speciality. Based on bacterial fermentation of carbohydrates, large volume products in the markets could be accessed by following cost reductions techniques that is delivered through its production. Currently, *E. coli* is the bacterial strain used for succinic acid production, but several companies toward other microorganisms such as *Coryne*-type bacteria and yeast for requirement for lowering costs in production. Succinic acid can be converted to 1,4-butanediol (BDO) and other products by further modifications [38]. BDO can serves as a raw material for a range of important chemicals including polymers polybutylene terephthalate (PBT) and polybutylene succinate (PBS). Approximately, 40% of BDO is consumed in tetrahydrofuran (THF) production [39].

Fumaric Acid: Fumaric acid is a C-4 chemical having a molecular formula $C_4H_4O_4$. It is presently produced by petrochemical synthesis through catalytic isomerization of maleic acid. The actual yield of fumaric acid from maleic acid can reach more than 90% with the theoretical yield of 100% [38]. Industrially, it is produced by catalytic isomerization of maleic acid derived from maleic anhydride, which is produced from n-butene *via* catalytic oxidation and it is used as food acidulant; feed additive; medicine; raw material for polyester resins, polyhydric alcohols, and others. Fumaric acid is mainly produced by *Rhizopus* sp. in the following way using various fermentors:

- Fumaric acid is produced by *Rhizopus arrhizus* in industrial process by using 20-L STR and shake flaskfermentor having substrate glucose and xylose where the yield is 0.70 g/g and 0.23 g/g and productivity is 1.22 g/L·h and 0.07 g/L·h, respectively [40, 41].
- *Rhizopus arrhizus* is also used in production using STR fermentor having wood as substrate giving a yield of 0.71 g/g [42].
- *Rhizopus oryzae* is also used in fumaric acid production by using STR fermentor with DO control and glucose as substrate, giving a yield of 0.54 g/g and productivity of 0.7g/L·h [43].
- *Rhizopus formosa* is also used in production using STR fermentor having cassava bagasse as substrate [44].

For production of fumaric acid, the pH of the medium remains ~6.0, so neutralizing agent such as $CaCO_3$ is used in the fermentor other than substrates which have high sources of carbon, nitrogen, phosphorous, and certain metals like $MgSO_4$, $ZnSO_4$, $FeCl_3$, and others. Fumaric acid is basically produced by filamentous fungi under aerobic conditions. So, maintenance of sufficient level of Dissolved Oxygen (DO) in the fermentation broth is a very important aspect.

Malic Acid: Malic acid is a C-4 chemical having a molecular formula $C_4H_6O_5$, which is primarily mass-produced through chemical synthesis *via* hydration of maleic or fumaric acid, producing a racemic mixture of d-(−) and l-(+) isomers [38]. Fermentation of malic acid can be carried out by various microorganisms. Various studies represented several microorganisms for high amount of l-malic acid production such as *Aspergillus flavus, Aspergillus oryzae, Aspergillus sojae, Aspergillus ochraceus, Aspergillus foetidus, and Aspergillus nidulans* [45]. Malic acid is mainly produced by certain fungus and bacteria as follows:

- Malic acid is produced by *Aspergillus flavus* in industrial process by using 16-STR fermentor having substrate glucose where the yield is 0.94 g/g and productivity is 0.59 g/L·h [46].
- *Aspergillus niger* is also used in production using Shake flask fermentor having thin still age as substrate giving a yield of 0.8 g/g and productivity is 0.09 g/L·h [87]
- *Zygosaccharomyces rouxii* is also used in production using flask fermentor having glucose as substrate giving a yield of 0.33 g/g and productivity is 0.54 g/L·h [88]
- *E. coli* is also used in production having shake flask or 3-L STR fermentorfor production using glucose as substrate giving a yield of 0.42–1.06 g/g and productivity is 0.69–74 g/L·h [89]
- *Saccharomyces cerevisiae* is also used in production having shake flask for production using glucose as substrate giving a yield of 0.31 g/g and productivity is 0.19 g/L·h [89]

Another technique for malic acid production is through enzymatic catalysis in which l-malic acid is produced from fumaric acid by immobilized bacterial cells (e.g., *Brevibacterium flavum*) containing the enzyme fumarase, a well-developed process used in Japan and China. The fermentative production of malic acid has also been demonstrated with *Aspergillus flavus*, attaining a high production and molar yield but it produces a toxic substance known as aflatoxin as a result of the accompanying production of malic acid [46]. Hence, *A. flavus* cannot be used as a producer of food-grade chemicals.

Butyric Acid: Butyric acid is a 4-carbon aliphatic fatty acid that is widely used in the food, perfume, and polymer industries [47]. It is the starting material for the formation of cellulose acetate butyrate, a polymer of butyric acid, which is used for the production of photographic films and eyeglass frames. Butyric acid is produced either by chemically oxidizing butane and butyraldehyde, or *via* sugar fermentation [48]. Several bacteria have shown to yield high amount of butyric acid belonging to the genera *Clostridium* sp. [49, 50], *Butyrivibriro* sp. [51], and *Enterococcus* sp. [52]. Butyric acid can be produced from the glycolytic pathway *via* condensation of acetyl-CoA to acetoacetyl-CoA [4]. To reduce by-product formation, researchers have used metabolic engineering strategies (e.g., deletion of the acetate-producing pathways) to enhance butyric acid yield and production titers. However, tolerance issues prevented continuous production of butyric acid as the titer increased. Thus, the butyric acid tolerance of *C. tyrobutyricum* should be improved to achieve the higher titers, yields, and productivities required for commercial production.

8.4 Production of C5 Chemicals

In this section, the bio-based production of representative C5 platform chemicals, including isoprene, xylitol, itaconic acid, levulinic acid, 5-aminovalaric acid, and glutamic acid, will be discussed.

Isoprene: Isoprene has a chemical name of 2-methyl-1,3-butadiene, which is a common C5 organic compound with the formula $CH_2=C(CH_3)CH=CH_2$. It is a colorless volatile liquid. Isoprene is the monomer used for the production of synthetic rubbers called polyisoprenes. Isoprene is commonly produced by extraction of the isoprene molecule from Crude C5 mixtures such as the C5 Diene Crude product. During the extraction process of isoprene, piperylene is also concentrated and collected as by-product. From this piperylene different C5-based resin are being produced.

Xylitol: Xylitol, a sugar alcohol, is a C5 chemical that has a molecular formula of $C_5H_{12}O_5$. It is a naturally occurring poly alcohol, basically a sugar alcohol found in certain fruits and vegetables in small amounts. It acts as an alternative to conventional sweeteners, i.e., sucrose. It also has many applications in the sectors of health, food, and drink. The chief raw materials for the microbial production of xylitol are xylans, which are a major part of plant biomass referring to lignocellulose biomass. Plant biomass is the most abundantly available in expensive energy source on earth. This source can be produced directly from plants, from agriculture and forest waste residues, or from crops. It is mainly composed of three polymers: cellulose, hemicellulose, and lignin. Among these polymers, the major hemicellulose part of lignocellulose biomass is composed of C-5 sugars such as xylans, which covers 11%–35% of plant biomass and in a lesser extent C-6 sugars are also present in most of the plant species. The composition and the variation of the hemicellulose content of lignocellulose are highly dependent on its source. The major component of hemicellulose is xylan, which is a heteropolysaccharide with a linear backbone of β-1, 4-linked xyloses [53]. Primarily, in a one-step process by chemical dehydrogenation of d-xylose, xylitol was produced. This process was expensive due to its high cost of the starting material, i.e., d-xylose. The d-xylose must be a pure form for chemical production. Xylitol should be a cost-effective material as it serves as a starting material for many industrial-scale chemicals.

Itaconic Acid: Itaconic acid (IA) is a C5 chemical with molecular formula $C_5H_6O_4$. It is predominantly produced by *Aspergillus terreus* [38]. In industrial process, different fermentors are used for its production utilizing different substrates as mentioned below:

- By using STR fermentor, IA is produced utilizing molasses as substrate where the productivity is 1 g/L·h [54].
- Porous disk bioreactor is also used for IA production using glucose as substrate having productivity is 0.73 g/L·h [55].
- Flasks and 2.5-L air lift bioreactor is also used for production where corn starch are used as substrates [56].
- 3-jar fermentor is another type of fermentor used for IA production using Sago starch hydrolysate as substrate having a yield of 0.34 g/g and productivity is 0.4 g/L·h [57].

The industrial fermentation for IA production was first realized by Charles Pfizer Co. in the United States in 1955, followed by many manufacturers in Europe, Russia, Japan, and China [58] and currently used in China. It is used in several industries for production of bioactive compounds in agriculture and medicine; polymer intermediate; coating, plasticizer, and others. IA itself has an addressable market in which it can be used as a substitute [38]. IA has the potential to replace petroleum-based products and other chemicals, such as acrylic acid, maleic anhydride, sodium tripolyphosphate, and acetone cyanohydrin.

Eventhough its position in the industry is not yet determined, a few applications with high-volume markets have been identified and it is in a developmental stage. The largest application of IA is for the production of methyl methacrylate [53].

Levulinic Acid: It is a 5-carbon chemical of molecular formula $C_5H_8O_3$. Levulinic acid (LVA) is one of the most commonly formed organic acids during biomass processing. The chemical synthesis of LVA from different feedstocks by using mineral acid as the catalyst has been known since 1840. One of the most studied areas for the synthetic production of LVA is the controlled degradation cellulose-containing biomass or lignocellulosic biomass at low cost. By this processdue to the by-product's formation, the practical yield of LVA was seen to be less than its theoretical yield. There were also problems in the yield and purity of the final product [53]. Many methods for production andprocessing for the separation of LVA from biomass and also for the prevention of by-product formation methods were developed and patented [59–61]. The costly chemical processes have been preventing a widespread adoption of LVA. There is no bioprocess that exists for the pure production of LVA, and undoubtedly, further research will definitely reduce the cost and open an addressable market for LVA. The presence of two highly reactive functional groups in LVA led to its increasing market value. They are: ketone and carboxylic acid groups, which allow this LVA to convert to a large number of chemical or biochemical compound.

5-aminovalaric Acid: It is a 5-carbon chemical of molecular formula $C_5H_{11}NO_2$. 5-aminovalerate (AMV) is a potential chemical for many downstream value-added chemicalderivatives and for polymer synthesis. AMV is known to be one of the natural degradation products formed during l-lysine degradation in "AMV pathway" of many *Pseudomonas* sp. Only a small amount of AMV is accumulated in the *Pseudomonas* sp. during l-lysine catabolism [62]. The direct fermentative production of 5-aminovaleric acid (AVA) at the industrial scale has not yet been demonstrated. Reconstruction of the AMV pathwayin an l-lysine overexpressing host will be the best way for the bio-based production of AMV. Enzymatic methods also yield a high conversion rate of l-lysine to AMV. An enzyme l-lysine α-oxidase was developed from microorganism *Trichoderma viride* by immobilization system on an epoxy-activated solid support. In AMV, production using efficient microorganisms is generally cellular factories of C-6monomer, which is a polyamide building block. The AMV monomer is a precursor of valerolactam, which is important for nylon-5 homopolymer synthesis [63]. This would drive the biopolyamide and biopolymer fields. The polymers, nylon-4,5 and nylon-5,5, can be produced by the polymerization of glutarate with putrescine and cadaverine, respectively. These polymers have additional piezoelectric and ferroelectric properties that are useful in sensors and electronics.

Glutamic Acid: Glutamic acid (Glu) is a 5-carbon molecule with molecular formula $C_5H_9NO_4$. Out of all the amino acids, Glu has the highest demand and industrial importance. The production of Glu exceeds 2.5million tons per year on a global basis [64]. Glu is now directly produced from microbial fermentation processes. *Corynebacterium glutamicum* was discovered in Japan as a natural producer of Glu in 1957. Apart from *C. glutamicum*, other bacterial strains of *Brevibacterium* sp. were also used as cost-effective bioconverters for Glu production [53]. Among the various *Corynebacterium* bacterial strains, the *C. glutamicum* strain has the ability to produce high amounts of Glu, which distinguishes it from others. Glucose is one of the most commonly used carbon sources for the production of Glu. Glucose is metabolized through the Embden-Meyerhof pathway and the TCA cycle.

The end metabolite of these pathways, 2-oxoglutarate, is aminated to glutamate by the action of glutamate dehydrogenase. The carbon source (glucose), ammonium, and biotic are the major limiting parameters for high Glu accumulation. Some researchers used other low-cost carbon sources, such as beet molasses [65], cassava residues [66], sugarcane bagasse [67], date waste, and syrup [68]. In addition to the exploitation of various kinds of raw materials for Glu biosynthesis, some studies used fermentation techniques, such as continuous fermentation, solid-state fermentation [69], temperature shift-up cultivation [70], and the immobilization of cells [71] for improvement in Glu production.

8.5 Production of C6 Chemicals

Six carbon (C6) sugar molecules can be retrieved by the hydrolysis of starch or cellulose or from sucrose to give glucose. Glucose used in fermentation processes provides access to a variety of important chemical building blocks and can also be converted by chemical processing to useful chemicals. Recent advancement in biotechnology allows industry to target fermentation products which may be new or previously abandoned to improve the economic and commercial potential of the fermented products. It is predicted that in the future abundant accessibility ofcarbohydrates obtained from lignocellulosic biomass may play a significant role. However, in recent years, there is steep increase in costs of carbohydrates. So, the use of carbohydrates for non-food products is under pressure. Fermentation in the industrial level gives access to new chemical molecules and its production which was previously inaccessible due to cost limits.

In this section, the bio-based production of representative C5 platform chemicals, including sorbitol, lysine, citric acid, adipic acid, and glucaric acid, will be discussed.

Sorbitol: Sorbitol is a C-6 chemical that has a chemical formula of $C_6H_{14}O_6$. It is produced on large industrial scale by catalytic hydrogenation of glucose which can be further activated by developing milder processing conditions and/or other catalysts to replace the nickel catalysts that are used nowadays [90]. This is done by the industrial implementation of a continuous process. Fermentative routes are also suggested [72] but are unlikely that these routes can replace the technically mature catalytic hydrogenation process. Besides food, sorbitol is also produced as the raw material for production of other products such as surfactants and polyurethanes. Sorbitol can also be further derivatized into ascorbic acid (80,000 ton/y by combined biotechnological/chemical process), Sorbitan (50,000 ton/y), Isosorbide (selective dehydration), and 1,2-propanediol by hydrogenolysis (900,000 ton/y) [73].

Lysine: Lysine is an amino acid being a C-6 chemical with molecular formula $C_6H_{14}N_2O_2$. Lysine is a high nitrogen containing amino acid where its production of nitrogen-containing bulk chemicals from biomass is in a less advanced state compared to oxygenated bulk chemicals such as glycols. Bio-based routes from lysine to caprolactam for the production of nylon have perhaps received the most attention [74]. In the 1950s, fermentation with *Corynebacterium glutamicum* was found to be a very efficient production route to L-glutamic acid. Since this time, biotechnological processes with bacteria of the species *Corynebacterium* developed to be among the most important in terms of tonnage and economical value. L-lysine is produced in bulk nowadays with development of genetically modified plants with elevated levels of certain amino acids such as lysine. In this way, amino acids that are naturally produced by plants can be produced at higher concentration levels by over-expression of certain structural genes [75].

Citric Acid: Citric acid is a C-6 chemical of molecular formula $C_6H_8O_7$. It is predominantly and commercially produced *via* submerged fermentation of molasses by *Aspergillus niger*. Early in 1893, citric acid production *via* fungal fermentation was first discovered [76]. In industrial process, different fermentors are used for its production utilizing different substrates and organisms for production as mentioned below:

- Citric acid is produced by *Aspergillus niger* in industrial process by using 120~250 m^3 STR/~900 m^3 airlift fermentor having substrates such as sucrose, molasses, or corn starch where the yield is 0.7~0.9 g/g and productivity is 0.5~0.8 g/L·h [77].
- It is also produced industrially by reciprocating jet bioreactor using substrate glucose having productivity 0.15 g/L·h [78].
- Using 1-bioreactor, it is industrially produced with anion-exchange resin using Beet molasses as substrates where the yield is 0.95 g/g and productivity is 0.54 g/L·h [79].
- By Solid State Fermentation technique using flasks or packed bed and corncob, cassava, bagasse as substrates, the yield of the product is 0.50 g/g [80, 81].
- *Candida oleophila* is also used in citric acid production by using 2-fermentor and glucose as substrate, giving a yield of 0.50 g/g and productivity of 1.28 g/L·h [82].
- *Yarrowialipolytica* produces citric acid utilizing 3.5-L STR fermentor having glycerol as substrate, giving a yield of 0.69 g/g and productivity of 1.16 g/L·h [83].

Citric acid is used in several industries such as food, beverages, pharmaceuticals, detergents; they are also used as buffering and chelating agents. Nowadays, the annual global citric acid production reaches over million tons, with China the largest citric acid-producing country accounting for ~40% of worldwide output [38]. However, in spite of the success of the commercial production by *A. niger*, several issues, including elucidating the mechanism of citric acid production and controlling cell morphology for acid production, are still not completely addressed.

Adipic Acid: Adipic acid have a chemical formula of $(CH_2)_4(COOH)_2$ is also known as hexanedioic acid or 1,4-butanedicarboxylic acid. It is the most important aliphatic dicarboxylic acid, a white crystalline powder. It is primarily used for the production of nylon. The current market trend for adipic acid production is close to 3 million tons per year, which approximately has a worth of $8 billion at current market prices. By fermentation of glucose, followed by catalytic hydrogenation to adipic acid, it can be biosynthesized in the form of cis, cis-muconic acid [84]. Besides optimization of production organisms, the recovery of adipic acid from aqueous medium at purity levels needed for polymer-grade products and catalytic conversion of muconic acid to adipic acid needs to be further investigated.

Glucaric Acid: Glucaric acid has a chemical formula of $C_6H_{10}O_8$. It is predominantly produced by an international company named Rivertop. It has developed a technique of catalytic oxidation to make glucaric acid from glucose. The current market for glucaric acid is very minor but the C-6 chemical has a huge potential. Glucaric acid has a drop-in replacement for phosphates in detergents. Phosphates are responsible for stimulating algae growth in the system, which in turn diminishes oxygen supply for other aquatic life. Glucaric acid also has corrosion inhibition properties but it can also be polymerized for higher utilization, which is yet to be commercialized as a polymer and may probably take 5–7 years.

8.6 Concluding Remarks

Almost all organic chemical products are currently produced from fossil oil, but the petroleum-based system is currently facing global crises, such as climate change and fossil resource depletion. Thus, there is an increasing demand for sustainable production of bio-based platform chemicals. In particular, microorganisms are considered attractive hosts for the production of platform chemicals from biomass sugars. Chemical products with a broad range of carbon lengths, including C4–C6 acids, alcohols, diols, and diamines, can be produced by the sugar fermentations of microorganisms. At present, these C-4, C-5, and C-6 chemicals are nearing large-scale commercialization as bio-based chemicals. However, intensive metabolic engineering is needed to enhance performance for the production of other chemicals, including the C-4, C-5, and C-6 platform chemicals reviewed in this work. Common aspects to be considered in strain development are: (1) selecting and optimizing metabolic pathways to produce the desired products, (2) utilizing a variety of carbon sources, (3) improving end-product tolerance, (4) minimizing nitrogen sources, and (5) expanding metabolic pathways to produce platform chemicals by one step reactions. Mixed six and five carbon products are formed from the hydrolysis of hemicelluloses. Hypothetically, the carbohydrate fermentation by microorganisms may produce fermented products such as six carbon sugar; but the technical block that still remains is the economic and biological one, which requires to be overcome before these opportunities can be utilized. Chemical manipulation and modifications of these things can provide us with a wide range of useful molecules for sustenance. The progression of fermentation processes in a cost-effective manner to production of succinic, itaconic, and glutamic acids promises the potential for novel chemical development.

The development of a fruitful bio-economy will depend on the development of robust biorefinery systems each with advanced technologies, which can process biological feedstocks into a variety of bio-based products with efficient and cost-effective processes. In a bio-economy, the basic building blocks for chemicals, materials, and energy are derived from renewable sources and are considered as an integral part of the development toward a more sustainable economy. Advanced biorefineries when developed at commercial scale will create and drive new business opportunities. However, the most advanced biorefineries are not yet commercialized and the development of these biorefineries depends on the progress made in the thermochemical and biochemical platform technologies [4, 91].

References

1. Kamm, B. and Kamm, M., Principles of biorefineries. *Appl. Microbiol. Biotechnol.*, 64, 137–145, 2004. [PubMed: 14749903].
2. Kamm, B., Gruber, P.R., Kamm, M., *Biorefineries-industrial processes and products*, Wiley-VCH Verlag GmbH & Co. K GaA, Weinheim, 2006.
3. Esposito, D. and Antonietti, M., Redefining biorefinery: The search for unconventional building blocks for materials. *Chem. Soc. Rev.*, 44, 5821–5835, 2015. [PubMed: 25907306].
4. Takkellapati, S., Li, T., Gonzalez, M.A., An Overview of Biorefinery Derived Platform Chemicals from a Cellulose and Hemicellulose Biorefinery. *Clean. Technol. Environ. Policy*, 20, 7, 1615–1630, 2018.
5. Hatti-Kaul, R., Tornvall, U., Gustafsson, L., Borjesson, P., Industrial biotechnology for the production of bio-based chemicals—A cradle to-grave perspective. *Trends Biotechnol.*, 25, 119–124, 2007.
6. Lee, J.W., Kim, T.Y., Jang, Y.S., Choi, S., Lee, S.Y., Systems metabolic engineering for chemicals and materials. *Trends Biotechnol.*, 29, 370–378, 2011.
7. Jang, Y.S., Kim, B., Shin, J.H., Lee, S.Y. *et al.*, Bio-based production of C2-C6 platform chemicals. *Biotechnol. Bioeng.*, 109, 10, 2437–2459, 2012.
8. Zhu, L., Biorefinery as a promising approach to promote microalgae industry: An innovative framework. *Renew. Sustain. Energy Rev.*, 41, 1376–1384, 2015.
9. Clark, J.H. and Deswarte, F.E.I., The Biorefinery Concept–An Integrated Approach, in: *Introduction to Chemicals from Biomass*, J.H. Clark and F.E.I. Deswarte (Eds.), pp. 1–20, John Wiley & Sons, Ltd, Chichester, UK, 2008.
10. Naik, S.N., Goud, V.V., Rout, P.K., Dalai, A.K., Production of first and second generation biofuels: A comprehensive review. *Renew. Sustain. Energy Rev.*, 14, 578–597, 2010.
11. Sarkar, N., Ghosh, S.K., Bannerjee, S., Aikat, K., Bioethanol production from agricultural wastes: An overview. *Renew. Energy*, 37, 19–27, 2012.
12. Talebnia, F., Karakashev, D., Angelidaki, I., Production of bioethanol from wheat straw: An overview on pre-treatment, hydrolysis and fermentation. *Bioresour. Technol.*, 101, 4744–4753, 2010. [PubMed: 20031394].
13. Zhao, J., Lu, C., Chen, C.C., Yang, S.T., Biological production of butanol and higher alcohols, in: *Bioprocessing Technologies in Biorefinery for Sustainable Production of Fuels, Chemicals, and Polymers*, First Edition, Yang, S.-T., El-Enshasy, H.A., Thongchul, N. (Eds.), 2013.
14. Madihah, M.S., Ariff, A.B., Sahaid, K.M., Suraini, A.A., Karim, M.I.A., Direct fermentation of gelatinized sago starch to acetone-butanol-ethanol by *Clostridium acetobutylicum*. *World J. Microbiol. Biotechnol.*, 17, 567–576, 2001.
15. Qureshi, N. and Maddox, I.S., Reduction in butanol inhibition by perstraction: Utilization of concentrated lactose/whey permeate by *Clostridium acetobutylicum* to enhance butanol fermentation economics. *Trans. I Chem. E. C.*, 83, 43–52, 2005.
16. Ezeji, T.C. and Blaschek, H.P., Fermentation of dried distillers grains and solubles (DDGS) hydrolysates to solvents and value-added products by solventogenic *Clostridia Bioresour. Technol.*, 99, 5232–5242, 2008.
17. Hipolito, C.N., Crabbe, E., Badillo, C.M., Zarrabal, O.C., Mora, M.A.M., Flores, G.P., Cortazar, M.A.H., Ishizaki, A., Bioconversion of industrial wastewater from palm oil processing to butanol by *Clostridium saccharoperbutylacetonicum* N1-4 (ATCC 13564). *J. Cleaner Prod.*, 16, 632–638, 2008.
18. Bruant, G., Levesque, M.J., Peter, C., Guiot, S.R., Masson, L., Genomic analysis of carbon onoxide utilization and butanol production by *Clostridium carboxidivorans* strain P7. *PLoS One*, 5, 9, e13033, 2010.

19. Qureshi, N., Maddox, I.S., Friedl, A., Application of continuous substrate feeding to the ABE fermentation: Relief of product inhibition using extraction, perstraction, stripping and pervaporation. *Biotechnol. Prog.*, 8, 382–390, 1992.
20. Maddox, I.S., Qureshi, N., Roberts-Thomson, K., Production of acetone-butanol-ethanol from concentrated substrates using *Clostridium acetobutylicum* in an integrated fermentation-product removal process. *Process. Biochem.*, 30, 209–215, 1995.
21. Qureshi, N. and Blaschek, H.P., Recovery of butanol from fermentation broth by gas stripping. *Renew. Energy*, 22, 557–564, 2001.
22. Ezeji, T.C., Qureshi, N., Blaschek, H.P., Acetone butanol ethanol (ABE) production from concentrated substrate: Reduction in substrate inhibition by fed-batch technique and product inhibition by gas stripping. *Appl. Microbiol. Biotechnol.*, 63, 653–658, 2004.
23. Dellomonaco, C., Clomburg, J.M., Miller, E.N., Gonzalez, R., Engineered reversal of the beta-oxidation cycle for the synthesis of fuels and chemicals. *Nature*, 476, 355–359, 2011.
24. Um, Y. and Kim, K.D., Biotechnological development for the production of 1,3-propanediol and 2,3-butanediol, in: *Bioprocessing Technologies in Biorefinery for Sustainable Production of Fuels, Chemicals, and Polymers*, First Edition, S.-T. Yang, H.A. El-Enshasy, N. Thongchul (Eds.), John Wiley & Sons, Inc., New York, United States, 2013.
25. Soltys, K.A., Batta, A.K., Koneru, B., Successful non freezing, subzero preservation of rat liver with 2,3-butanediol and type I antifreeze protein. *J. Surg. Res.*, 96, 30–34, 2001.
26. Celinska, E. and Grajek, W., Biotechnological production of 2,3-butanediol—Current state and prospects. *Biotechnol. Adv.*, 27, 715–725, 2009.
27. Ji, X.J., Huang, H., Zhu, J.G., Ren, L.J., Nie, Z.K., Du, J., Li, S., Engineering *Klebsiella oxytoca* for efficient 2,3-butanediol production through insertional inactivation of acetaldehyde dehydrogenase gene. *Appl. Microbiol. Biotechnol.*, 85, 1751–1758, 2010.
28. Sun, L.H., Wang, X.D., Dai, J.Y., Xiu, Z.L., Microbial production of 2,3-butanediol from Jerusalem artichoke tubers by *Klebsiella pneumoniae. Appl. Microbiol. Biotechnol.*, 82, 847–852, 2009.
29. Wu, K.J., Saratale, G.D., Lo, Y.C., Chen, W.M., Tseng, Z.J., Chang, M.C., Tsai, B.C., Su, A., Chang, J.S., Simultaneous production of 2,3-butanediol, ethanol and hydrogen with a Klebsiella sp strain isolated from sewage sludge. *Bioresour. Technol.*, 99, 7966–7970, 2008.
30. Syu, M.J., Biological production of 2,3-butanediol. *Appl. Microbiol. Biotechnol.*, 55, 10–18, 2001.
31. Tran, A.V. and Chambers, R.P., The dehydration of fermentative 2,3-butanediol into methyl ethyl ketone. *Biotechnol. Bioeng.*, 29, 343–351, 1987.
32. Harden, A. and Walpole, G.S., Chemical action of Bacillus lactis aerogenes (*Escherich*) on glucose and mannitol: Production of 2,3-butyleneglycol and acetylmethylcarbinol. *Proc. R. Soc. B.*, 77, 399–405, 1906.
33. Donker, H., *Bijdrage tot de kinnis der boterzuurbutylalkohol in acetongistigen*, PhD Thesis, Delft, 1926.
34. Fulmer, E.I., Christensen, L.M., Kendali, A.R., Production of 2,3-butylene glycol by fermentation. *Ind. Eng. Chem.*, 25, 798–800, 1933.
35. Blackwood, A.C., Wheat, J.A., Leslie, J.D., Ledingham, G.A., Simpson, F.T., Production and properties of 2,3-butanediol: XXXI. Pilot plant studies on the fermentation of wheat by *Aerobacillus polymyxa. Can. J. Res.*, 27 F, 199–210, 1949.
36. Ledingham, G.A. and Neish, A.C., Fermentative production of 2,3-butanediol, in: *Industrial fermentations*, L. Under kofler and R. Hickey (Eds.), pp. 27–93, Chemical Publishing Co, New York, 1954.
37. Garg, S.K. and Jain, A., Fermentative production of 2,3-butanediol—A review. *Bioresour. Technol.*, 51, 103–109, 1995.
38. Zhang, K., Zhang, B., Yang, S.T., Production of citric, itaconic, fumaric, and malic acids in filamentous fungal fermentations, in: *Bioprocessing Technologies in Biorefinery for Sustainable*

Production of Fuels, Chemicals, and Polymers, First Edition, Yang, S.-T., El-Enshasy, H.A., Thongchul, N. (Eds.), 2013.

39. Jong, E., Higson, A., Walsh, P., Wellisch, M., Bio-based Chemicals, Value Added Products from Biorefineries, *IEA Bioenergy—Task42 Biorefinery*, NNFCC, York, United Kingdom, 2013.
40. Rhodes, R.A., Lagoda, A.A., Jackson, R.W., Misenhei, T.J., Smith, M.L., Anderson, R.F., Production of fumaric acid in 20 liter fermentors. *Appl. Environ. Microbiol.*, 10, 9–15, 1962.
41. Kautola, H. and Linko, Y.Y., Fumaric acid production from xylose by immobilized *Rhizopus arrhizus* cells. *Appl. Microbiol. Biotechnol.*, 31, 448–452, 1989.
42. Rodríguez-López, J., Antonio, J.S., Diana, M.G., Aloia, R., Juan, C.P., Fermentative production of fumaric acid from *Eucalyptus globulus* wood hydrolyzates. *J. Chem. Technol. Biotechnol.*, 87, 1036–1040, 2012.
43. Fu, Y.Q., Li, S., Chen, Y., Xu, Q., Huang, H., Sheng, X.Y., Enhancement of fumaric acid production by *Rhizopus oryzae* using a two-stage dissolved oxygen control strategy. *Appl. Biochem. Biotechnol.*, 162, 1031–1038, 2010.
44. Carta, F.S., Soccol, C.R., Ramos, L.P., Fontana, J.D., Production of fumaric acid by fermentation of enzymatic hydrolysates derived from cassava bagasse. *Bioresour. Technol.*, 68, 23–28, 1999.
45. Bercovitz, A., Peleg, Y., Battat, E., Rokem, J.S., Goldberg, I., Localization of pyruvate carboxylase in organic acid producing *Aspergillus* strains. *Appl. Environ. Microbiol.*, 56, 1594–1597, 1990.
46. Battat, E., Peleg, Y., Bercovitz, A., Rokem, J.S., Goldberg, I., Optimization of L-malic acid production by *Aspergillus flavus* in a stirred fermentor. *Biotechnol. Bioeng.*, 27, 1108–1116, 1991.
47. Zhang, A. and Yang, S.T., Engineering *Propionibacterium acidipropionici* for enhanced propionic acid tolerance and fermentation. *Biotechnol. Bioeng.*, 104, 766–773, 2009.
48. Lah, S.L., *Encyclopaedia of petroleum science and engineering*, Gyan Publishing House, India, 2003.
49. van Andel, J., Zoutberg, G., Crabbendam, P., Breure, A., Glucose fermentation by *Clostridium butyricum* grown under a self generated gas atmosphere in chemostat culture. *Appl. Microbiol. Biotechnol.*, 23, 21–26, 1985.
50. Wu, Z. and Yang, S.T., Extractive fermentation for butyric acid production from glucose by *Clostridium tyrobutyricum*. *Biotechnol. Bioeng.*, 82, 93–102, 2003.
51. Bryant, M.P. and Small, N., The anaerobic monotrichous butyric acid producing curved rod-shaped bacteria of the rumen. *J. Bacteriol.*, 72, 16–21, 1956.
52. Centeno, J.A., Menendez, S., Hermida, M., Rodriguez-Otero, J.L., Effects of the addition of *Enterococcus faecalis* in Cebreiro cheese manufacture. *Int. J. Food Microbiol.*, 48, 97–111, 1999.
53. Pulicharla, R., Lonappan, L., Brar, S.K., Verma, M., Production of Renewable C5 Platform Chemicals and Potential Applications. *Platform Chemical Biorefinery*, First Edition, S. K. Brar, S. J. Sarma, K. Pakshirajan (Eds.), Future Green Chemistry, Elsevier, 201–216, 2016.
54. R.C. Nubel and E.D. Ratajak, Process for producing itaconic acid. US Patent 3044941 (to Pfizer), 1964.
55. Ju, N. and Wang, S.S., Continuous production of itaconic acid by *Aspergillus terreus* immobilized in a porous disk bioreactor. *Appl. Microbiol. Biotechnol.*, 23, 311–314, 1986.
56. Yahiro, K., Shibata, S., Jia, S.R., Yahiro, K., Shibata, S., Jia, S.R., Park, Y., Okabe, M., Efficient itaconic acid production from raw corn starch. *J. Ferment. Bioeng.*, 84, 375–377, 1997.
57. Dwiarti, L., Otsuka, M., Miura, S., Yaguchi, M., Okabe, M., Itaconic acid production using sago starch hydrolysate by *Aspergillus terreus* TN484-M1. *Bioresour. Technol.*, 98, 3329–3337, 2007.
58. Willke, T. and Vorlop, K.D., Biotechnological production of itaconic acid. *Appl. Microbiol. Biotechnol.*, 56, 289–295, 2001.
59. Rackemann, D.W. and Doherty, W.O., The conversion of lignocellulosics to levulinic acid. *Biofuel. Bioprod. Bior.*, 5, 2, 198–214, 2011.

60. Weingarten, R., Conner, W.C., Huber, G.W., Production of levulinic acid from cellulose by hydrothermal decomposition combined with aqueous phase dehydration with a solid acid catalyst. *Energy Environ. Sci.*, 5, 6, 7559–7574, 2012.
61. Muranaka, Y., Suzuki, T., Sawanishi, H., Hasegawa, I., Mae, K., Effective production of levulinic acid from biomass through pre-treatment using phosphoric acid, hydrochloric acid, or ionic liquid. *Ind. Eng. Chem. Res.*, 53, 29, 11611–11621, 2014.
62. Revelles, O., Espinosa-Urgel, M., Fuhrer, T., Sauer, U., Ramos, J.L., Multiple and interconnected pathways for L-lysine catabolism in *Pseudomonasputida* KT2440. *J. Bacteriol. Res.*, 187, 21, 7500–7510, 2005.
63. Pukin, A.V., Boeriu, C.G., Scott, E.L., Sanders, J.P., Franssen, M.C., An efficient enzymatic synthesis of 5-aminovaleric acid. *J. Mol. Catal.*, 65, 1, 58–62, 2010.
64. Dutta, S., Ray, S., Nagarajan, K., Glutamic acid as anticancer agent: An overview. *Saudi Pharm. J.*, 21, 4, 337–343, 2013.
65. Yoshikiro, T., Hiroe, Y., Yoshio, R., Fermentative production of L-glutamic acid. *Japan KokaiTokkyoKoho JP*, 78, 1994, 1979.
66. Jyothi, A.N., Sasikiran, K., Nambisan, B., Balagopalan, C., Optimisation of glutamic acid production from cassava starch factory residues using *Brevibacterium divaricatum. Process. Biochem.*, 40, 11, 3576–3579, 2005.
67. Amin, G.A. and Al-Talhi, A., Production of L-glutamic acid by immobilized cell reactor of the bacterium *Corynebacterium glutamicum* entrapped into carrageenan gel beads. *World Appl. Sci. J.*, 2, 1, 62–67, 2007.
68. Ahmed, Y.M., Khan, J.A., Abulnaja, K.A., Al-Malki, A.L., Production of glutamic acid by *Corynebacterium glutamicum* using dates syrup as carbon source. *Afr.*, 7, 19, 2071–2077, 2013.
69. Nampoothiri, K.M. and Pandey, A., Solid state fermentation for L-glutamic acid production using *Brevibacterium* sp. *Biotechnol. Lett.*, 18, 2, 199–204, 1996.
70. Choi, S.U., Nihira, T., Yoshida, T., Enhanced glutamic acid production of *Brevibacterium* sp. with temperature shift-up cultivation. *J. Biosci. Bioeng.*, 98, 3, 211–213, 2004.
71. Nampoothiri, M. and Pandey, A., Immobilization of *Brevibacterium* cells for the production of L-glutamic acid. *Bioresour. Technol.*, 63, 1, 101–106, 1998.
72. Akinterinwa, O., Khankal, R., Cirino, P.C., Metabolic engineering for bioproduction of sugar alcohols. *Curr. Opin. Biotech.*, 19, 461–467, 2008.
73. Harmsen, B. and Annevelink, D 2.1 Background information and biorefinery status, potential and sustainability—Task 2.1.2 Market and Consumers; Carbohydrates, *Star Colibri project*, Wageningen Food & Biobased Research (WUR FBR), Wageningen, Netherlands, 2010. http://www.star-colibri.eu/files/files/Deliverables/D2.1-Report-19-04-2010.pdf
74. Haveren, J., Scott, E.L., Sanders, J., Bulk chemicals from biomass. *Biofuel. Bioprod. Bior.*, 2, 1, 41–57, 2008.
75. Gibson, L., Verdezyne proves adipic acid production process. *Biomass Bioenerg.*, 4, 25, 2010.
76. Papagianni, M., Advances in citric acid fermentation by *Aspergillus niger*: Biochemical aspects, membrane transport and modelling. *Biotechnol. Adv.*, 25, 244–263, 2007.
77. Max, B., Salgado, J.M., Rodriguez, N., Cortes, S., Converti, A., Dominguez, J.M., Biotechnological production of citric acid. *Braz. J. Microbiol.*, 41, 862–875, 2010.
78. Wieczorek, S. and Brauer, H., Continuous production of citric acid with recirculation of the fermentation broth after product recovery. *Bioprocess Eng.*, 18, 75–77, 1998.
79. Wang, J., Wen, X., Zhou, D., Production of citric acid from molasses integrated with *in-situ* product separation by ion-exchange resin adsorption. *Bioresour. Technol.*, 75, 231–234, 2000.
80. Hang, Y.D. and Woodams, E.E., Production of citric acid from corncobs by *Aspergillus niger. Bioresour. Technol.*, 65, 251–253, 1998.

81. Vandenberghe, L.P.S., Soccol, C.R., Prado, F.C., Pandey, A., Comparison of citric acid production by solid-state fermentation in flask, column, tray and drum bioreactor. *Appl. Biochem. Biotechnol.*, 118, 1–10, 2004.
82. Anastassiadis, S. and Rehm, H.J., Citric acid production from glucose by yeast *Candida oleophila* ATCC 20177 under batch, continuous and repeated batch cultivation. *Electron. J. Biotechnol.*, 9, 26–39, 2006.
83. Rywinska, A., Rymowicz, W., Larowska, B., Wojtatowicz, M., Biosynthesis of citric acid from glycerol by acetate mutants of *Yarrowialipolytica* in fed-batch fermentation. *Food Technol. Biotechnol.*, 47, 1–6, 2009.
84. Mac Gillavry, C.H., The crystal structure of adipic acid. *Recueil des Travaux Chimiques des Pays-Bas*, 60, 8, 605–617, 2010.
85. Roffler, S.R., Blanch, H.W., Wilke, C.R., *In-situ* recovery of butanol during fermentation. Part 2: Fed-batch extractive fermentation. *Bioprocess Eng.*, 2, 181–190, 1987.
86. Yang, X. and Tsao, G.T., Enhanced acetone-butanol fermentation using repeated fed-batch operation coupled with cell recycle by membrane and simultaneous removal of inhibitory products by adsorption. *Biotechnol Bioeng.*, 47, 4, 444–450, 1995.
87. West, T. P., Malic acid production from thin stillage by Aspergillus species. *Biotechnol. Lett.*, 33 12, 2463–2467, 2011.
88. Taing, O. and Taing, K., Production of malic and succinic acids by sugar-tolerant yeast *Zygosaccharomyces rouxii. Eur. Food Res. Technol.*, 224, 3, 343–347, 2007.
89. Zelle, R. M., de Hulster, E., van Winden, W. A., de Waard, P., Dijkema, C., Winkler, A. A., Geertman, J. M., van Dijken, J. P., Pronk, J. T., and van Maris, A. J., Malic acid production by Saccharomyces cerevisiae: Engineering of pyruvate carboxylation, oxaloacetate reduction, and malate export. *Appl Environ Microbiol.*, 74, 9, 2766–2777, 2008.
90. Patel, M., Crank, M., Dornburg, V., Hermann, B., Roes, L., Hüsing, B., van Overbeek, L., Terragni, F., Recchia, E., Medium and long-term opportunities and risks of the biotechnological production of bulk chemicals from renewable resources - *The BREW Project. Final report by European Commission's GROWTH Programme (DG Research)*, Utrecht University, Netherlands, 2006.
91. Pandey, M. P. and Kim, C. S., Lignin Depolymerization and Conversion: A Review of Thermochemical Methods. *Chem. Eng. Technol.*, 34, 1, 29–41, 2010.

9

Value-Added Products from Guava Waste by Biorefinery Approach

Pranav D. Pathak[1,2], Sachin A. Mandavgane[1]* and Bhaskar D. Kulkarni[3]

[1]*Department of Chemical Engineering, Visvesvaraya National Institute of Technology, Nagpur, India*
[2]*MIT School of Bioengineering Sciences and Research, Pune, India*
[3]*CSIR- National Chemical Laboratories, Pune, India*

Abstract
Guava (*Psidium guajava*) is cultivated in many tropical and subtropical countries owing to its ability to produce fruits year around. Besides, the fruits are cheap, easy to transport, and has high consumer demand. The fruit is highly nutritious and has medicinal value. Most people consume the fruit afresh. Commercially, it is used in the production of juice, jams, jelly, beverages, canned slices, etc., which leaves behind huge amounts of guava wastes in the form of peels, eaves, bark, seeds, and pomace. However, studies on utilization of wastes have confirmed that guava waste can be a valuable resource for the production of several value-added products. This article critically reviews the recent developments in the utilization of guava waste. We include a summary of the physicochemical characterization of guava wastes, which is important for understanding and identifying the multitude of ways in which the guava waste can be used. The waste is commonly used as an ingredient in animal feeds, essential oils, food ingredients, medicine and nano-material synthesis, and as a water adsorbent, with no or negligible toxicity.

Keywords: Guava waste, valorization, agricultural, value-added products, polyphenol

9.1 Introduction

Food-processing industries are increasing rapidly due to the high demand for processed foods. A serious problem resulting from these setups is the food waste generated during processing of fruits and vegetables for preparing jams, jellies, juices, wines, ice cream, sweets, etc. These wastes generated also cause ecological problems associated with proliferation of insects and rodents, as well as an economical burden due to their transportation to depositories [1, 2]. However, the food manufacturing costs may be reduced if the manufacturer gets some value from these wastes. The wastes from fruit- and vegetable-processing industries are a promising source of bioactive compounds such as dietary fiber (DF), essential

**Corresponding author*: sam@che.vnit.ac.in

Arindam Kuila and Mainak Mukhopadhyay (eds.) Biorefinery Production Technologies for Chemicals and Energy, (163–196)

fatty acids, antioxidants, antimicrobials, and minerals. Thus, the recovery of these valuable compounds is economically beneficial to both the farmers and the manufacturers [3].

Guava (*Psidium guajava* Linn, Family: Myrtaceae) is a semideciduous tropical tree widely grown worldwide for its fruits. Guava is a tropical berry that consists of fleshy pulp and numerous small seeds. The fruit is small and is pear-shaped that turns from green to reddish-yellow upon ripening. Several portions of the fruit are potentially nutritious, including the pulp, peel, and seed [4, 5]. It is the fifth most significant fruit crop in India, comprising about 3.38% of the total land area for fruit cultivation [6]. Guava is cultivated in many tropical and subtropical countries and mainly consumed afresh. Guava fruits (GFs) are usually 4–12 cm long. The fruit is a berry, which holds a fleshy pericarp and a seed cavity containing fleshy pulp and many small seeds. GFs have both nutritional and medicinal values due to the high level of antioxidants, such as vitamin C (40.1 mg g^{-1} to 180.5 mg 100 g^{-1}, which is higher than that in an orange by three to six times), carotenoids (β-carotene and lycopene), and phenolic compounds (anthocyanin and ellagic acid) [7–9].

GFs are mainly used in the production of beverages, syrup, ice cream, cheese, toffee, jams, jellies, juice, wine, and dehydrated and canned products [8, 10, 11]. Following their utilization, a heterogeneous mixture of peels (GPs), seeds (GSs), and residual pulp (GFP) is generated, which accounts for 4%–30% of the total mass (w/w) [12]. Guava leaves (GLs) and tree bark (GB) are the other products obtained from guava tree. In GF processing industries, these products are mainly generated during the stages of crushing, refining, and sieving, namely, guava refiner, guava siever, and guava decanter, respectively. The first portion, guava refiner contains green-brownish fraction that mostly includes GPs and GSs. The guava siever and guava decanter mostly comprise the GFP [8, 10, 12–14].

In most cases, the wastes generated are generally thrown into landfills, which may increase the environmental load. Furthermore, these require additional handling, which increases the total product cost. Thus, it is essential to reuse this waste to produce some valuable products that will increase the profitability of the manufacturing industries as well as that of the farmers, in turn adding value to the waste and subsequently lowering the environmental burden. We herein report the process/methodology to produce some value-added products from GF wastes obtained in various forms. This process/methodology not only enhances the value of waste by producing profitable products but also reduces their burden on the ecological system. We also provide a detailed characterization report of various forms of GF.

9.2 Physicochemical Characterization

The characterization plays a significant role in explaining the morphological (surface) structure of any material in depth as well as provides an insight into its chemical structure.

GF is very nutritional in nature. On an average, GF contains water, dry matter, ash, fat, protein, carbohydrates, minerals, and vitamins. The detailed food value of GF is presented in Table 9.1. Furthermore, GF contains alkaloids, glycosides, flavonoids, steroids, tannins, and saponins [16].

GP contains minerals such as Mg (206.65 ppm), Ca (17.31 ppm), Na (2.04 ppm), phenolic compounds (596.67 mg/L) [4], ascorbic acid [17], and polyphenols. In addition, GS

Table 9.1 Food value of GF [15].

Parameter	Value	Parameter	Value
Calories (g)	77–86	Phosphorous (mg)	0.30–0.70
Moisture (g)	2.8–5.5	Iron (I.U.)	200–400
Crude fiber (g)	0.9–1.0	Carotene (Vitamin A) (mg)	0.046
Protein (g)	0.1–0.5	Thiamin (mg)	0.03–0.04
Fat (g)	0.43–0.7	Riboflavin (mg)	0.6–1.068
Ash (g)	9.5–10	Niacin (I.U.)	40
Carbohydrate (mg)	9.1–17	Vitamin B3 (I.U.)	35
Calcium (mg)	17.8–30	Vitamin G4 (mg)	36–50

contains 14% oil (8.83 ± 0.75), nitrogen-free extract (on a dry basis), phenolic compounds, flavonoids [8, 17], cellulose (31.4%), hemicellulose (14.3%), and lignin (40.2%) [18]. The detailed values of minerals, fatty acid contents, vitamin, and other bioactive contents in GF are shown in Table 9.2.

GL contains fixed oil (6%), volatile oil (0.365%), resin (3.15%), tannin (8.5%) [20] and various antioxidants, whereas GB contains resins, polyphenols, and calcium oxalate crystals. The roots are a source of tannin, leucocyanidins and sterols, carbohydrates, gallic acid, salts, and tannic acid, and the twigs contain various minerals, including calcium, magnesium, iron, potassium, phosphorus, sodium, copper, fluoride, zinc, manganese, and lead [16]. Table 9.4 presents various compounds that are available and can be extracted from various parts of GF using different methodologies. Thus, a wide range of chemicals from GP and GL can be extracted for various purposes.

9.3 Valorization of GW

Because of their varied physicochemical properties, various parts of GW (GP, GS, GFP, GL, and GB) can be used for generating a variety of value-added products. Valorizations of GW is discussed in depth in the following sections and depicted schematically in Figure 9.1.

9.3.1 Medicinal Uses

The extracts of guava plants are a good source of medicine for many human ailments. These medicines have gained increased attention due to their biological and pharmacological properties [21]. Traditionally, guava has been used as a folklore medicine in the treatment of many disorders. In many countries in the world, the extracts of bark, roots, and leaves are used in curing infections caused by bacteria and virus, dysentery, diarrhea, and stomach upsets.

Table 9.2 Physicochemical characcterization of GS [19].

Parameters (g/100 g)	Results	Minerals	Value	Fatty acid	Value
Moisture	6.68 ± 0.00	Calcium	0.05 ± 0.14	Lauric acid	0.07 ± 0.00
Ash (g/100 g)	1.18 ± 0.02	Magnesium	0.13 ± 0.02	Myristic acid	0.10 ± 0.00
Total Lipids (g/100 g)	13.93 ± 0.03	Sulfur	0.09 ± 0.27	Palmitic acid	8.00 ± 0.04
Protein (g/100g)	11.19 ± 0.28	Iron	13.8 ± 2.95	Hepta-decanoic acid	0.07 ± 0.00
Carbohydrate (g/100 g)	3.08	Manganese	0.44 ± 0.47	Stearic acid	4.48 ± 0.17
Pectin (g/100 g)	0.58 ± 0.01	Zinc	3.31 ± 2.52	Oleic acid (n-9)	9.42 ± 0.26
Frutose (g/100 g)	0.29 ± 0.01	Sodium	0.05 ± 0.02	Linoleic acid (n-6)	77.35 ± 0.35
Starch (g/100 g)	0.17 ± 0.00	Potassium	0.20 ± 0.02	Arachidic acid	0.12 ± 0.00
Total Dietary Fiber (g/100 g)	63.94 ± 0.10	Phosphorus	0.30 ± 0.45	Gondoic acid	0.14 ± 0.00
Total Calories (Kcal/100 g)	182			Linolenic acid	0.15 ± 0.00
				Behenic acid	0.10 ± 0.00
Bioactive Compounds					
Vitamin C (mg ascorbic acid/100 g)	87.44 ± 1.70	Carotenoids Totals (mg/100 g)	1.25 ± 0.14		
Soluble Dietary Fiber (g/100 g)	0.39 ± 0.02	Insoluble Dietary Fiber (g/100 g)	63.55 ± 0.12		

Values on % dry basis.

9.3.1.1 GL, GB, and GF in Medicines

GL has good potential to prevent several cardiovascular and neurodegenerative diseases. In addition, the aqueous extracts of GL and GB are effective as a cough suppressant, central nervous system depressant, sedative, hypotensive (lowering blood pressure), and analgesic (pain reliever). These free radical scavenging, immunostimulatory, and antioxidative

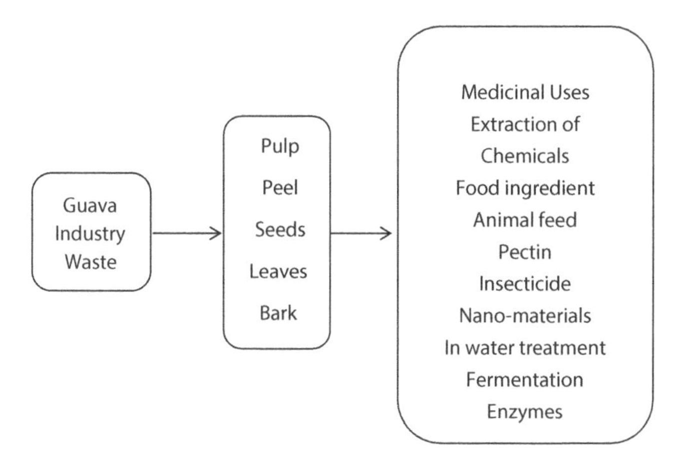

Figure 9.1 Value added products from guava industry waste demonstrate the impact of considered factors on the production of value added products.

properties are due to the presence of polyphenols, flavonoids, and terpenoids [21–23]. GL tea is commonly used as a local medicine against gastroenteritis and diarrhea in children [24]. When boiled with rice, GL and GB (rice ORS) offer an alternative to antibiotics in treating cholera and support the recovery process (convalescence period) by functioning as an electrolyte and fluid replacement. This combined effect helps in saving the lives of those suffering from cholera and diarrhea, especially those living in remote areas with limited access to vaccines [25]. People from Oriental countries like Japan and Taiwan boil the water with GL and drink the extract as a folk medicine for diabetes. According to Cheng *et al.* (2009), quercetin present in the aqueous extract of GLs enhances glucose uptake by liver cells and contributes to the improvement of hypoglycemia [26].

Moreover, guava is also used in the treatment of gastrointestinal disturbances and respiratory disorders. Compounds β-caryophyllene (17.6%) and limonene (11.0%) from GF oil plus β-caryophyllene (16.9%) and selin-7(11)-en-4α-ol (8.3%) from GL oil are common ingredients in various medicines. El-Ahmady *et al.* (2013) extracted various oils from GF and GL by hydrodistillation and determined their antioxidant and anti-inflammatory actions along with their cytotoxicity to HepG2 and MCF-7 carcinoma cells to expound the ethnopharmacological uses of guava [27].

Ademiluyi *et al.* (2015) investigated the antihypertensive and antioxidant activities of different varieties (giant white, small white, stripped, and pink) of GF and GL extracts [extraction by methanol/1 M HCl (20:1 v/v)]. The extracts significantly ($P < 0.05$) inhibited scavenging radicals (DPPH, ABTS•+, nitric oxide, and hydroxyl), ACE activity, chelated Fe^{2+} along with inhibition of Fe^{2+} and SNP-induced lipid peroxidation in the heart of rats (*in vitro*). Rosmarinic acid, carvacrol, eugenol, catechin, and caffeic acid were identified as the active phenolic compounds in these extracts [28], studied the antimicrobial effect of GL (ethanol, water, and acetone-diluted) extracts using the bacterial strains *Staphylococcus aureus* and *Escherichia coli* isolated from fresh shrimps and concluded that extracts of guava sprout can be used in the possible treatment of diarrhea caused by *Staphylococcus aureus* and *Escherichia coli*–produced toxins [29]. Joseph and Priya (2016) reported that the methanolic extract of GL has a cytotoxic effect on human cervical cancer cells and provides a scientific basis for using crude extracts of guava in herbal medicine [30]. The GL also exhibits anticancer activity (anti–DU-145 bioactivity) due to the presence of large amounts of polyphenolic compounds, includinggallic acid (348 mg/g), catechin (102 mg/g), epicatechin (60 mg/g), rutin (100 mg/g), quercetin (102 mg/g), and rutin (100 mg/g) [22]. GL extracts shows antifatty liver effects induced by ethanol in rats [31]. The GL extracts show preventive effects on inflammatory biomarkers (NO and PGE2 production and iNOS and COX-2 expression) in LPS-stimulated RAW264.7 cells [21]. Im *et al.* (2012) investigated the molecular mechanism responsible for the possible antimetastatic and anti-invasive effects of butanol extracts of GL. The extraction was done in a sonication bath for 3 days at room temperature. The butanol fraction of GL extracts was found to reduce the metastasis of lung cancer cells. D-Glucuronic acid, loganin, quercetin 3-glucuronide, and xanthyletin are the major compounds in identified GL extracts [32]. The yield of hydrodistilled oil-from GL was 1.6 g/kg on fresh weight basis; limonene (54.70%) was the major compound identified, with 1,8-cineole identified as the major oxygenated monoterpenoid (32.14%) in common GL. The hydrodistilled oil shows antibacterial activity against *Bacillus subtilis, Staphylococcus aureus, Escherichia coli, Streptococcus faecalis, Pseudomonas aeruginosa,* and *Neisseria gonorrhoeae*. Furthermore, it showed antifungal activity against *Candida albicans* and *Aspergillus flavus* [33]. GL extracts (hexane, methanol, ethanol, and water) exhibit antimicrobial effect by preventing the growth of Gram-negative bacteria (*Escherichia coli* and *Salmonella enteritidis*) and Gram-positive bacteria (*Staphylococcus aureus, Bacillus subtilis,* and *Bacillus cereus*). Besides, GL displays high scavenging effect (90%) as determined by DPPH [34, 35]. Tannin extracts (ethanol) from GLs exhibit inhabitation effects against *Escherichia coli, Staphylococcusaureus, Pseudomonas aeruginosa, Aspergillus niger,* and *Candida albicans* [36]. The ethanolic extracts of GL have significant antidiabetic and antidiarrheal activities in Wister rats. An oral dose of 1.00 and 0.50 g/kg of GL extract meaningfully ($P < 0.05$) reduced their blood glucose levels [37]. The methanoic extracts of GL exhibited effective antidiabetic activity in a type 2 diabetic mice model by inhibiting PTP1B activity. Using an extract dose of 10 mg/kg in 1- and 3-month-old Leprdb/Leprdb mice, substantial reduction in blood glucose level was observed after intraperitoneal injection [38]. The aqueous and ethanol extracts of GL show a significant drop in the blood sugar level in diabetic rats. Long-term use of GL extracts increases the plasma insulin level and glucose utilization in diabetic rats [39].

Moreover, the aqueous extracts of GLs reduced the frequency of cough induced by capsaicin aerosol by 35% and 54% in rats and guinea pigs, respectively. Jaiarj *et al.* (1999) reported that the extract directly stimulated muscle contraction and synergized by the stimulatory result of pilocarpine, which was antagonized by an atropine. This water, methanol, and chloroform extract of GL can also inhibit the growth of *Staphylococcus aureus* and β-streptococcus group A. Thus, the leaf extract can be used as a remedial measure for cough [40]. Gonçalves (2008) studied the antimicrobial effect of essential oils extracted from GL as well as that of methanol, hexane, ethyl acetate extracts against diarrhea-causing bacteria (*Salmonella* spp., *Staphylococcus aureus,* and *Escherichia coli*). These bacterial strains were isolated from seabob shrimp, *Xiphopenaeuskroyeri* (Heller), along with laboratory-type strains. It was concluded that, among the strains studied, *Staphylococcus aureus* was the most inhibited by the extracts, with the methanol extract being the most effective. Moreover, the essential oils extracted show good inhibitory effects against *S. aureus* and *Salmonella* spp. These results show that GL extracts and essential oils inhibit very active bacterial strains, thereby making them a potential source for developing new antimicrobial compounds [24].

GL extracts have potential antiglycative and anticoagulant agents, which demonstrate preventive properties against glycation-associated cardiovascular diseases in diabetes due to the presence of phenolic compounds such as ferulic acid, gallic acid, and quercetin [41]. Methanol extracts of GL contain quercetin, morin, and quercetin-3-O-glucopyranoside, which are responsible for the antioxidant activity. Nantitanon (2012) concluded that quercetin was the most active antioxidant from GL; importantly, addition of quercetin to chitosan nano-particles significantly improved its stability [42].

Uduak (2012) concluded that the methanol extracts of GL are safe in the treatment of ethanol-produced gastric ulcer in adult Wistar rats [43]. Irondi *et al.* (2016) reported the presence of flavonoids (quercetin >kaempferol>catechin>quercitrin>rutin>luteolin>epicatechin) and phenolic compounds (caffeic acid >chlorogenic acid >gallic acids) in the GL extracts. These extracts efficiently withdrew XO-, ACE-, and Fe^{2+}-induced lipid peroxidation in a dose-dependent manner. The inhibition of xanthine oxidase and angiotensin 1-converting enzyme activity along with scavenging of free radicals may be underlying mechanisms for the antigout and antihypertensive effects of GL extracts, which are attributed to the presence of flavonoids and phenolic acids. Therefore, GL can be a potential food ingredient for the treatment of gout and hypertension as well as in the reduction of oxidative stress related to these two ailments [44]. Ethanol extracts (75%) from GL were further extracted using CH_2Cl_2, EtOAc, and n-BuOH to obtain four different fractions: CH2Cl2-soluble, EtOAc-soluble, n-BuOH-soluble, and remaining fractions. Among these, n-BuOH-soluble and EtOAc-soluble fractions displayed extraordinary inhibitory activity toward α-glucosidase and α-amylase. This supports the use of GL in in the treatment of diabetes (traditional medicine) [45]. Athikomkulchai *et al.* (2008) extracted volatile oils from GL by hydrodistillation and analyzed their properties. γ-Terpinene and α-pinene were the main chemical constituents in these oils. The extracted volatile oil shows good antimicrobial activity against *Propionibacterium* acnes (MIC = 9.38 mg/ml, MBC = 37.50 mg/ml for guava oil). Prepared oil-in-water creams incorporating 4% w/w guava oils show good permanence after kept at −4°C and 45°C alternately (freeze-thawing) for four cycles. The monoterpenes (α-pinene, sylvestrene), oxygenated monoterpenes (1,8-cineol), sesquiterpenes (E-caryophyllene, aromadendrene, α-humulene, β-bisabolene), oxygenated sesquiterpenes [E-nerodiol, caryophylleneoxide, caryophylla-4(12), 8(13)-dien-5-alpha-ol] are the other chemical constituents of GL volatile oil obtained by hydrodistillation [45].

Saraya *et al.* (2008) prepared chewable tablets against *Streptococcus mutans* using GL extracts (boiling in water for 15 minutes and lyophilized to be the powder). *Streptococcus mutans* is a usual floral bacterium in human oral cavity and is known for causing dental caries and bad breath odor. Crude extract, PVP K30 (as 10% solution), mannitol, AEROSIL, magnesium stearate, peppermint, and menthol are the ingredients of guava chewable tablets. These tablets show excellent killing effect against *S. mutans* [47].

It was noted that the nutrients in GL vary with season and cultivators. Very high levels of nutrients were observed just after leaf appearance and relatively stable levels were observed at leaf age of 180–240 days, which is also the best time for sample collection [48].

9.3.1.2 GP in Medicines

The antioxidant activities of GP, guava pulp, and GS were compared using FRAP assay and the values obtained were as follows: GP, 10.24 ± 0.24 mmol/100g wet weight; pulp, 6.07 ±

0.69 mmol/100 g wet weight; and GS, 4.71 ± 0.24 mmol/100 g wet weight [49]. Besides, the aqueous extracts of GP show hypoglycemic and antidiabetic effect on blood glucose level in rats [5]. Budin *et al.* (2013) suggested that the aqueous extracts of GP have the potential to decrease oxidative stress in the pancreas of diabetic rats [50]. At optimal conditions (60% methanol at 55°C and 120 minutes) the extracts obtained from GP exhibit the highest antioxidant activity (1,021.00 μmol/L) and phenolic content (596.67mg/L). In addition, GP contains minerals such as Mg (206.65 ppm), Ca (17.31 ppm), and Na (2.04 ppm) [4]. According to Abdelmalek (2016), gallic acid and ferulic acid in the aqueous extracts of GP exhibit antimicrobial activity against *S. aureus* and bacteria such as *E. coli*, *P. aeruginosa*, and *L. monocytogenes* [51]. The supplementation of ripe GF with GP reduces BMI and blood pressure ($p < 0.05$) while the levels of FPG, total cholesterol, andtriglycerides was significantly reduced ($p < 0.05$) in the first group; by contrast, the second group showed successful reduction in BMI and blood pressure as well as significant falls ($p < 0.05$) in the levels of serum total cholesterol, triglycerides, and low-density lipoprotein cholesterol. These results suggest that GF without peel is more effective in reducing blood sugar along with serum total cholesterol, triglycerides, and low-density lipoprotein cholesterol [52]).

Pharmacological studies on GW (leaves, peel, bark, seeds) demonstrate its diverse applications for the treatment of diarrhea, gastroenteritis and rotavirus enteritis, wounds, acne, malaria, allergies, coughs, diabetes, degenerative muscular diseases, cardiovascular disorder, dental plaque, inflammatory ailments including rheumatism, menstrual pain, cancer, liver diseases, bacterial/fungal infections, etc. However, guava extracts and derived phytochemicals (phenolics, carotenoids, flavonoids, triterpenes, and essential oil constituents) have low toxicity. Some of the compounds in guava leaves, pulp, seed, skin, and bark and their pharmacological effects are presented in Table 9.3.

Table 9.3 Compounds in guava leaves, pulp, seed, skin, and bark and their pharmacological effects [53].

Skin	Phenolic compounds	Endothelial progenitor cells and improvement of their intestinal absorption
Seed	Glycosids; Carotenoids, phenolic compounds	Antimicrobial
Leaves	Phenolic compounds, isoflavonoids, gallic acid, catechin, epicathechin, rutin, naringenin, kaempferol	Hepatoprotection, antioxidant, anti-inflammatory, anti-spasmodic, anti-cancer, antimicrobial, anti-hyperglycemic, analgesic
Pulp	Ascorbic acid, carotecoids (lycopene, β-carotene, β-cryptoxanthin	Antioxidant, anti-hyperglycemic, anti-neoplasic
Bark	Phenolic compounds	Strong antibacterial activity (against multi-drug- resistant Vibrio cholera); stomachache and diarrhea

9.3.2 Extraction of Chemicals

GF and the various wastes obtained from it contain various compounds such as tannins, polyphenols, flavonoids, terpenoids, glycosides, among others. These compounds can add value to the waste if extracted efficiently. Various studies have tried to extract many chemicals from available guava waste using different methodologies. Phenolic compounds are very important chemicals as they exhibit antioxidant, antiatherogenic, antiallergenic, antimicrobial, anti-inflammatory, antithrombotic, cardioprotective, and vasodilatory properties [8]. Thus, their extraction from plant waste by-products is economically beneficial.

The ethanol extracts of guava pomace containphenolic acids such as gallic acid, isovanillic acid, m-coumaric acid, epicatechin, and quercetin [17]. About 50 compounds were isolated from essential oil obtained by hydrodistillation of GF, with the major constituents being β-caryophyllene (27.7%), α-pinene (14.7%), and 1,8-cineole (12.4%; see Table 9.4 [54]. Sukeksi *et al.* (2016) extracted the polyphenols from pink GW using methanol, ethanol, acetone, acetonitrile, and water as a solvent and suggested that 60% methanol/water was the best solvent for polyphenol extraction [55]. The response surface methodology (RSM) was successfully implemented for the extraction of phenolic compounds from pink GW from puree industry by Kong *et al.* (2010). The maximum yield of phenolic content was obtained at pH 2, temperature 60°C, and for 5 hours [13].

9.3.2.1 Extraction from GL

Ethanol extracts of GL contain bioactive compounds such as gallic acid, (+)-catechin, chlorogenicacid, rutinhydrate, caffeicacid, q-coumaricacid, luteolin7-glucoside, myricetin, quercetin, luteolin, and kaempferol [21]. Ethanol/water (80:20, v/v) is the best solvent for the extraction of phenolic compounds, especially for flavonols and flavan-3-ols, from GL [56]. Other solvents such as n-hexane, ethyl acetate, n-butanol, and methanol can also be used for extraction of polyphenolic compounds from GL [57]. Tannin is another important polyphenolic compound and can be extracted from GL using ethanol, acetone, water, and n-hexane as solvents [58]. Among these, 30% ethanol is the best solvent to extract tannins from GL (yield achieved, 2.351 mg/g) [36].

The pre-treatment, drying process, extraction method, and leaf maturity are important factors that should be considered when evaluating the antioxidant power of bioactive compounds from GL extracts. Ultrasonication is the best method for extraction of polyphenols from GL followed by Soxhletextraction and maceration. Ultrasound bath and ethanol:water [80/20 (v/v)] as extractant solvent are the best systems for extraction. Díaz-de-Cerio *et al.* (2015) isolated 72 phenolic compounds from GL (Table 9.4). Moreover, Nantitanon *et al.* (2010) added that the maturity stages of GL and solvent for extraction are some other essential parameters. Newly grown GLs are best for extraction followed by old and middle leaves, respectively [56, 59]. Moura *et al.* (2012) obtained greater yields of essential oils, flavonoids, antioxidant compounds, and β-carotene from GL by supercritical extraction (temperature: 30°C and 50°C, pressure: 300 bar) compared with other process such as Soxhlet extraction, low-pressure solvent extraction, hydrodistillation, and ultrasound extraction [60]. According to Tuan *et al.* (2016), microwave (ethanol as solvent) increases the phenolic and flavonoid content by 12.7% and 36.5%, respectively, compared with the solid-liquid extraction of biochemicals from GLs. Microwave extraction also increases the efficiency of

Table 9.4 The chemicals available in various parts of GF.

Part of guava	Compounds	Reference
Guava pomace	Bioactive phenolic compounds: Phosphoric acid, tristrimethylsilyl, Beta-caryophyllene, Malic acid (TMS), Alpha-selinene, Trimethylsilyl 3-phenyl-2-propenoate, Isovanilic acid, Tris (trimethylsilyl) 2-[(trimethylsilyl)oxy]-1,2,3-propanetricarboxylate, Beta-l-mannofuranose, 6-deoxy-1,2,3,5-tetrakis-o-(trimethylsilyl), Glucofuranoside, methyl 2,3,5,6-tetrakis-o-(trimethylsilyl)-, alpha-d-, Gallic acid, Hexadecanoic acid, trimethylsilyl ester, Oleic acid, trimethylsilyl ester, Epicatechin, Myricetin, Quercetin	[74]
Pink GP	Phenolic acid derivatives: Galloyl-hexoside, Galloyl-hexoside, Gallic acid, Galloyl-pentoside, Hydroxybenzoyl-galloylglucoside, Dimethoxycinnamoyl-hexoside Flavones: Chrysin-C-hexoside Ellagitannins: Valoneic acid bilactone, valoneic acid Flavonols: Quercetin-galloyl-hexoside, Quercetin-hexoside, Quercetin-glucuronide, Quercetin-pentoside, Quercetin-pentoside, Quercetin-galloyl-pentoside (guavinoside C), Quercetin-deoxyhexoside-hexoside, Quercetin Monomeric Flavonols: Gallocatechin, Epigallocatechin, Catechin, Epicatechin, Gallocatechingallate, Epigallocatechin gallate, Catechingallate, Epicatechingallate Proanthocyanidins Dihydrochalcones: Phloretin-C-glucoside (nothofagin), Phloretin-O-glucoside:phlorizin, Stilbenes: Piceatannol-O-Glucoside (astringin) Acetophenones: Myrciaphenone B Benzophenones: Guavinoside A, Guavinoside B - isomer, Guavin B - isomer, Glucopyranosyl-benzophenone Other polar compounds: Cinnamoyl-hexoside, Abscisic acid - hexoside, Abscisic acid Anthocyanidins: Cyanidin-3-O-glucoside	[65]

(Continued)

Table 9.4 The chemicals available in various parts of GF. (*Continued*)

Part of guava	Compounds	Reference
Constituents identified from volatile oil of the GL	α-Pinene (1.53%), Benzaldehyde (0.83%), p-Cymene (0.52%), Limonene (54.7%), 1,8-Cineole (32.14%), β-cis-Ocimene (0.28%), γ -Terpinene (0.38%), α-Terpineol (1.79%), β-Caryophyllene (2.91%), α-Humulene (0.77%) Hydrocarbons: Monoterpenes (58.24%), Sesquiterpenes (3.68%) Oxygenated compounds: Monoterpenes (33.93%)	[33]
Leaf oil of *Psidium guajava*	Aliphatic Alcohols: 3-hexenol, 2-hexenol Aliphatic Aldehydes: hexanal, 2-hexenal, benzaldehyde Aliphatic Esters: 3-hexenyl acetate Aliphatic Ketones: 6-methyl-5-hepten-2-one Terpene Alcohols: linalool, α-fenchol, trans-pinocarveol, borneol, terpinen-4-ol, α-terpineol, (E)-nerolidol, spathulenol, globulol, ledol, α-cadinol Terpene Alcohols: caryophyllenol, farnesol Terpene Esters: α-fenchyl acetate, bornyl acetate, neryl acetate, geranyl acetate Terpene Ketone: pinocarvone Terpene oxides: 1,8-cineole, caryophyllene oxide Turpene Hydrocarbons: α-thujene, α-pinene, camphene, b-pinene, myrcene, α-phellandrene, α-terpinene, ρ-cymene, limonened, (Z)-β-ocimene, γ-terpinene, terpinolene, allo-ocimene, α-muurolene, α-copaene, β-caryophyllene, aromadendrene, α-humulene, allo-aromadendrene, (Z)-α-bisabolene, β-bisabolene, δ-cadinene, (E)-γ-bisabolene	[54]

(*Continued*)

Table 9.4 The chemicals available in various parts of GF. (*Continued*)

Part of guava	Compounds	Reference
Phenolic compounds in Psidium guajava L. leaves	HHDP glucose Isomer, HHDP glucose Isomer, HHDP glucose Isomer, Prodelphinidin B2 Isomer, Gallic acid, Pedunculagin/Casuariin Isomer, Prodelphinidin Dimer Isomer, Gallocatechin, Vescalagin/Castalagin Isomer, Prodelphinidin Dimer Isomer, Uralenneoside, Geraniin Isomer, Pedunculagin/Casuariin Isomer, Geraniin Isomer, Procyanidin B Isomer, Galloyl(epi) catechin-(epi)gallocatechin, Procyanidin B Isomer, Tellimagrandin I Isomer, Pterocarinin A, Pterocarinin A Isomer, Stenophyllanin A, Procyanidin trimer Isomer, Catechin, Procyanidin tetramer, Procyanidin trimer Isomer, Guavin A, Casuarinin/Casuarictin Isomer, Galloyl(epi) catechin-(epi)gallocatechin, Procyanidinpentamer, Galloyl-(epi)catechin trimer Isomer, Gallocatechin, Tellimagrandin I Isomer, Vescalagin, Stenophyllanin A Isomer, Galloyl-(epi) catechin trimer Isomer, Myricetinhexoside Isomer, Stachyuranin A, Procyanidingallate Isomer, Myricetinhexoside Isomer, Vescalagin/castalagin Isomer, Myricetin-arabinoside/xylopyranoside Isomer, Myricetin-arabinoside/xylopyranoside Isomer, Procyanidingallate Isomer, Myricetin-arabinoside/xylopyranoside Isomer, Myricetinhexoside Isomer, Myricetin-arabinoside/xylopyranoside Isomer, Quercetin-galloylhexoside Isomer, Ellagic acid deoxyhexoside, Quercetin-galloylhexoside Isomer, Myricetin-arabinoside/xylopyranoside Isomer, Morin, Myricetin-arabinoside/xylopyranoside Isomer, Ellagic acid, Hyperin, Quercetin glucuronide, Isoquercitrin, Procyanidingallate Isomer, Reynoutrin, Guajaverin, Guavinoside A, Avicularin, Quercitrin, Myrciaphenone B, Guavinoside C, Guavinoside B, Guavinoside A Isomer, Guavinoside B Isomer, 2,6-dihydroxy-3-methyl-4-O-(6″-O-galloyl-β-D-glucopyranosyl)-benzophenone, Guavin B, Quercetin, Naringenin	[56]

(*Continued*)

Table 9.4 The chemicals available in various parts of GF. (*Continued*)

Part of guava	Compounds	Reference
Guava leaves ultrasound aqueous extract	HHDP glucose isomer, HHDP glucose isomer, HHDP glucose isomer, Gallic acid, Prodelphinidin B2 isomer, Pedunculagin/casuariin isomer, Prodelphinidin dimer isomer, Gallocatechin, Prodelphinidin dimer isomer, Geraniin isomer, Pedunculagin/casuariin isomer, Geraniin isomer, Procyanidin B isomer, Procyanidin B isomer, Tellimagrandin I isomer, Catechin, Casuarinin/casuarictin isomer, Tellimagrandin I isomer, Gallocatechin, Myricetin-arabinoside/xylopyranoside isomer, Procyanidin B isomer 2, Myricetinhexoside isomer, Myricetinhexoside isomer, Myricetin-arabinoside/xylopyranoside isomer, Quercetin-galloylhexoside isomer, Quercetin-galloylhexoside isomer, Myricetin-arabinoside/xylopyranoside isomer, Morin, Myricetin-arabinoside/xylopyranoside isomer, Ellagic acid, Hyperin, Quercetin glucuronide, Isoquercitrin, Reynoutrin, Guajaverin, Guavinoside A, Avicularin, QuercitrinGuavinoside C, Guavinoside B, Guavinoside A isomer, Prodelphinidin B2 isomer 2, Guavinoside B Isomer, Guavinoside B isomer, Guavinoside B isomer, Guavin B, Quercetin, Naringenin	[75]
Supercritical fluid extraction of GL	n-Octane(0.36 %), m-Xylene (0.40 %), 1,8-Cineole (0.82 %), β-Phellandrene (2.98 %), α-Copaene (0.88 %), β-Caryophyllene (18.77 %), Aromadendrene (1.89 %), Gcrmacrene-D (0.55 %), β-Selinene (1.91 %), α-Selinene (23.70 %), δ-Sclinene (18.28 %), (E)-Nerolidol (0.71 %), Spathulenol (1.76 %), Caryophyllene epoxide (5.16 %), δ-Cadinol (0.47 %), *allo*-Aromadendrene (6.03 %), Ethyl hexadecanoate (1.43 %)	[63]
Sulfur Volatiles from GL	2-ethylfuran, 3-pentanone, methyl 2-methylbutyrate, methyl 3-methylbutyrate, α-pinene ethyl vinyl ketone, hexanal, E-2-pentenal, Z-3-hexenal, isobutyl 2-methylbutyrate, isobutyl 3-methylbutyrate, isoamyl butyrate, isoamylisobutyrate, limonene, E-2-hexenal, (Z)-β-ocimene, isoamyl 2-methylbutyrate, 2-methylbutyl 2-methylbutyrate, isoamyl 3-methylbutyrate, Z-2-pentenol, unidentified 3-methyl-3-butenyl 3-methylbutyrate, Z-3-hexenol, neo-allo-ocimene, (E,Z)-2,4-hexadienal, (E,E)-2,4-hexadienal, α-p-dimethylstyrene, Z-3-hexenyl 2-methylbutyrate, Pentadecane, Z-3-hexenyl 3-methylbutyrate, α-copaene, cyclohexyl 3-methylbutyrate, benzaldehyde, β-elemene, unidentified β-caryophyllene, β-farnesene, 3-methylbutyric acid, methyl geranate, (Z,E)-α-farnesene, β-bisabolene, (E,E)- α-farnesene, geranyl acetate, curcumene, 5-ethyl-2(5H)-furanone, geranyl propionate, geranyl butyrate, geranyl isovalerate, isoamyl benzoate, unidentified	[76]

extraction, reduces extraction time, and increases the antioxidant activity, as this method damages plant cells more substantially [61]. Compounds such as gallic acid (0.87 ± 0.02%), caffeic acid (0.11 ± 0.01%), chlorogenicacid (0.62 ± 0.05%), catechin (2.25 ± 0.29%), epigallocatechin gallate (0.47 ± 0.06%), epicatechin (1.45 ± 0.13%), and quercetin (0.83 ± 0.06%) are obtained from GL by ultrasound-assisted extraction [62]. Sagrero-Nieves *et al.* (1994) extracted volatile compounds such as n-octane, m-xylene, 1,8-cineole, β-phellandrene, α-copaene, β-caryophyllene, germacrene-D, aromadendrene, β-selinene, α-selinene, δ-sclinene, spathulenol, (E)-nerolidol, caryophylleneepoxide, δ-cadinol, do-aromadendrene, and ethyl hexadecanoate from GLs using supercritical fluid extraction [63]. According to Rodriguez (1967), P, Ka, and N content in the GL reduces definitely with age while Ca, Mg, Mn, Na, and Al content increases with age. The other nutrients are variants other than B, it tends to increase in the middle position and decreases at beginning there is not a large difference between fruiting and non-fruiting terminals was observed. Chemical composition was found to be steady within the fourth and seventh positions in the terminals as age is an important parameter affecting parameter [64].

9.3.2.2 *Extraction from GP*

Among polyphenols, cinnamoyl-O-hexoside is the main polar compound in pink guava followed by benzophenones, guavin B isomer, and abscisic acid. Among flavonoids, quercetin-O-hexoside is the important compound found in GP followed by quercetin-O-pentoside and adimethoxycinnamoyl-O-hexoside. The higher concentration of polyphenols in the GP is due to its more increased to sunlight. Rojas-Garbanzo *et al.* (2017) found about 42 more compounds in GP (see Table 9.4) [65].

9.3.2.3 *Extraction from GS*

Roasting is a thermal process by which the seed gets the typical color, flavor, and taste. Roasting also changes the chemical composition and alters the nutritional values and shelf life. Roasting of GSs reduces the moisture content, crude protein, crude fiber, arginine, isoleucine, glutamic and total aromatic and sulfur amino acids, antinutritional factors (tannins and phytic acid), flavonoids, ash, and mineral contents though the oil content [66]. Amino acid composition (g/100 g protein) of GS is as follows: isoleucine (4.05), leucine (5.93), lysine (0.91), cystine (2.13), methionine (3.52), total sulfur amino acids (5.65), tyrosine (4.12), phenylalanine (2.61), total aromatic amino acids (6.73), threonine (3.93), tryptophan (2.64), valine (5.35), total essential amino acids (35.19), histidine (2.02), arginine (8.52), aspartic acid (5.49), glutamic acid (9.06), serine (5.98), proline (8.65), glycine (18.75), alanine (6.34), and total non-essential amino acids (64.81) [66]. According to Castro-Vargas *et al.* (2010), supercritical fluid extraction (using CO_2) with ethyl acetate and ethanol as co-solvents is the best method for the extraction of phenolic fraction from GS. The temperature and pressure applied affect the extraction yield [67]. Glutelin is the main protein fraction from GS, which can be extracted using Na_2B4O_7, 2-mercaptoethanol, 2-mercaptoethanol/NaOH. The extracted glutins show high values for several properties such as surface hydrophobicity, solubility at pH 10, water-holding capacity at pH 3.6, emulsifying activity index, and foaming properties [68, 69].

In the oil extracted from GS using solvent $CHCI_3$:MeOH (2:1), the total phenolic contents were found to be 9.73% of the dry weight. This also confirms the presence of 15 amino acids in the protein of GS (in ascending order): lysine, threonine, proline, isoleucine, histidine, alanine, phenylalanine, tyrosine, serine, valine, leucine, glycine, aspartic acid, glutamic acid, and arginine. Among these arginine, glutamic acid, aspartic acid, glycine, and leucine are the major amino acids in guava protein contributing about 67% of the total amino acids [70]. Castro-Vargas *et al.* (2011) extracted the GS oil using a homemade supercritical fluid extraction system in which pure CO_2 and CO_2 with co-solvents were used. Supercritical fluid extraction system with ethanol as co-solvent gives the highest yield of GS oil (17.30% w/wat 30MPa pressure, 313 K temperature, and 30 minutes), which is close to the oil yield achieved with Soxhlet extraction (20.2% w/w) [71].

The pre-treatment step plays an important role in the determination of phenolic compounds (i.e., the antioxidants in the fruit also regulates the ability of sample storage) [4]. Besides, compared with solar drying, oven drying produces a large amount of bioactive substances [72]. In addition to the pre-treatment method, factors such as maturity of plant material, particle size, solvent polarity, extraction procedures, and conditions also affect the extraction of phenolic compounds from any plant material [8, 59, 73].

9.3.3 Food Supplements

The wastes obtained from guava industries have good nutritional value (protein, 4.8%; ash, 2.4%; fat, 1.4%; carbohydrates, 22.2%; and moisture, 9.3). They comprise about 50% of DFs and so can be utilized as raw material to produce DF-rich nutritional food products [77]. One example is DF-rich yogurt, which is prepared by mixing GS powder (1.4%) and yogurt culture (3.86%) at an incubation temperature of 39.45°C. Similarly, a mixture of GS powder (17.65 g), wheat flour (62g), and sugar (20g) can be used to produce biscuits rich in DF [78]. GS flour can also be used for producing supplemented pasta. This resulted in increased amounts of protein, fat, ash, moisture, and crude fiber, with lowered carbohydrate values. The replacement upto 10% of GS flour does not responsible for the cooking losses. Also, the physical assessment showed that, with replacement level up to 30%, 20% and 10% does not affects the stickiness, appearance, flavor, and tenderness, respectively [79].

The GP flour can replace upto 30% wheat flour in the preparation of cookies, which is the degree to which various attributes (e.g., aroma, flavor, and texture) remain unaffected. This replacement shows good nutritional advantages such as increased amounts of fiber and protein and decreased levels of fat and carbohydrates. GP flour shows large amounts of total polyphenols, lycopene, and β-carotene, which remain conserved during the whole process. Addition of increased amount of GP flour enhances the levels of phenol and β--carotene in the product while it also does not show in differences in terms of color, flavor, and appearance. However, when wheat flour is replaced up to 50% and 70%, only the aroma is retained. Thus, GP flour can significantly increase the nutritional value of cookies without affecting the quality of product [2].

Jellies can be prepared from GP. According to Tavares *et al.* (2015), the formulations prepared using higher amounts of the extract were accepted better in relation to their sensory attributes such as color, appearance, tasty, aroma, texture, and global impression [77].

Shams El-Din and Yassen prepared cookies from GS meal and recommended wheat flour replacement of upto 9%, which is acceptable and has a positive effect on specific volume, cookie volume, diameter, and thickness; however, compared with the control sample this is of low grade. The GS meal is also a good source of fiber and iron and the GS oil is a good source of linoleic acid, an essential fatty acid [80].

Tangirala *et al.* (2012) suggested addition of guava pomace up to 10% to ready-to-eat snacks along with rice flour. The addition of guava pomace resulted in rise in hardness, water absorption index, and lowered expansion, bulk density, and water solubility index [3].

The addition of guava pulp powder to gluten-free bread is advantageous for patients with celiac disease who require gluten-free diet. The addition of guava pulp powder non-considerably affects the protein content and calorific value of gluten-free bread. According to Arslan *et al.* (2017), the guava pulp powder supplementation enhances the nutritional composition of the gluten-free bread. Inclusion of GP powder upto 5% does not show any adverse effects on the quality attributes of gluten-free bread, but after this level, it shows adverse effects on its color, volume, textural, and sensorial properties [81].

9.3.4 Extraction of Pectin

Pectin is an important chemical that is widely used in the food industry as thickening, emulsifying, and gelling agent in the manufacture of jam, sauce, preservatives, jellies, dairy products etc. Besides, pectin has a wide range of applications in cosmetics, pharmaceutical preparations, pastes, etc. Guava press contains the highest amount of calcium pectate (13.03%). The variety of fruit also affects the quantity of calcium pectate content; for example, seedless guava has maximal amount of calcium pectate (13.43%) [82].

Thongsombat *et al.* (2007) extracted pectin from peel, pulp, and seed cakes using sodium hexametaphosphate, which was followed by pectin precipitation by acidified ethanol. They achieved a yield of 30.50 ± 0.34% of crude pectin. This extracted crude pectin contains moisture (4.71 ± 0.18%), protein (0.34 ± 0.21%), ash (0.68 ± 0.00%), and soluble DFs (20.70 ± 0.16 g (%dwb) with pH (3.06 ± 0.02) [83].

Bhat and Singh (2014) extracted pectin from GP powder using HCl and citric acid at different conditions. The amount of pectin extracted using HCl and citric acid as solvents varies from 3.87% to 16.8% and 2.65% to 11.12%, respectively. Pectin extracted from HCl and citric acid gives higher yield at 85°C, 60 min, and pH 2.0. The properties of pectin extracted using HCl are as follows: equivalent weight, 685.3; methoxyl content, 4.25%; anhydrouronic acid, 67.4%; degree of esterification, 52.85%, whereas those using citric acid are as follows: equivalent weight, 345.4; methoxyl content, 3.50%; an hydrouronic acid, 82.1%; degree of esterification, 44.3%. The representative pectin obtained is low methoxyl pectin [14].

9.3.5 Animal Feed

Various agroindustrial wastes are used as a nonconventional food source for feeding animals. The main advantages of using such a type of feed is that it reduces the nutritional deficiency requirements among animals while also reducing the feeding costs, as more than 70% of the total production costs in any livestock is related to the feed resource. The use of agro industrial waste products in animal feedings signifies a way of nutrient recycling and should be considered as a preferable route of nutritious by-product elimination. In this regard, the waste obtained from guava processing industry can be used as partial replacement in livestock feedings.

GW can be successfully used as a feed ingredient in the diet of growing rabbits and Ossimi lambs. Replacing traditional food with GW up to 20% does not show any negative effects on the productive and economic efficiency. Besides, this partial replacement does not show any negative effects on growth, digestibility, health, and carcass characteristics of the rabbit as well as economizes feed costs [84, 85]. Denny *et al.* (2013) revealed the presence of bioactive phenolic compounds in guava pomace. The extracted compounds are listed in Table 9.4. These extracted compounds show that guava pomace can be a good source of anti-inflammatory and analgesic substances in animal models [74]. GLs are also used as a feed for Black Bengal goats [86]. Inclusion of GW (up to 12%) in the feeds for broiler chickens enhances the performance and carcass yield, which is similar to that achieved with corn- and soybean-based meal [87]. Ethanol extracts of guava pomace (GS and GL) exhibit a good protective effect against lipid oxidation in chicken meal even after 14 days of storage in aerobic packaging at 4 ± 1°C [17].

Sun-dried guava by-products (GBP, pulp, GP, GS, and inedible fruits) can be replace upto 15% of normal feed for laying hen diets at ages of 32–42 weeks without any adverse effect on productive performance and egg quality [88].

9.3.6 As Insecticide

The Asiatic citrus psyllid is a disturbing disease of citrus caused by phloem-limited bacteria. From multiple experiments, Zaka *et al.* (2010) concluded that GL possessed a repellent effect toward the adult citrus psyllids. By covering the guava shoots with net cloth, it was exposed that the repellent effect of GL is due to the presence of some volatile compounds effective against adult psyllids instead of some physical factors. These volatile compounds are mainly aldehydes and alcohols, which show repellent effects on insects. Barman and Zeng (2014) studied the effect of GL on psyllids, *Diaphorinacitri* Kuwayama, in various concentrations (100 mg/L to 10,000 mg/L). The adult psyllid significantly affected to citrus in both cage and Y-olfactometer tests at higher concentrations. After the treatments of 12 and 24 h and at the concentrations of 10,000 mg/L extract solution significant highest reduction (44.2% and 50%), followed by 5,000 mg/L GL extract solution (41.9% and 47.6%) and 1,000 mg/L GL extract solution (32.6% and 35.7%), respectively was obtained in choice cage tests. Moreover, the effect of GL extracts on adult psyllids was dosage dependent. Between the fractions studied, only petroleum ether had a substantial effect in lowering the psyllid settlement on citrus shoots [for the 12- and 24-hour treatment with 1,000 mg/L GL extract (35.7% and 43.8%) and with 500 mg/L GL extract (28.9% and 39.5%), respectively] in cage test. In the Y-Tube test, the repellent effects were 54.7% in petroleum ether fraction >37.0% in ethyl acetate fraction >31.5% in n-butyl fraction. The oral acute toxicity of GL extract to mice was also very low (LD50 of more than 20.0 g/kg), indicating that GL extracts can be safely used in living beings [89, 90]. Rouseff *et al.* (2008) reported the presence of sulfur volatiles, mainly hydrogen sulfide, sulfur dioxide, methanethiol, dimethyl sulfide, dimethyl disulfide, methional, and dimethyl trisulfide, etc. (see Table 9.4). Among these, dimethyl disulfide is toxic to insects, which is formed as a defensive volatile compound only by wounded guava but not

Table 9.5 Nanoparticles from GF.

Type of waste	Nanoparticle	Size (nm)	Reference
pink GW extract	Ag	23–105	[92]
GF extract	Cu	15–30	[95]
pink GL extract	Ag	10–90	[93]
pink GL extract	NiO	30	[94]
pink GL extract	Ni	103	[94]
pink GL extract	Au	27	[96]

citrus leaves, and thus may be responsible for the protective effect of guava against Asian citrus psyllid (Diaphorinacitri Kuwayama) [76].

9.3.7 Synthesis of Nanomaterials

The biosynthesis of nanoparticles using plant extracts is an emerging area of research. This route is known to be safe, cost-effective, sustainable, and environment friendly. In addition, the biosynthetic route provides better size and structure to the produced nano-particles [91].

In this regard, agrowastes from guava industries have been used in the production of nanomaterials. Silver nanoparticles were produced using pink GW [92] and GL extracts [93]. NiO and Ni nanoparticles (NPs) were successfully synthesized by boiling GLs [94]. The aqueous extract of GW was successfully used for the extraction of copper nanoparticles [95]. Microwave-exposed aqueous extract of GL is used to produce gold nanoparticles from a solution of gold chloride. In this case, the flavonoids separated during the microwave heating of microcellular solution of GL are responsible for the biosynthesis of gold nanoparticles [96]. Table 9.5 presents the various nanoparticles synthesized from the different parts of guava tree along with the particle sizes produced. In most cases, GL extract is used for extracting various types of metal nanoparticles.

9.3.8 In Fermentations

GS flour along with dry mycelium of *Aspergillus niger* is a good source of nitrogen for the production of alcohol by fermentation. Using GS flour has some other advantages too; for example, it contains lipids, which show antifoaming effects during fermentation due the presence of linoleic acid (79%), palmitic acid (8%), stearic acid (5%), oleic acid (7%), and triglycerides (60% trilinolein). In addition, the vitamin and mineral contents in GS favor the growth and sustainability of the yeast, adding more value to the generated waste. According to Cock *et al.* (2013), the use of such a substrate as a nitrogen source can reduce the operational costs uptoabout 35%, compared with current sources. The use of GS flour as a nitrogen source yields 43.58% more ethanol than other similar sources [97].

9.3.9 As a Water Treatment Agent

Various parts of guava (e.g., GL, GS, GP, and GW) were successfully used as an adsorbent for wastewater treatment. These materials were used either in their natural form or treated by physical/chemical methods prior to their use. The physical/chemical modification of adsorbents enhances the surface area or activates the active sites on their surface, which results in higher adsorption efficiency. Different parts of guava were used for the successful removal of heavy metal, dyes, and organic compounds from the wastewater with more than 90% efficiency. According to the literature sources, various parts of guava were mostly used for the treatment of dye containing wastewater with the majority using them following physical/chemical modifications. An in-depth literature overview of adsorption using guava waste is given in Table 9.6.

GS has an acidic character due to the high concentration of bulk functional groups (C=O); however, their acidic character may change due to the carbonization [98]. Amri *et al.* (2013) applied response surface mechanism (RSM) to study the variables for preparing activated carbons from GW with the aim of optimizing GW activated carbon synthesized as well as its NaOH:char impregnation ratio to enhance the removal of MB dye from aqueous solution [99]. Elizalde-González and Hernández-Montoya (2009) applied the Taguchi method was used for experimental design of acid orange 7 removal using GS and achieved >99% removal [100]. Dávila-Jiménez *et al.* (2015) impregnated the TiO_2 nano-particles on GS and utilized these for the adsorption of photoproducts obtained by decomposition of the herbicide isoproturon [101] .

Settheeworrarit *et al.* (2005) investigated the extract of GL as an alternative natural indicator for iron quantification by the flow injection method. In Thailand, the change in the color of water to a darker color after addition of GL extract helps local people to identify iron-containing water [102].

9.3.10 Production of Enzymes

Ismail and Faizal (2016) used GP to extract proteases and ammonium sulfate precipitation to partially purify protease from guava peel. Based on the data obtained, the maximum enzyme activity was found at pH 5 and 500°C [16]. Pectinase is the other enzyme that can be extracted from GP using ultrasound. RSM using a four-factor central composite design (CCD) was applied to define the optimal conditions for extraction of pectinase in an ultrasound environment. Amid *et al.* (2015) reported the optimized conditions as follows: sonication time of 20 min, temperature 40°C, and pH 5.0 using a solvent-to-sample ratio of 4:1 ml/g for achieving maximum pectinase yield (96.2%) [113].

9.4 Sustainability of Value-Added Products From GW

From these discussions, it can be concluded that GWs can be a potential source for a variety of applications; however, their utilization on an industrial scale is not an easy task, due to some factors that restrict their direct use. First, it is difficult to procure GW (peels, bark, seeds, leaves) throughout the year. Because these are lignocellulosic materials, they cannot be stored for long periods and should be utilized as collected; otherwise, they start to

Table 9.6 Adsorption parameters using GW (peel, leaves, seeds).

Adsorbate	Adsorbent	Modification process	Adsorption capacity (mg/g)	% removal	Reference
Congo Red	GP	Powdered GP was soaked in 0.1NKOH [GP: KOH is 1:1.5 (wt.%)] for 10 min. to obtain thick syrupy solution. The solution is then placed in microwave oven (input power: 600 W for 6 min). Then, the activated GP were treated with 0.5 N HCl to remove any remaining KOH. Finally, after filtration, the obtain carbon was dried at 80°C for 3 hours.	120.62	85	[10]
Congo Red	GP	Collected GP was washed thoroughly with distilled water to remove dust and other impurities. The dried GP was milled and sieved to obtain powder for further uses.	61.42	45	[10]
Chromium	GL	Collected GL were washed with deionized water, dried, ground and sieved (<1 mm). The sieved GL was used the extraction of tannins form them using acetone, ethyl alcohol and deionized water as extractant. The tannin extracted GL was used as an adsorbent.	–	95.3	[103]
Auramine Dye	GL	–	7.76	92-94	[104]
Methylene Blue	GL	–	295	95.81	[105]

(Continued)

Table 9.6 Adsorption parameters using GW (peel, leaves, seeds). (*Continued*)

Adsorbate	Adsorbent	Modification process	Adsorption capacity (mg/g)	% removal	Reference
Congo Red Dye	GL	25 g of GL was mixed to 500 ml of (0.3 mol) H_3PO_4 and heated to form apaste. The formed paste was then placed in the furnace and heated to 300°C for 30 min. Finally, it was washed with distilled water to a pH of 6.8 and oven dried at 105°C for 4 hours and then grinded to the particle size of 106 μm.	47.62	97.6	[106]
Cadmium	GL	GL was firstly washed several times with tap water followed by distilled water to remove particulate material from their surface. Then, dried at 100°C for 24 hours. Finally, it was ground and sieved to a particle size of about 0.187 mm.	10.53	–	[20]
Cadmium	GL	GL was soaked in 1 M NaOH solution for 24 hours. Then, it was filtered, washed with distilled water and dried in oven at 100°C.	18.66	–	[20]
2,4-Dichlorophenol	GS	GS was semi-carbonized by pyrolysis was carried out in muffle furnace at 300 °C for 1 hour. After pyrolysis the material was activated using ZnCl2 with impregnation ratios (ZnCl2:AC) were altered from 3:1 (w/w). Finally, it was dried at 85 °C. Both semi-carbonization and activation of GL were done in self-generated atmosphere.	20.88	–	[9]

(*Continued*)

Table 9.6 Adsorption parameters using GW (peel, leaves, seeds). (*Continued*)

Adsorbate	Adsorbent	Modification process	Adsorption capacity (mg/g)	% removal	Reference
Acid Orange 7	GS	The Taguchi method was used for experimental design. The chosen experimental factors and their ranges were: pH (2–12), temperature (15–35°C), specific surface area (50–600 $m^2 g^{-1}$) and adsorbent dosage (16–50 mg ml^{-1}).	–	>99%	[100]
Methylene Blue Dye	GS	The seeds (from the species *Psidium guajava* L.) were collected from a wet market in Sabah (Malaysia). The seeds were washed and dried in an oven at 110°C for 24 hours. Carbonization by pyrolysis was carried out in a muffle furnace (Carbolite, RHF 1500). The sample was placed on a pyrex petri dish and placed in the hot zone of the furnace. The samples were heated at 200°C for 15 minutes. The carbonized samples were taken out of the furnace and stored in a desiccator until the furnace reached the set activation temperature (500°C, 45 min before being put back in for activation. After activation at the required temperature, the activated samples were repetitively washed with 0.1 M phosphorus acid to reduce the ash and mineral content of the samples. The washing was carried out in a water bath shaker at 70°C for 4 hours. Finally, hydroxide. The end product was the carbon was neutralized with 0.1 M sodium hydroxide.	198.12	84.75	[107]

(*Continued*)

Table 9.6 Adsorption parameters using GW (peel, leaves, seeds). (*Continued*)

Adsorbate	Adsorbent	Modification process	Adsorption capacity (mg/g)	% removal	Reference
Nickel(II)	GS	The crushed GS were soaked in 50% ZnCl2 solution for 48 hours in a ratio of 1:2 (W/W). After drying it was then carbonized at 700°C for 1 hour. After cooling the excess ZnCl2 present in the carbonized GS was leached out using dilute HCl solution. Finally, the obtained carbon was washed several times to remove of ZnCl2 and HCl and dried.	18.05	99.8	[18]
Nickel(II)	GS	GS was soaked in the solution of monomethylolthiourea for 15 min and dried at 30°C for 20 min. Then, thermal treatment was given at 130°C–140°C for 15 min. Then, it was washed several times with distilled water followed by boiling with distilled water for 1 hour and finally washed with methanol several times and dried.	32.05	99.7	[18]
Acid Blue 80	GS	The GS was washed with deionized water, dried at 70°C for 24 h and Sieved. Then, 30 g of GS were heated in tubular furnace with a quartz reactor from room temperature to 600 or 1,000°C in the atmosphere of the products released as a result of the carbonization. The temperature (from room temperature to 70°C with 8°C min^{-1}, from 70 to 600 or 1,000°C with 5°C min^{-1}). The maximum heat treatment temperature was 4 hours.	6.6 X 10^{6}mol/g	–	[108]

(*Continued*)

Table 9.6 Adsorption parameters using GW (peel, leaves, seeds). (*Continued*)

Adsorbate	Adsorbent	Modification process	Adsorption capacity (mg/g)	% removal	Reference
Acid Green 27	GS	Four samples of carbon were obtained from GS at 1,000°C for different sizes. The Taguchi method was used for the experimental design to determine the optimum conditions for the adsorption of acid orange 7 in batch experiments. The selected experimental conditions were: temperature (15–35°C), pH (2–12), specific surface area (50–600 m^2/g) and adsorbent dosage (16–50 mg/ml).	–	100	[108]
Fluoride	GS	GS was washed and rinsed many times with distilled water to eliminate traces of fruit pulp and impurities. Then, dried at room temperature, sieved to obtain the 2-mm size and stored for further use.	116.50	–	[109]
Chromium (VI)	GS	GS were rinsed with distilled water and then dried in an oven for 72 hours at 105°C. Finally, it were ground to a particle size of 550 μm.	10.50	100	[110]
Methylene Blue	GW	GW was washed with water to remove any impurities and contaminants. Then, it was dried in oven at 110°C for 48 hours. The dried GW was grinded and sieved (<500 μm). Finally, it was carbonized at 400°C for 1 hour in a muffle furnace.	250.00	–	[111]

(*Continued*)

Table 9.6 Adsorption parameters using GW (peel, leaves, seeds). (*Continued*)

Adsorbate	Adsorbent	Modification process	Adsorption capacity (mg/g)	% removal	Reference
Amoxicillin	GS	7 g of GS (particle size: 425–590 μm) were placed in a horizontal stainless steel reactor and heated in muffle furnace (N_2 flow of 100 cm^3/min) at rate of 20°C/min to 500°C, and maintain this temperature for 2 hours. The obtained carbon is then activated with NaOH with impregnation ratio of 3:1 (wt:wt) NaOH:GS carbon. Finally, it was washed with (0.1 mol/L) HCl followed by hot distilled water (until pH:6.5).	570.48	–	[112]
Isoproturon	GS	In first stage, 5 g of GS were pre-activated with H_3PO_4 by stirring at 323 k for 1 hour. Then, the carbonization was carried out at 1,000°C in quartz reactor in an atmosphere of combustion gases. Finally, it was impregnated with TiO_2 in the ratio of 5:1 (weight of carbon: Weight of TiO_2).	10.2	99	[101]

Table 9.7 Unit operation and processes of GW products.

Products	Unit operation & processes
Essential oil, nano-materials	Steam distillation/flashing
Pectin & starch	Hydrolysis
Polyphenols, antioxidants, insecticide	Solvent extraction
Cattle feed, dietary fiber, adsorbents	Size reduction & blending

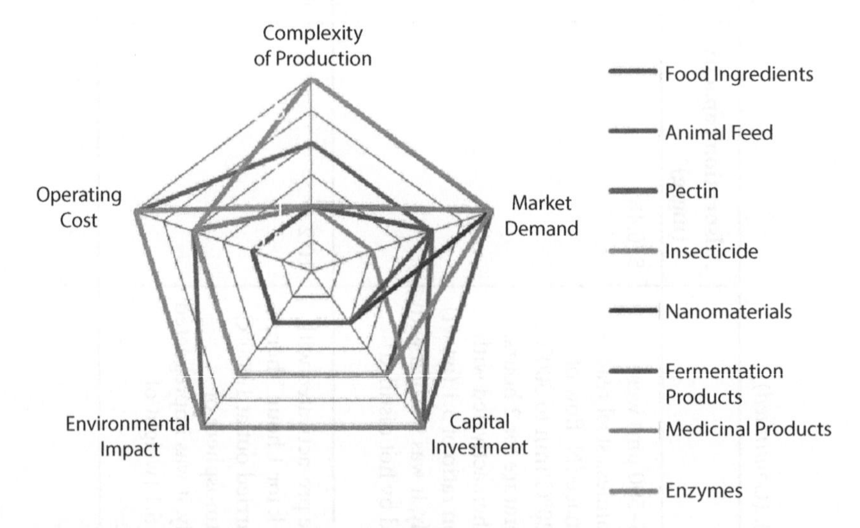

Figure 9.2 Plot depicting the indicators for sustainability assessment.

degrade, which is also favored by their high moisture content. Thus, on-site drying of GW is necessary to ensure its long-term storage, but this is quite difficult. Therefore, it is necessary to identify sustainable, environment benign, low-cost, and highly advanced technologies that can convert available GW into more value-added products within the available period.

Other important areas to be considered are ease of production of products and the capital investment. Thus, an ideal conversion process design should consider the environmental impact, overall demand of products in market, the simplicity of production, and capital investment required. The various products obtained from GW as discussed are compared based on five constraints: market demand, complexity of production, capital investment, environmental impact, and operating cost. Table 9.7 discusses various unit operations/processes and possible products derived from FPW.

The antioxidants and polyphenols obtained from GW require sophisticated devices for separation, and therefore, it is difficult to produce these on a large scale. The obtained polyphenols and antioxidants have enough demand in the market, but they have a comparatively higher environmental impact. However, products such as cattle feed and adsorbents (in raw form) tend to have minimum environmental impacts due to low cost and simple processing Figure 9.2.

9.5 Conclusion

Guava waste (pomace, peel, leaves, bark, seeds) contains many nutritionally and economically valuable components that have significant uses and thus need to be exploited. Various products are obtained from these GWs, including animal feeds, essential oils, food ingredients, nano-materials, insecticides, water adsorbents, medicines, etc. The economic valorization of GW is possible more than ever by adapting biotechnological tools and techniques. Therefore, by applying these techniques, the waste generally discarded can be converted into value-added products with high economic potential. However, for this to become a reality, novel, advanced, and economic technologies should be developed on the commercial level. Furthermore, ensuring availability of raw material throughout the year, gaining in-depth knowledge of the physicochemical characterization of waste obtained, formulating simple and economic designs of conversion process that is flexible according to raw material (as chemical composition changes with variety, seasonal changes, maturity, location, etc.), and performing an integrated study of utilization of wastes generated in different steps can make this biorefinery approach a reality.

References

1. Bernardino-Nicanor, A., Scilingo, A.A., Anon, M.C., Davila-Ortız, G., Guava seed storage protein Fractionation and characterization. *LWT*, 39, 902–910, 2006.
2. Bertagnolli, S.M.M., Silveira, M.L.R., Fogaça, A.D.O., Umann, L., Penna, N.G., Bioactive compounds and acceptance of cookies made with Guava peel flour. *Food Sci. Technol, Campinas*. 34, 2, 303–308, 2014.
3. Tangirala, S., Sarkar, B.C., Sharma, H.K., Kumar, N., Modeling and Characterization of Blended Guava Pomace and Pulse Powder Based Rice Extrudates. *Int. J. Food Eng.*, 8, 1, 1–24, 2012.
4. Rejal, S.Z.B., Extraction of antioxidant acivity, phenolic content and mineral content from guava peel, in: *Chemical Engineering (Biotechnology)*, Faculty of Chemical & Natural Resources Engineering Universiti Malaysia Pahang, Malaysia Pahang, 2010.
5. Rai, P.K., Jaiswal, D., Mehta, S., Watal, G., Anti-hyperglycaemic potential of Psidium guajava raw fruit peel. *Indian J. Med. Res.*, 129, 561–565, 2009.
6. Shiva, B., Nagaraja, A., Srivastav, M., Kumari, S., Goswami, A.K., Singh, R., Arun, M.B., Characterization of guava (Psidium guajava) germplasm based on leaf and fruit parameters. *Indian J. Agric. Sci.*, 87, 5, 634–638, 2017.
7. Rai, P.K., S. Mehta, and G. Watal, Hypolipidaemic & hepatoprotective effects of Psidium guajava raw fruit peel in experimental diabetes. *Indian J. Med. Res.*, 131, 820–824, 2010.
8. Mohamed, G.F., S.S. Mohamed, and F.S. Taha, Antioxidant, Antimicrobial, and Anticarcinogenic Properties of Egyptian Guava Seed Extracts. *Nat. Sci.*, 9, 11, 32–41, 2011.
9. Anisuzzaman, S.M., Joseph, C.G., Krishnaiah, D., Bono, A., Suali, E., Abang, S., Fai, L.M., Removal of chlorinated phenol from aqueous media by guava seed (Psidium guajava) tailored activated carbon. *Water Resour. Ind.*, 16, 29–36, 2016.

10. Singh, P., Raizada, P., Pathania, D., Sharma, G., Sharma, P., Microwave induced KOH activation of guava peel carbon as an adsorbent fot congo red dye removal from aqueous phase. *Indian J. Chem. Technol.*, 20, 305–311, 2013.
11. Jimenez-Escrig, A., Rinco, M., Pulido, R., Saura-Calixto, F., Guava Fruit (Psidium guajava L.) as a New Source of Antioxidant Dietary Fiber. *J. Agric. Food Chem.*, 49, 5489–5493, 2001.
12. Sousa, B.A. and R.T.P. Correia, Phenolic content, antioxidant activity and antiamylolytic activity of extracts obtained from bioprocessed pineapple and guava wastes. *Braz. J. Chem. Eng.*, 29, 1, 25–30, 2012.
13. Kong, K.-W., Ismail, A.R., Tan, S.-T., Prasad, K.M.N., Ismail, A., Response surface optimisation for the extraction of phenolics and flavonoids from a pink guava puree industrial by-product. *Int. J. Food Sci. Technol.*, 45, 1739–1745, 2010.
14. Bhat, S.A. and E.R. Singh, Extraction And Characterization of Pectin From Extraction And Characterization of Pectin From guava fruit peel. *Int. J. Res. Eng. Adv. Technol.*, 2, 3, 1–7, 2014.
15. Kamath, J.V., Rahul, N., Kumar, C.K.A., Lakshmi, S.M., Psidium guajava L: A review. *Int. J. Green Pharm.*, 2, 1, 9–12, 2008.
16. Ismail, N. and Faizal, M., Characterization and Purification of Protease extracted from Guava (Psidium guajava) peel. *Sci. Lett.*, 10, 1, 2016.
17. Packer, V.G., S.Melo, P., Bergamaschi, K.B., Selani, M.M., Villanueva, N.D.M., Alencar, S.M.d., Contreras-Castillo, C.J., Chemical characterization, antioxidant activity and application of beetroot and guava residue extracts on the preservation of cooked chicken meat. *J. Food Sci. Technol.*, 52, 11, 7409–7416, 2015.
18. Zewail, T.M. and El-Garf, S.A.M., Preparation of agriculture residue based adsorbents for heavy metal removal. *Desalin. Water Treat.*, 22, 1–3, 363–370, 2010.
19. Uchôa-Thomaz, A.M.A., Sousa, E.C., Carioca, J.O.B., Morais, S.M.D., Lima, A.D., Martins, C.G., Alexandrino, C.D., Ferreira, P.A.T., Rodrigues, A.L.M., Rodrigues, S.P., Thomaz, J.C.D.A., Silva, J.D.N., Rodrigues, L.L., Chemical composition, fatty acid profile and bioactive compounds of guava seeds (Psidium guajava L.). *Food Sci. Technol. Campinas*, 34, 3, 485–492, 2014.
20. Abdelwahab, O., Fouad, Y.O., Amin, N.K., Mandor, H., Kinetic and thermodynamic aspects of cadmium adsorption onto raw and activated guava (Psidium guajava) leaves. *Environ. Prog. Sustainable Energy*, 34, 2, 351–358, 2014.
21. Jang, M., Jeong, S.W., Cho, S.K., Ahn, K.S., Lee, J.H., Yang, D.C., Kim, J.C., Anti-inflammatory effects of an ethanolic extract of guava (Psidium guajava L.) leaves *in vitro* and *in vivo*. *J. Med. Food*, 17, 6, 678–85, 2014.
22. Chen, K.-C., Hsieh, C.-L., Huang, K.-D., Ker, Y.-B., Chyau, C.-C., Peng, R.Y., Anticancer Activity of Rhamnoallosan against DU-145 Cells Is Kinetically Complementary to Coexisting Polyphenolics in Psidium guajava Budding Leaves. *J. Agric. Food Chem.*, 57, 6114–6122, 2009.
23. Laily, N., Kusumaningtyas, R.W., Sukarti, I., Rini, M.R.D.K., The Potency of Guava Psidium guajava (L.) Leaves as a Functional Immunostimulatory Ingredient. *Procedia Chem.*, 14, 301–307, 2015.
24. Gonçalves, F.A., Neto, M.A., Bezerra, J.N.S., Macrae, A., Sousa, O.V.D., Fonteles-Filho, A.A., Vieira, R.H.S.F., Antibacterial activity of GUAVA, Psidium guajava Linnaeus, leaf extracts on diarrhea-causing enteric bacteria isolated from Seabob shrimp, Xiphopenaeus. *Rev. Inst. Med. trop. S. Paulo*, 50, 1, 11–15, 2008.
25. Rahim, N., Gomes, D.J., Watanabe, H., Rahman, S.R., Chomvarin, C., Endtz, H.P., Alam, M., Antibacterial Activity of Psidium guajava Leaf and Bark against Multidrug-Resistant Vibrio cholerae Implication for Cholera Control. *Jpn. J. Infect. Dis.*, 63, 271–274, 2010.
26. Cheng, F.C., S.C. Shen, and J.S. Wu, Effect of guava (Psidium guajava L.) leaf extract on glucose uptake in rat hepatocytes. *J. Food Sci.*, 74, 5, H132–H138, 2009.

27. El-Ahmady, S.H., Ashour, M.L., Wink, M., Chemical composition and anti-inflammatory activity of the essential oils of Psidium guajavafruits and leaves. *J. Essent. Oil Res.*, 25, 6, 2013.
28. Ademiluyi, A.O., Oboh, G., Ogunsuyi, O.B., Oloruntoba, F.M., A comparative study on anti-hypertensive and antioxidant properties of phenolic extracts from fruit and leaf of some guava (Psidium guajava L.) varieties. *Comp. Clin. Pathol.*, 25, 363–374, 2015.
29. Vieira, R.H.S.D.F., Rodrigues, D.D.P., Gonçalves, F.A., Menezes, F.G.R.D., Aragão, J.S., Sousa, O.V., Microbicidal effect of medicinal plant extracts (Psidium guajava Linn. and Carica papaya Linn.) upon bacteria isolated from fish muscle and known to induce. *Rev. Inst. Med. trop. S. Paulo,* 43, 3, 145–148, 2001.
30. Joseph, B. and R.M. Priya, Preliminary Phytochemicals of Psidium guajava L. Leaf of Methanol Extract and its Cytotoxic Study on hela Cell Lines. *Inventi Rapid: Ethnopharmacol.,* 1, 2, 1–3, 2016.
31. Abozid, M.M. and Farid, H.E.A., The anti-fatty liver effects of guava leaves and pomegranate peel extracts on ethanol-exposed rats Medhat M. Abozid, Hoda E. A. Farid. *J. Biol. Chem. Environ. Sci.*, 8, 3, 83–104, 2013.
32. Im, I., Park, K.-R., Kim, S.-M., Kim, C., Park, J.H., Nam, D., Jang, H.-J., Shim, B.S., Ahn, K.S., AshikMosaddik, Sethi, G., Cho, S.K., Ahn, K.S., The Butanol Fraction of Guava (Psidium cattleianum Sabine) Leaf Extract Suppresses MMP-2 and MMP-9 Expression and Activity Through. *Nutr. Cancer*, 64, 2, 255–266, 2012.
33. Soliman, F.M., Fathy, M.M., Salama, M.M., Saber, F.R., Comparative study of the volatile oil content and antimicrobial activity of Psidium guajava L. and Psidium cattleianum Sabine leaves. *Bull. Faculty Pharm. Cairo University*, 54, 2, 219–225, 2016.
34. Zahidah, W.Z.W.N., Noriham, A., Zainon, M.N., Antioxidant and antimicrobial activities of pink guava leaves and seeds. *J. Trop. Agric. Fd. Sc*, 41, 1, 53–62, 2013.
35. Biswas, B., Rogers, K., McLaughlin, F., Daniels, D., Yadav, A., Antimicrobial Activities of Leaf Extracts of Guava (Psidium guajava L.) on Two Gram-Negative and Gram-Positive Bacteria. *Int. J. Microbiol.*, 2013, 746165, 2013.
36. Mailoa, M.N., Mahendradatta, M., Laga, A., Djide, N., Antimicrobial Activities of Tannins Extract From Guava Leaves (Psidium Guajava L) On Pathogens Microbial. *Int. J. Sci. Technol. Res.*, 3, 1, 236–241, 2014.
37. Mazumdar, S., Akter, R., Talukder, D., Antidiabetic and antidiarrhoeal effects on ethanolic extract of Psidium guajava (L.) Bat. leaves in Wister rats. *Asian Pac. J. Trop. Biomed.*, 5, 1, 10–14, 2015.
38. Oh, W.K., Lee, C.H., Lee, M.S., Bae, E.Y., Sohn, C.B., Oh, H., Kim, B.Y., Ahn, J.S., Antidiabetic effects of extracts from Psidium guajava. *J. Ethnopharmacol.*, 96, 3, 411–5, 2005.
39. Shen, S.C., Cheng, F.C., Wu, N.J., Effect of guava (Psidium guajava Linn.) leaf soluble solids on glucose metabolism in type 2 diabetic rats. *Phytother. Res.*, 22, 11, 1458–64, 2008.
40. Jaiarj, P., Khoohaswan, P., Wongkrajang, Y., Peungvicha, P., Suriyawong, P., Saraya, M.L.S., Ruangsomboon, O., Anticough and antimicrobial activities of Psidium guajava Linn. leaf extract. *J. Ethnopharmacol.*, 67, 203–212, 1999.
41. Hsieh, C.-L., Lin, Y.-C., Yen, G.-C., Chen, H.-Y., Preventive effects of guava (Psidium guajava L.) leaves and its active compounds against α-dicarbonyl compounds-induced blood coagulation. *Food Chem.*, 103, 528–535, 2007.
42. Nantitanon, W., Comparison of antioxidant activity of compounds isolated from guava leaves and a stability study of the most active compound. *Drug Discoveries Ther.*, 6, 1, 38–43, 2012.
43. Uduak, E.U., Timbuak, J.A., Musa, S.A., Ikyembe, D.T., Abdurrashid, S., Hamman, W.O., Ulceroprotective Effect of Methanol Extract of Psidium guajava Leaves on Ethanol Induced Gastric Ulcer in Adult Wistar Rats. *Asian J. Med. Sci.*, 4, 2, 75–78, 2012.

44. Irondi, E.A., Agboola, S.O., Oboh, G., Boligon, A.A., Athayde, M.L., Shode, F.O., Guava leaves polyphenolics-rich extract inhibits vital enzymes implicated in gout and hypertension *in vitro*. *J. Intercult Ethnopharmacol.*, 5, 2, 122–30, 2016.
45. Wang, H., Du, Y.-J., Song, H.-C., α-Glucosidase and α-amylase inhibitory activities of guava leaves. *Food Chem.*, 123, 6–13, 2010.
46. Athikomkulchai, S., Watthanachaiyingcharoen, R., Tunvichien, S., Vayumhasuwan, P., Karnsomkiet, P., Sae-Jong, P., Ruangrungsi, N., The development of antI-acne products from Eucalyptus globulus and Psidium guajava Oil. *J. Health Res.*, 22, 3, 109–113, 2008.
47. Saraya, S., Kanta, J., Sarisuta, N., Samranri, K., Temsiririrkkul, R., Chumnumwat, S., Development of Guava Extract Chewable Tablets for Anticariogenic Activity against Streptococcus mutans. *Mahidol Univ. J. Pharm. Sci.*, 35, 1–4, 18–23, 2008.
48. Chetri, K., Sanyal, D., Kar, P.L., Changes in nutrient element composition of guava leaves in relation to season, cultivar, direction of shoot, and zone of leaf sampling. *Commun. Soil Sci. Plant Anal.*, 30, 1–2, 121–128, 2008.
49. Guo, C., Yang, J., Wei, J., Li, Y., Xu, J., Jiang, Y., Antioxidant activities of peel, pulp and seed fractions of common fruits as determined by FRAP assay. *Nutr. Res.*, 23, 1719–1726, 2003.
50. Budin, S.B., Ismail, H., Chong, P.L., Psidium guajava Fruit Peel Extract Reduces Oxidative Stress of Pancreas in Streptozotocin-induced Diabetic Rats. *Sains Malaysiana*, 42, 6, 707–713, 2013.
51. Abdelmalek, S., Mohsen, E., Awwad, A., Issa, R., Peels of Psidium guajava fruit possess antimicrobial properties. *Int. Arabic J. Antimicrob. Agents*, 6, 3, 1, 2016.
52. Kumari, S., Rakavi, R., Mangaraj, M., Effect of Guava in Blood Glucose and Lipid Profile in Healthy Human Subjects A Randomized Controlled Study. *J. Clin. Diagn. Res.*, 10, 9, BC04–BC07, 2016.
53. Barbalho, S.M., Farinazzi-Machado, F.M.V., Goulart, R.d.A., Brunnati, A.C.S., Ottoboni, A.M.M.B., Nicolau, C.C.T., Psidium Guajava (Guava) A Plant of Multipurpose Medicinal Applications. *Med Aromat Plants*, 1, 4, 1–6, 2012.
54. Chen, H.-C., Sheu, M.-J., Lin, L.-Y., Wu, C.-M., Chemical Composition of the Leaf Essential Oil of Psidium guajava L. from Taiwan. *J. Essent. Oil Res.*, 19, 4, 345–347, 2007.
55. Sukeksi, L., Hassan, C.R.C., Sulaiman, N.M.N., Rashidi, H., Emami, S.D., Determination of guava (Psidium guajava L.) leaf phenolic compounds using HPLC-DAD-QTOF-MSPolyphenols Recovery from Tropical Fruits (Pink Guava) Wastes via Ultra - Filtration Membrane Technology Application by Optimum Solvent Selection. *Iran. J. Chem. Chem. Eng.*, 35, 3, 53–63, 2016.
56. Díaz-de-Cerio, E., Gómez-Caravaca, A.M., Verardo, V., Fernández-Gutiérrez, A., Segura-Carretero, A., Determination of guava (Psidium guajava L.) leaf phenolic compounds using HPLC-DAD-QTOF-MS. *J. Funct. Foods*, 22, 376–388, 2016.
57. Tachakittirungrod, S., Okonogi, S., Chowwanapoonpohn, S., Study on antioxidant activity of certain plants in Thailand Mechanism of antioxidant action of guava leaf extract. *Food Chem.*, 103, 381–388, 2007.
58. Mailoa, M.N., Mahendradatta, M., Laga, A., Djide, N., Tannin extract of guava leaves (psidium guajava l) variation with concentration organic solvents. *Int. J. Sci. Technol. Res.*, 2, 9, 106–110, 2013.
59. Nantitanon, W., Yotsawimonwat, S., Okonogi, S., Factors influencing antioxidant activities and total phenolic content of guava leaf extract. *LWT - Food Sci. Technol.*, 43, 7, 1095–1103, 2010.
60. Moura, P.M., Prado, G.H.C., Meireles, M.A.A., Pereira, C.G., Supercritical fluid extraction from guava (Psidium guajava) leaves Global yield, composition and kinetic data. *J. Supercrit. Fluids*, 62, 116–122, 2012.

61. Tuan, P.M., Anh, H.T.V., Cam, L.T.H., Chi, V.N.Q., Oanh, D.T.B., Khue, D.B., Minh, P.D.T., Tri, M.H., Sang, N.T.T., Mai, D.S., Nghia, T.T., Extraction and encapsulation of polyphenols from guava leaves. *Ann. Food Sci. Technol.*, 17, 1, 34–40, 2016.
62. Liu, C.-W., Wang, Y.-C., Lu, H.-C., Chiang, W.-D., Optimization of ultrasound-assisted extraction conditions for total phenols with anti-hyperglycemic activity from Psidium guajava leaves. *Process Biochem.*, 49, 1601–1605, 2014.
63. Sagrero-Nieves, L., Bartley, J.P., Provis-Schwede, A., Supercritical fluid extraction of the volatile components from the leaves of Psidium guajava L.(guava). *Flavour Fragrance J.*, 9, 135–137, 1994.
64. Rodriguez, S.J., Variation in Chemical Composition of Guava Leaves (Psidium guajava L.) as Affected by Position in the Terminals. *J. Agric. Univ. Puerto Rico*, 51, 3, 252–259, 1967.
65. Rojas-Garbanzo, C., Zimmermann, B.F., Schulze-Kaysers, N., Schieber, A., Characterization of phenolic and other polar compounds in peel and flesh of pink guava (Psidiumguajava L. cv. 'Criolla') by ultra-high performance liquid chromatography with diode array andmass spectrometric detection. *Food Res. Int.*, 100, 445–453, 2017.
66. Anany, A.E., Nutritional composition, antinutritional factors, bioactive compounds and antioxidant activity of guava seeds (Psidium Myrtaceae) as affected by roasting processes. *J. Food Sci. Technol.*, 52, 4, 2175–2183, 2015.
67. Castro-Vargas, H.I., Rodríguez-Varela, L.I., Ferreira, S.R.S., Parada-Alfonso, F., Extraction of phenolic fraction from guava seeds (Psidium guajava L.) Using supercritical carbon dioxide and co-solvents. *J. Supercrit. Fluids*, 51, 319–324, 2010.
68. Fontanari, G.G., Souza, G.R., Batistuti, J.P., Neves, V.A., Pastre, I.A., Fertonani, F.L., DSC Studies on Protein Isolate of Guava Seeds Psidium Guajava. *J. Therm. Anal. Calorim.*, 93, 2, 397–402, 2008.
69. Bernardino-Nicanor, A., Anoän, M.C., Scilingo, A.A., Vila-Ortiäz, G.D., Functional Properties of Guava Seed Glutelins. *J. Agric. Food Chem.*, 53, 9, 3613–3617, 2015.
70. Habib, M.A., Studies on the lipid and protein composition of guava seeds (Psidium guajava). *Food Chem.*, 22, 7–16, 1986.
71. Castro-Vargas, H.I., Rodríguez-Varela, L.I., Parada-Alfonso, F., Guava (Psidium guajava L.) Seed oil obtained with a homemade supercritical fluid extraction system using supercritical CO2 and co-solvent. *J. Supercrit. Fluids*, 56, 238–242, 2011.
72. Khalifa, I., Barakat, H., El-Mansy, H.A., Soliman, S.A., Optimizing Bioactive Substances Extraction Procedures from Guava, Olive and Potato Processing Wastes and Evaluating their Antioxidant Capacity. *J. Food Chem. Nanotechnol.*, 2, 4, 170–177, 2016.
73. Hurtado, J.G.F., Piedra, M.A.A., Cuenca, E.L.P., Effect of particle size on the antioxidant capacity of guava (Psidium guajava L.) Byproduct. *Rev. Venezolana de Ciencia y Tecnología de Alimentos*, 3, 2, 202–209, 2012.
74. Denny, C., Melo, P.S., Franchin, M., Massarioli, A.P., Bergamaschi, K.B., Alencar, S.M.d., Rosalen, P.L., Guava pomace a new source of anti-inflammatory and analgesic bioactives. *BMC Complement. Altern. Med.*, 13, 235–241, 2013.
75. Díaz-de-Cerio, E., Verardo, V., Gómez-Caravaca, A.M., Fernández-Gutiérrez, A., Segura-Carretero, A., Determination of Polar Compounds in Guava Leaves Infusions and Ultrasound Aqueous Extract by HPLC-ESI-MS. *J. Chem.*, 2015, 1–9, 2015.
76. Rouseff, R.L., Onagbola, E.O., Smoot, J.M., Stelinski, L.L., Sulfur volatiles in guava (Psidium guajava L.) leaves possible defense mechanism. *J. Agric. Food Chem.*, 56, 8905–8910, 2008.
77. Pereira, A.G.T., Pereira, P.A.P., Borges, S.V., Dias, M.V., Figueiredo, L.P., Valente, W.A., Physicochemical characterization and sensory evaluation of jellies made with guava peels (Psidium guajava L.). *Int. J. Agric. Policy Res.*, 3, 11, 396–401, 2015.

78. Maurya, A.K. and Pandey, R.K., Development of dietary fibre rich yoghurt and biscuits using guava seed powder, in: *Food Science & Technology*, Banaras Hindu University, Varanasi, India, 2015.
79. Hussein, A.M.S., Kamil, M.M., Mohamed, G.F., Physicochemical and Sensorial Quality of Semolina-Defatted Guava Seeds Flour Composite Pasta. *J. Am. Sci.*, 7, 6, 2011.
80. El-Din, M.H.A.S. and Yassen, A.A.E., Evaluation and utilization of guava seed meal (Psidium guajava L.) in cookies preparation as wheat flour substitute. *Nahrung*, 41, 6, 344–348, 1997.
81. Arslan, M., Rakha, A., Khan, M.R., Zou, X., Complementing the dietary fiber and antioxidant potential of gluten free bread with guava pulp powder. *Food Meas.*, 11, 1959–1968, 2017.
82. Begum, R., Yusof, Y.A., Aziz, M.G., Uddin, M.B., Screening of Fruit Wastes as Pectin Source. *J. Environ. Sci. Nat. Resour.*, 10, 1, 65–70, 2017.
83. Thongsombat, W., Sirichote, A., Chanthachum, S., The production of guava juice fortified with dietary fiber. *Songklanakarin J. Sci. Technol.*, 29, 1, 187–196, 2007.
84. Kamel, E.R., Abdel-Fattah, F., El-Qaliouby, H.S., A, M.E.A., Response of New Zealand Rabbits to Diet Containing Guava Waste (Psidium Guaijava L.). *Alexandria J. Vet. Sci.*, 50, 1, 24–35, 2016.
85. Hassan, T.M.M., Abdel-Fattah, F.A.I., Farid, A.S., Kamel, E.R., Effect of feeding guava waste on growth performance, diet digestibility, carcass characteristics and production profitability of ossimi lambs. *Egypt. J. Nutr. Feeds*, 19, 3, 463–475, 2016.
86. Kibria, S.S., Nahar, T.N., Mia, M.M., Tree leaves as alternative feed resource for Black Bengal goats under stall-fed conditions. *Small Ruminant Res.*, 13, 217–222, 1994.
87. Lira, R.C., Rabello, C.B.-V., Ferreira, P.V., Lana, G.R.Q., Lüdke, J.V., Junior, W.M.D., Inclusion of guava wastes in feed for broiler chickens. *R. Bras. Zootec.*, 38, 12, 2401–2407, 2009.
88. El-Deek, A.A., Hamdy, S.M., Attia, Y.A., El-Shahat, A.M., Guava By-Product Meal Processed in Various Ways and Fed in Differing Amounts as a Component in Laying Hen Diets. *Int. J. Poult. Sci.*, 8, 9, 866–874, 2009.
89. Zaka, S.M., Zeng, X.-N., Holford, P., Beattie, G.A.C., Repellent effect of guava leaf volatiles on settlement of adults of citrus psylla, Diaphorina citri Kuwayama, on citrus. *Insect Sci.*, 17, 39–45, 2010.
90. Barman, J.C. and Zeng, X., Effect of Guava Leaf Extract on Citrus Attractiveness to Asian Citrus Psyllid, Diaphorina citri Kuwayama. *Pakistan J. Zool.*, 46, 4, 1117–1124, 2014.
91. Madhumitha, G. and Roopan, S.M., Devastated Crops Multifunctional Efficacy for the Production of Nanoparticles. *J. Nanomater.*, 2013, 1–13, 2013.
92. Zamanhuri, N.A., Alrozi, R., Osman, M.S., Biosynthesis of Silver Nanoparticles using Pink Guava Waste Extract (PGWE), in: *IEEE Business, Engineering & Industrial Applications Colloquium (BEIAC)*, 2012.
93. Bose, D. and Chatterjee, S., Biogenic synthesis of silver nanoparticles using guava (Psidium guajava) leaf extract and its antibacterial activity against Pseudomonas aeruginosa. *Appl. Nanosci.*, 6, 6, 895–901, 2015.
94. Mariam, A.A., Kashif, M., Arokiyaraj, S., Bououdina, M., Sankaracharyulu, M.G.V., Jayachandran, M., Hashim, U., Bio-Synthesis of Nio And Ni Nanoparticles And Their Characterization. *Dig. J. Nanomater. Biostructures*, 9, 3, 1007–1019, 2014.
95. Caroling, G., Priyadharshini, M.N., Vinodhini, E., Ranjitham, A.M., Shanthi, P., Biosynthesis of copper nanoparticles using aqueous guava extract –characterisation and study of antibacterial effects. *Int. J. Pharm. Bio Sci.*, 5, 2, 25–43, 2015.
96. Raghunandan, D., Basavaraja, S., Mahesh, B., Balaji, S., Manjunath, S.Y., Venkataraman, A., Biosynthesis of Stable Polyshaped Gold Nanoparticles from Microwave-Exposed Aqueous Extracellular Anti-malignant Guava (Psidium guajava) Leaf Extract. *NanoBiotechnology*, 5, 1–4, 34–41, 2009.

97. Cock, L.S., Ayala, J.D.M., López, J.E.A., Schouben, A.L.G., Kinetics of Alcoholic Fermentation Using Guava (Psidium Guajava) Seed Flour And Dry Mycelium of Aspergillus Niger as Nitrogen Sources. *Dyna*, 80, 180, 113–121, 2013.
98. Elizalde-González, M.P. and Hernández-Montoya, V., Guava seed as an adsorbent and as a precursor of carbon for the adsorption of acid dyes. *Bioresour. Technol.*, 100, 2111–2117, 2009.
99. Amri, N., Alrozi, R., Osman, M.S., Nasuha, N., Optimization of Pink Guava (Psidium guajava) Waste-based Activated Carbon Preparation Conditions for Methylene Blue Dye Removal. *J. Mater. Chem. Eng.*, 1, 1, 32–43, 2013.
100. Elizalde-González, M.P. and Hernández-Montoya, V., Removal of acid orange 7 by guava seed carbon. *J. Hazard. Mater.*, 168, 515–522, 2009.
101. Dávila-Jiménez, M.M., Elizalde-González, M.P., García-Díaz, E., Santes-Aquino, A.M., Assessment of the effectiveness of combined adsorption and photocatalysis for removal of the herbicide isoproturon. *Phys. Chem. Earth*, 91, 77–86, 2015.
102. Settheeworrarit, T., Hartwell, S.K., Lapanatnoppakhun, S., Jakmunee, J., Christian, G.D., Grudpan, K., Exploiting guava leaf extract as an alternative natural reagent for flow injection determination of iron. *Talanta*, 68, 2, 262–7, 2005.
103. Kumtabtim, U., Sornsathian, P., Pichrugul, O., Yimwasana, N., Chromium Removal from Wastewater from COD Analysis Using Tannin Adsorbent from Guava Leaves. *UTK*, 10, 1, 18–22, 2016.
104. Gaikwad, R.W. and Misal, a.S.A., Studies on auramine dye adsorption on psidium guava leaves. *Korean J. Chem. Eng.*, 26, 1, 102–107, 2009.
105. Ponnusami, V., Vikram, S., Srivastava, S.N., Guava (Psidium guajava) leaf powder Novel adsorbent for removal of methylene blue from aqueous solutions. *J. Hazard. Mater.*, 152, 276–286, 2008.
106. Ojedokun, A.T. and Bell, O.S., Kinetic modeling of liquid-phase adsorption of Congo red dye using guava leaf-based activated carbon. *Appl. Water Sci.*, 7, 1965–1977, 2017.
107. Joseph, C.G., Bono, A., Krishnaiah, D., Soon, K.O., Sorption Studies of Methylene Blue Dye in Aqueous Solution by Optimised Carbon Prepared from Guava Seeds (Psidium guajava L.). *Mater. Sci. (Medžiagotyra)*, 13, 1, 2007.
108. Elizalde-González, M.P. and Hernández-Montoya, V., Fruit Seeds as Adsorbents And Precursors Of Carbon For The Removal of Anthraquinone Dyes. *Int. J. Chem. Eng.*, 1, 2–3, 243–253, 2008.
109. Valencia-Leal, S.A., Cortés-Martínez, R., Alfaro-Cuevas-Villanueva, R., Evaluation of Guava Seeds (Psidium Guajava) As a LowCost Biosorbent for the Removal of Fluoride from Aqueous Solutions. *Int. J. Eng. Res. Dev.*, 5, 4, 69–76, 2012.
110. Abdelwahab, O., Sikaily, A.E., Khaled, A., Nemr, A.E., Mass-transfer processes of chromium (VI) adsorption onto guava seeds. *Chem. Ecol.*, 23, 1, 73–85, 2007.
111. Amri, N., Alrozi, R., Osman, M.S., Nasuha, N., Aman, N.S., Removal Of Methylene Blue Dye From Aqueous Solution Using Pink Guava (Psidium Guajava) Waste-based Activated Carbon, in: *IEEE Symposium on Humanities, Science and Engineering Research*, 2012.
112. Pezoti, O., Cazetta, A.L., Bedin, K.C., Souza, L.S., Martins, A.C., Silva, T.L., Júnior, O.O.S., Visentainer, J.V., Almeida, V.C., NaOH-activated carbon of high surface area produced from guava seeds as a high-efficiency adsorbent for amoxicillin removal Kinetic, isotherm. *Chem. Eng. J.*, 288, 778–788, 2016.
113. Amid, M., Murshid, F.S., Manap, M.Y., Sarker, Z., Optimisation of Ultrasound-Assisted Extraction of Pectinase Enzyme from Guava (Psidium guajava) Peel Enzyme Recovery, Specific Activity, Temperature and Storage Stability. *Prep. Biochem. Biotechnol.*, 46, 1, 91–99, 2016.

10

Case-Studies Towards Sustainable Production of Value-Added Compounds in Agro-Industrial Wastes

Massimo Lucarini[1]*, Alessandra Durazzo[1], Ginevra Lombardi-Boccia[1], Annalisa Romani[2], Gianni Sagratini[3], Noemi Bevilacqua[4], Francesca Ieri[2], Pamela Vignolini[2], Margherita Campo[2] and Francesca Cecchini[4]

[1]CREA-Research Centre for Food and Nutrition, Via Ardeatina, Rome, Italy
[2]PHYTOLAB, University of Florence, Sesto Fiorentino, Firenze, Italy
[3]School of Pharmacy, University of Camerino, Via Sant'Agostino 1, Camerino, Italy
[4]CREA-Research Centre for Vitinculture and Enology, Velletri, Roma, Italy

Abstract
Currently, different strategies are proposed in different Countries in order to tackle the food waste valorization: particularly, a scheme addressing, particularly, towards high value products, involve biorefinery as integrated approach of the characterization of collected biowaste. This chapter describes some case-studies towards sustainable production of value-added compounds in agro-industrial wastes, applied on chestnut, soy, olive oil wastewater and *Olea europaea* L. leaves. The experimental pilot plants are here reported and discussed.

Keywords: Bio-based compounds, experimental pilot plant, chestnut, soy, olive oil wastewater, *Olea europaea* L. leaves

10.1 Introduction

Food waste is currently generated in great quantities worldwide; Food and Agricultural Organization (FAO) [1] reported that one-third of food produced globally for human consumption (nearly 1.3 billion tons) is lost along the food supply chain. For each food, different percentages of food waste are produced along different stages of food chain and industrial processing [2].

Currently, different strategies are proposed in different countries in order to tackle the food waste valorization: a scheme addressing, particularly, toward high value products, involve biorefinery as integrated approach of the characterization of collected biowaste.

Nowadays, the meaning of "waste" is based upon the perspective of circular economy and is directed toward the agrofood chain and the lifestyle toward a "zero waste" model.

**Corresponding author*: massimo.lucarini@crea.gov.it

Arindam Kuila and Mainak Mukhopadhyay (eds.) Biorefinery Production Technologies for Chemicals and Energy, (197–220) © 2020 Scrivener Publishing LLC

A shift from linear approach to circular economy will move beyond singular point solutions: transition to a more circular economy requires changes in the agricultural and industrial organization, stepping up efforts to increase resource productivity.

Sustainability and innovation are some of the keywords of a novel economic concept, named the circular economy, based on legislative proposals suggested and adopted by the European Community to increase global competitiveness, foster economic growth, and create new jobs by saving resources and energy. The aim of the circular economy is to "close the loop" of product lifecycles through greater recycling and re-use and create benefits for both the environment and the economy.

This approach can be applied in almost all manufacturing fields. For example, the agro-industrial field offers a good opportunity when considering the large amounts of waste and by-products produced every year during fruit and vegetable processing. In this area, a circular economy process can be achieved through efficient small and industrial scale bioenergy plants, biorefineries, and environmentally friendly processes to obtain bioactive compounds that can be used as active ingredients for agronomic, cosmetic, food, feed, and pharmaceutical formulations.

A biorefinery is a facility that integrates efficient and flexible conversion of biomass feedstocks through a combination of physical, chemical, biochemical, and thermochemical processes and equipment to produce multiple products: chemicals, materials, value-added chemicals at the core, together with the production of power, energy, biofuels using a range of sustainable and low environmental impact combined technologies [2, 3]. The biorefinery is defined as the conversion of all kinds of biomass (e.g., organic residues, energy crops, aquatic biomass) into different bio-based products, such as food and feed, chemicals, materials, fuels, power, and heat [4, 5]. The biorefinery goal could be simplified and described into the separation of the components of the biomass in its constituents with a specified use.

As well summarized by Lucarini *et al.* [6], and well-schematized biorefinery outputs are classified into a biovalor hierarchy that indicates the value of biomass transformations based on new circular agricultural or economic models: at the top of the pyramid, there are substances for fine and pharmaceutical chemistry, useful for the synthesis of vaccines, antibiotics, and immunotherapy proteins; further down the hierarchy of biorefinery products, there are food and feed products, followed by substances for the chemical industries, such as bioplastics, lubricants, solvents, adhesives, fibers, and dyes; at the bottom of the pyramid, there are all the substances for the production of biogas by fermentation and biofuels in the field of energy sector [6]. The more suitable technologies for separation, fermentation, gasification, and chemical conversion must be identified, as well as for biomasses' pre-treatment and storage and at the same time safety aspects should be taking into account.

The potential of new technologies can strengthen and enhance the "Green Economy" in agriculture and agro-industry, redefining the agro-production model where each component of the production cycle is "a resource", with economic, environmental, and social value which contributes to a more integrated and sustainable local environment. The central instruments of this model will be by-product, co-products, and waste (BCW) conversion systems, which will turn agribusiness BCW into new products. These systems will be tailored to the agricultural and agro-industrial framework of the regional context, involving all the actors of the production chain, from farm to industry and service sectors.

The importance of the cooperation of different sectors and sciences, i.e., chemistry, biology, environmental sciences, economics, statistics, and engineering, and includes research institutes and industries, from both food and non-food field is marked by Fava *et al.* [7]. Moreover, in many countries, synergistic networks are being developed to improve resource efficiency by using waste from one industry as a resource in another sector. The key to this new approach of raw materials, waste recovery and integration is achieving through efficient small and industrial scale of biogas plants, biorefineries, and environmentally friendly process for the production of enriched extracts in biomolecules from vegetal sources. At territorial level, the availability of biomass requires the development of integration processes within the relationship with suppliers, by allowing to access to a limited supply range to qualified materials to be included in the production processes.

Finally, it is worthy to mention the work of Accardi *et al.* [8] that underlined the primary importance of informing consumers about the benefits that can arise from the valorization of biomass as well as the safety aspects.

10.2 Experimental Pilot Plant

10.2.1 Chestnut

The use of products from sweet chestnut tree has recently been rediscovered and promoted as a natural and sustainable one in several economic and productive sectors, in geographic areas (such as Tuscany) where both chestnut and industries exploiting its products are widespread. The many different uses of sweet chestnut products concern chestnut wood, but also the fiber and natural extracts rich in hydrolyzable tannins, polyphenolic compounds typical of this vegetal species with interesting and useful properties. The presence of a high number of phenolic groups and their complex structure are what give tannins their main characteristics: they can interact both with polar groups of biologic macromolecules and with other substrates, such as metals. Therefore, extracts from sweet chestnut wood can be used as tanning agents for leather, as mordants or dying products for fabrics, paper, wood etc., to clarify wines and stabilize their organoleptic properties; but they also have specific biological properties that suggest more targeted uses as antioxidant, radical scavenging and antimicrobial agents [9–11]. Moreover, according to recent studies, they also have anti-inflammatory, antitumor, cholesterol-lowering, antiviral, nematostatic properties, and biostimulant activity on plant tissues [9, 12–15].

The so-called tannic acid, generally obtained from sweet chestnut (*Castanea sativa* Mill.) aqueous extract, is a typical product containing hydrolyzable tannins and it is known for its ability to have beneficial effects on human health through the expression of some biological activities, related to its antimutagenic, anticancer, and antioxidant properties. In addition, its ability to reduce serum cholesterol and triglycerides and to suppress lipogenesis by insulin has been documented [16]. Tannins are present in buds, stems, roots, seeds, bark, and leaves, commonly with a weight ranging between 5% and 10% on dry plant material, so it's possible to extract them as a co-product from the waste of other productive industrial processes such as the one of furniture. This allows to integrate the extraction of tannins as a step of production in biorefinery plants.

Concerning the industrial sustainable exploitation, thorough considerations on extraction and fractionation processes are important to isolate and stabilize noble fractions high in active molecules, because of the complex composition of the extracts and the low stability of hydrolyzable tannins in water at high temperature during extraction. On the other hand, hot water is a suitable solvent to extract efficiently polyphenolic compounds, hydrolyzable tannins in particular, from wood matrices, and to keep the process eco-friendly at the same time. The partial hydrolysis in water of high weight chestnut tannins yields lower weight polyphenols, ellagic acid, gallic acid, and glucose, and,without prejudice to the need for standardization and stabilization of the finished products, the low stability of high weight tannins during extraction does not necessarily compromise their biological activity as demonstrated by previous studies [17, 18]. On the contrary, it is essential to consider the stability of the finished products to evaluate their shelf life and the possibility of commercialization and use for new product formulation.

Figure 10.1 shows the operating diagram of the industrial extraction and fractionation/ concentration plant operating in Radicofani (Siena), Tuscany, for Mauro Saviola Holding [6, 11]. This process is complementary to that for the production of MDF (Medium Density

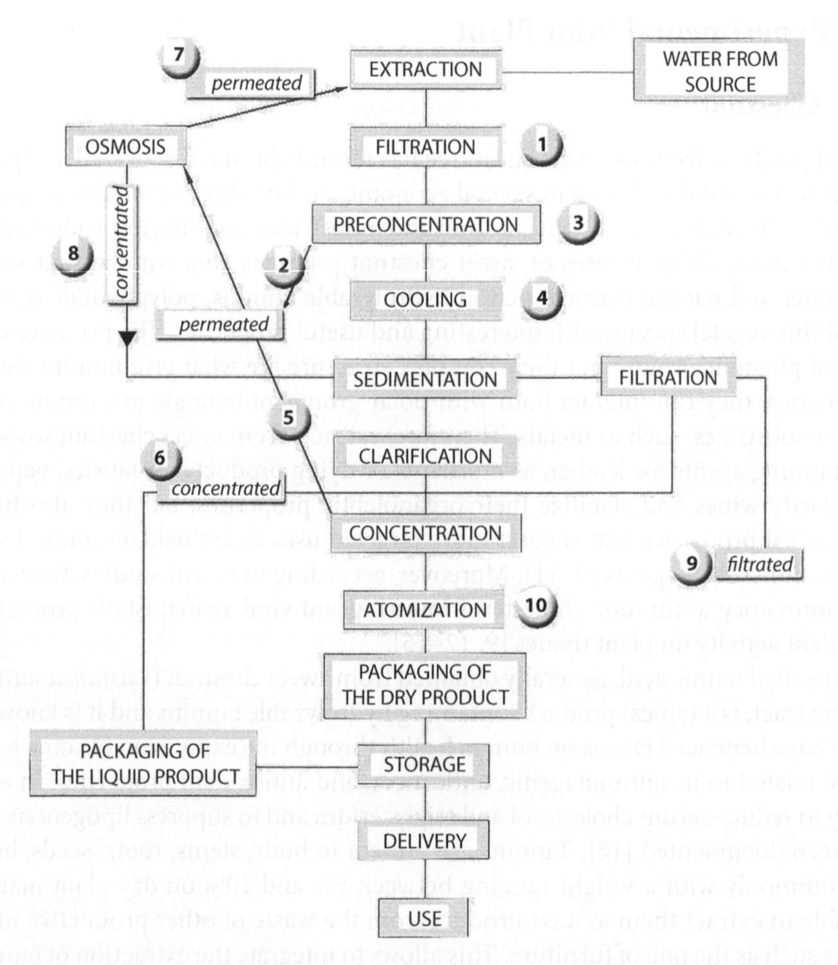

Figure 10.1 Operating diagram of the Mauro Saviola Holding extraction and fractionation plant: 1) filtered tannin broths; 2) permeate from nanofiltration step-1; 3) concentrate from nanofiltration step-1; 4) concentrate from nanofiltration step-2; 5) permeate from nanofiltration step-2; 6) concentrate from nanofiltration step-3; 7) osmosis permeate; 8) osmosis concentrate; 9) settled fraction from clarification step; 10) spray-dried obtained from fraction 6 [6, 11].

Fiberboard) panels that requires the removal of tannins from the wood before introducing the glue, so that the panels can be modeled before hardening.

The plant is suitable for the production of purified and concentrated aqueous extracts rich in hydrolyzable tannins from sweet chestnut and, in general, for the extraction of water-soluble active principles from plant matrices, using low percentages of ethanol if necessary. The final refined products are obtained via a membrane separation technology system coupled to the extraction system. The extraction chamber is fed with 20 m^3 of biomass and the extraction solvent can reach a temperature of 80°C. The boiler is fed by using the exhausted biomass that has already undergone the polyphenol extraction process, so that the plant can be entirely powered by cogeneration according to the specific process parameters and calorific values of the different matrices (wood matrices have an average calorific value of 2,000–5,000 Kcal/Kg; the boiler can exploit 10 tons of biomass each batch) [11]. The purification and concentration steps are performed by membrane technology, avoiding the use of the organic solvents usually employed in industrial purification processes. This methodology also allows, when necessary, for a selective concentration of specific subclasses of compounds with different chemical and biological properties. The final product was a concentrated extract or spray-dried powder obtained with an average yield of 5% with respect to the vegetal material [6, 11].

Fractions 6 (liquid, concentrate from nanofiltration) and 10 (spray-dried obtained from fraction 6) are commercially available; the other fractions, not yet marketed, are reintroduced into the process to be further refined or added to the extraction water. The use of only water as solvent makes ecologically and economically sustainable the described extraction and purification method, whereas the industrial processes are usually based on purification through extraction with organic solvents. Moreover, the use of a co-product for obtaining differentiated and refined marketable fractions eliminates many of the ecological and economic issues associated with the disposal of production waste.

The two commercial fractions of sweet chestnut have been analyzed and characterized by chromatographic, spectrophotometric, and spectrometric (HPLC-DAD-ESI-MS) methods in order to identify and quantify the secondary tannic metabolites. In Table 10.1, the individual compounds quali-quantitative analysis is reported [11].

The analyses performed on all fractions from the described process (data not shown) demonstrated that the membrane filtration steps are quite effective for concentrating the extracts and separating compounds according to their molecular weights and chemical characteristics, in particular to separate gallotannins from ellagitannins and gallic acid from the other compounds (mainly gallic and ellagic esters of glucose). The osmosis permeate (fraction 7), very diluted, contains 2.68 mM polyphenols, 100% of gallic acid. Tests for further optimization of the process, based on variations of the membranes molecular weight cut-offs and filtration/concentration times and pressures, leaded to obtaining osmosis permeates more concentrated in gallic acid with comparable purities. This allows to hypothesize the production of a concentrated fraction of gallic acid, marketable for use in various sectors.

The dry commercial fraction was stable at the 6-month follow-up analytical control and at all successive analytical controls through time; for the liquid fraction some stability issues were detected, in particular vescalagin and castalagin were partially hydrolyzed yielding vescalin and castalin. This is not currently a problem because the liquid fraction is used essentially in the tanning sector where the product performances are not affected by slight

Table 10.1 Individual compounds quali-quantitative analysis of sweet chestnut commercial fractions: liquid (fraction 6) and dry (fraction 10). Results in mg/g of sample. Data from Campo *et al.* [11].

Compound	Content in commercial liquid fraction	Content in commercial dry fraction
	mg/g	mg/g
Vescalin	3.57	8.68
Castalin	3.21	8.83
Pedunculagin I	3.31	10.1
Monogalloyl glucose I	2.51	4.69
Gallic acid	1.53	13.6
Monogalloyl glucose II	2.68	5.03
Roburin D	4.03	9.61
Vescalagin	26.7	45.2
Dehydrated tergallic-C-glucoside	1.77	6.03
Castalagin	21.0	39.7
Digalloyl glucose	3.79	12.2
O-galloyl-castalagin isomer	20.8	32.0
Trigalloyl glucose	5.03	12.1
Tetragalloyl glucose	4.72	9.18
Ellagic acid	2.50	7.80
Pentagalloyl glucose	3.67	0.00
Total tannins	111	225

Table 10.2 Measured values of: GAE. Gallic Acid Equivalents (g gallic acid/100g sample); EC_{50}. extract concentration inhibiting DPPH· activity to 50% (g/L sample); TE. Trolox Equivalents from ORAC assay (μmol/g sample), for commercial sweet chestnut fractions. Data from Campo *et al.* [11].

	GAE	EC_{50}	μmol TE/g
Liquid fraction	30.991	0.010	450.4
Dry fraction	56.990	0.002	3050.1

variations in the concentration of the individual tannin compounds. For applications in different sectors, new formulations will be tested to stabilize the tannins in solution.

Data concerning antioxidant and radical-scavenging activities of commercial fractions, evaluated by *in vitro* assays with Folin-Ciocalteu reagent, stable radical DPPH· (1,1-diphenyl-2-picrylhydrazyl), and ORAC assay, are shown in Table 10.2.

GAE (Gallic Acid Equivalents) values and EC_{50} and TE values are the results of Folin-Ciocalteu, DPPH and ORAC assays, respectively. The Folin-Ciocalteu assay results (GAE), expressed with respect to the sample weight, depend on the content of antioxidant compounds (hydrolyzable tannins) in the extracts and should give an idea of their antioxidant capacity. The EC_{50}s is the extract concentration inhibiting DPPH· activity to 50% in aqueous solutions; TEs are expressed as μmol Trolox with respect to the sample weights.

Evaluating the antioxidant activity of the extracts is important not only because many of the applications of sweet chestnut tannins are based on this biological property, but also because the presence itself of antioxidant compounds within commercial formulations is useful to preserve and extend the shelf-life of products, thus avoiding the use of chemicals.

The biological properties and green agricultural applications of sweet chestnut tannins were tested under the project LIFE+ "Environmentally friendly biomolecules from agricultural wastes as substitutes of pesticides for plant diseases control - EVERGREEN" (LIFE13 ENV/IT/000461) over the period 01/10/2014 to 30/09/2016. The EVERGREEN UE project consisted of several actions related to each other, able to demonstrate the validity and effectiveness of innovative and standardized preparations of high quality polyphenolic-based molecules from agricultural vegetable biomass and waste plant pathogenic bacteria and nematodes and to control the diseases they cause on plants. The EVERGREEN polyphenolic extracts, co-formulated with chestnut tannins, are active in plant protection against phytopathogenic Gram-negative bacteria and nematodes, as demonstrated by applying four optimized formulations on model systems at pilot-scale and field screening, with beneficial effects on soil microflora [18–20].

The studies carried out in the EVERGREEN UE project were based on previous scientific results concerning the optimization of the extraction and purification process, the chemical characterization and standardization of the sweet chestnut fractions under study, and the biological activities such as antioxidant, antimicrobial, and biostimulant [9, 21, 22]. The antimicrobial activity of diluted sweet chestnut extracts was also successfully exploited in the curing process of tobacco, to minimize the fermentation processes leading to the formation of nitrosamines, compounds responsible for part of the toxic effects of smoking [23]. Further studies are in progress to evaluate the possibility of exploitation of sweet chestnut hydrolyzable tannins for different purposes in agronomy and environment sectors, as biostimulant and antimicrobial agents for partial or total replacement of the traditional chemicals, and also for the formulations of active substrates for a possible phytoremediation action on soils contaminated by organic pollutants derived from industrial activities [24–26].

For the leather industry, the chestnut extract is a natural and very high quality alternative compared to the use of chemical products for tanning with a higher environmental impact. Within the activities of the R&D project POR FESR 2014-2020 of Region Tuscany "Green for Fashion", ("G4F"), sweet chestnut extracts are in a test phase for the development of new bags and accessories completely made with environmentally sustainable green methods of vegetal tanning and dyeing, with beneficial effects on the environment but also for the users.

For the zootechnical industry, the chestnut extract feed is an astringent and natural antiseptic that drastically reduces the use of antibiotics and can have a positive effect on derived products [27–29].

Sweet chestnut is an official species with regard to leaves, cupules, and bark, so that these parts of the plant can be used in the cosmetic, food, and oenological sectors to prolong the shelf lives of the products thanks to the antioxidant and antimicrobial activities, allowing for a drastic reduction or the elimination of the traditional chemicals. The same biological properties can also have positive effects on the health of the users. Sweet chestnut extracts are also in a test phase in the wine sector to optimize new natural green formulations with other vegetal products to clarify wines and to stabilize their characteristics [30–32].

Further studies are in progress to assess the biological properties of hydrolyzable tannins from both sweet chestnut and other vegetal species and to test the possible applications in different sectors [33].

Nowadays, new and innovative applications for sweet chestnut extracts were found, and the traditional uses have been improved by applying the latest technical and scientific results and the modern criteria proposed by the circular economy.

10.2.2 Soy

Soy is one of the few plants that provides a complete protein as it contains all eight amino acids essential for human health and the oil has a unique fatty acid composition that is relatively unaffected by environmental factors. Soybeans are the dominant oilseeds worldwide and they make up well over one-half of world oilseeds production. About 85% of the world's soybeans are processed annually into soybean meal and oil. Approximately, 98% of the soybean meal that is crushed is further processed into animal feed with the balance used to make soy flour and protein for food use. Of the oil fraction, 95% is consumed as edible oil; the rest is used for industrial products such as fatty acids, soaps, and biodiesel. Food uses of soybeans include traditional soy foods such as tofu and soymilk. Soy ingredients have become staples in the food manufacturing industry. Lecithin is widely used as an emulsifier; since the 1970s, partially hydrogenated soybean oil has been a mainstay in the production of snacks, baked goods, salad dressings, and other foods. Soy protein ingredients play functional roles in baked foods, processed meats, and other products. In addition to being used for their functional characteristics, soy ingredients are used to add nutrition to processed foods; some isolated soy proteins, for instance, are specifically designed to be used in acidic or clear beverages—products that could not, until recently, be protein-fortified. Soybeans are also processed into many industrial products such as biodiesel. In addition, soybeans are processed into hydraulic oil, grease, solvent, ink, plastics, and other products. In all, soybean shows a remarkable record of achievement, resulting from a combination of science and technology with development of a new crop and industry.

Several constituents of medical interest have been isolated from soybeans, including isoflavones, phytoestrols, protease inhibitors, inositol hexaphosphate, and saponins. Isoflavonoids are a category of polyphenols, mainly found in legumes such as soybeans [34], which health benefits have been extensively investigated. The isoflavones genistein and daidzein are among the most abundant phytoestrogens in human diets and are found predominantly in legumes like soy.

In recent years, due to the importance of isoflavonoids especially in the diet of oriental people, most research on the flavonoid content of soybean has been devoted to this particular group of polyphenols [35]. Isoflavones are structurally similar to the mammalian oestrogen, oestradiol-17β, and exhibit oestrogenic properties [36]. The role of isoflavones in the diet is claimed to be as a protective compound in cardiovascular diseases, osteoporosis, menopausal symptoms, and breast and prostate cancers [37–40].

The structure backbone of isoflavones is show in Figure 10.2. The main rapresentatives components are genistein, daidzein, Biochanin A, and glycitein.

The study of Romani A. *et al.* [34] investigates the polyphenol content in different plant parts (roots, stems, leaves, cotyledons, pods, and seeds) of three not exhaustively characterized soybean (*Glycine max* L.) varieties, grown under natural conditions, and inoculated with *Rhizobium japonicum* (*Bradyrhizobium japonicum*) that fixes N2 into a form that the plant can use. After the inoculation, growth of the seedlings was followed for three months in order to obtain the fully developed parts of the plant, including seeds.

In Table 10.3, the quantitative data for the "Emiliana" cultivar from sampling III are reported together with the identified compounds. The results show that the highest amount of isoflavones is found in the roots, with a great predominance of daidzein derivatives relative to genistein derivatives, while in the leaves, only genistein derivatives were found. The leaves are the part where the most flavonols were found, while smaller quantities were detected in pods and stem. The same evaluation was performed for the three cultivars at the three sampling dates; the polyphenol content of all three cultivars were similar. To our knowledge, this is the first investigation which analyzed both quercetin and kaempferol glycosides from a qualitative and quantitative point of view. In leaves of all three cultivars, irrespective of the sampling age, only genistein and its derivatives were found.

Indeed, isoflavones were present in most of the investigated parts (except pods) and the amounts change with the sampling period and the cultivar. Flavonols of the leaves were characterized and their amount was evaluated so that, it may be possible to extract polyphenols from the different parts in order to exploit the whole plant also for its food use [34].

Many papers have dealt with the separation of phytoestrogens in soybeans, soy foods, and human biological fluids. Among the different separation methods, HPLC plays an important role: the main improvements concern the kind of material from which isoflavones must be extracted and the detection limits. Also, the time needed to perform the complete analysis plays an important role; in fact, the presence of polyphenols in food increases both quality and commercial value since the food may subsequently be regarded as "health food". It is worth mentioning the study of Heimler D. *et al.* [41] was to optimize a rapid chromatographic method to characterize and quantify the isoflavones content of

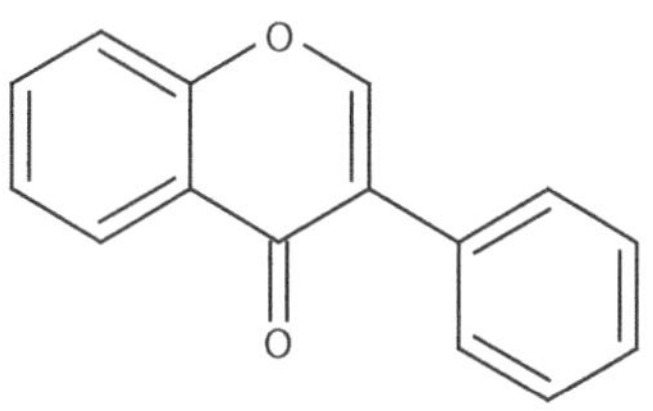

Figure 10.2 The structure backbone of isoflavones subclass.

Table 10.3 Composition of different soy parts (cv. Emiliana) collected after 77 days. Data are expressed as mg/kg of dry weight. Average value ± SD of three soy samples . Data from Romani *et al.* [34].

Compounds (Rt)	Stem	Leaves	Roots	Pods
Genistein-glycoside (29.2 min)	86.4 ± 1.7	55.1 ± 1.1	n.d.	n.d.
Genistein-7-*O*-glucoside (44.9 min)	115.3 ± 1.8	1649.3 ± 20.0	traces	n.d.
Genistein-malonylglycoside (54.6 min)	17.3 ± 0.5	55.1 ± 1.1	n.d.	n.d.
Genistein-7-*O*-malonylglucoside (57.9 min)	691.6 ± 9.6	202.9 ± 3.5	805.7 ± 11.5	n.d.
Genistein acetylglycoside (64.9 min)	17.3 ± 0.5	17.4 ± 0.3	n.d.	n.d.
Genistein (74.3 min)	34.6 ± 0.7	n.d.	238.8 ± 4.2	n.d.
Daidzein-7-*O*-glucoside (30.7 min)	n.d.	n.d.	719.7 ± 10.1	n.d.
Daidzein-malonylglucoside (40.7 min)	n.d.	n.d.	471.3 ± 7.7	n.d.
Daidzein-malonylglucoside (42.9 min)	n.d.	n.d.	592.3 ± 8.4	n.d.
Daidzein-7-*O*-malonylglucoside (51.8 min)	890.5 ± 10.1	n.d.	3,665.6 ± 36.4	n.d.
Daidzein (59.8 min)	n.d.	n.d.	3,710.2 ± 32.2	n.d.
Coumestrol-7-*O*-glucoside (52.6 min)	n.d.	n.d.	703.8 ± 9.9	n.d.
Coumestrol-malonylglucoside (60.5 min)	n.d.	n.d.	2,496.8 ± 29.9	n.d.
Coumestrol (76.1 min)	n.d.	n.d.	875.8 ± 10.2	n.d.
Quercetin-triglycoside (30.0 min)	n.d.	1,098.5 ± 14.9	n.d.	1,011.6 ± 13.9
Quercetin-triglycoside (30.6 min)	n.d.	771.0 ± 9.8	n.d.	n.d.
Quercetin diglycoside (32.0 min)	n.d.	201.0 ± 3.0	n.d.	n.d.
Quercetin diglycoside (32.9 min)	n.d.	371.0 ± 4.05	n.d.	n.d.
Quercetin diglycoside (33.6 min)	n.d.	199.1 ± 2.95	n.d.	n.d.
Methyl-quercetin-triglycoside (38.7 min)	n.d.	510.1 ± 7.1	n.d.	n.d.
Methyl-quercetin-diglycoside (41.9 min)	n.d.	327.5 ± 4.7	n.d.	n.d.
Quercetin-diglycoside (43.1 min)	n.d.	452.2 ± 7.2	n.d.	n.d.

(*Continued*)

Table 10.3 Composition of different soy parts (cv. Emiliana) collected after 77 days. Data are expressed as mg/kg of dry weight. Average value ± SD of three soy samples . Data from Romani *et al.* [34].

Compounds (Rt)	Stem	Leaves	Roots	Pods
Quercetin diglycoside (44.6 min)	n.d.	501.2 ± 7.9	n.d.	n.d.
Quercetin diglycoside (45.3 min)	n.d.	452.2 ± 7.4	n.d.	n.d.
Methyl-quercetin-diglycoside (52.5 min)	n.d.	411.6 ± 5.7	n.d.	n.d.
Methyl-quercetin-diglycoside (52.9 min)	n.d.	307.2 ± 4.4	n.d.	n.d.
Kaempferol-triglycoside (36.2 min)	n.d.	559.4 ± 8.1	n.d.	n.d.
Kaempferol-triglycoside (37.9 min)	n.d.	626.1 ± 9.3	n.d.	n.d.
Kaempferol diglycosides (39.6 min)	n.d.	307.2 ± 4.7	n.d.	n.d.
Kaempferol-diglycoside (40.6 min)	89.3 ± 1.7	220.3 ± 3.7	n.d.	110.4 ± 2.3
Kaempferol-diglycoside (48.9 min)	106.6 ± 1.8	115.9 ± 2.1	n.d.	866.3 ± 10.9
Kaempferol glucoside (50.3 min)	n.d.	46.4 ± 1.0	n.d.	n.d.
Kaempferol-diglycoside (53.0 min)	n.d.	153.6 ± 3.7	n.d.	n.d.
Kaempferol (63.6 min)	n.d.	153.6 ± 3.3	n.d.	n.d.
Caffeic acid derivatives	54.7 ± 1	1927.5 ± 28	79.6 ± 2	58.1 ± 1

soybean seeds, one of soy's edible parts. The method was then applied to five different soy cultivars grown under natural conditions.

Moving into soybean oil, soybean oil processing has three major steps: preprocessing, extraction and separation, and post-processing. Practically, all soybean oil is produced by solvent extraction, except in some mechanical extraction facilities in small-scale plants throughout the world [42]. Mechanical extraction has low capital cost and no solvent requirements; however, the efficiency of mechanical oil extraction is much less than that of the solvent extraction method.

The extraction mechanism can be described in three the following steps: 1) penetration of solvent into the soybean solids; 2) solubilization of oil in the solvent; 3) oil diffusion from the solid matrix into the liquid solvent phase. The oil concentration gradient from the solid matrix to the liquid phase is the driving force of diffusion.

There are wide prospects in the innovative uses of conventional products (oil production to fair trade and protein meal for animal and human use at high commercial value), but especially in the active compounds from husk panel of exhausted and vegetable tissues. The versatility and potential usage of soybean promotes the enhancement of comprehensive and integrated plan.

From Romani *et al.* [43], the possibility to consider soybeans waste matter as source of bio-product possessing a high added value: they are aiming at the development of new products, at exhibiting enhanced functional and bioactive properties to be implemented in the sector of phytoterapy, cosmetic, and agro-food industry (isoflavones production).

For this reason, they consider all input/output data based on the energy and material balanced calculation at each stage of soybean oil production to conduct a simplified Life Cycle Assessment (LCA) study. Such interdisciplinary approach may allow a preliminary study of environmental evaluation (through involving of local industries) and results could be utilized in the future as inputs to conduct an Initial Environmental Review in ISO 14001, covering the aspects of an Environmental Management System (EMS). LCA preliminary results could be also used as inputs to obtain environmental product labels (such as Ecolabel or EPD—Environmental Product Declaration) for finished products.

The plant provides the integration of the extraction and refining process of soybean oil with extraction and fractionation of active principals of biological/alimentary interest. Thus, it is possible to optimize the use of all vegetable waste (biomass) and produce electricity and heat using vegetable oil as fuel (Table 10.4).

The plant was organized on two main production lines:

1. Oil line: line of extraction of soybean oil for food, for bio-lubricants, for bio-fuels (bio jet fuel) and for energy; co-production of protein meal for animal and human use;
2. Pharma line: line of extraction of bioactive compounds (particularly isoflavones) from soybean waste matter (tissues plants and waste processing) possessing a high added value.

Table 10.4 Raw materials, products, and applications. Data from Romani *et al.* [43].

Raw material	Products	Applications
Dried soybean 11%	Oil for food	Food consumption
	Bio-oil for energy	Cogeneration Bio Jet fuel – fuel for aviotransportation
	Bio-lubricants	Textile industry, food, pharmaceutical, and cosmetic
	Soy meal	Food consumption and lifestock breeding
Waste processing and plant tissues	Isoflavones	Pharmaceutical Food supplements Food advanced Cosmetics Basic food

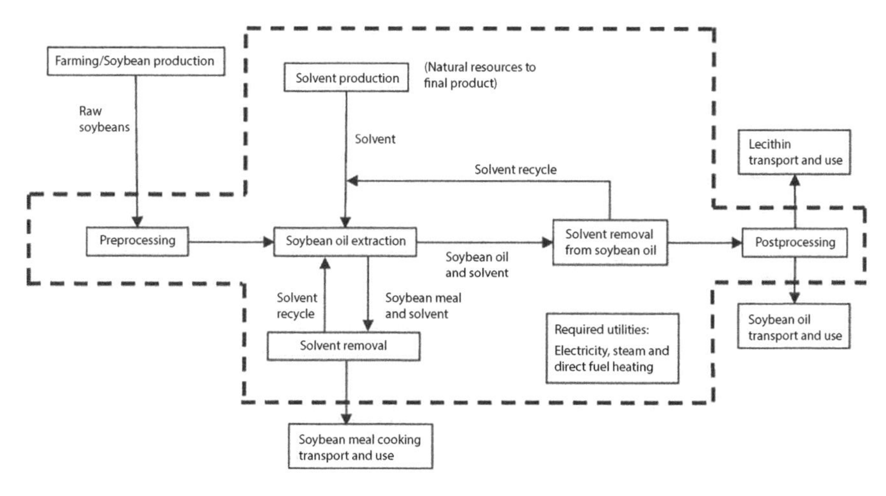

Figure 10.3 Overview of the soybean oil production process [43].

The oil line and the subsequent production of protein meal, working continuously 24 hours daily, consists on the following phases (Figure 10.3):

A) Soybean preparation and oil extraction line:

1. trucks unloading
2. preliminary storage and selection of raw materials (division between the seed and the leaf and stalk)
3. Industry acceptance, storage, and preparation of biomass (seeds, leaves, and residue processing)
4. precleaning and seeds dryer
5. seeds storage
6. continuous soybean conditioning and dehulling plant: seeds cleaning, seeds cracking, seeds dehulling, seeds flaking, and seeds expander
7. continuous soybean solvent extraction plant: solvent extraction, solvent distillation, meal desolventizing, drying and cooling section
8. continuous soybean oil water degumming plant and recovery of lecithin
9. control board and automation section

B) Soybean oil physical refining line (post-processing):

1. continuous multipurpose plant for special degumming or hot chemical neutralizing
2. pneumatic bleaching earths handling plant
3. continuous dry degumming and bleaching plant
4. fully automatic filtration section
5. continuous deodorizing/physical refining plant for vegetable oil

In this integrated process, plant will also be combined:

- Bio fuel for aviotransportation and bio-lubricants production
- Cogeneration plant to produce electricity and heat for use in both production lines
- Storage of co-products and additives
- Hangar for meal and refined oil tanks
- Storage of dried plant material (already processed)
- Expediting area
- General services area of plant and guardian
- Auxiliary facilities protection (fire, water, and wastewater treatment, etc.)

LCA is an objective evaluation procedure to examine the energetic and environmental impact related to a product, process, or activity. LCA applications can evaluate the effects of the complete life cycle product [44], in terms of potential impact, on human health, ecosystem quality, depletion of resources, and climate change. For the Life Cycle Impact Assessment (LCIA) phase, the Impact 2002+ method [45] is applied. The Impact 2002+ evaluation method is one of the latest methods; it represents the evolution and the update of the widespread Eco-indicator 99 method. Impact 2002+ allows to calculate the mid-point categories and the end-point categories, it comprehends four damage categories: Human Health, measured in DALY (Disability Adjusted Life Year); Ecosystem Quality, measured in PDF*m^2*yr (Potentially Disappeared Fraction); Climate Change, measured in kg of CO_2 equivalent in air, that derives from impact category Global warming; Resources, measured in MJ; Non-renewable energy and mineral extraction. For the study, the Sima Pro code will be used and the database Ecoinvent [46].

In regards to the Pharma line, the attention is focused at the isolation, characterization, and quali-quantitative evaluation of natural compounds present in these vegetal matrices.

The exploitation of all chemical components comes through preliminary processing, shredding, hydrolysis, and chemical separations (membrane techniques and chromatography). In particular, the isoflavones extraction plant is based on membrane techniques as microfiltration (MF), ultrafiltration (UF), nanofiltration (NF), and reverse osmosis (RO). The plant has a capacity to treat 10 of m^3/day of water solution (85°C), resulting from decoction of plant tissues and waste processing.

Soy extract, obtained from soy plant, used for the production of food supplement is 40% in isoflavones. Table 10.5 lists the amounts of the different isoflavones from HPLC in the two commercial products.

As a preliminary step of the study, the standard LCA methodology was used for the identification of the environmental impacts of the first part of the process: from soybean cultivation to soya oil extraction. The functional unit (FU) is 1 kg of soya oil produced. The system boundaries include: the soybean cultivation and the soybean production at farm (the processes of soil cultivation, sowing, weed control, fertilization, pest and pathogen control, harvest and drying of the grains, machine infrastructure, and a shed for machine sheltering); the transports of soybean from farms to soy oil producers; the soybean conditioning; the soybean solvent extraction. The filtration membrane technologies, the post-processing phase and the soybean meal cooking are not included. The 77% of impact assessment is due to soybeans production at farm (Figure 10.4), the kg/CO_2eq. value is 1.4 for 1 kg of soy oil produced, and the Ecosystem Quality are the damage category most affected.

Table 10.5 Amount of different isoflavones (mg/g) from HPLC data. D = daidzein, Gly = glycitein, G = genistein, gluc = glucoside, acgluc = acethylglucoside. Data are the mean of three determinations (standard deviation < 3%).

Compounds	**Sample 1**	**Sample 2**
G der	0.057	0.069
Gly der	0.112	0.163
D 7-O-gluc	9.178	13.091
Gly gluc	0.575	0.941
G 7-O-gluc	32.011	35.781
D der	traces	traces
D aclgluc	0.247	0.561
G der	1.610	1.189
D	4.213	3.775
Gly	0.779	1.212
G	4.266	4.115

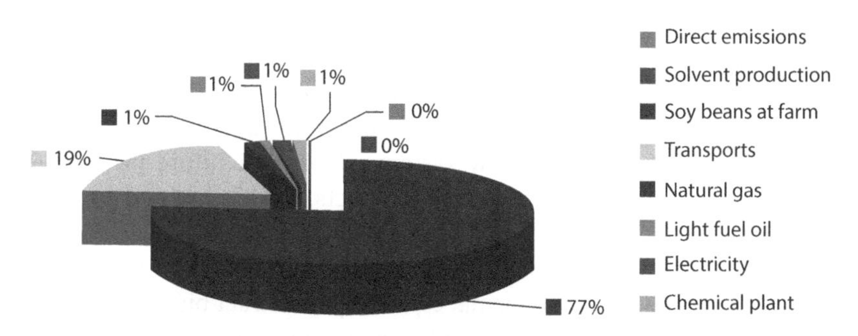

Figure 10.4 The processes contribution of preliminary LCA soybean oil production process (FU: 1 kg). Data from Romani *et al.* [43].

The results obtained from the standard LCA methodology could be utilized as inputs to conduct an Initial Environmental Review [44], following the requirement of standard ISO 14001, to implement an EMS of the proposed platform. At present, the results focused the environmental impacts of the first part of the process: from soybean cultivation to soya oil extraction. Further feasibility studies could be made to use LCA preliminary results to determinate technical requirements in order to obtain environmental (such as Ecolabel or EPD) labels for finished bio-based products.

One of the main targets it's to develop new biodegradable products derived from soybeans oil to be applied in many different industrial and manufacturing applications. This new bio-based products can directly replace traditional petroleum based lubricants and fuels.

Vegetable oil, refined and blanched, can be easily fueled into combined power system's engine to produce electric and thermal energy. Vegetable oils are environment-friendly fuels, which do not contribute to global greenhouse effect (they are considered a renewable energy source) and contain low levels of other pollutants. The result is an integrated agricultural and energetic system, which uses locally available plantations to produce low-emissions and relatively low-cost electricity. The power station is based on diesel reciprocating engines able to convert the chemical energy of fuels into mechanical energy. The thermal energy cascade can be reused to supply heat and cool system for greenhouses. When compared to conventional lubricating products, bio-lubricants offers superior corrosion protection, is biodegradable, and is safe to use. It is recommended for use in a wide variety of applications including mould release, a general-purpose lubricant, and preservative oil for corrosion prevention. In addition, the nature of the soybean oil used to formulate bio-lubricants gives it excellent penetrating properties, useful in loosening frozen nuts and bolts. This kind of products can be used for industrial, shop, or home use; it can be used indoors or outdoors and will not harm most plastics.

Regarding on biofuels, starting from the soybean oil based and selecting many different chemical process and bio-addictives (extracted and refined from soya vegetable tissues), it's possible to design renewable fuels for aviotransportation.

Our environmental strategy has long been focused on designing fuel-efficient airplanes. Better airplane fuel efficiency means reduced airplane emissions of carbon dioxide, the primary gas linked to climate change concerns. Thematically, these are new sources of fuel, but chemically, they contain the same types of molecules that are in traditional petroleum jet fuel. The new crops can include algae and camelina too but chemical compatibility is essential, because a new fuel won't be commercially viable if it can't be dropped into the existing aviation infrastructure.

Last but not least, target is to have a complete self-reliance production system able to co-generate electric and thermal energy to be use into agro-industrial platform. The agro-industrial platform will be the new system where agriculture potential meets the industrial needs to cooperate into renewable products production and to proposed new added value goods.

This platform will be integrated with greenhouses to develop new vegetable kinds oriented to nutraceuticals and natural pharma, especially to extract bioactive compounds.

The greenhouses will be equipped with photovoltaic modules installed on the roof to produce electric energy only. The thermal energy (necessary to the plant during the winter period to have a rapid growth) will be provided by engines fuelled by vegetable oil (liquid biomass) that it is able to produce electric energy recovering the downstream thermal energy (renewable power genset).

Thanks to the European Regulation for renewable energy, Romanian government supported the development and the expansion of solar photovoltaic in Romania. The feed-in tariff had a great success, and like in Italy, Spain, and Germany, big size photovoltaic plants have been built, especially in the south of the country. The south of Romania presents high irradiation (up to 1,300 equivalent hours) and economically agriculture is an important local resource. This aspect led to the concept of the solar greenhouses, a way to match the electricity production by PV modules with an improvement of the cultivation possibilities. Solar greenhouses include integrated PV modules mounted on the roof oriented to south and his design is new and still has to be evaluated in details. In particular, important parameters like the luminance, the type of cultivations and the temperature of the PV modules must be carefully analyzed to have a real good match between the agriculture and the electricity production purpose.

The greenhouse system combined with PV modules and bio-oil fuelled genset can provide to agriculture a new tools to fight the winter period improving the economic performance of a in crisis sector like agriculture is.

The management of biomass will be performed under a "zero discharge" cycle where starting from the soybeans, it's possible to produce vegetable oil, meals, bio jet fuels and bio-lubricant, powering engines, and solar greenhouses to produce renewable energy and recovery thermal energy to sustains the whole process energy consumption.

10.2.3 Olive Oil By-Products Case Studies

10.2.3.1 Olive Oil Wastewater

The olive oil processing produces large amount of waste, including olive mill wastewaters, leaves and olive pulp, which represent an environmental problem for their high toxicity and a cost of disposal. At the same time, they are a precious source of bioactive compounds including low-molecular weight phenols. The extraction of the new biomolecules from the olive oil milling can be considered an integrated step of the circular economy process applied to the olives processing. The use of a by-product for obtaining differentiated and refined marketable fractions eliminates many ecological and economic problems associated with the disposal of EVOO's production waste. Olives' aqueous fraction is a high interest raw material for the food, feed, and pharmaceutical industry. Rodis *et al.* [47] reported that 53% of the olives' phenolic compounds are lost in the vegetable waters and 45% in the pomace, whereas only 2% are found in olive oil. Olive mill wastewaters are the main waste produced from three-phase olive processing.

Pomace and vegetable water composition are affected mainly by the olives harvesting time and the extraction method. Bitler *et al.* [48] reported the anti-inflammatory activity of hydrolyzed olive vegetation water in mice.

The work of Quaglia *et al.* [49] has led through a series of technical and experimental studies to the creation and characterization of functional ingredients, obtained from residues of two- and three-phase mills' wastes, which have been eventually used in food preparations with high nutraceutical value. In the work of Quaglia *et al.* [49] samples were: vegetable waters from three-phase extractors, wet olive pomace from two-phase extractors, dry pomace without kernels from two-phase extractor with Jambo 2 separator to separate the kernels.

After freeze-drying, three different heat treatments were performed, simulating respectively the pasteurization, cooking, and sterilization treatments.

The authors [49] concluded that mill by-products does not show any critical issue about respecting hygiene requirements but the high concentration of active ingredients imposes strict control on their use. For the realization of these products can be advantageously used both the vegetable water and the dry pomace without kernels. Even the wet pomace, after removing the pits, could be used to make these products too. Kernels removing process can be done whether at the time of pressing, or on the dehydrated product.

10.2.3.2 Olea europaea L. leaves

Olea europaea L. leaves, a typical herbal drug of the Mediterranean region, have been widely used like traditional remedy as extract, infusion, herbal tea, and powder in countries such as Greece, Spain, Italy, France, Turkey, Israel, Morocco, Albania, and Tunisia. Olive leaves are the source of many bioactive compounds, the main of which is oleuropein, a secoiridoid, which can constitute up to 6%–9% of leaf dry matter. Oleuropein and its derivatives exhibit specific biological activities as antioxidant, antihypertensive, antiatherogenic, anti-inflammatory, hypoglycemic, hypocholesterolemic, antiproliferative, and antifungal. The composition of leaves extract has been studied, and active compounds were identified such as secoiridoids, flavonoids, and triterpenes. Olive leaves may be regarded as a by-product in the cultivation of olives both for olive oil and table olives during pruning operations and/or during olive harvest; leaves extract is used to prepare commercial affordable dietary supplements. Extraction process in order to obtain commercial supplements needs quite constant starting material while it has been pointed out that leaf polyphenols content depends on cultivar, geographic production zone, and time of olive leaf harvesting.

From the quantitative determination of flavonoids and secoiridoid derivatives of leaves, subjected to different treatments, the final product, i.e., dietary supplements and/or dry leaves, or extracts used for pharmaceutical purposes, can be achieved with a quite constant content of bioactive compounds. Romani *et al.* [50] set up a method, which was tested to characterize and quantify secondary metabolites (oleuropein and its derivatives, flavonoids, hydroxycinnamic acids, hydroxytyrosol, and elenolic acid derivatives) in *Olea europaea* leaves extracts. The aim of the study was the characterization of fresh, refrigerated, frozen, dried, and lyophilized *Olea* leaves of different cultivars under various extraction conditions. The identification of the best operating conditions, which may help in obtaining a high and almost constant bioactive products yield when *Olea* leaves are used in the achievement of commercial food supplements, is the further goal of the study.

Twenty-five compounds, among which flavonoids and secoiridoids, were separated and quantified after extraction from *Olea europaea* L. leaves. Differences were found in total polyphenols content and in oleuropein depending on cultivar, production area, sampling time (pruning or harvest time), and state of leaves (fresh, refrigerated, dried, frozen, or lyophilized). Polyphenols content in fresh leaves ranged from 34.21 to 7.87 mg/g, while oleuropein content changes from 21.03 to 2.79 mg/g in fresh leaves of different cultivars and decreases after the drying process. In addition, five commercial food supplements from olive leaves were analyzed, and their total polyphenol, secoiridoids, and flavonoid contents were detected by HPLC/DAD analysis.

The commercial products analyzed are used as antioxidants and/or as arterial blood pressure modulators. Oleuropein content and stability has been demonstrated as related to both the drying process and the extraction temperature; this occurrence has never been pointed out before. The bioactive compounds content variability, which was demonstrated, does not allow a proven efficacy and biological efficiency. However, from the knowledge of raw material composition, harvest time, drying conditions and extraction procedures, commercial products with a constant and standardized content of active ingredients could be obtained [50].

Oleuropein (Ole), a secoiridoid glucoside present in olive leaves, gained the interest of many scientists thanks to its several biological properties, including the anti-cancer one. So, Ruzzolini *et al.* [51] verified whether Ole might potentiate cytotoxicity of some conventional drugs used to treat (murine sarcoma viral oncogene homolog B) BRAF melanoma, disclosing new potential therapeutic strategy.

Ole was able, at a dose of 500 μM, to stimulate apoptosis in melanoma cells, while at a non-toxic dose of 250 μM, it affected cell proliferation and induced the downregulation of protein kinase B (pAKT)/ribosomal protein (pS6) pathway. 250 μM Ole did not potentiate the effect of Vemurafenib, but it succeeded in increase the cytotoxic effect of Dacarbazine. The major effect was found in the association between Ole and Everolimus, also on Vemurafenib-resistant BRAF melanoma cells, possibly cooperating in the inhibition of pAKT/pS6 pathway. Of interest, an olive leaf extract enriched in equimolar Ole was able to improve Dacarbazine and particularly Everolimus efficacy on BRAF melanoma cells in a rather comparable way of Ole, extending the health benefits of this olive leaf product.

Ole represents a natural product able to potentiate a wide array of chemotherapeutics against BRAF melanoma cells affecting pAKT/pS6 pathway.

In conclusion, Ole exhibited a promising potential as adjuvant in conventional anticancer therapies. Furthermore, it may reverse drug resistance of cancer cells to chemotherapeutics and reduce adverse effects of conventional therapies on nontarget cells.

The limited *in vivo* animal studies, as recently summarized by two reviews and the paucity of in human studies, in particular randomized controlled clinical trials, still represents the major drawback [52, 53]. So, preclinical evidence needs to be substantiated by an evidence-based approach to determine effective dose and best route of Oleuropein administration, on the basis of its biodisponibility, and any side effects related to chronic administration [51].

Bernini *et al.* [54] studied the effects of fractions from *Olea europaea* L. by-products on a model of colorectal cancer cells. A hydroxytyrosol (HTyr)-enriched fraction containing HTyr 6% w/w, derived from *Olea europaea* L. by-products and obtained using an environmentally and economically sustainable technology, was lipophilized under green chemistry conditions. The effects of three fractions containing hydroxytyrosyl butanoate, octanoate, and oleate, named, respectively, lipophilic fractions 5, 6, and 7, and unreacted HTyr on the human colon cancer cell line HCT8-β8 engineered to overexpress estrogen receptor β (ERβ) were evaluated and compared to those of pure HTyr. The experimental data demonstrated that HTyr and all fractions showed an antiproliferative effect, as had been observed by the evaluation of the cellular doubling time under these different conditions (mean control, 32 ± 4 h; HTyr 1, 65 ± 9 h; fraction 5, 64 ± 11 h; fraction 6, 62 ± 14 h; fraction 7, 133 ± 30 h). As evidenced, fraction 7 containing hydroxytyrosyl oleate showed the highest activity. These results were related to the link with ER-β, which was assessed through simultaneous treatment with an inhibitor of ERβ.

The data obtained have shown the antiproliferative effects of HTyr 1 on HCT8-β8-expressing cells, comparable to the antiproliferative effect induced by 17β-E2. This is in agreement with observations from other studies on CRC.

Taken together, data suggested not only that the inhibition of cell growth depends on the binding of these polyphenolic molecules to ER-β but also that the HTyr fraction enriched with HTyr oleate (fraction 7) has a greater antiproliferative effect than the other fractions. This effect could be related to the facility with which this molecule passes through the cell membrane, thanks to the major lipophilicity given to the molecule by the presence of the long chain of the oleate [54].

References

1. Food and Agriculture Organization (FAO), 2011. Global food losses and food waste – Extent, causes and prevention. Rome. UN FAO. Accessible at: http://www.fao.org/food-loss-and-food-waste/en/.
2. Lin, C.S.K., Koutinas, A.A., Stamatelatou, K., Mubofu, E.B., Matharu, A.S., Kopsahelis, N., Pfaltzgraff, L.A., Clark, J.H., Papanikolaou, S., Kwan, T.H., Luque, R., Current and future trends in food waste valorization for the production of chemicals, materials and fuels: A global perspective. *Biofuels Bioprod. Bioref.*, 8, 686–715, 2014.
3. Cherubini, F., Jungmeier, G., Wellisch, M., Willke, T., Skiadas, I., Van Ree, R., Toward a common classification approach for biorefinery systems. *Biofuels Bioprod. Bioref.*, 3, 534–546, 2009.
4. Ragauskas, A.J., Williams, C.K., Davison, B.H., Britovsek, G., Cairney, J., Eckert, C.A., The path forward for biofuels and biomaterials. *Science*, 311, 484–489, 2006.
5. Demirbas, A., *Biorefineries. For Biomass Upgrading Facilities*, Springer, London, UK, 2010.
6. Lucarini, M., Durazzo, A., Romani, A., Campo, M., Lombardi-Boccia, G., Cecchini, F., Bio-Based Compounds from GrapeSeeds: A BiorefineryApproach. *Molecules*, 23, 1888, 2018.
7. Fava, F., Totaro, G., Diels, L., Reis, M., Duarte, J., Carioca, O.B., Poggi-Varaldo, H.M., Ferreira, B.S., Biowaste biorefinery in Europe: Opportunities and research & development Needs. *New Biotechnol.*, 32, 100–108, 2015.
8. Accardi, D.S., Bubbico, R., Di Palma, L., Pietrangeli, B., Environmental and Safety Aspects of Integrated BioRefineries (IBR) in Italy. *Chem. Eng. Trans.*, 32, 169–174, 2013.
9. Buzzini, P., Arapitsas, P., Goretti, M., Branda, E., Turchetti, B., Pinelli, P., Ieri, F., Romani, A., Antimicrobial and antiviral activity of hydrolyzabletannins. *Mini-Rev. Med. Chem.*, 8, 1179–1187, 2008.
10. Romani, A., Campo, M., Pinelli, P., HPLC/DAD/ESI-MS analyses and anti-radical activity of hydrolyzable tannins from different vegetal species. *Food Chem.*, 30, 214–221, 2012.
11. Campo, M., Pinelli, P., Romani, A., Hydrolyzable Tannins from Sweet Chestnut Fractions Obtained by a Sustainable and Eco-friendly Industrial Process. *Natural Product Comm.*, 11, 3, 409–415, 2016.
12. Okuda, T., Systematic effects of chemically distinct tannins in medicinal plants. *Phytochemistry*, 66, 2012–2031, 2005.
13. Lupini, C., Cecchinato, M., Scagliarini, A., Graziani, R., Catelli, E., *In vitro* antiviral activity of chestnut and quebracho woods extracts agaist avian reovirus and metapneumovirus. *Res. Veter. Sci.*, 87, 482–487, 2009.
14. Lee, C.J., Chen, L.G., Liang, W.L., Wang, C.C., Anti-inflammatory effects of Punica granatum Linne *in vitro* and *in vivo*. *Food Chem.*, 118, 315–322, 2010.
15. Bargiacchi, E., Miele, S., Romani, A., Campo, M., Biostimulant activity of hydrolyzable tannins from sweet chestnut (Castanea sativa Mill.), in: *Proceedings of the Ist World Congress on the Use of Biostimulants in Agriculture, Acta Horticolturae ISHS*, vol. 1009, pp. 111–116, 2013.
16. Ong, K.C., Khoo, H.E., Das, N.P., Tannic acid inhibits insulin-stimulated lipogenesis in rat adipose tissue and insulin receptor function *in vitro*. *Cell. Mol. Life Sci.*, 51, 577–584, 1995.

17. Sorrentino, E., Succi, M., Tipaldi, L., Pannella, G., Maiuro, L., Sturchio, M., Coppola, R., Tremonte, P., Antimicrobial activity of gallic acid against food-related Pseudomonas strains and its use as biocontrol tool to improve the shelf life of fresh black truffles. *Int. J. Food Microbiol.*, 266, 183–189, 2018.
18. Romani, A., Campo, M., Scardigli, A., Biancalani, C., Cerboneschi, M., Tegli, S., Natural Standardized Polyphenolic Fractions for Innovative Green formulations in Agronomics, in: *Polyphenols Communications 2016 - XXVIII International Conference on Polyphenols 2016 Acts*, pp. 482–483, 2016.
19. Bargiacchi, E., Campo, M., Milli, G., Miele, S., LIFE+2013 EVERGREEN Identified Polyphenol Botanical Biostimulants as Potential Substitutes of Agrochemicals to Increase Plant Resistance to Meloidogyne arenaria Chit, in: *Polyphenols Communications 2016 - XXVIII International Conference on Polyphenols 2016 Acts*, pp. 122–123, 2016.
20. Biancalani, C., Cerboneschi, M., Tadini-Buoninsegni, F., Campo, M., Scardigli, A., Romani, A., Tegli, S., Global analysis of Type Three Secretion System and Quorum Sensing inhibition of Pseudomonas savastanoi by polyphenols extracts from vegetable residues. *PLoS ONE*, 11, 3, 409–415, 2016.
21. Romani, A., Campo, M., Pinelli, P., Turchetti, B., Buzzini, P., Hydrolyzable tannins from different vegetal species, fractionation, HPLC/DAD/MS analyses, and anti-yeast activity, in: *XXVth International Conference on Polyphenols, Polyphenols Communication, 2010*, vol. II, pp. 611–61, 2010.
22. Romani, A., Romani, M., Vignolini, P., Olivieri, G., Borsacchi, L., Feasibility and life cycle assessment of soybean oil, isoflavon and energy production, 17th IGWT Symposium "Facing the Challenges of the Future: Excellence in Business and Commodity Science". Bucharest, Romania, September 21st-25rd, pp. 1–3, 2010.
23. PCT (WO2014024020). Method for making low nitrosamine contents tobacco. Applicants: Gruppo Mauro Saviola S.R.L.; Fattoria Autonoma Tabacchi S.C.A.R.L.; Inventors: Miele S., Romani A., Bargiacchi E., Campo M. February 13th, 2014.
24. Simone, G., Moncini, L., Bernini, R., Campo, M., Romani, A., Tannini idrolizzabili da scarti della lavorazione del castagno: caratterizzazione chimica e valutazione *in vitro* dell'attività inibitoria verso funghi fitopatogeni, in: *Acts of the XXVIII National Congress on Commodity Sciences*, Florence, Italy, February 21st-23rd, pp. 456–461, 2018.
25. Vona, T., Innovative Green Active Compost from pruning and urban solid waste, in: *Acts of the XXVIII National Congress on Commodity Sciences*, Florence, Italy, February 21st-23rd, pp. 2–6, 2018.
26. Romani, A., Bernini, R., Ieri, F., Campo, M., Ciani Scarnicci, M., Villanelli, F., Masci, C., Bova, D.M., Maggino, F., Uso innovativo delle tecnologie di substrato attivo e di fitodepurazione per l'impatto ambientale chimico del suolo e valutazione di nuovi indicatori BES, in: *Acts of the V National Congress of the Italian Association for the Studies on Life Quality (AIQUAV)*, Fiesole (FI), December 13th-15th, pp. 115–121, 2018.
27. Buccioni, A., Pauselli, M., Viti, C., Minieri, S., Pallara, G., Roscini, V., Rapaccini, S., TrabalzaMarinucci, M., Lupi, P., Conte, G., Mele, M., Milk fatty acid composition, rumen microbial population, and animal performances in response to diets rich in linoleic acid supplemented with chestnut or quebracho tannins in dairy ewes. *J. Dairy Sci.*, 98, 2, 1145–1156, 2015.
28. Messini, A., Buccioni, A., Minieri, S., Mannelli, F., Mugnai, L., Comparini, C., Venturi, M., Viti, C., Pezzati, A., Rapaccini, S., Effect of chestnut tannin extract (*Castanea sativa Miller*) on the proliferation of Cladosporium cladosporioides on sheep cheese rind during the ripening. *Int. Dairy J.*, 66, 6–12, 2017.

29. Minieri, S., Buccioni, A., Serra, A., Galigani, I., Pezzati, A., Rapaccini, S., Antongiovanni, M., Nutritional characteristics and quality of eggs from laying hens fed on a diet supplemented with chestnut tannin extract (*Castanea sativa Miller*). *Brit. Poultry Sci.*, 57, 6, 824–832, 2016.
30. Scardigli, A., Vita, C., Masci, C., Vignolini, P., Romani, A., Ingredienti alimentari innovativi ottenuti da sottoprodotti del settore agronomico con tecnologia green, in: *Acts of the XXVIII National Congress on Commodity Sciences*, Florence, Italy, February 21st-23rd, pp. 498–504, 2018.
31. Bargiacchi, E., Campo, M., Romani, A., Miele, S., Hydrolysable Tannins from Sweet Chestnut (*Castanea sativa* Mill.) to improve Tobacco and Food/Feed Quality, in: *XXVII International Conference on Polyphenols 2014Acts*, 2014.
32. Bargiacchi, E., Bellotti, P., Costa, G., Miele, S., Pinelli, P., Romani, A., Zambelli, P., Scardigli, A., Uso di estratto di tannini di castagno come additivo antiossidante, antimicrobico e per ridurre nitrosammine e micotossine. Patent number 0001422367 Gruppo Mauro Saviola S.R.L, https://www.grupposaviola.com/en/, 2014.
33. Imperatori, F., Barlozzari, G., Scardigli, A., Romani, A., Macrì, G., Polinori, N., Bernini, R., Santi, L., Leishmanicidal activity of green tea leaves and pomegranate peel extracts on L. infantum. *Natural Product Comm.*, 4, 1–7, 2018.
34. Romani, A., Vignolini, P., Galardi, C., Aroldi, C., Vazzana, C., Heimler, D., Polyphenolic content in different plant parts of soy cultivars grown under natural conditions. *J. Agric. Food. Chem.*, 51, 18, 5301–5306, 2003.
35. Xiao, Y., Zhang, S., Tong, H., Shi, S., Comprehensive evaluation of the role of soy and isoflavone supplementation in humans and animals over the past two decades. *Phyt. Res.*, 32, 3, 384–394, 2018.
36. Chadha, R., Bhalla, Y., Jain, A., Chadha, K., Karan, M., Dietary soy isoflavone: A mechanistic insight. *Nat. Prod. Commun.*, 12, 4, 627–634, 2017.
37. Ahn-Jarvis, J., Clinton, S.K., Riedl, K.M., Vodovotz, Y., Schwartz, S.J., Impact of food matrix on isoflavone metabolism and cardiovascular biomarkers in adults with hypercholesterolemia. *Food Funct.*, 3, 10, 1051–1058, 2012.
38. Zhang, X., Gao, Y.T., Yang, G., Li, H., Cai, Q., Xiang, Y.B., Ji, B.T., Franke, A.A., Zheng, W., Shu, X.O., Urinary isoflavonoids and risk of coronary heart disease. *Int. J. Epidemiol.*, 41, 5, 1367–1375, 2012.
39. Varinska, L., Gal, P., Mojzisova, G., Mirossay, L., Mojzis, J., Soy and breast cancer: Focus on angiogenesis. *Int. J. Mol. Sci.*, 16, 5, 11728–11749, 2015.
40. Yu, D., Shu, X.O., Li, H., Yang, G., Cai, Q., Xiang, Y.B., Ji, B.T., Franke, A.A., Gao, Y.T., Zheng, W., Zhang, X., Dietary isoflavones, urinary isoflavonoids, and risk of ischemic stroke in women. *Am. J. Clin. Nutr.*, 102, 3, 680–686, 2015.
41. Heimler, D., Vignolini, P., Galardi, C., Pinelli, P., Romani, A., Simple Extraction and Rapid Quantitative Analysis of Isoflavones in Soybean Seeds. *Chromatographia*, 59, 361–365, 2004.
42. Wan, P.J. and Wakelyn, P.J., *Technology and solvent for estracting oilseeds and nonpetroleum oils*, pp. 36–37, AOCS Press, Champaign, IL, 1998.
43. Romani, A., Romani, M., Neri, P., Olivieri, G., Tacconi, D., Vignolini, P., Deasibility and life cycle assessment of soybean oil and soy-isoflavon production, in: *16th IMEKO TC4 Symposium Exploring New Frontiers of Instrumentation and Methods for Electrical and Electronic Measurements*, Florence, Italy, Sept, pp. 22–24, 2008.
44. Olivieri, G., Romani, A., Neri, P., Environmental and economic analysis of aluminium recycling through Life Cycle Assessment. *Int. J. Sustainable Develop. World Ecol.*, 13, 4, 269–276, August 2006. Sapiens Publishing.

45. Jolliet, O., Margni, M., Charles, R., Humbert, S., Payet, J., Rebitzer, G., Rosenbaum, R., IMPACT 2002+: A new life cycle impact assessment methodology. *Int. J. Life Cycle Assess.*, 8, 6, 324–330, 2003.
46. Frischknecht, R., Jungbluth, N., Althaus, H.J., Doka, G., Dones, R., Heck, T., Hellweg, S., Hischier, R., Nemecek, T., Rebitzer, G., Spielmann, M., The ecoinvent Database: Overview and Methodological Framework. *Int. J. LCA*, 10, 3–9, 2005.
47. Rodis, P.S., Karathanos, V.T., Mantzavinou, A., Partitioning of olive oil antioxidants between oil and water phases. *J. Agric. Food Chem.*, 50, 596–601, 2002.
48. Bitler, C.M., Viale, T.M., Damaj, B., Crea, R., Hydrolyzed olive vegetation water in mice has anti-inflammatory activity. *J. Nutr.*, 135, 1475–1479, 2005.
49. Quaglia, G.B., Bucarelli, F.M., Lucarini, M., Lupattelli, A., Mingoia, E., Salustri, C., Cimaroli, L., New functional ingredients from olive millswastes: Feasibility study with experimental activity. *La Rivista della Scienza dell'Alimentazione. J. Food Sci. Nutr.*, 2, 11–25, 2016.
50. Romani, A., Mulas, S., Heimler, D., Polyphenols and secoiridoids in raw material (*Olea europaea* L. leaves) and commercial food supplements. *Eur. Food Res. Technol.*, 243, 429–435, 2017.
51. Ruzzolini, J., Peppicelli, S., Andreucci, E., Bianchini, F., Scardigli, A., Romani, A., La Marca, G., Nediani, C., Calorini, L., Oleuropein, the Main Polyphenol of Olea europaea Leaf Extract, Has an Anti-Cancer Effect on Human BRAF Melanoma Cells and Potentiates the Cytotoxicity of Current Chemotherapies. *Nutrients*, 10, 12, 1950, 2018.
52. Shamshoum, H., Vlavcheski, F., Tsiani, E., Anticancer effects of oleuropein. *Biofactors*, 43, 517–528, 2017.
53. Fabiani, R., Anti-cancer properties of olive oil secoiridoid phenols: A systematic review of *in vivo* studies. *Food Funct.*, 7, 4145–4159, 2016.
54. Bernini, R., Carastro, I., Palmini, G., Tanini, A., Zonefrati, R., Pinelli, P., Brandi, M.L., Romani, A., Lipophilization of Hydroxytyrosol-Enriched Fractions from Olea europaea L. byproducts and evaluation of the *in vitro* effects on a model of colorectal cancer cells. *J. Agric. Food Chem.*, 65, 1–7, 2017.

11

Biorefining of Lignocellulosics for Production of Industrial Excipients of Varied Functionalities

UpadrastaLakshmishri Roy[1]*, DebabrataBera[2], Sreemoyee Chakraborty[2] and Ronit Saha[1]

[1]Department of Food Technology, Techno Main Salt Lake, Kolkata, India
[2]Department of Food Technology and Biochemical Engineering, Jadavpur University, Kolkata, India

Abstract

Lignocellulosic (LC) biomass resources are available as potential candidates that are convertible into high value bio-products by biotechnological approaches. In general, LCs are complex matrices because of their phenolic content. A wide range of biologically important compounds comprising of various sub-classes like phenylpropanoids, phenylethanoids, flavonoids, benzofurans, anthraquinones, coumarins, tannins, neolignans, etc., constitute the phenolic compounds. The global interest in phenolic compounds has rapidly increased due to their possible role as natural antioxidants. Additional attributes include: the recognition of their wide ranging health promoting biological and pharmacological activities such as antibacterial, anti-inflammatory, antifungal, and anticancer activities. In addition, phenolics are also being widely employed as flavoring and nutraceutical agents. The present chapter is a comprehensive overview of types of phenolics available in LCs, strategies employed for their recovery, and plausible avenues of their diverse applications. In the concluding section, an overall summary of the current practices adopted in this area have been included.

Keywords: Biorefining, lignocellulose, functional components, antioxidant, polyphenols

11.1 Introduction

Lignocellulosic (LC) biomass comprises biomass from agricultural and agro-industrial wastes. These are opulent, renewable, and cheap energy sources [1]. Such wastes embrace a range of materials corresponding to saw dust, poplar trees, sugarcane pulp, paper, brewer's spent grains, switchgrass, and straws, stems, stalks, leaves, husks, shells, and peels from cereals like wheat, rice, corn, barley, and sorghum, among others. This biomass is presently undergoing an in-depth analysis and intensive study as an alternative substrate for bio-based chemicals and energy production [2–5]. Despite the benefits in endurability and accessibility, materialistic use of lignocellulose remains problematic. Hydrolysis of hemicellulose and cellulose into five- and six-carbon sugars prior to fermentation and chemicals recovery process becomes imperative due to the chemical constituency of LC materials.

**Corresponding author*: drlakshmi1371@gmail.com; lakshmi1371@gmail.com

Arindam Kuila and Mainak Mukhopadhyay (eds.) Biorefinery Production Technologies for Chemicals and Energy, (221–240) © 2020 Scrivener Publishing LLC

Lignocellulose wastes are accumulated each year in massive quantities, inflicting environmental issues. Globally, USA, Canada, China, and New Zealand top in production of large amount of agricultural and forest residues annually [1, 6–9]. However, because of their chemical composition, supported sugars, and alternative compounds of interest, they might be utilized for the assembly of variety of added products, corresponding to ethyl alcohol, food additives, enzymes, and others. Therefore, besides the environmental issues caused by their accumulation in nature, the non-use of these materials constitutes a loss of probably valuable sources. Thus, if LC resources are to become universal raw materials, an industrially feasible efficient refining process and technology needs to be developed [10]. In future, LC Green conversion techniques are likely to be most opted for as they can keep energy consumption at minimum and achieve optimum conversion efficiency [11].

11.2 Structure and Composition

Lignocellulose is universally composed of two basic components—polymeric carbohydrates (cellulose and hemicellulose) and aromatic polymer lignin—bound together forming a complex network. Irrespective of their sources, their basic structure remains same [12]. Cellulose, hemicellulose, and lignin constitute 30%–50%, 20%–40%, and 15%–25% of total LC feedstock (LCF) dry matter, respectively [13, 14]. Depending on the source, lignocellulose may also contain varied but smaller amounts of pectin, proteins, extractives, and inorganic compounds. The cellulose fragment is composed of monomeric units linked by 1-4-β glycosidic bonds. Degree of polymerization in cellulose (10,000 or higher) is the highest among all three of the LC polymers, thus attributing to its low flexibility and solubility in water and most solvents, followed by hemicellulose [15]. The hemicellulose backbone consists mainly of different types of sugar residues of five- or six-carbon sugars, i.e., arabinose, xylose, galactose, glucose, and mannose. Hemicellulose polymers are branched containing functional groups such as acetyl, methyl, cinnamic, glucuronic, and galacturonic acids [13, 15]. Lignin which is produced as the plant growth ceases is water insoluble and aromatic polymer provide the strength and rigidity to the plant tissue [15].

Among the three components, hemicellulose is relatively easier to hydrolyze than cellulose and lignin. Lignocellulose is used for extraction of different bioactive chemicals as a part of biorefinery [13, 14]. The complex structures formed by hemicellulose and lignin around the cellulose platform make them resistant to biological and chemical degradation [16, 17]. Thereby necessitating preliminary pre-treatment conversion processes for degradation in order to make the polysaccharides accessible. This conversion step is the most important part in bio-conversion of lignocellulose [18, 19].

11.3 Lignocellulosic Residues: A Bioreserve for Fermentable Sugars and Polyphenols

Cellulose and hemicellulose portions of the LC residues (LCRs) act as a source of fermentable sugars which can be further transformed to bioethanol wherein unutilized lignin is a hindrance for the conversion process [20]. Recent researches have yielded strategies

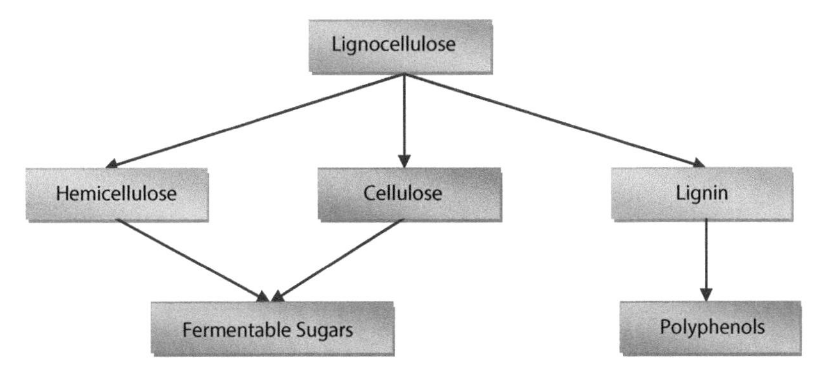

Figure 11.1 Plausible products from lignocellulosic substrates.

for conversion of lignin to value-added products, thereby making the complete utilization of LCR and achieve a zero-waste process [21]. Bioactive molecules or phenolic compounds can be obtained from acid-soluble lignin or by hydrolysis of hemicellulose [22]. Additionally, lignin is capable of producing antioxidants like p-coumaric acid, ferulic acid, syringaldehyde, and vanillin [23].

Among various secondary metabolites extracted from agro wastes, polyphenols are the largely studied antioxidants for multidisciplinary application. These compounds are basically made of phenolic acids and flavonoids with antioxidant and anti-inflammatory effects [24]. They have been found to reduce the threat of degenerative diseases by decreasing the oxidative stress and prevent or decrease the rate of oxidation [25]. Furthermore, lignin-derived phenolics can be applied in place of synthetic antibiotics and chemical food preservatives due to their antimicrobial and antioxidant nature [26]. Progress has been made in the area of extraction of polyphenols using the crops, fruits, and vegetables [27–29]. Use of LC wastes as a substrate for bioactive compounds is likely to mitigate the dependency on global plant based food reserves. Studies have shown that extraction of the polyphenols through various processes, namely, solvent extraction, alkali treatment, and acid treatment, has been successful in the past [23, 30, 31]. However, the low yield is a hurdle in its commercial production. Moreover, use of methanol for the extraction of polyphenols becomes unsuitable for human consumption as it possesses toxicity [32].

11.3.1 Biorefining of Lignocellulosic Residues

Biorefining is a process that combines biomass conversion processes and equipment to produce green fuels, power, and chemicals from biomass with a variety of new fermentation and thermo-chemical processes [33]. It is industrial symbiosis—where reutilization of by-products and wastes lead to generation of newer value-added resources that have both environmental sustainability and economic benefits [34]. Three types of biorefineries known as phase I, II, and III have been described by Kamm and Kamm [35] and van Dyne *et al.* [36]. While phase I and II biorefineries are only capable of using a single category of feedstock for production of ethanol and other bio-extractives, phase III biorefineries can utilize a mix of biomass feedstocks and yield an array of products by using different processing methods and a combination of technologies [35]. Phase III biorefineries can

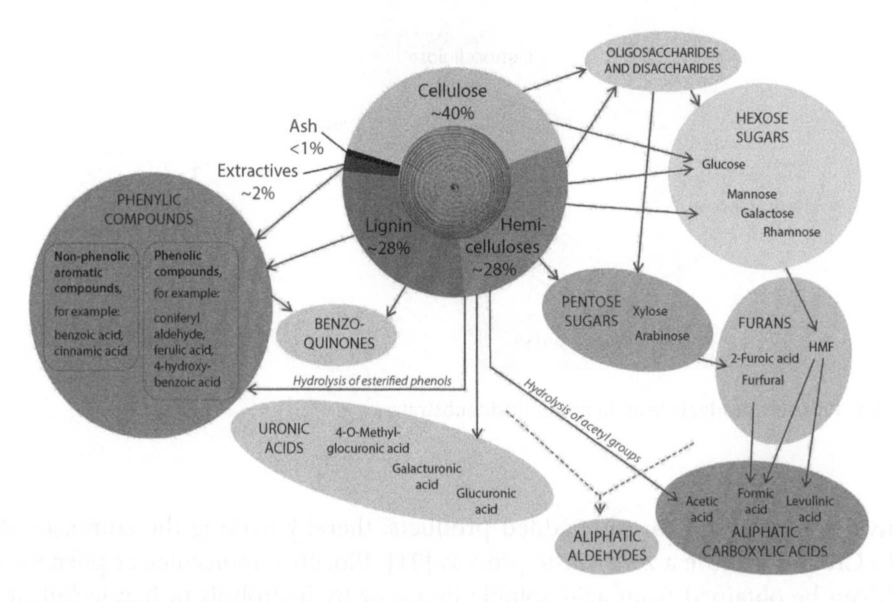

Figure 11.2 Lignocellulosic biorefining.

produce a wide range of high valued chemicals as minor products of their processes along with fermentable sugars as major product.

For complete conversion of feedstock, the step which is utmost crucial is that of the pre-treatment of LC residues.

11.4 Pre-Treatment of Lignocellulosics

Pre-treatment is required to alter the structure of the biomass as well as its chemical composition and structure so that hydrolysis of the carbohydrate fraction to monomeric sugars can be achieved more rapidly and with greater yields. The pre-treatment step which is aimed at separation of lignocellulose into cellulose, hemicellulose, and lignin can be thermo-physical, physico-chemical, and/or biological processes.

11.4.1 Physico-Chemical Process

11.4.1.1 Acid Refining

Luo *et al.* [37] designed a LCF biorefinery producing ethanol, succinic acid, acetic acid, and electricity from corn stover, and next analyzing the refinery from an economic and environmental point of view. They selected dilute acid pre-hydrolysis process for the pre-treatment of corn stover as shown by Aden *et al.* [38]. In the pre-treatment step, LCF was treated with dilute sulphuric acid at a very high temperature of 190°C. Most of the hemicellulose portion is converted to soluble sugars—primarily xylose, arabinose, mannose, and galactose. Simultaneously, glucan in the hemicellulose and a small portion of the cellulose are converted to glucose. Following the pre-treatment step, the pentose sugars are fermented by recombinant strains of *Zymomonas mobilis* to produce ethanol and the glucose fraction is fermented by *C. glutamicum*to produce succinic acid and acetic acid. This method is cost effective and can be easily scaled up for commercial use; however, the only drawback is that during downstream processing of the fermentation broth the filtration is adversely effected due to accumulation of divalent cations over the membrane. However, by employing proper neutralization, separation, and electrodialysis methods high levels of purity for both succinic acid (99.5%) and acetic acid (99.9%) can be achieved eventually.

Jones and Semrau [39] studied the effect of using concentrated acid to increase the yield of fermentable sugars. The temperature required was considerably lower; however, the hydrolysis rate was slower and the hexoses decomposed to form furfural groups which had inhibitory effect on the subsequent fermentation of sugars. Other disadvantages include equipment corrosion, toxicity to environment, and difficulties in acid recovery.

Recent researches have been focused on combined treatment of dilute and concentrated acids which ensures an efficient conversion under lower temperatures and lesser time along with reduction in formation of corrosive and toxic by products. At first, concentrated sulphuric acid was used to hydrolyze hemicellulose and decrystallize cellulose at a temperature of 40°C followed by treatment with dilute acid at a temperature around 90°C, finally, sulphuric acid is recovered using a ion exchange column. This process is presently under way for commercial application.

In a bid to reduce instrument corrosion and lower energy demand use of organic acids as a replacement for inorganic acids have been proposed. Efficiency of maleic, succinic, oxalic, fumaric, and acetic acid has been evaluated. But it is seen that while organic acids can be as effective as sulphuric acid in case of catalyzing cellulose, the rate hydrolysis of hemicellulose is much lower [40].

11.4.1.2 Alcohol Refining

Another method of biorefining of LCR was proposed by Garcia *et al.* [41] which involved the application of ethanol organosolv process. Their study was aimed at fractionating biomass feedstock into solid cellulose and liquid hemicellulose sugars containing some fractions of lignin followed by purification to obtain high valued excipients with potential industrial applications. The milled LCF was treated with aqueous solution of ethanol at high temperature 160°C, and the final solid and liquid fraction were separated using membrane filtration. Finally, the chemical structure of lignin was studied for its feasibility in industrial applications.

11.4.1.3 Alkali Refining

Alkaline pre-treatment of LCs was applied in the method of soda pulping patented in 1,854 (Watt) using sodium hydroxide (NaOH), potassium hydroxide (KOH), aqueous ammonia (NH4OH), calcium hydroxide (Ca(OH)2), and oxidative alkali as reagents. Alkaline hydrolysis can be performed at a much lower temperature but the treatment takes comparatively longer time than acid hydrolysis [42]. The alkaline reagent cleaves the 4-O-methyl-D-glucuronic linkages resulting in easy removal of lignin and hemicellulose. Biomass pretreated with NaOH has a greater porosity and cellulose accessibility. However, alkaline hydrolysis is more effective on LCF residues which have lower lignin content [43]. Hence, in order to increase lignin removal and improve efficiency alkaline treatment can be coupled with application of high pressure like Ammonia-fiber expansion (AFEX) and Ammonia recycle percolation (ARP). Pressure increased the surface area of the fibers and helps in better accessibility of enzymes during fermentation and improved digestibility. The combined effect of alkali and high pressure creates a physico-chemical change in the spatial arrange of the fibers. Moreover, there is almost no sugar loss and the pre-treatment results in decrystallization of cellulose, partial depolymerization of hemicellulose, removal of acetyl group from hemicellulose and cleavage of lignin-xylan bonds.

11.4.2 Thermo-Physical Process

Physico-chemical processes produce considerable amount of inhibitory products, and the reagent recovery system is a time-consuming and complex process. To prevent this hydrothermal pre-treatment processes like steam explosion, supercritical/subcritical water and hot-compressed water treatments were introduced to fractionate the LCF residues.

11.4.2.1 Steam Explosion Process

It is a thermo-physical process where high pressure saturated steam at about 0.69–4.83 MPa with temperature ranged between 160°C and 260°C is applied to the biomass for several seconds. The heat and pressure subjects the cell wall to rapid explosion and partial hydrolysis followed by instant reduction in pressure to atmospheric levels. Acetic acid is released from xylan and catalyzes the hydrolysis. Temperature, residence time, and moisture content are the three deciding factors of conversion efficiency in this process. Steam explosion is a cost-effective process; however, the drawbacks include incomplete disruption of lignin carbohydrate matrix subsequently leading to formation of inhibitory compounds during fermentation or enzymatic treatment [44–46].

11.4.2.2 Supercritical and Subcritical Water Treatment

This method is one of the most environmentally benign, non-toxic, non-flammable, non-mutagenic, and thermodynamically stable processes. When water crosses its critical temperature (374°C) and critical pressure (22.1 MPa), it becomes supercritical and represents a non-condensable gas-like state. When the temperature and pressure of water does not cross the critical point but is near its vicinity it is known as subcritical water. In subcritical and supercritical states, the ionic product and dielectic constant of water can be changed accordingly to increase its hydrolytic capacity. Ehara *et al.* [47] showed how cellulose is hydrolyzed to produce glucose, hexose sugars, and low-molecular-weight organic acids. First cellulose is broken down to form polysaccharides which degree of polymerization (DP) between 13 and 100. Then, it is further fractionated and dehydrated to form oligosaccharides (DP 2–12). The reducing ends are hydrolyzed and fragmented to hexose sugar and non-reducing ends to glucose and fructose. Further dehydration and fragmentation of hexose residues can form low-molecular-weight organic acids such as pyruvic, lactic, formic, and acetic acids. The reaction time was as low as 0.19–0.22 seconds and the conversion rate was 65% of cellulosic hexoses. Moreover, combined subcritical and super critical treatments can be applied to realize shorter reaction time, increased yield with very few inhibitory by-products [47].

11.4.2.3 Hot-Compressed Water Treatment

Hot compressed water at temperature above 200°C and varying high pressure was able to hydrolyze LC biomass in two stages due to the difference in hydrolysis rate for cellulose and hemicellulose. The first stage is performed at low severity (230°C/10 MPa/15 min) to breakdown hemicelluloses while the second stage, at a higher severity (270°C–280°C/10 MPa/15 min) aims to hydrolyze cellulose and increase the yield of saccharides. Hydrolyzed products obtained at the end of the first stage are xylo-oligosaccharides, xylose, glucuronic acid, and acetic acid, and in the second stage, glucose is obtained. Lignin is partially liquified in the process and can be easily separated by further processing in biorefineries [48, 49].

11.4.3 Biological Process

Most physical and chemical pre-treatments use acid, alkali, microwave heating, steam explosion, ionizing radiation, or combined two or more of the above processes. These methods require lot of technical support and chemical treatments sometimes generate inhibitors which affect enzymatic hydrolysis and fermentation [50]. Hydrolysis of lignocellulose without any pre-treatment can yield maximum 20% of total sugars, while after pre-treatment, yield can be as 90%. Biological pre-treatment of lignocellulose is done using either enzymes directly or by inoculating a particular strain of polysaccharide degrading microorganism or a combination of different strains. While these organisms sometimes might produce secondary metabolites which may act as inhibitors to the hydrolytic process, enzymes on the other hand, may produce by-products which actually act synergistically during refining of lignocellulose. Table 11.1 shows different biological pre-treatment strategies involved for hydrolysis of LC biomass.

Utilizing microbial consortium for fractionation of LCF is one of the most efficient processes of pre-treatment. However, this also has two major drawbacks. Firstly, production of unwanted metabolites which can be toxic to the organism or inhibit efficient hydrolysis and secondly a considerable portion of polysaccharide residues are consumed by the organism itself for its growth and sustainability. This requires continuous removal of toxic metabolites and adding substantial biomass. Hence, this has become one of the major limiting factors of successful commercialization of LCF as substrate for production of high valued industrial excipients.

Recent studies have shown that supplementation of accessory enzymes can improve the hydrolytic efficiency. Some proteins like swollenin play an important role in non-hydrolytically loosening the cellulosic fibril network so the hydrolytic enzymes can effectively act on them. Factors which affect rate of hydrolysis include enzyme concentration, enzyme adsorption, end-product inhibition, thermal inactivation, and unproductive binding to lignin. Other factors include cellulose crystallinity, degree of polymerization, accessible surface area, particle size, as well as structural arrangement of hemicelluloses and lignin [61].

11.4.3.1 Lignin Degrading Enzymes

Lignin is an aromatic polymer consisting of both phenolic and non-phenolic structures. Some lignin digesting enzymes include laccases (E.C. 1.10.3.2.), lignin peroxidase

Table 11.1 Some methods of biological treatment of lignocellulosics [51].

Microorganism	Biomass	Reaction conditions & minor observations	Major effects	References
Punctualaria sp. TUFC20056	Bamboo culms		50% of lignin removal	[52]
Irpexlacteus	Corn stalks	28 days reaction time; by products improved efficiency	82% of hydrolysis yield	[53]
Fungal consortium	Straw	Co-culturing the fungal consortium	Seven-fold increase in hydrolysis	[54]
P. ostreatus/P. pulmonarius	Eucalyptus grand is saw dust	Structural changes in the saw dust; selective degradation of lignin	Twenty-fold increase in hydrolysis	[55]
P. chrysosporium	Rice husk	18-day treatment time; cellulase, xylanase, lignin peroxidase, glyoxidase, and aryl alcohol oxidase were produced		[56]
Fungal consortium	Corn stover	42-day pre-treatment time	43.8% lignin removal/ seven-fold increase in hydrolysis	[57]
Ceriporiopsis subvermispora	Wheat straw	10-week reaction time	Minimal cellulose loss; 44% sugar yield	[58]
Ceriporiopsis subvermispora	Corn stover	Addition of glucose and malt extract improved cellulose digestibility of wheat straw	2–3-fold increase in reducing sugar yield	[59]
Fungal consortium	Plant biomass	Simultaneous pre-treatment and saccharification; laccase-detoxifying agent	Complete elimination of use of hazardous chemicals	[60]

(E.C. 1.11.1.7), and manganese peroxidase (E.C. 1.11.1.7). Laccases are copper containing enzymes capable of complete degradation of lignin. It catalyzes the oxidation of phenolic units and aromatic amines to radicals while the phenolic compounds like 3-hydroxyanthranilic acid, 2,2 P-azino-bis (3-ethylthiazoline-6-sulfonate) present in act as redox mediators [62]. Lignin peroxidase acts only on non-phenolic units while manganese peroxidase can act on both phenolic and non-phenolic units [61]. These three enzymes are synergistically capable of degrading the lignin fraction of LCF to prepare for it for efficient fermentation. However, due to its high phenolic content, lignin residues are also treated to extract high valued polyphenols; some of the methods are discussed further in the chapter.

11.4.3.2 Cellulose Degrading Enzymes

Conversion of cellulose into glucose is generally achieved by endoglucanase, cellobiohydrolases, and b-glucosidase. Endglucanases hydrolyze beta-1, 4 glycosidic linkages in the cellulose chain; cellobiohydrolase cleaves off cellobiose units from the end of the chain and beta-glucosidase coverts cellobiose to glucose [63]. Cellulases have a carbohydrate binding module which binds the enzyme to the crystalline cellulose enhancing cellulase activity [64].

11.4.3.3 Hemicellulose Degrading Enzymes

Hemicelluloses are composed of pentoses like xylose, and complete hydrolysis requires the action of multiple xylanases with different specificities and action [61]. Hemicellulases and other accessory enzymes are crucial for the improved hydrolysis efficiency of LC biomass. Hence, efficient hydrolysis of hemicellulose fractionation becomes crucial and supplementation of accessory enzymes increase hydrolysis yields and thereby reduces enzyme costs and dosages.

The factors which play an important role in determining the efficiency of biological pretreatment of LCF include:

i. Biomass type: Cellulose, hemicelluloses, and lignin and small amount of other organic and non-organic components like proteins, lipids, and extractives form the LC structure. Composition of LCF varies with maturation, growth condition, and species of the feedstock.
ii. Type of organism/source of enzyme: Selection of a source or an enzyme/enzyme conglomerate depends on the end product required as rate of hydrolysis, product formation, degree of purity, formation of inhibitors, and consumption of substrates varies with respect to species. Also, sometimes, using a mixture of different organisms yield better efficiency than using a single strain Asiegbu *et al.* [65].
iii. Incubation temperature: It is important to optimally control the temperature for lignification and hydrolysis. Because the reactions involved in hydrolysis are mostly exothermic, accumulation of heat can inhibit microbial growth or destroy the reactive enzymes.

iv. pH: Optimal enzyme activity is achieved only by optimally controlling the pH of the system.
v. Incubation time: As shown in Table 11.1, the incubation time required for biological pre-treatment depends upon the composition of the biomass and the strain or enzyme used for pre-treatment.
vi. Inoculums concentration: Time required for colonization and multiplication has a direct effect on the efficiency of the process. Higher inoculum volume seems to generate relatively shorter pre-treatment time.
vii. Moisture content: Initial moisture content is a very important factor in case of microbial pre-treatment as optimal microbial growth and enzyme production is directly related to the moisture content of the substrate. Moreover, high degree of dry matter may lead to formation of high concentration of inhibitory products. Studies conducted by Reid [66] revealed that an initial moisture content of 70%–80% was optimal for lignin degradation and ligninase activities of most white rot fungi.
viii. Aeration rate: Proper aeration induces effective heat dissipation, CO2 removal, oxygenation, humidity maintenance, as well as proper distribution of nutrients and substrate. Hence, controlled aeration is required for efficient delignification of substrates.
ix. Production of Reactive Oxygen Species by the organism.

11.4.4 Phenols as By-Products of Lignocellulosic Pre-Treatment Process

During pre-treatment of LCR, lignin is mostly liquified and separated out. This liquified lignin fraction contains high percentages of phenolic compounds which are actually inhibitors of enzyme hydrolysis of cellulose and hemicellulose. However, these aromatic units are also high valued chemicals with potentials industrial applications. For instance, the presence of vanillin at 10 mg/ml decreased the cellulose conversion of lignin-free cellulose (Avicel) by 26%, which was almost a half conversion yield when compared to the control (53%, without vanillin) [67]. Also, phenolics recovered from the pretreated biomass had a negative impact on enzyme performance. Michelin *et al.* observed that phenolics from liquid hot water pretreated sugarcane bagasse (log R0 = 3.83, 3.5 mg phenolic/mg protein enzyme) led to a 20% lower cellulose (SolkaFloc) conversion compared to a control while the phenolic compounds recovered at higher severity condition (log R0 = 4.42, 6.2 mg phenolic/mg protein enzyme) resulted in a 45% lower yield [68]. Another study showed that phenolics obtained from the liquid hot water pretreated hardwood (log R0 = 4.25, 2 mg phenolic/mg protein enzyme) decreased conversion yield of by about 50% when they incubated with consortium of enzymes hydrolysis of Spezyme CP and Novozyme 188 [69]. Phenolics can also increase the fluidity of the cell membrane, possibly causing intracellular potassium levels to drop significantly. Furthermore, phenolic compounds are able to promote a loss of structural integrity in biological membranes, inhibit cell growth and further sugar assimilation, as well as can cause breakdown of DNA, resulting in the inhibition of RNA and protein synthesis [70, 71]. Hence, it is important to remove the phenolic compounds from the biomass for complete utilization and effective conversion of the residues.

11.5 Methods of Extraction of Polyphenols From Lignocellulosic Biomass

Sugar present in the hydroxyl groups tend to reduce the yield of polyphenols and also diminish their antioxidant properties. Hence, efforts have been made to extract phenolic compounds from the hydroxyl groups using different enzymatic treatments [72, 73]. Two such successfully established methods of extraction are enzyme affiliated extraction (EAE) (consortia of lignocellulolytic enzymes) and solvent affiliated extraction (SAE) (ethanol and methanol) utilizing the LCR, namely, coffee cherry husk (CCH), pomegranate peel (PP), and watermelon rind (WM).

11.5.1 Solvent Affiliated Extraction

The polyphenols extraction was carried out using 50% (v/v) of ethanol and methanol at 10% (w/v) of the LCRs at room temperature for 24 h under static conditions [74]. After the incubation period, the contents were filtered through and the filtrate was centrifuged at 10,000 rpm for 10 min and the clear supernatant was obtained. Further, the concentration of the sample was carried out in rotary evaporator, and the dry matter was expressed as yield percentage.

11.5.2 Enzyme Affiliated Extraction

The polyphenols were extracted from the various LCRs at pH 7, 40°C, and 10% (w/v) concentration under static condition for 24 h by the consortia of lignocellulolytic enzymes (50 mg of protein in the crude enzyme hydrolysate) secreted by Sphingobacterium sp. ksn [75]. After incubation, the constituents of the flasks were centrifuged at 10,000 rpm for 10 min to get the clear supernatant. Further, concentration and yield percentage were estimated as previously done for SAE section.

Shankar *et al.* [76] found that EAE was a better process than SAE. The cell wall hydrolyzing enzymes were far more effective in fractionating and separating the phenolic from the plant cells. Moreover, utilizing a combination of different cell wall hydrolyzing enzymes, the researches have developed a cost-effective and novel method of extraction of phenolics from the LCR. However, since cellulase and hemicellulase enzymes actively participated in extraction of phenolicsits is obvious that the hydrolysate will contain sugar which may decrease the antioxidant activity. Hence, the extract was purified by subjecting it to absorption by activated charcoal wherein the phenolics were separated and trapped on charcoal bed followed by desorption in 50% ethanol. After flash evaporation of ethanol the hydrolysate was found to contain gallic acid, caffeic acid, p-coumaric acid, 1-hydroxybenzoic acid, 2,5-dihydroxybenzoic acid, phydroxybenzaldehyde, syringic acid, quercetin, kaemferol, and epicatechin. Gallic acid displayed considerable antioxidant and anti-inflammatory activities, quercitin, kaempferol, and catechin displayed antibacterial, antiviral, antineoplastic bioactivities [77].

Kurakake *et al.* [78] demonstrated that treatment with 25% ammonia water at 120°C for 20 min followed by vacuum drying (to remove ammonia) efficiently produced a thermostable antioxidant from lignin of selected herbaceous and soft wood species. The antioxidants produced by ammonia water treatment likely includes phenolic compounds, equivalent in activity to Vitamin E, gallic acid, and ascorbic acid, released from lignin because the antioxidant capacity of lignin is known to be closely related to its total phenolic content [79]. The treatment of LC wastes with ammonia water is an attractive new method to produce antioxidants which can be successfully used in food.

Xavier *et al.* [80] performed solid-liquid extraction of LC biomass from forest residues with different concentrations of ethanol. The extraction was carried out at a temperature of 75°C for 180 mins with varying ethanol concentrations (10%–90%). Highest phenol yield was obtained at 75 °C, 55 min, and ethanol concentration of 30%. An abundance of several phenolic compounds including gallic acid, ascorbic, and ellagic acid with high antioxidant activity was found at high levels.

Akl *et al.* [81] used acid and alkali hydrolysis followed by extraction with organic solvents to extract the bioactive material from the hydrolyzed LCRs of jatropha and jojoba hulls. The unique feature of this study is that the research was conducted using ultrasound waves (38.5 kHz) and low temperature (room temperature) aimed at minimizing the depletion of antioxidant activity. Bioactive compounds extracted in their natural form had high antioxidant activities exceeding that of BHT. The extraction of phytochemical compounds with elevated antioxidant activities from jojoba was successful by using NaOH, while (1, 0.5 N) HCl with ethanol (30%–70%) concentrations was used to extract phytochemicals with high antioxidant activity from jatropha. The main advantage of this method is that it uses a small amount of solvent and less energy. The bioactive extracts can be used in food biotechnology and pharmaceutical industry. Introduction of different structure of bioactive compost extracts from jojoba and jatropha hulls through hydrolyzes of lignocellulolytic hulls can provide a multi-functional product with multiple uses in industrial purposes.

A sustainable process of valorization of LC biomass for extraction and separation of high valued phenolic compounds was developed by Lopes *et al.* [82]. In the first stage, a pre-treatment and fractionation process of biomass was carried out by an ionic liquid (IL). The main objective of this work was to separate and purify the phenol rich liquid. The phenolic extraction was studied using adsorption methodology with specific polymeric resins like Amberlite XAD-7, Silica C18 resins, PVPP, etc. Among the examined resins, Amberlite XAD-7 demonstrated the highest degree of phenolic compounds extraction in comparison to IL. A residual quantity of IL was extracted by applying supercritical CO_2 extraction approach. Proposed methodology lead to the production of high purity main biomass fractions free of phenolic compounds in a green and sustainable way.

11.5.3 Advanced Technological Methods Adopted for Recovery of Phenolics: (Pulsed-Electric-Field Pre-Treatment)

Pulsed-electric-field (PEF) pre-treatment involves application of a brief burst of high voltage to a sample placed between two electrodes. PEF pre-treatment has serious effects on the structure of plant tissues. Once a high-intensity, external field of force is applied, an important voltage is elicited across the semipermeable membrane, that ends up in fast electrical breakdown and native structural changes of the semipermeable membrane, the plasma membrane, and thus the plant part. The electrical field leads to a dramatic increase in mass porosity and, in some cases, mechanical rupture of the plant part. The electrical field pulses most ordinarily applied are within the style of exponential-decay or square waves.

Application of high-intensity field of force pulses from nanoseconds to microseconds in length ends up in the permeabilization of biological membranes.

In applying PEF pre-treatment to plant process, the electrical intensity (voltage/distance), the amount of pulses, and therefore the treatment time are the foremost necessary factors. Typically, the plant part is placed or transported between two electrodes, and therefore, the electrical discharges are applied within the style of pulses. In biomass-to-fuel conversion, the biomass has to be treated in order that the cellulose within the plant fibers is exposed. Pre-treatment with PEFs will facilitate this method. By utilization of high field strengths that vary from 5- to 20 kV/cm, plant cells may be considerably damaged. By applying electrical pulses with high field strengths, PEF pre-treatment will produce permanent pores within the semipermeable membrane and thus facilitate the entry of acids or enzymes to break down the cellulose into its constituent sugars.

11.5.4 Catalytic Microwave Pyrolysis

Catalytic microwave shift of biomass by utilization of carbon (AC) is performed simply to analyze and verify the consequences of transmutation conditions on the yields of phenol and phenolics. Bio-oils with high concentrations of phenol (38.9%) and phenolics (66.9%) were obtained. These levels were beyond those obtained by shift while not AC addition and were closely associated with the decomposition of polymer. A high concentration of esters (42.2% within the upgraded bio-oil) was obtained within the presence of metallic element powder as catalyst and formic acid/ethanol as reaction medium. Most of the esters known by GC–MS were long chain acetic acid esters. The high content of phenols and esters obtained during this study are often used as partial replacement of crude oil fuels when separation of oxygenates or as feedstock for organic synthesis within the industry when purification. Bio-oil is taken into account as an alternative to petroleum-based sources for a large vary of solvents, fuels, and chemicals and alternative merchandise. Potential bio-based chemicals from bio-oil embrace phenolics and cyclic ketones for resins and solvents, levoglucosan and levoglucosenone for polymers, and aromatic hydrocarbons for fuels and solvents. Lignins are way more thermally stable than polysaccharide and hemicellulose throughout biomass shift. The compounds obtained from decomposition of polymer became one of the most important challenges throughout upgrading of bio-oils. Though variety of studies are conducted on the preparation of phenolic like phenol-formaldehyde.

11.5.5 Multifaceted Applications of Phenolics

The main property attributed to phenolic compounds is their antioxidant activity against reactive molecule species involved in provoking diverse human diseases. Phenols can be important components of the human diet due to their potential antioxidant activity, their capacity to diminish oxidative stress induced tissue damage resulted from chronic diseases and their potentially important properties such as anticancer activities. Promising results have been achieved in human disease prevention studies increasing, for instance, the demand on food commodities with functional properties. Less polar phenolic substances, such as tocopherols, lignans, or resins, may be added to fats and oils increasing their stability on storage and heating. More polar phenols are advantageous for stabilization of food dispersions. Rosemary and sage resins were found efficient in frying fats. Particularly high content of phenols is found in spices, tea leaves, roasted coffee, and cocoa beans, and in red wine. Attempts to increase the antioxidant levels in blood stream were not very efficient as most phenols are inactivated before or immediately after the resorption through the intestine wall. Phenolic antioxidants have been found to be capable to protect polyunsaturated fatty acids in oil against autoxidation in plants naturally present. During oilseed processing, less polar antioxidants are co-extracted with oil, while more polar antioxidants, insoluble in the extraction solvent, remain in the extracted meal. Plant seeds are important sources of phytochemicals for nutritional, industrial, and pharmaceutical applications.

Chlorogenic acid and many other polyphenol compounds are extensively used in medicine and industries such as in consumer chemicals and food industries. Chlorogenic acid is used as various additives in beverage, cosmetics, tea products, and foods, as well as medical substances. Chlorogenic acid has antibacterial and antiviral properties. Chlorogenic acid, the most potent functional inhibitor of the microsomal glucose-6-phosphate translocase (G6PT), is thought to possess cancer chemopreventive properties. It is also a promising precursor compound for the development of medicine that can resist AIDS virus HIV.

Ferulic acid is a ubiquitous plant constituent that arises from the general phenylpropanoid pathway in plants. This pathway originates from phenyalanine and tyrosine and is responsible for the biosynthesis of a very large number of diverse secondary metabolites such as lignin and lignin precursors, including feruloyl CoA and pcoumaroyl CoA. Ferulic acid is a phenolic acid of low toxicity; it can be absorbed and easily metabolized in the human body. Due to its phenolic nucleus and an extended side chain conjugation, it readily forms a resonance stabilized phenoxy radical which accounts for its potent antioxidant potential. By virtue of effectively scavenging deleterious radicals and suppressing radiation-induced oxidative reactions, ferulic acid may serve an important antioxidant function in preserving physiological integrity of cells exposed to both air and impinging UV radiation. Ferulic acid has been reported to have many physiological functions, including protection against coronary disease, lowers cholesterol, and increases sperm viability. In same time, because of photoprotective properties and low toxicity, ferulic acid is now widely used in the food and cosmetic industries. The photoprotective properties are afforded to skin by ferulic acid dissolved in cosmetic lotions. Its incorporation into a topical solution of 15% L-ascorbic acid and 1% α-tocopherol improved chemical stability of the vitamins (C + E) and doubled photoprotection to solar-simulated irradiation of skin from four-fold to approximately eight-fold as measured by both erythema and sunburn cell formation. Inhibition of apoptosis was associated with reduced induction of caspase-3 and caspase-7. This antioxidant formulation efficiently reduced thymine dimer formation. This combination of pure natural low molecular weight antioxidants provides meaningful synergistic protection against oxidative stress in skin and should be useful for protection against photoaging and skin cancer. Ferulic acid is addition to foods inhibits lipid peroxidation and subsequent oxidaive spoilage. By the same mechanism ferulic acid may protect against various inflammatory diseases like other representatives of hydroxycinnamates group. The interest in hydroxycinnamates as bioactive components of the diet, as structural and functional components of plant cell walls, and as precursors for favors in the food industry has expanded rapidly in the last 5–10 years.

P-Coumaric acid (4-hydroxycinnamic acid) is a hydroxyl derivative of cinnamic acid. P-coumaric acid has antioxidant properties and is believed to reduce the risk of stomach cancer by reducing the formation of carcinogenic nitrosamines. A further analysis of several p-coumaric acid analogs suggests that the para positioning of the hydroxyl group in the phenyl ring and the double bond of p-coumaric acid may be important for its biological activity.

One of the most important chemical properties of tanins is the ability to form soluble and insoluble complexes with macromolecules, such as protein, fiber and starch. Their content in foods is affected by many factors that influence phenolic stability, biosynthesis, and degradation. The beneficial effects observed following the consumption of condensed tannins are mainly due to their protein-binding ability, which protects dietary protein from degradation in the rumen and thus increases protein availability in the lower digestive tract. Interestingly, the protein-binding capacity of tannins has been also considered responsible for causing adverse effects on livestock production. Most phenolic antioxidants are flavonoids, such as catechins, of different structures and antioxidant activities. Some of these used for flavoring certain types of vodka, was also found very effective as an antioxidant. Certain scatechins, flavonoids are applied to food and are widely used in (fermented) tea. Content of catechins is particularly high in green tea, but its infusions are less concentrated.

11.6 Conclusion

For LC biorefinery to be a cost-attractive value proposition calls for innovative solutions to the technological, environmental, and economic challenges that mainly revolve around the feedstock processing and pre-treatment. A combinatorial approach involving process integration, intensification, and optimization for conversion of LC biomass into fuels and value added products can be the most sustainable approach toward making LC biorefinery concept a success.

References

1. Zhang, Y.P., Reviving the carbohydrate economy *via* multi-product lignocellulose biorefineries. *J. Ind. Microbiol. Biotechnol.*, 35, 5, 367–375, 2008.
2. Eggeman, T. and Elander, R.T., Process and economic analysis of pre-treatment technologies. *Bioresour. Technol.*, 96, 18, 2019–2025, 2005.
3. Jin, S. and Chen, H., Fractionation of fibrous fraction from steam-exploded rice straw. *Process Biochem.*, 42, 2, 188–192, 2007.
4. Kim, T.H. and Lee, Y., Fractionation of corn stover by hot-water and aqueous ammonia treatment. *Bioresour. Technol.*, 97, 2, 224–232, 2006.
5. Sun, F. and Chen, H., Enhanced enzymatic hydrolysis of wheat straw by aqueous glycerol pretreatment. *Bioresour. Technol.*, 99, 14, 6156–6161, 2008.
6. Zhu, J.Y. and Pan, X.J., Woody biomass pre-treatment for cellulosic ethanol production: Technology and energy consumption evaluation. *Bioresour. Technol.*, 101, 4992–5002, 2010.
7. Kumar, R., Singh, S., Singh, O.V., Bioconversion of lignocellulosic biomass: Biochemical and molecular perspectives. *J. Ind. Microbiol. Biotechnol.*, 35, 377–391, 2008.
8. Kurian, J.K., Nair, G.R., Hussain, A., Raghavan, G.S.V., Feedstocks, logistics and pre-treatment processes for sustainable lignocellulosicbiorefineries: A comprehensive review. *Renew. Sust. Energ. Rev.*, 25, 205–219, 2013.

9. FAOSTAT, 2006. FAO statistical databases. http://faostat.fao.org/.
10. Chen, H.Z., Qiu, W.H., Xing, X.H., Xiao, X., Key process of biomass refining for the new generation of biological and chemical industry. *Chin. Basic Sci.*, 11, 5, 32–37, 2009.
11. Chen, H.Z., *Biotechnology of Lignocellulose: Theory and Practice*, Springer Press, Berlin, 2014.
12. Taherzadeh, M.J. and Karimi, K., Pre-treatment of lignocellulosic wastes to improve ethanol and biogas production: A review. *Int. J. Mol. Sci.*, 9, 1621–1651, 2008.
13. Menon, V. and Rao, M., Trends in bioconversion of lignocellulose: Biofuels, platform chemicals &biorefinery concept. *Prog. Energ. Combust. Sci.*, 38, 522–550, 2012.
14. Cherubini, F., The biorefinery concept: using biomass instead of oil for producing energy and chemicals. *Energy Convers. Manage.*, 51, 1412–1421, 2010.
15. Brandt, A., Gräsvik, J., Halletta, J.P., Welton, T., Deconstruction of lignocellulosic biomass with ionic liquids. *Green Chem.*, 15, 550–583, 2013.
16. Hong, J., Ye, X., Zhang, Y.H.P., Quantitative determination of cellulose accessibility to cellulase based on adsorption of a nonhydrolytic fusion protein containing CBM and GFP with its applications. *Langmuir*, 23, 12535–12540, 2007.
17. Pan, X., Xie, D., Gilkes, N., Gregg, D.J., Saddler, J.N., Strategies to enhance the enzymatic hydrolysis of pretreated softwood with high residual lignin content. *Appl. Biochem. Biotechnol.*, 124, 1069–1080, 2005.
18. Alvira, P., Tomás-Pejó, E., Ballesteros, M., Negro, M.J., Pre-treatment technologies for an efficient bioethanol production process based on enzymatic hydrolysis: A review. *Bioresour. Technol.*, 101, 4851–4861, 2010.
19. Ragauskas, A.J., Beckham, G.T., Biddy, M.J., Chandra, R., Chen, F., Davis, M.F., Davison, B.H., Dixon, R.A., Gilna, P., Keller, M., Langan, P., Naskar, A.K., Saddler, J.N., Tschaplinski, T.J., Tuskan, G.A., Wyman, C.E., Lignin valorization:Improving lignin processing in the biorefinery. *Science*, 344, 6185, 709–719, 2014.
20. Vijayalaxmi, S., Appaiah, K.A.A., Jayalakshmi, S.K., Mulimani, V.H., Sreeramulu, K., Production of Bioethanol from Fermented Sugars of Sugarcane Bagasse Produced by Lignocellulolytic Enzymes of Exiguobacterium sp. VSG-1. *Appl. Biochem. Biotechnol.*, 171, 1, 246–260, 2013.
21. Vazquez-Olivo, G., Lopez-Martinez, L.X., Contreras-Angulo, L., Heredia, J.B., Antioxidant Capacity of Lignin and Phenolic Compounds from Corn Stover. *Waste Biomass Valor.*, 10, 95–102, 2019.
22. Conde, E., Moure, A., Domınguez, H., Carlos, J., Production of Antioxidants by Non-Isothermal Autohydrolysis of Lignocellulosic Wastes. *LWT Food Sci. Technol.*, 44, 436–442, 2011.
23. Um, M., Shin, G., Lee, J., Extraction of Total Phenolic Compounds from Yellow Poplar Hydrolysate and Evaluation of Their Antioxidant Activities. *Ind. Crop. Prod.*, 97, 574–581, 2017.
24. Xiao, H., Wang, Y., Xiang, Q., Xiao, C., Yuan, L., Liu, Z., Liu, X., Novel Physiological Properties of Ethanol Extracts from EremurusChinensisFedtsch. Roots: In Vitro Antioxidant and Anticancer Activities. *Food Funct.*, 3, 1310–1318, 2012.
25. Oliveira, I., Sousa, A., Ferreira, I., Bento, A., Estevinho, L., Pereira, J.A., Total Phenols, Antioxidant Potential and Antimicrobial Activity of Walnut (Juglansregia L.) Green Husks. *Food Chem. Toxicol.*, 46, 2326–2331, 2008.
26. Oliveira, I., Sousa, A., Valentao, P., Andrade, P., Ferreira, I., Ferreres, F., Bento, A., Seabra, R., Estevinho, L., Pereira, J.A., Hazel (Corylusavellana L.) Leaves as Source of Antimicrobial and Antioxidative Compounds. *Food Chem.*, 105, 1018–1025, 2007.
27. Karamac, M., Orak, H.H., Amarowicz, R., Orak, A., Phenolic Contents and Antioxidant Capacities of Wild and Cultivated White Lupin (LupinusAlbus L.) Seeds. *Food Chem.*, 258, 1-7, 2018.
28. Nobre, C.B., Sousa, E.O., Camilo, C.J., Machado, J.F., Silva, J., Filho, J.R., Coutinho, H.D.M., Costa, J.G.M., Antioxidative Effect and Phytochemical Profile of Natural Products from the

Fruits of Babaçu (Orbigniaspeciose) and Buriti (Mauritiaflexuosa). *Food Chem. Toxicol.*, 121, 423–429, 2018.
29. Fang, L., Meng, W., Min, W., Phenolic Compounds and Antioxidant Activities of Flowers, Leaves and Fruits of Five Crab apple Cultivators (MalusMill.species). *Sci. Hortic.*, 235, 460–467, 2018.
30. Fu, D., Farag, S., Chaouki, J., Jessop, P.G., Extraction of Phenols from Lignin Microwave-Pyrolysis Oil Using a Switchable Hydrophilicity Solvent. *Bioresour. Technol.*, 154, 101–108, 2014.
31. Abussaud, B., Asmaly, H.A., Saleh, A., Kumar, V., Ali, M., Sorption of Phenol from Waters on Activated Carbon Impregnated with Iron Oxide, Aluminum Oxide and Titanium Oxide. *J. Mol. Liq.*, 213, 351–359, 2016.
32. Malviya, S. and Jha, A., Antioxidant and Antibacterial Potential of Pomegranate Peel Extracts. *J. Food Sci. Technol.*, 51, 4132–4137, 2014.
33. NREL Biomass Research: http://www.nrel.gov/biomass/biorefinery.html(assessed June 2009).
34. Realff, M.J. and Abbas, C., Industrial symbiosis – refining the biorefinery. *J. Ind. Ecol.*, 7, 3–4, 5–9, 2004.
35. Kamm, B. and Kamm, M., Principles of biorefinery. *Appl. Microbiol. Biotechnol.*, 64, 137–145, 2004.
36. van Dyne, D.L., Blase, M.G., Clements, L.D., A strategy for returning agriculture and rural America to long-term full employment using biomass refineries, in: *Perspectives on New Crops and New Uses*, J. Janick (Ed.), ASHS Press, Alexandria, VA, 1999.
37. Luo, L., van der Voet, E., Huppes, G., Biorefining of lignocellulosic feedstock - Technical, economic and environmental considerations. *Bioresour Technol.*, 101, 5023–5032, 2010.
38. Aden, A., Ruth, M., Ibsen, K., Jechura, J., Neeves, K., Sheenhan, J. *et al.*, Lignocellulosic Biomass to Ethanol Process Design and Economics Utilizing Cocurrent Dilute Acid Prehydrolysis and Enzymatic Hydrolysis for Corn Stover, 2002. NREL/TP-510-32438, NREL.
39. Jones, J.L. and Semrau, K.T., Wood Hydrolysis for ethanol production-previous experience and the economics of selected processes. *Biomass*, 5, 2, 109–135, 1984.
40. Rabemanolontsoa, H. and Saka, S., *in Zero-Carbon Energy Kyoto 2011*, T. Yao (Ed.), pp. 135–140, Springer Japan, Tokyo, 2012.
41. García, A., Egues, I., Ana, T., Alriols, M., Serrano, L., Labidi, J., Biorefining of Lignocellulosic residues using Ethanol Organosolv Process. *Chem Eng Trans.*, 18, 911–916, 2009.
42. McMillan, J.D., Pre-treatment of lignocellulosic biomass, in: *enzymatic conversion of biomass for fuel production, ACS Symposium series*, vol. 566, M.E. Himmel, J.O. Baker, R.P. Overend (Eds.) pp. 292–324, ACS, Washington, DC, 1994.
43. Millett, M.A., Effland, M.J., Caulfield, D.P., Influence of fine grinding on the hydrolysis of cellulosic materials, acid versus enzymes. *Adv. Chem. Series*, 181, 71–89, 1979.
44. Grous, W.R., Converse, A.O., Grethlein, H.E., Effect of steam explosion pre-treatment on pore size and enzymatic hydrolysis of poplar. *Enzyme Microb Technol.*, 8, 274–280, 1986.
45. Wright, J.D., Ethanol from biomass by enzymatic hydrolysis. *Chem Eng Prog.*, 84, 62–74, 1988.
46. Jacquet, N., Vanderghem, C., Danthine, S., Quiévy, N., Blecker, C., Devaux, J., Paquot, M., Influence of steam explosion on physicochemical properties and hydrolysis rate of pure cellulose fibers. *Bioresour Technol.*, 121, 221–227, 2012.
47. Ehara, K. and Saka, S., A comparative study on chemical conversion of cellulose between the batch-type and flow-type systemsin supercritical water. *Cellulose.*, 9, 301–331, 2002.
48. Nakahara, Y., Yamauchi, K., Saka, S., MALDI-TOF/MS analysis of decomposition behavior of beech xylan as treated by semi flow hot compressed water. *J. Wood Sc*, 60, 3, 225–231, 2014.

49. Takada, M. and Saka, S., Characterization of lignin derived products from Japanese cedar as treated by semi flow hot compressed water. *J. Wood Sc*, 61, 1–9, 2015.
50. Mosier, N., Wyman, C., Dale, B., Elander, R., Lee, Y.Y., Holtzapple, M., Ladisch, M., Features of promising technologies for pre-treatment of lignocellulosic biomass. *Bioresour. Technol.*, 96, 673–686, 2005.
51. Sindhu, R., Binod, P., Pandey, A., Biological pre-treatment of lignocellulosic biomass: An overview. *Bioresource Technol.*, 199, 76–82, 2015.
52. Suhara, H., Kodama, S., Kamei, I., Mawekawa, N., Screening of lignin degrading basidiomycetes and biological pre-treatment for enzymatic hydrolysis of bamboo culm. *Int Biodeter biodegrad*, 75, 176–180, 2012.
53. Du, W., Yu, H., Song, L., Zhang, J., The promoting effect of byproducts from Irpex lacteous on subsequent enzymatic hydrolysis of bio pretreated corn stalks. *Biotechnol Biofuels*, 4, 1, 37, 1–8, 2011.
54. Taha, M., Shahasavari, E., AI-Hothali, K., Mouradov, A., Smith, A.T. *et al.*, Enhanced biological strw saccharification through coculturing of lignocellulose degrading microorganisms. *Appl. Biochem. Biotechnol.*, 175, 3709–3728, 2015.
55. Castoldi, R., Bracht, A., de Morais, G.R., Baesso, M.L., Correa, R.C.G. *et al.*, Biological pre-treatment of Eucaliptus grandis sawdust with white rot fungi: Study of degradation patterns and saccharification kinetics. *Chem Eng J.*, 258, 240–246, 2014.
56. Potumarthi, R., Baadhe, R.R., Nayak, P., Jetty, A., Simultaneous pre-treatment and saccharification of rice husk by Phanerochete chrysosporium for improved production of reducing sugars. *Bioresour Technol.*, 128, 113–117, 2013.
57. Song, L., Yu, H., Ma, F., Zhang, X., Biological pre-treatment under nonsterile conditions for enzymatic hydrolysis of cornstover. *Bioresour*, 8, 3, 3802–3816, 2013.
58. Cianchetta, S., Di Maggio, B., Burzi, P.L., Galletti, S., Evaluation of selected white-rot fungal isolates for improving the sugar yield from wheat straw. *Appl. Biochem. Biotechnol.*, 173, 2, 609–623, 2014.
59. Wan, C. and Li, Y., Effectiveness of microbial pre-treatment by Cariporiopsis subvermispore on different biomass feed stocks. *Bioresour Technol.*, 102, 7507–7512, 2011.
60. Dhiman, S.S., Haw, J.R., Kalyani, D., Kalia, V.C., Kang, Y.C. *et al.*, Simultaneous pre-treatment and saccharification: Green technology for enhanced sugar yields from biomass using a fungal consortium. *Bioresour Technol.*, 179, 50–57, 2015.
61. Binod, P., Janu, K.U., Sindhu, R., Pandey, A., Hydrolysis of lignocellulosic biomass for bioethanol production, in: *Chapter 10 in Biofuels*, pp. 229–250, Academic Press, Oxford, UK, 2011.
62. Saloheimo, M., Paloheimo, M., Hakola, S., Pere, J., Swanson, B. *et al.*, Swollenin, a Trichoderma reesei protein with sequence similarity to the plant expansins, exhibits disruption activity on cellulosic materials. *Eur J Biochem.*, 269, 4202–4211, 2002.
63. Himmel, M.E., Ruth, M.F., Wyman, C.E., Cellulase for commodity products from cellulosic biomass. *Curr. Opin. Biotechnol.*, 10, 4, 358–364, 1999.
64. Bayer, E.A., Morag, E., Lamed, R., Yaron, S., Shoham, Y., Cellulosome structure: Fourpronged attack using biochemistry, molecular biology, crystallography and bioinformatics. *J. Struc. Biol.*, 124, 221–234, 1998.
65. Asiegbu, F.O., Paterson, A., Smith, J.E., The effects of co-fungal cultures and supplementation with carbohydrate adjuncts on lignin biodegradation and substrate digestibility. *World J. Microbiol. Biotechnol.*, 12, 3, 273–279, 1996.
66. Reid, I.D., Solid-state fermentations for biological delignification. *Enz. Microb. Technol.*, 967, 11, 786–803, 1989.
67. Qin, L., Li, W.C., Liu, L., Zhu, J.Q., Li, X., Li, B.Z., Yuan, Y.J., Inhibition of lignin-derived phenolic compounds to cellulase. *Biotechnol. Biofuels*, 9, 70, 2016.

68. Michelin, M., Ximenes, E., de Lourdes, de MoraesPolizeli, T.M., Ladisch, M.R., Effect of phenolic compounds from pretreated sugarcane bagasse on cellulolytic and hemicellulolytic activities. *Bioresour. Technol.*, 199, 275–278, 2016.
69. Kim, Y., Ximenes, E., Mosier, N.S., Ladisch, M.R., Soluble inhibitors/deactivators of cellulase enzymes from lignocellulosic biomass. *Enzym. Microb. Technol.*, 48, 408–415, 2011.
70. Fitzgerald, D.J., Stratford, M., Gasson, M.J., Ueckert, J., Bos, A., Narbad, A., Mode of antimicrobial of vanillin against Escherichia coli, Lactobacillus plantarum and Listeria innocua. *J. Appl. Microbiol.*, 97, 104–113, 2004.
71. Klinke, H.B., Thomsen, A.B., Ahring, B.K., Inhibition of ethanol-producing yeast and bacteria by degradation products produced during pre-treatment of biomass. *Appl. Microbiol. Biotechnol.*, 66, 10–26, 2004.
72. Ajila, C.M., Gassara, F., Brar, S.K., Polyphenolic Antioxidant Mobilization in Apple Pomace by Different Methods of Solid- State Fermentation and Evaluation of Its Antioxidant Activity. *Food Biopro Tech.*, 5, 7, 1–11, 2011.
73. Dulf, F.V., Vodnar, D.C., Dulf, E.H., Pintea, A., Phenolic Compounds, Flavonoids, Lipids and Antioxidant Potential of Apricot (Prunusarmeniaca L.) Pomace Fermented by Two Filamentous Fungal Strains in Solid State System. *Chem. Cent. J.*, 11, 1, 92, 2017.
74. Vijayalaxmi, S., Jayalakshmi, S.K., Sreeramulu, K., Polyphenols from Different Agricultural Residues: Extraction, Identification and Their Antioxidant Properties. *J. Food Sci. Technol.*, 52, 2761–2769, 2015.
75. Neelkant, K.S., Shankar, K., Jayalakshmi, S.K., Sreeramulu, K., Optimization of Conditions for the Production of Lignocellulolytic Enzymes by Sphingobacterium sp. ksn-11 Utilizing Agro-Wastes under Submerged Condition. *Prep. Biochem. Biotechnol.*, 49, 9, 927–934, 2019.
76. Shankar, K., Neelkant, S., Kulkarni, S., Jayalakshmi, K., Kuruba, S., Comparative assessment of solvents and lignocellulolytic enzymes affiliated extraction of polyphenols from the various lignocellulosic agro-residues: Identification and their antioxidant properties. *Prep Biochem Biotechnol*, 50, 7, 1–8, 2019.
77. Malviya, S. and Jha, A., Antioxidant and Antibacterial Potential of Pomegranate Peel Extracts. *J. Food Sci. Technol.*, 51, 4132–4137, 2014.
78. Kurakake, M., Hirotsu, S., Shibata, M., Kubota, A., Makino, A., Lignin antioxidant extracted from lignocellulosic biomasses by treatment with ammonia water. *Ind Crops Prod.*, 77, 1028–1032, 2015.
79. Tangkhavanich, B., Kobayashi, T., Adachi, S., Properties of rice straw extract after subcritical water treatment. *Biosci. Biotechnol. Biochem.*, 76, 1146–1149, 2012.
80. Xavier, L., Freire, M.S., González-Álvarez, J., Modeling and optimizing the solid–liquid extraction of phenolic compounds from lignocellulosicsubproducts. *Biomass Conv. Bioref.*, 9, 737–747, 2019. https://doi.org/10.1007/s13399-019-00401-9.
81. Akl, E.M., Taha, F.S., Mohamed., S.S., Wagdy, S.M., Hamid, S.M.A., Effective treatments of jojoba and jatropha hulls to obtain phytochemical compounds for industrial, nutritional, and pharmaceutical uses. *Bull Nat. Res Centre.*, 43, 21, 1–14, 2019.
82. da Costa Lopes, A.M., Brenner, M., Falé, P., Roseiro, L.B., Bogel-Łukasik, R., Extraction and Purification of Phenolic Compounds from Lignocellulosic Biomass Assisted by Ionic Liquid, Polymeric Resins, and Supercritical CO2. *ACS Sust Chem. Eng.*, 4, 6, 3357–3367, 2014.

12

Bioactive Compounds Production from Vegetable Biomass: A Biorefinery Approach

Shritoma Sengupta[1,2], Debalina Bhattacharya[3] and Mainak Mukhopadhyay[1*]

[1]Department of Biotechnology, JIS University, Kolkata, India
[2]Department of Biochemistry, University of Calcutta, Kolkata, India
[3]Department of Microbiology, Maulana Azad College, Kolkata, India

Abstract
"Bioactive compounds" means various molecules that have some biological activity. New generations of biorefinery combine innovative biomass waste resources from different origins, chemical extraction and/or synthesis of biomaterials, biofuels, and bioenergy via green and sustainable processes. Utilization of biomass as a sustainable renewable resource is the only way to replace carbon from fossil sources for the production of the carbon-based products such as chemicals, materials, and liquid fuels with a remarkable reduction of CO_2 releases into the atmosphere. Biorefinery utilization for production of bioactive compounds provides perceptions on the recent development, trends, innovations, explanations, and practical challenges that come across in the fields of enzyme technology and nanobiotechnology for the production of bioactive materials with extra health benefits. These wastes from the vegetable biomass are converted to different bioactive compounds with an excellent nutraceutical functions and with high adding value. In this chapter, it is discussed on the updated survey of different bioactive compounds from organic materials and vegetable biomass utilizing biorefinery techniques for its extraction that can be converted for bioenergy production and the various extraction techniques of the bioactive compounds.

Keywords: Bio-active compounds, biomass degradation, biorefinery, conventional solvents extraction, supercritical fluid extraction, pulsed-electric field extraction

12.1 Introduction

"Bioactive compounds" means various molecules that have some biological activity. Thus, a definition of bioactive compounds in plants is secondary plant metabolites prompting pharmacological or toxicological effects in men and animals. Bioactive compounds are the phytochemicals that present naturally to lesser extents in plants as well as foodstuffs [1] and have the potential to amend metabolic processes for the promotion of better health. The typical bioactive compounds produced in plants are secondary metabolites not required for the circadian functioning of the plant [2]. Bioactive compounds are extremely heterogeneous

**Corresponding author*: mainak@jisuniversity.ac.in; m.mukhopadhyay85@gmail.com

Arindam Kuila and Mainak Mukhopadhyay (eds.) Biorefinery Production Technologies for Chemicals and Energy, (241–258) © 2020 Scrivener Publishing LLC

class of compounds that includes plant growth factors, alkaloids, mycotoxins, food-grade pigments, antibiotics, flavonoids, and phenolic acids, etc. with dissimilar chemical structures (hydrophilic or lipophilic), specific to ubiquitous distribution in nature, significant amount present in foods and human body, efficient against oxidative species, and possess the potential biological action [3, 4]. There is another approach where the wastes from the vegetable biomass are converted to different bioactive compounds with an excellent nutraceutical function and with high adding value. Researches are being focused on causes for clean renewable energy production due to the drastic exhaustion of non-renewable energy, and the various environmental pollutions caused by the use of these non-renewable fossil energies [5]. Biomass resources are plentiful, renewable in nature, and also eco-friendly that plays a vital role in the supply of energy [6]. The conversion of biomass resources to renewable fuels, chemicals, or important bioactive compounds by various techniques is thus a very demanding and trending research topic.

The term biorefinery is widely used for describing such a concept [7] and was recently redefined in the project Biorefinery Euroview as follows: "Biorefineries could be described as integrated biobased industries using a variety of technologies to make products such as chemicals, biofuels, food, and feed ingredients, biomaterials, fibers and heat, and power, aiming at maximizing the added value along the three pillars of sustainability (Environment, Economy and Society)". Energy counts a great deal for both economic and social development. Along with the development of the world's economy through industrialization, urbanization, and modernization, the global energy supply has been increasing dramatically. According to the International Energy Agency (IEA), it is estimated that the global energy consumption will observe a 53% increase by 2030 [8], and the total energy consumption of developing countries will exceed that of the developed countries in 2030 [9]. For large-scale production of bioactive materials with extra health benefits, utilization of biorefinery techniques provides insights on the recent development, inclinations, novelties, elucidations, and practical challenges that come across in the fields of nanobiotechnology and enzyme technology.

The majority of bioactive compounds belong to the secondary metabolites' families, having particular structural characteristics ascending from the way in which they are biosynthesized in nature [10]. Secondary metabolites are classified into three main classes namely (a) alkaloids, (b) terpenes/terpenoids, and (c) phenolic compounds which are all important naturally obtained bioactive compounds. There are four major biosynthetic pathways for the production of these bioactive secondary metabolites. The pathways are: (1) shikimic acid pathway, (2) malonic acid pathway, (3) mevalonic acid pathway, and (4) non-mevalonate (MEP) pathway [11]. The isolated bioactive compounds include terpenes, phenolics, alkaloids, lipids, carbohydrates, peptides, and proteins. These important bioactive compounds are extracted from numerous sources which include microorganisms, algae, plant tissues, and marine organisms which are naturally obtained or they are produced by metabolic engineering of genes [10].

The useful biological activities of bioactive compounds, plant, as well as microbial secondary metabolites have received significant consideration especially for their beneficial effects on human health. Utilizing metabolic engineering and industrial biotechnology for the biosynthesis of these bioactive compounds or secondary metabolites basically offers a significant advantage over conventional methods for extraction from biomass [12]. Among the microorganisms, different types of marine bacteria and actinomycetes yield unique and novel secondary metabolites. These organisms display interesting biological activities and also play a widespread role in the pharmaceutical and medical industry for their capacity to produce secondary metabolites with diverse chemical structures and biological activities [10, 12]. Various bioactive compounds produced from agro-industrial wastes are polyphenols, pigments, peptides, biocomposites, etc. [13]. The absorption of bioactive compounds does not take place by simple diffusion processes. They also have the inability to pass the lipid-rich outer membrane of the small intestine [14]. Various technologies like nanocarriers, phytosome, etc., are being developed to increase the bio-accessibility as well as bio-availability of bioactive compounds. They also help to protect the active substances from oxidation or other degradation reactions in the gastrointestinal tract [15].

Sustainability and feasibility of biorefinery utilizing biowaste such as food and vegetable biomass mostly depend on the availability of biomass feedstock. Agricultural wastes are accumulated seasonably. Hence, the production of some of these food processing by-products from the biomass and waste from them are also produced seasonally. Seasonal processing in biorefinery chains results in futile use of the equipment and irregular supply of yields to the market. Thus, to recover from this situation, the requirement of R&D is beneficial for identifying the cause. They have certain objectives which include:

- The utilization of several feedstocks and innovative cultivations for round the year use of various amenities such as the utilization of third-generation biomass, i.e., algal biomass;
- Different conditions to increase the storage life of the feedstocks and their by-products;
- The advancement of new sorting and collection techniques (automation) together with the development of sustainable techniques for the integrated characterization of collected biowaste and successive decentralized technologies for their stabilization, fractionation, transportation within new logistic value chains are necessary;
- Optimization of biowaste and biomass conditions in relation to the availability and content of the target biomolecules that need to be recovered or bio-converted into more sophisticated compounds with respect to the concept of biorefinery.

12.2 Production of Bioactive Compounds

During the process of metabolic breakdown in fermentation, they also release several additional compounds other than the usual products of fermentation, such as carbon dioxide and alcohol. These additional compounds are called secondary metabolites. Secondary metabolites vary from numerous peptides to antibiotics and enzymes to growth factors [16–18]. These secondary metabolites are also termed as "bioactive compounds" as they hold biological activity. Hence, fermentation is one such technique for the production of bioactive compounds that has been widely used to manufacture an extensive variation of substances that are highly beneficial to both individuals and industry. Over the years, fermentation techniques have gained enormous status due to their environmental and economic advantages. Early practices have been additionally improvised and refined for the maximum level of production of beneficial compounds, which has also involved the advancement of new machinery and processes.

In recent years, excessive attention has been paid toward bioactive compounds because of their ability in dealing with human health, such as reducing the rate of cardiovascular diseases and other progressive diseases such as cancer, diabetes, etc. [2]. The naturally occurring bioactive compounds or the secondary metabolites are biosynthesized through the following pathways: (1) shikimic acid pathway, (2) malonic acid pathway, (3) mevalonic acid pathway, and (4) non-mevalonate (MEP) pathway [11]. On the contrary, the extraction and bioactivity screening of these natural bioactive compounds includes quite a few steps. They are:

- Firstly, the raw materials are selected based on their nutritional or medicinal effects were using standard protocols, their toxicity levels are checked [19].
- In the next step, various elemental analyses are carried out for determining their chemical composition and potential bioactivities of the materials [20].
- The compounds are extracted from the crude extracts and their potential activities are checked in both conditions: *in vitro* and *in vivo* [21].
- Finally, the bioactive compounds are commercialized into medicinal products and proved to be fruitful in curbing various diseases [10].

Bioactive peptides are another form of bioactive compounds that perform diverse functions that range from specific hormone-like tasks to functions relating to cellular processes [22]. They are also functional food components that are involved in human being regulatory activities [23] and are vastly encoded inside bioactive proteins [24]. Till now, cheese [25], dairy products [26], and bovine milk [27–30] have been considered as the greatest sources of bioactive proteins and peptides derived from foods. They are also obtained from various animal sources such as bovine blood [31], gelatin [32], meat, eggs, and fish species such as tuna, sardine, herring, and salmon [33]. Some bioactive peptides and proteins are obtained from vegetable sources such as maize, rice [34], soy [35], mushrooms, pumpkin, sorghum [36], and amaranth [37].

Worldwide, a huge number of by-products or wastes are formed from various food industries. These wastes and by-products released out cause a serious problem to dispose of the environment. So, all over the world, people are working out with different approaches which can be used as an alternative way to use these wastes because these by-products are an outstanding source of numerous bioactive compounds such as peptides, flavonoids, polyphenols, carotenoids, caffeine, polysaccharides, and others, which are advantageous for human health [2]. Various bioactive compounds produced from agro-industrial wastes are polyphenols, pigments, peptides, biocomposites, etc. [13].

Bioactive compounds from natural sources not only provide a platform for drug finding but also remediate the food crisis. The naturally obtained bioactive compounds may interact with different biomolecules such as proteins, DNA, and others to produce the desired outcome. Hence, both the pharmaceutical and food industries have strong attention to characterizing new bioactive compounds for developing therapeutic agents as well as nutraceuticals and functional foods. Natural compounds are biosynthesized in conjugation or in combination with another form in small quantities [38]. Thus, the concentration and purification of these conjugated products require the repeated separation of complex extracts into individual bioactive compounds and this process is time-consuming and tedious which makes it unprofitable for industries [10]. Various modern techniques have been developed for the extraction and production of bioactive compounds from natural sources such as plants and marine algae. For example, natural compounds can be efficiently purified with much lower cost using metabolic engineering and other molecular biological techniques [39].

Recently, various studies show that researches demonstrating the utilization of several secondary metabolites which have both industrially and economically important can be harnessed using biorefinery techniques. These technologies have been utilized in a variety of industries such as pharmaceuticals and food, especially in the field of probiotics [40] and prebiotics [41]. The development of these industries has led to the amplification of several techniques used in the laboratory on a large scale, which has exhibited a huge number of problems. Adverse circumstances may also affect in the production of unwanted compounds instead of the bioactive compound of interest. Two broad fermentation techniques have emerged as a result of this rapid development which has gained immense popularity are: Submerged Fermentation (SmF) and Solid State Fermentation (SSF) for the basal and industrial level of production of bio-active compounds [18]. These SSF and SmF techniques have been more polished based on several features such as the environmental parameters, substrates used, and the organisms used for fermentation of biomass and others. Various research works mentioned that in SSF, certain bioactive compounds are found to be produced in higher quantities, whereas the other compounds have been extracted using SmF. Hence, a fermentation process may be classified into SSF and SmF mainly based on the type of substrate used during fermentation [17, 18]. There has been little information regarding the relative study of these two fermentation techniques with respect to the production of bioactive compounds. This analysis is necessary as utilizing these fermentation processes in the industrial level using specialized biorefinery techniques, productivity must be maximized to meet demands. Several bioactive compounds are extracted such as enzymes [42], antioxidants [43], pigments [40], bioactive peptides [22, 25], biosurfactants, and other medicinal compounds such as antibiotics [44, 45], antihypertensive agents [46], antitumor agents [47], and hypercholesterolemic agents [48] that have been extracted using these two fermentation techniques.

SSF utilizes solid substrates, like bran, bagasse, paper pulp, vegetable, and fruit biomass, and others for the production of several compounds that have potent bioactivity. In this fermentation practice, the substrates are used very slowly and steadily by the microorganisms, so the initial substrate added to the system can be utilized for long periods of fermentation. Hence, this technique supports the controlled release of nutrients. The main advantage of this process is using these substrates which are nutrient-rich waste materials and can be easily recycled as substrates [49]. Wheat bran, rice and rice straw, hay, fruit and vegetable waste, paper pulp, bagasse, coconut coir, and synthetic media are some of the well-known and regularly used substrates for SSF [50].

SmF also known as Liquid Fermentation (LF) utilizes free-flowing liquid substrates, such as molasses and broths. In this process of fermentation, the bioactive compounds produced are released into the fermentation broth. The substrates in this process are used up quite rapidly; so, it requires to have a constant flow of replaced or new supplement of nutrients. Soluble sugars, molasses, liquid media, fruit and vegetable juices, and even sewage/wastewater are some common substrates used in SmF.

12.3 Bioactive Compounds From Vegetable Biomass

Agro-industrial processes are on the rising path and expectedly would continue to increase as the world population escalates that needs sustenance. Various agro-industrial activities have led to the growth of numerous kinds of by-products either as waste or as secondary products with recyclable value. In addition, most waste products from these agro-industries are dumped into the environment; thus, constituting an environmental irritation. Sustainable biotechnology, on the other hand, initiates for zero waste and purposefully boosts the valorization of waste to high-value products such as bio-active compounds where microbes and microbial products with unique properties are employed in the bio recycling processes for high-value products [51]. Sustainability and feasibility of biorefinery utilizing biowastes such as food and vegetable biomass mostly depend on the availability of biomass feedstock.

Bioactive compounds such as bioactive peptides are extracted from soybean and the extracted soy milk has been widely used in various food products and also extensively studied [22, 35]. Bioactive peptides are produced from soybean and soymilk that are obtained from hydrolytic digestion, which showed that antimicrobial properties and the endogenous proteases are also used to investigate the presence of peptides [35]. Peptides found in soy milk samples could be formed during food processing [52]. Soy hydrolysate and the soy-fermented foods such as natto and tempeh are processed with a variety of endoproteases such as pronase, trypsin, and others to produce oligopeptides, which are mainly derived from a soy protein known as glycinin. These digests showed angiotensin-converting enzyme inhibitory activity and surface-active properties [53]. Cereal grains are other rich sources of bioactive peptides, where grains such as rice, wheat, barley, millet, rye, sorghum, oat, and corn have shown substantial qualitative production of active peptides. The highest diversity and plenty of peptides with potential biological activity among the cereal proteins are generally showed by Grains of wheat and barley [54]. Wheat grains and oats have shown the presence of a specific peptide, i.e., angiotensin-converting enzyme inhibitory peptides, as well as peptides with hypotensive, antioxidant, antithrombotic, and opioid activities [22]. Various studies showed bioactive peptides obtained from rice have also shown anticancer activity [22, 54].

The cyclic dipeptides such as 2, 5-diketopiperazines (DKPs) are present in various foods, particularly in cocoa, roasted coffee, roasted malt, and fermented foods such as distillation residue of awamori, beer, and others that have received considerable attention as bioactive compounds. Their nature of bioactive compound is due to the formation of N-terminal amino acid residues of a linear peptide or protein that have shown antioxidant activities [2, 22].

Numerous fruit and vegetable processing industries are based on various industries of fruits as well as vegetables such as juice industries, pickle industries, oil industries, etc. These industries processed the substrate to increase their shelf life by using canning, drying, freezing, and preparation of juices, jams, and jellies, etc. [2]. The fruit and vegetable industries usually produce a huge amount of effluents as well as solid waste. The main solid waste constitutes organic materials, including discarded fruits vegetables, peel/skin, seeds, stones, etc., whereas the effluents contain liquid waste of juice and wash waters. In India, Asia's largest vegetable, fruit, and flower market, i.e., the Koyambedu market, Chennai spread over an area of 60 acres and produced approximately 80 tons of solid waste per day [55]. Subsequently, there is a major issue in regard to squander transfer, which can prompt issues with flies and rats around the preparing room, if not accurately managed. In most of the Asian countries, there is shortage of feed for livestock e.g., in India, a deficiency of 25, 159, and 117 million tons of concentrates, green forages, and crop residues [56], and China has a shortage of 10, 30, and 20 million tons of protein feed, energy feed, and aquatic feed, respectively [57]. To overcome this problem, the fruit and vegetable waste has been used as an alternate source of livestock feed. Various studies showed that vegetable biomass used for bioactive compound extraction are:

- Onion skin used for Quercetin extraction by subcritical solvent extraction (SWE) using water at 100–190°C for 5–30 min at pressure 90–131 bar which is used in anticancer, antivirus, and anti-inflammatory responses [58].
- Carrot leaves are used for extraction of phenolics, luteolin for agriculture purpose and food industry at 110–230°C for time 0–114 min, having a solid-liquid ratio (15 and 35 g/L) [59].
- Potato peel is used for extraction of glycoalkaloids, aglycone alkaloids by pulsed-electric field (PEF) extraction technique having a field strength of 0.75 kV/cm and 600 μs of treatment time. Potato peels are also extracted by solid-liquid extraction (SLE) technique using methanol which is used in the food industry and other industrial purposes [60]
- Avocado peel is used for extraction of phenolic compounds which are commercially used in food, cosmetic, and pharmaceutical industries by the extraction process of Pressurized liquid extraction (PLE) at 200°C, where ethanol/water ratio is used in the ratio of 1:1 v/v [61]. Orange peel is used for extraction of hesperidin, narirutin for antioxidants and food industry by extracting at the temperature of 110–150°C and water flow having rate 10–30 ml/min [62]
- Cocoa bean shell and coffee silverskin are used for the extraction of polyphenols and methylxanthines by PEF technique for industrial productions [63].
- Mango peels are used for the extraction of some bioactive compounds by PEF technique using 13.3 kV/cm pulses which are used in several food and industrial purposes [64].
- Winery grape seeds are used for catechins and proanthocyanidins extraction and in pharmaceutical and food industry by SWE using water at 50, 100, and 150°C and 1,500 psi [65]
- Litchi seeds are used for extraction of polysaccharides for antitumoral, antioxidant, and hypoglycaemic properties by ultrasonic-assisted extraction (UAE) technique using water for 45 min at 222 W [66]
- Apple and peach pomaces (plant fruit) are used for polyphenols extraction by SWE using carbon dioxide and ethanol at 40–60°C with 14–20% ethanol for 10–40 min [67]
- Tomatoes are used for extraction of carotenoids by PEF technique of varying 5, 18, and 30 pulses (40, 120, and 200 kV m^{-1}) at 4°C for 24 h which are used in food and pharmaceutical industries [68]

- Peaches are used for the extraction of phenols, flavonoids, and antioxidant compounds by PEF technique using 80% methanol in food and pharmaceutical industries [69]
- Lemons are used for the extraction of total phenols, flavanones hesperidin, and eriocitrin by PEF technique using 30 pulses of 30 ms, 7 kV/cm, which have importance in food and industrial uses [70].
- Grapes are used for the extraction of anthocyanins, phenolic acids, flavanols, and flavonols by PEF technique for industrial uses.
- The cocoa bean is used for the extraction of catechin, epicatechin, procyanidin B2, caffeine, and theobromine by PLE technique for utilization in fertilizers and animal feed [71].
- Wheat straw, germ, and bran are used for the extraction of policosanols by PLE technique, using n-hexane, ethanol, petroleum ether, chloroform at 80–125°C and pressure 1,500 psi [72].
- Eggplant peel is used for the extraction of anthocyanins by solvent extraction using acidified ethanol containing tartaric acid and malic acid at 40°C for 60–80 min which is commercially used for natural food colorings [73].
- *Mangifera pajang* peels are used for the extraction of phenolics by solvent extraction using 68% ethanol at 55°C [74].
- Oregano leaves (plant herb) are used for the extraction of flavone, flavanone, and flavonols by supercritical fluids extraction using 0–7% EtOH and CO2 at 40–60°C and pressure of 150–360 bar [75].
- Cherry (plant fruit) is used for the extraction of phenols and perillyl alcohol by supercritical fluids extraction technique using CO2 at 50°C for 60 min at pressure 25 MPa [76].
- Tomato skins are used for the extraction of lycopene by supercritical fluid extraction (SFE) technique using CO2 at 40–100°C and pressure of 20–40 MPa [77].
- Coriander seeds are used for the extraction of antioxidants such as DPPH by supercritical fluids extraction technique using CO2 at the pressure of 116–180 bar [78].
- Rice bran is used for the extraction of endoproteases mixture through enzymatic hydrolysis having a system at temperature 60°C, pH 8 [79].

12.4 Role of Biorefinery in Production of Bioactive Compounds

As a multifunctional process, biorefinery applies a varied range of raw materials to sustainably and concurrently generate a range of diverse intermediates and products including food and feed. The biorefinery chain consists of the pre-treatment and separation of biomass components (primary biorefinery) and the subsequent conversion (secondary biorefinery) [80]. Using green and sustainable processes, new generations of biorefinery techniques combine various types of biomass waste resources from different origins, following chemical extraction and finally synthesis of biomaterials, biofuels, or bioenergy. Utilization of biomass as a sustainable renewable resource is the only way to replace carbon from fossil sources for the production of carbon-based products such as chemicals, materials, and liquid fuels with a remarkable reduction of CO_2 releases into the atmosphere. Utilization of biorefinery aspect for production of various bioactive compounds provides insights on the current development, trends, innovations for bioactive compounds production, as well as it explains the practical challenges that come across in the fields of enzyme technology and nanobiotechnology for the production of bioactive materials with extra health benefits.

The extraction and bioactivity screening of natural compounds comprises of several steps. First, raw materials are selected based on their nutritional or medicinal effects. The selected materials are checked for toxicity with standard protocols. Then, elemental analysis is carried out to determine the chemical composition and potential bioactivities of the materials [20]. The compounds are isolated from the crude extracts and their potential activities are tested *in vitro* and *in vivo* [21]. Finally, the bioactive compounds are commercialized into medicinal products and proved to be fruitful in curbing the various diseases.

From various studies, it was observed that in recent years, solvent and supercritical fluids assisted extraction techniques were the most common methods of extracting compounds from plants [5]. The different methods of extraction of bioactive compounds for its production are:

- Conventional solvents extraction: It is widely used for its easy availability and low cost in extracting. This process is one of the conventional methods for extracting compounds from bacteria, algae, fungi, and plants where raw materials from the organisms are often ground into powder form to enhance the extraction efficacy. In this process, both polar and nonpolar solvents such as ethanol, ether, chloroform, hexane, benzene, and water have been used for the extraction in different combinations and ratios [81].
  Solvent extraction is an advantageous process compared to other methods due to low processing cost and ease of operation. On the other side, this technique has certain drawbacks of time-consuming and labor-intensive procedure which is tried to overcome using advanced methods of soxhlet, ultrasound, and microwave extraction [10].
- Non-conventional methods: The limitations of conventional extraction methods led to the development of more advanced methods such as microwave-assisted extraction (MAE), PLE, UAE, SFE, PEF extraction and others [82]. These new methods have decreased the extraction time of the process and have mostly diminished the thermal decomposition of the extracted bioactive compounds, retaining their efficacy [83].
- Pressurized liquid extraction (PLE): Of late, new extraction systems have been projected for replacing common solvent extraction by new PLE technique, which is lately gaining importance and has been widely employed in various extraction processes from natural sources [84, 85]. This technology was firstly described by Richter and known as pressurized fluid extraction (PFE) and enhanced solvent extraction. This extraction technique is referred to as accelerated solvent extraction or pressurized solvent extraction technique. In this process, organic liquid solvents are used at high temperatures (50°C to 200°C) and pressure (1,450 to 2,175 psi) to confirm the swift and speedy extraction rate of compounds [72]. This technique allows a faster extraction

in which less amount of solvents is used and higher yields are obtained in comparison with conventional solvent extraction. The high pressure helps in the extraction of cells to be filled faster and forces liquid into the solid matrix [86]. This technique has a lot of advantages over conventional methods, but this method is not suitable for thermolabile compounds as high temperatures can have lethal effects on the compounds' structure and functional activity [87].

- Subcritical water extraction (SWE): The use of water under high temperature and pressure below supercritical conditions in extraction processes is generally referred to as SWE which was developed as a useful tool to replace traditional extraction methods [86]. It represents a sequence of significant advantages over the traditional extraction techniques such as being a faster technique, producing high yields, where the use of different solvents can be greatly abridged [85] and it is an environmentally friendly technique compared to conventional organic solvent utilizing liquid solvent extraction techniques. This system the function is carried out using hot water having temperatures between 100°C and 374°C. The latter temperature (374°C) is considered to be the critical the temperature of water under high pressure (ranging in between 10 and 60 bar) that sustains water in its liquid state only [88]. Dielectric constant, solubility, and temperature are considered to be certain parameters of the solvent that is affected when the liquid state is maintained.
- Supercritical fluid extraction (SFE): Various studies showed in recent years that it is the most common method of extracting compounds from plants. This SFE was first introduced by Hannay and Hogarth in 1879, as an alternative extraction method to the conventional ones [10]. The detailed SFE study was conducted in around 1960s [89] and the elaborative study of the extraction procedure was done by the SFE process, which was used to decaffeinate the coffee beans [90]. Studies showed that the supercritical state of any solvents was accomplished when temperature and pressure are beyond its critical point at which the solvent possesses both gas and liquid-like properties such as diffusion, surface tension, viscosity, density, and salvation [91]. This principle of extraction is based on various parameters such as fluid, density, the viscosity of the solvent above their critical points. The yield of SFE technique depends upon numerous operational conditions and parameters, such as raw materials, solvents, pressure, and temperature that need to be optimized for the extraction of several components.

  SFE is an eco-friendly and highly selective method and this technique has been utilized to extract several bioactive compounds from macroalgae, cyanobacteria, and marine invertebrates which includes organisms such as crustacean, crawfish, crab, shrimp, squid, urchin, starfish, and others [92]. Carbon dioxide (CO_2) is the most common solvent for SFE and other solvents are also used, such as ethylene, methane, nitrogen, xenon, or fluorocarbons [93]. There are varying other applications of SFE such as coffee decaffeination, phenol, and flavonoid extraction, fatty acid refining, nutraceutical, and functional food preparation. SFE was less used frequently for the extraction of some secondary metabolites such as phenolics and isoflavones. To expand the applicability of SFE, later, a new hyphenated technique was developed for the extraction of isoflavones from macroalgae by supercritical CO_2 extraction followed by fast chromatography using 3% (v/v) of $MeOH/H_2O$ as solvent [10, 94].

- Microwave-assisted extraction (MAE): MAE was first explained in 1986 where using microwave energy, products from solvents or solutions were extracted [95]. MAE technique utilizes electromagnetic radiations with a wavelength from 0.001 m to 1 m and the system is composed of two oscillating fields which are perpendicular to the magnetic and electric field [10, 96]. In the MAE technique, microwave energy is converted into thermal energy which causes direct impacts on polar materials by exerting pressure on cell walls [97]. This MAE technique is a comparatively new extraction technology that is being widely used in a variety of industries with many advantages over conventional extraction practices including lesser environmental pollution, advanced extraction efficacy, and shorter time for extraction. In order to be considered in industrial applications at least two important limitations must be improved which includes i) the recovery of non-polar compounds and ii) the modification of the chemical structure of target compounds which may alter their bioactivity and limit their application [86].
- Green extraction techniques: "Green Extraction is based on the discovery and design of extraction processes which will reduce energy consumption, allows the use of alternative solvents and renewable natural products that guarantees a harmless and high-quality extract/product" [98]. This technique employs principles of green chemistry which promotes the sustainability of the products and allows them to exploit nature without exhausting the natural environment. The main aspects behind green chemistry are waste, energy, and hazard which dominate among the 12 principles laid by Anastas and Warner [99]. This thought has been introduced in various chemical procedures which include synthesis, catalysis, separation, and monitoring of the extracted compounds [10].
- Ultrasonic- or ultrasonic-assisted extraction (UAE): UAE practices sound wave at 20 kHz–100 MHz for creating compression and expansion of the cells to extract the required bioactive compounds which are chemical components [10]. This technique is based on the mechanism of dispersion across cells and breach of the cells due to mass transfer [98]. This technique has been utilized for the extraction of sugars [100], polysaccharides, proteins [101], polysaccharide-protein complex [102], oils [103], phenolics, and carotenoids [104]. This technique has two phases of cycles: the expansion cycle and the compression cycle. When the ultrasound intensity is sufficient, the expansion cycle can create cavities or microbubbles in the liquid. Specifically, ultrasonic creates voids or microbubbles which captivates energy and inflate in size in the expansion cycle and later recompress in the compression cycle where the cycles cause a sequence of production, growth, and collapse of bubbles in cells. The collapsing process generates sound waves of high pressure

(1,000 atm) and high temperature (about 5,000 K) at a heating and cooling rate above 1,010 K/s [105]. The explosion of the bubbles results in fragmentation of the cells that release out the bioactive compounds. Compared to conventional methods, the ultrasonic consumes less solvent and is more cost-effective and efficient to extract polyphenols and other compounds [10]. The UAE process has been proposed as a small alternative to conventional solvent extraction, providing greater recovery of the targeted bioactive compounds with lower solvent consumption and faster analysis and bioactivity of the extracted compounds. Better extraction efficiency is related to a phenomenon known as acoustic cavitation [86]. However, this technique should be cautiously used in the extraction of unstable compounds such as carotenoids because a major denaturation has been observed in the compounds extracted in this technique when compared with other technologies [106].

- Pulsed-electric field (PEF) extraction: PEF extraction is a non-thermal technique that uses short electric field pulse to extract the bioactive compounds and improve the quality of food materials. During the PEF technique, the cell membrane is ruptured by the electric field and the electric potential is transferred to the cells [107]. This electric potential generates charge in the cell membrane and the bioactive compounds are separated from the cells when the electric potential exceeds 1V [10]. The effectiveness of the extraction is dependent on various operational parameters such as field strength, energy input, temperature, the material used, and pulse number [108, 109]. Technical improvements have enabled the use of high electric field intensity (up to 80 kV/cm) in pulse electric and suggestively increased the extraction efficiency. PEF technique has been used for various procedures changing from pressing, extraction, drying, to diffusion. This technology reduces the time of extraction and increases mass transfer by disrupting the membrane of the raw material. The disruption or the damage of the cell membrane is important to the enrichment of penetrability of the membrane and demonstrates to be advantageous over conventional methods of extraction [10, 109].

12.5 Concluding Remarks

This chapter focuses on various biorefinery techniques and the technologies used for the production and extraction of bioactive compounds from natural sources and vegetable biomass and its utility in several other industries such as food and pharmaceutical. Nowadays, people are more aware of the components contained in the foods they consume, preferring those obtained from natural sources due to the often imagined negative effects of some compounds that are attained by chemical synthesis provoke. Moreover, several diseases predominant all over the world have caused high mortality. So, pharmaceutical industries have been searching for novel bioactive compounds for developing more effective treatments. Bioactive compounds are produced by using two techniques at the industrial level: SSF and SmF. The extraction of bioactive compounds using conventional methods has been seen to be monotonous and time-consuming. Several biorefinery techniques having advanced extraction methods such as SFE, UAE, PLE, MAE, and others have enhanced the extraction efficacy and the yield of the bioactive compound. The increasing demand for natural products in the food and pharmaceutical industry requires more efficient, productive, and environmental-friendly extraction techniques. More research is required in the near future to demonstrate the effectiveness of these bioactive compounds with more *in vivo* studies to advance more rapidly the design of new functional foods and nutraceuticals, which might be beneficial for human welfare.

References

1. Martins, S., Mussatto, S.I., Martínez-Avila, G., Montañez-Saenz, J., Aguilar, C.N., Teixeira, J.A., Bioactive phenolic compounds: Production and extraction by solid-state fermentation: A review. *Biotechnol. Adv.*, 29, 365–373, 2011.
2. Sadh, P.K., Kumar, S., Chawla, P., Duhan, J.S., Fermentation: A Boon for Production of Bioactive Compounds by Processing of Food Industries Wastes (By-Products). *Molecules*, 23, 2560–2593, 2018.
3. Porrini, M. and Riso, P., Factors influencing the bioavailability of antioxidants in foods: A critical appraisal. *Nutr. Metab. Cardiovasc. Dis.*, 18, 647–650, 2008.
4. Carbonell-Capella, J.M., Buniowska, M., Barba, F.J., Esteve, M.J., Frígola, A., Analytical methods for determining bioavailability and bioaccessibility of bioactive compounds from fruits and vegetables: A review. *Compr. Rev. Food Sci. Food Saf.*, 13, 155–171, 2014.
5. Wang, Y.P., Ke, L.Y., Yang, Q., Peng, Y.J., Hu, Y.Z., Dai, L.L. *et al.*, Biorefinery process for production of bioactive compounds and bio-oil from *Camellia oleifera* shell. *Int. J. Agric. Biol. Eng.*, 12, 5, 190–194, 2019.
6. Doddapaneni, T.R.K.C., Konttinen, J., Hukka, T.I., Moilanen, A., Influence of torrefaction pretreatment on the pyrolysis of Eucalyptus clone: A study on kinetics, reaction mechanism and heat flow. *Ind. Crops Prod.*, 92, 244–254, 2016.
7. Ohara, H., Biorefinery. *Appl. Microbiol. Biotechnol.*, 62, 474–477, 2003.
8. Ong, H.C., Mahlia, T.M.I., Masjuki, H.H., A review on energy scenario and sustainable energy in Malaysia. *Renew. Sust. Energ. Rev.*, 15, 639–647, 2011.
9. Saito, S., Role of nuclear energy to a future society of shortage of energy resources and global warming. *J. Nucl. Mater.*, 398, 1–9, 2010.
10. Gill, B.S., Navgeet, Qiu, F., Technologies for extraction and production of bioactive compounds. *Biotechnol. Prod. Bioact. Compd.*, Chapter-1, 1–36, 2019.
11. Taiz, L., and Zeiger, E., Stress physiology, in: *Plant Physiology, 4th edition,* L. Taiz, E. Zeiger (Eds.). pp. 671–681, Sinauer Associates, Inc., Sunderland, MA, 2006.
12. Janardhan, A., Kumar, A.P., Viswanath, B., Saigopal, D.V.R., Narasimha, G., Production of Bioactive Compounds by Actinomycetes and Their Antioxidant Properties. *Biotechnol. Res. Int.*, Article ID 217030, 2014, 1-8, 2014.
13. Ramírez, F.B., Tamayo, D.O., Corona, I.C., González-Cervantes, J.L.N., Esparza-Claudio, J. de J., Rodríguez, E.Q., Agro-Industrial Waste Revalorization: The Growing Biorefinery. *Biomass Bioenergy Recent Trends Future Challenges*, Chapter-5, 1-20, 2019.
14. Awasthi, R., Kulkarni, G.T., Pawar, V.K., Phytosomes: An approach to increase the bioavailability of plant extracts. *Int. J. Pharm. Pharm. Sci.*, 3, 1–3, 2011.
15. De Souza, R.J., Mente, A., Maroleanu, A., Cozma, A.I., Ha, V., Kishibe, T., Anand, S.S., Intake of saturated and trans unsaturated fatty acids and risk of all cause mortality, cardiovascular disease, and type 2 diabetes: Systematic review and meta-analysis of observational studies. *BMJ*, 351:h3978, 2015.

16. Balakrishnan, K. and Pandey, A., Production of Biologically Active Secondary Metabolites in Solid State Fermentation. *JSIR*, 55, 365–372, 1996.
17. Robinson, T., Singh, D., Nigam, P., Solid-state fermentation: a promising microbial technology for secondary metabolite production.. *Appl. Microbiol. Biotechnol.*, 55, 284–289, 2001.
18. Subramaniyam, R. and Vimala, R., Solid State and Submerged Fermentation for the production of Bioactive Substances: A comparative study. *Int. J. Sci. Nat.*, 3, 3, 480–486, 2012.
19. Gill, B.S. and Kumar, S., Ganoderic acid targeting multiple receptors in cancer: in silico and in vitro study. *Tumor Biol.*, 37, 10, 14271–14290, 2016a.
20. Gill, B.S. and Kumar, S., Triterpenes in cancer: significance and their influence. *Mol. Biol. Rep.*, 43, 9, 881–896, 2016b.
21. Gill, B.S. and Kumar, S., Ganoderic acid modulating TNF and its receptors: *In silico* and *in vitro* study. *Med. Chem. Res.*, 26, 6, 1336–1348, 2017.
22. Chauhan, V. and Kanwar, S.S., Bioactive peptides: Synthesis, functions and Biotechnological applications, *Biotechnol. Prod. Bioact. Compd.*, Chapter-4, 107–137, 2019.
23. Gobbetti, M., Stepaniak, L., De Angelis, M., Corsetti, A., Di Cagno, R., Latent bioactive peptides in milk proteins: Proteolytic activation and significance in dairy processing. *Crit. Rev. Food Sci. Nutr.*, 42, 223–239, 2002.
24. Meisel, H. and Bockelmann, W., Bioactive peptides encrypted in milk proteins: Proteolytic activation and thropho-functional properties; Antonie Van Leeuwenhoek. *Int. J. Gen. Mol. Microbiol.*, 76, 207–215, 1999.
25. Pritchard, S.R., Phillips, M., Kailasapathy, K., Identification of bioactive peptides in commercial Cheddar cheese. *Int. Food Res. J.*, 43, 1545–1548, 2010.
26. Choi, J., Sabikhi, L., Hassan, A., Anand, S., Bioactive peptides in dairy products. *Int. J. Dairy Technol.*, 65, 1–12, 2012.
27. Korhonen, H., Milk-derived bioactive peptides: from science to applications. *J. Func. Foods*, 1, 177–187, 2009.
28. Léonil, J., Milk bioactive peptides: Their interest for cardiovascular diseases and metabolic syndrome prevention. *Medecine Des Maladies Metaboliques*, 8, 495–499, 2014.
29. Mohanty, D.P., Mohapatra, S., Misra, S., Sahu, P.S., Milk derived bioactive peptides and their impact on human health: A review. *Saudi J. Biol. Sci.*, 23, 577–583, 2015.
30. Mohanty, D., Jena, R., Choudhury, P.K., Pattnaik, R., Mohapatra, S., Saini, M.R., Milk derived antimicrobial bioactive peptides: A review. *Int. J. Food Prop.*, 19, 837–846, 2016.
31. Przybylski, R., Firdaous, L., Châtaigné, G., Dhulster, P., Nedjar, N., Production of an antimicrobial peptide derived from slaughterhouse by-product and its potential application on meat as preservative. *Food Chem.*, 211, 306–313, 2016.
32. Lassoued, I., Mora, L., Barkia, A., Aristoy, M.C., Nasri, M., Toldra, F., Bioactive peptides identified in thornback ray skin's gelatin hydrolysates by proteases from Bacillus subtilis and *Bacillus amyloliquefaciens*. *J. Proteomics*, 128, 8–17, 2015.
33. Kouhdasht, M.A. and Nasab, M.M., Bioactive peptides derived from fish by-product collagen. *Int. J. Environ. Sci. Nat. Res.*, Juniper Publishers Inc., 13, 2, 47–50, 2018.
34. Selamassakul, O., Laohakunjit, N., Kerdchoechuen, O., Ratanakhanokchai, K., A novel multi-biofunctional protein from brown rice hydrolysed by endo/endo-exoproteases. *Food Funct.*, 7, 2635–2644, 2016.
35. Singh, B.P., Vij, S., Hati, S., Functional significance of bioactive peptides derived from soybean. *Peptides*, 54, 171–179, 2014.
36. Moller, N.P., Scholz-Ahrens, K.E., Roos, N., Schrezenmeir, J., Bioactive peptides and proteins from foods: indication for health effects. *Eur. J. Nutr.*, 47, 171–182, 2008.

37. Silva-Sanchez, C., de la Rosa, A.P.B., Leon-Galvan, M.F., de Lumen, B.O., de LeonRodriguez, A., de Mejia, E.G., Bioactive peptides in amaranth (*Amaranthus hypochondriacus*) seed.. *J. Agric. Food Chem.*, 56, 1233–1240, 2008.
38. Altemimi, A., Lakhssassi, N., Baharlouei, A., Watson, D.G., Lightfoot, D.A., Phytochemicals: Extraction, isolation, and identification of bioactive compounds from plant extracts. *Plants*, 6, 4, 42–65, 2017.
39. Sweetlove, L.J., Nielsen, J., Fernie, A.R., Engineering central metabolisme: A grand challenge for plant biologists. *Plant J.*, 90, 4, 749–763, 2017.
40. Dharmaraj, S., Marine Streptomyces as a novel source of bioactive substances. *World J. Microb. Biot.*, 26, 2123–2139, 2010.
41. Wang, Y., Prebiotics: Present and future in food science and technology. *Int. Food Res. J.*, 42, 8–12, 2009.
42. Kokila, R. and Mrudula, S., Optimization of culture conditions for amylase production by thermohilic *Bacillus sp.* in submerged fermentation. *AJMBES* 12, 3, 653–658, 2010.
43. Tafulo, P.K.R., Queirós, R.B., Delerue-Matos, C.M., Ferreira, M.G., Control and comparison of the antioxidant capacity of beers. *Int. Food Res. J.*, 43, 1702–1709, 2010.
44. Ohno, A., Ano, T., Shoda, M., Production of a lipopeptide antibiotic, surfactin, by recombinant *bacillus subtilis* in solid state fermentation. *Biotechnol. Bioeng.*, 47, 209–214, 1995.
45. Maragkoudakis, P.A., Mountzouris, K.C., Psyrras, D., Cremonese, S., Fischer, J., Cantor, M.D., Tsakalidou, E., Functional properties of novel protective lactic acid bacteria and application in raw chicken meat against *Listeria monocytogenes* and *Salmonella enteritidis*. *Int. J. Food Microbiol.*, 130, 219–226, 2009.
46. Nakahara, T., Sano, A., Yamaguchi, H., Sugimoto, K., Chilkata, H., Kinoshita, E., Uchida, R., Antihypertensive effect of peptide-enriched soy sauce-like seasoning and identification of its angiotensin i-converting enzyme inhibitory activity. *J. Agr. Food. Chem.*, 58, 821–827, 2010.
47. Ruiz-Sanchez, J., Flores-Bustamante, Z.R., Dendooven, L., Favela-Torres, E., Soca-Chafre, G., Galindez-Mayer, J., Flores-Cotera, L.B., A comparative study of taxol production in liquid and solid state fermentation with *Nigrospora* sp., a fungus isolated from Taxus globosa. *J. Appl. Microbiol.*, 109, 2144–2150, 2010.
48. Xie, X. and Tang, Y., Efficient synthesis of simvastatin by use of whole-cell biocatalysis. *Appl. Environ. Microbiol.*, 73, 7, 2054–2060, 2007.
49. Babu, K.R. and Satyanarayana, T., Production of bacterial enzymes by solid state fermentation. *JSIR*, 55, 464–467, 1996.
50. Pandey, A., Selvakumar, P., Soccol, C.R., Singh, N.N., Poonam, Solid state fermentation for the production of industrial enzymes. *Curr. Sci.*, 77, 1, 149–162, 1999.
51. Ahuja, S.K., Ferreira, G.M., Moreira, A.R., Utilization of enzymes for environmental applications. *Crit. Rev. Biotechnol.*, 24, 2–3, 125–154, 2004.
52. Capriotti, A.L., Cavaliere, C., Piovesana, S., Samperi, R., Aldo Laganà, A., Recent trends in the analysis of bioactive peptides in milk and dairy products. *Anal. Bioanal. Chem.*, 408, 2677–2685, 2016.
53. Gibbs, B.F., Zougman, A., Masse, R., Mulligan, C., Production and characterization of bioactive peptides from soy hydrolysate and soy-fermented food. *Food Res. Int.*, 37, 123–131, 2004.
54. Malaguti, M., Dinelli, G., Leoncini, E., Bregola, V., Bosi, S., Cicero, A., Hrelia, S., Bioactive peptides in cereals and legumes: Agronomical, biochemical and clinical aspects. *Int. J. Mol. Sci.*, 15, 21120–21135, 2014.
55. Kameswari, S.B., Velmurugan, B., Thirumaran, K., Ramanujam, R.A., Biomethanation of Vegetable Market Waste-Untapped Carbon Trading Opportunities. *Proceedings of the International Conference on Sustainable Solid Waste Management*, Chennai, India, September, pp. 415–420, 2007, 5–7.

56. Ravi Kiran, G., Suresh, K.P., Sampath, K.T., Giridhar, K., Anandan, S., *Modeling and Forecasting Livestock and Fish Feed Resources: Requirements and Availability in India; Ph.D. Thesis*, National Institute of Animal Nutrition and Physiology, Bangalore, India, 2012.
57. Chen, J., Aquatic feed industry under tension in world and China's grain supply and demand. *Chin. Fish*, 6, 32–34, 2012.
58. Ko, M.J., Cheigh, C.-I., Cho, S.-W., Chung, M.-S., Subcritical water extraction of flavonol quercetin from onion skin. *J. Food Eng.*, 102, 4, 327–333, 2011.
59. Song, R., Ismail, M., Baroutian, S., Farid, M., Effect of subcritical water on the extraction of bioactive compounds from carrot leaves. *Food Bioprocess Technol.*, 11, 10, 1895–1903, 2018.
60. Hossain, M.B., Aguiló-Aguayo, I., Lyng, J.G., Brunton, N.P., Rai, D.K., Effect of pulsed electric field and pulsed light pre-treatment on the extraction of steroidal alkaloids from potato peels. *Innov. Food Sci. Emerg. Technol.*, 29, 9–14, 2015.
61. Figueroa, J.G., Borrás-Linares, I., Lozano-Sánchez, J., Quirantes-Piné, R., Segura-Carretero, A., Optimization of drying process and pressurized liquid extraction for recovery of bioactive compounds from avocado peel by-product. *Electrophoresis*, 39, 1908–1916, 2018.
62. Lachos-Perez, D., Baseggio, A.M., Mayanga-Torres, P., Junior, M.R.M., Rostagno, M., Martínez, J. *et al.*, Subcritical water extraction of flavanones from defatted orange peel. *J. Supercrit. Fluids*, 138, 7–16, 2018.
63. Barbosa-Pereira, L., Guglielmetti, A., Zeppa, G., Pulsed electric field assisted extraction of bioactive compounds from cocoa bean shell and coffee silverskin. *Food Bioprocess Technol.*, 11, 4, 818–835, 2018.
64. Parniakov, O., Barba, F.J., Grimi, N., Lebovka, N., Vorobiev, E., Extraction assisted by pulsed electric energy as a potential tool for green and sustainable recovery of nutritionally valuable compounds from mango peels. *Food Chem.*, 192, 842–848, 2016.
65. Duba, K.S. and Fiori, L., Supercritical CO_2 extraction of grape seed oil: Effect of process parameters on the extraction kinetics. *J. Supercrit. Fluids*, 98, 33–43, 2015.
66. Chen, Y., Luo, H., Gao, A., Zhu, M., Ultrasound-assisted extraction of polysaccharides from litchi (Litchi chinensis Sonn.) seed by response surface methodology and their structural characteristics;Innov. *Food Sci. Emerg. Technol.*, 12, 3, 305–309, 2011.
67. Adil, I.H., Cetin, H., Yener, M., Bayındırlı, A., Subcritical (carbon dioxide + ethanol) extraction of polyphenols from apple and peach pomaces, and determination of the antioxidant activities of the extracts. *J. Supercrit. Fluids*, 43, 1, 55–63, 2007.
68. González-Casado, S., Martín-Belloso O., Elez-Marínez, P., Soliva-Fortuny, R., Enhancing the carotenoid content of tomato fruit with pulsed electric field treatments: Effects on respiratory activity and quality attributes. *Postharvest Biol. Technol.*, 137, 113–118, 2018.
69. Redondo, D., Venturini, M.E., Luengo, E., Raso, J., Arias, E., Pulsed electric fields as a green technology for the extraction of bioactive compounds from thinned peach byproducts. *Innov. Food Sci. Emerg. Technol.*, 45, 335–343, 2018.
70. Peiró, S., Luengo, E., Segovia, F., Raso, J., Almajano, M.P., Improving polyphenol extraction from Lemon residues by pulsed electric fields. *Waste Biomass Valori.*, 10, 4, 1–9, 2017.
71. Okiyama, D.C., Soares, I.D., Cuevas, M.S., Crevelin, E.J., Moraes, L.A., Melo, M.P. *et al.*, Pressurized liquid extraction of flavanols and alkaloids from cocoa bean shell using ethanol as solvent. *Food Res. Int.*, 114, 20–29, 2018.
72. Dunford, N., Irmak, S., Jonnala, R., Pressurised solvent extraction of policosanol from wheat straw, germ and bran. *Food Chem.*, 119, 3, 1246–1249, 2010.
73. Todaro, A., Cimino, F., Rapisarda, P., Catalano, A.E., Barbagallo, R.N., Spagna, G., Recovery of anthocyanins from eggplant peel. *Food Chem.*, 114, 2, 434–439, 2009.

74. Nagendra, P.K., Hassan, F.A., Yang, B., Kong, K.W., Ramanan, R.N., Azlan, A., Ismail, A., Response surface optimisation for the extraction of phenolic compounds and antioxidant capacities of underutilised *Mangiferapajang* Kosterm peels. *Food Chem.*, 128, 4, 1121–1127, 2011.
75. Cavero, S., García-Risco, M., Marín, F., Jaime, L., Santoyo, S., Señoráns, F., Reglero, G., Ibañez, E., Supercritical fluid extraction of antioxidant compounds from oregano: Chemical and functional characterization via LC-MS and *in vitro* assays. *J. Supercrit. Fluid*, 38, 1, 62–69, 2006.
76. Serra, A., Seabra, I., Braga, M., Bronze, M., de Sousa, H., Duarte, C., Processing cherries (Prunus avium) using supercritical fluid technology; Part 1: Recovery of extract fractions rich in bioactive compounds. *J. Supercrit. Fluid*, 55, 1, 184–191, 2010.
77. Yi, C., Shi, J., Xue, S., Jiang, Y., Li, D., Effects of supercritical fluid extraction parameters on lycopene yield and antioxidant activity. *Food Chem.*, 113, 4, 1088–1094, 2009.
78. Yepez, B., Espinosa, M., López, S., Bolaños, G., Producing antioxidant fractions from herbaceous matrices by supercritical fluid extraction. *Fluid Phase Equilib.*, 194, 879–884, 2002.
79. Wang, T., Jonsdottir, R., Kristinsson, H.G., Hreggvidsson, G.O., Jonsson, J.O., Thorkelsson, G., Olafsdottir, G., Enzyme-enhanced extraction of antioxidant ingredients from red algae *Palmariapalmata. LWT-Food Sci. Technol.*, 43, 9, 1387–1393, 2010.
80. Zhu, L., Biorefinery as a promising approach to promote microalgae industry: An innovative framework. *Renew. Sust. Energ. Rev.*, 41, 1376–1384, 2015.
81. Negi, A. and Gill, B.S., Success stories of enolate form of drugs. *Pharma. Tutor.*, 1, 2, 45–53, 2013.
82. Gill, B.S. and Kumar, S., Differential algorithms-assisted molecular modeling-based identification of mechanistic binding of ganoderic acids. *Med. Chem. Res.*, 24, 9, 3483–3493, 2015.
83. De Castro, M.L. and Garcıa-Ayuso, L., Soxhlet extraction of solid materials: An outdated technique with a promising innovative future. *Anal. Chim. Acta*, 369, 1–2, 1–10, 1998.
84. Miron, T., Plaza, M., Bahrim, G., Ibáñez, E., Herrero, M., Chemical composition of bioactive pressurized extracts of Romanian aromatic plants. *J. Chromatogr.*, 218, 30, 4918–4927, 2010.
85. Plaza, M., Santoyo, S., Jaime, L., García-Blairsy, R.G., Herrero, M., Señoráns, F., Ibáñnez, E., Screening for bioactive compounds from algae.. *J. Pharm. Biomed. Anal.*, 51, 2, 450–455, 2010.
86. Gil-Chávez G.J., Villa, J.A., Ayala-Zavala, J.F., Heredia, J.B., Sepulveda, D., Yahia, E.M., González-Aguilar, G.A., Technologies for Extraction and Production of Bioactive Compounds to be Used as Nutraceuticals and Food Ingredients: An Overview. *Comprehen. Rev. Food Sci. Food Safety*, 12, 1, 5–23, 2013.
87. Ajila, C.M., Brar, S.K., Verma, M., Tyagi, R.D., Godbout, S., Valero, J.R., Extraction and analysis of polyphenols: Recent trends. *Crit. Rev. Biotechnol.*, 31, 3, 227–249, 2011.
88. Herrero, M., Jaime L., Martín-Álvarez P. J., Cifuentes A., Ibáñez E., Optimization of the Extraction of Antioxidants from *Dunaliella salina* Microalga by Pressurized Liquids; *J. Agric. Food Chem.*, 54(15): 5597–5603, 2006.
89. Hosikian, A., Lim, S., Halim, R., Danquah, M.K., Chlorophyll extraction from microalgae: A review on the process engineering aspects. *Int. J. Chem. Eng.*, 2010, Article ID 391632, 1–11, 2010.
90. Zosel, K., *Process for the Decaffeination of Coffee, US Patent*, US4260639A, 2014.
91. Sihvonen, M., Järvenpää, E., Hietaniemi, V., Huopalahti, R., Advances in supercritical carbon dioxide technologies. *Trends Food Sci. Technol.*, 10, 6–7, 217–222, 1999.
92. Wang, L. and Weller, C.L., Recent advances in extraction of nutraceuticals from plants. *Trends Food Sci. Technol.*, 17, 6, 300–312, 2006.
93. Daintree, L., Kordikowski, A., York, P., Separation processes for organic molecules using SCF technologies. *Adv. Drug Deliv. Rev.*, 60, 3, 351–372, 2008.
94. Klejdus, B., Lojková, L., Plaza, M., Snóblová, M., Sterbová, D., Hyphenated technique for the extraction and determination of isoflavones in algae: Ultrasound-assisted supercritical fluid

extraction followed by fast chromatography with tandem mass spectrometry. *J. Chromatogr., A.* 1217, 51, 7956–7965, 2010.

95. Ganzler K., Salgó A., Valkó K., Microwave extraction: A novel sample preparation method for chromatograph, J. Chromatogr. A, 1986, 371: 299–306.
96. Zhang, H.F., Yang, X.H., Wang, Y., Microwave assisted extraction of secondary metabolites from plants: Current status and future directions. *Trends Food Sci. Technol.*, 22, 12, 672–688, 2011.
97. Letellier M., Budzinski H., Microwave assisted extraction of organic compounds, Analusis, 1999, 27(3): 259–270.
98. Mason, J., Chemat, F., Vinatoru, M., The extraction of natural products using ultrasound or microwaves. *Curr. Org. Chem.*, 15, 2, 237–247, 2011.
99. Anastas, P.T. and Warner, J.C., Principles of green chemistry. *Green Chem. Theor. Pract.*, Chapter-4, 29–56, 1998.
100. Karki, B., Lamsal, B.P., Jung, S., Van Leeuwen, J.H., Pometto, A.L., Grewell, D., Khanal, S.K., Enhancing protein and sugar release from defatted soy flakes using ultrasound technology. *J. Food Eng.*, 96, 2, 270–278, 2010.
101. Qu, W., Ma, H., Jia, J., He, R., Luo, L., Pan, Z., Enzymolysis kinetics and activities of ACE inhibitory peptides from wheat germ protein prepared with SFP ultrasound-assisted processing. *Ultrason. Sonochem.*, 19, 5, 1021–1026, 2012.
102. Cheung, Y.C., Siu, K.C., Liu, Y.S., Wu, J.Y., Molecular properties and antioxidant activities of polysaccharide–protein complexes from selected mushrooms by ultrasound-assisted extraction. *Process Biochem.*, 47, 5, 892–895, 2012.
103. Adam, F., Abert-Vian, M., Peltier, G., Chemat, F., Solvent-free" ultrasound-assisted extraction of lipids from fresh microalgae cells: A green, clean and scalable process. *Bioresour. Technol.*, 114, 457–465, 2012.
104. Shi, M., Yang, Y., Hu, X., Zhang, Z., Effect of ultrasonic extraction conditions on antioxidative and immunomodulatory activities of a *Ganoderma lucidum* polysaccharide originated from fermented soybean curd residue. *Food Chem.*, 155, 50–56, 2014.
105. Herrera M. C., Luque de Castro, M. D.; Ultrasound-assisted extraction of phenolic compounds from strawberries prior to liquid chromatographic separation and photodiode array ultraviolet detection, J. Chromatogr. A, 2005, 1100(1): 1–7.
106. Zhao, L., Zhao, G., Chen, F., Wang, Z., Wu, J., Hu, X., Different effects of microwave and ultrasound on the stability of (all-E)-astaxanthin. *J. Agric. Food Chem.*, 54, 21, 8346–8351, 2006.
107. Bryant, G. and Wolfe, J., Electromechanical stresses produced in the plasma membranes of suspended cells by applied electric fields. *J. Membr. Biol.*, 96, 2, 129–139, 1987.
108. Heinz, V., Toepfl, S., Knorr, D., Impact of temperature on lethality and energy efficiency of apple juice pasteurization by pulsed electric fields treatment;Innov. *Food Sci. Emerg. Technol.*, 4, 2, 167–175, 2003.
109. Chemat, F., Vian, M.A., Cravotto, G., Green Extraction of Natural Products: Concept and Principles. *Int. J. Mol. Sci.*, 13, 8615–8627, 2012.

Part 3

BIOREFINERY FOR PRODUCTION OF ALTERNATIVE FUEL AND ENERGY

13

Potential Raw Materials and Production Technologies for Biorefineries

Shilpi Bansal[1], Lokesh Kumar Narnoliya[2] and Ankit Sonthalia[3]*

[1]*ICAR – Indian Agricultural Research Institute, Pusa Road, New Delhi, India*
[2]*Regional Centre for Biotechnology, NCR Biotech Science Cluster, Gurgaon Faridabad Expressway, Faridabad, India*
[3]*Department of Automobile Engineering, SRM Institute of Science and Technology, NCR Campus, Ghaziabad, India*

Abstract
Biomass use has increased exponentially in the last decade. Its use is not only limited to biofuel production but it has expanded to chemicals and direct bioenergy, among others. First- and second-generation biomass has been used lately for biofuels production. However, the biomass used in these generations is from edible crops, which threatens the food security and competes with the use of productive land. Therefore, it is necessary to look for biomass that is readily available and has the potential of obtaining value added materials without risking the food security. Efforts have been made in converting the biomass using efficient, sustainable, and cost-effective processes into useful products. Although, processes are available on large scale but improvements are still required to make the product market competitive. This chapter highlights the potential raw materials and biomass conversion technologies for moving toward a bio-based economy.

Keywords: Bioresources, biomass, biochemicals, biofuels, production technologies

13.1 Introduction

With increase in population and rise in economy, the world's energy demand is increasing. The global primary energy consumption of the world was 13511.2 Mtoe (million tons of oil equivalent) in 2017. Coal, natural gas, and oil are the major energy sources comprising of 85.18% of the total energy consumption. Whereas, only 3.6% of the energy consumption comprised of renewable sources such as solar, wind, geothermal, biomass, and waste [1]. As compared to 2015, the energy consumption from renewable sources increased by 14% [2]. This shows that the world is slowly moving toward energy sources that are sustainable and environment friendly.

Biofuels derived from biomass, which is biodegradable and environment friendly, are considered the best form of renewable energy. They can be derived from biomass which is biodegradable and environment friendly. In 2017, 84.12 Mtoe of biofuels was produced in

**Corresponding author*: ankit_sont@yahoo.co.in

Arindam Kuila and Mainak Mukhopadhyay (eds.) Biorefinery Production Technologies for Chemicals and Energy, (261–288) © 2020 Scrivener Publishing LLC

the world, an increment of 124.7% in one decade. Moreover, the biofuels production in India increased by 335% in one decade [1]. The combustion of biofuels is carbon neutral, i.e., they absorb carbon dioxide from the atmosphere, while growing and after burning, they release the same carbon dioxide back to the atmosphere [3]. Their combustion produces less toxic compounds [4]. Thus, biofuels can replace conventional fuels for internal combustion (IC) engines. However, issues like its cost-competitiveness with fossil-based fuels inhibit its use in IC engines. This can be solved by the intervention of the governments and implementing mandates that makes the use of biofuel compulsory. India in 2018 approved the National Policy on Biofuels that allowed ethanol production from sugarcane juice, sweet sorghum, sugar beet, and corn and cassava (starch containing materials). The policy allowed damaged food grains deemed unfit for human consumption to be converted to biofuels. Also, during surplus production years, due to high supply, the cost of the produce may be less, so to safeguard the farmers, they can sell their crops for biofuel production [5].

Organic chemicals can also be produced from the biomass: 10% of the organic chemicals produced in the world are bio-based [6]. These products in terms of chemical structure are highly diverse; they require highly sophisticated methods for production. However, they are mostly intermediates in a multi-step processing chain resulting in final consumer product [7]. Currently, compounds made from biomass are commercially successful as these compounds are available from biomass only or their cost is lower than fossil-based counterparts. Cosmetics, pharmaceuticals, and some high performance polymers are the best examples of compounds, which can be developed only using biomass. Bio-based basic chemicals are not used on a large scale mainly due to the cost competitiveness [8]. It is forecasted that by 2020, due to high consumer demand, bio-polymers will capture a market share of nearly 3% [9]. Moreover, by using the biorefinery concept and with advanced bio-based catalysts [10], the sustainable chemicals cost structure will improve and a steady 22% growth in sales is expected by 2020 [11]. Until the government provides incentives for producing bio-based chemicals, the production volume of these chemicals will be much smaller than biofuels [12].

The world today is based upon linear economy, which uses petroleum, natural gas, and coal. The flow of the linear economy is (i) virgin resources utilization for producing goods, (ii) using the products, and (iii) throwing away of unusable wastes. Non-renewability and limitation in supply are the major problems in using the fossil resources. It was estimated that oil, coal, and natural gas would be available for the next 33, 105, and 35 years, respectively. Therefore, it is expected that till 2112 coal will be available, and after 2042, it will be the only fossil fuel available. Therefore, for meeting the population requirements in the long run, the linear economy is unsustainable and ineffective [13]. The world should move toward a circular economy wherein the resources are properly managed. Also, the virgin resources are not used and the yields are optimized by manufacturing reusable products with less wastes [14]. To close the loop, the goods can be fixed during their lifetime, and at the end of their life, the whole product is reused again [15]. This model allows realizing the concept of reduce, reuse, and recycle which prevents and minimizes wastes thus solving the problem of landfills and shortage of fossil resources [16].

The concept of biorefinery fits well with the circular economy. The biorefinery is similar to petroleum refinery, wherein many products are produced using biomass as the feedstock. The valuable products include fuel, power, food ingredients, and chemicals [17]. The biorefinery concept applies technologies from varied fields like bioengineering, polymer chemistry, and agriculture [18]. The biorefinery system can be named on the basis of the main product such as product-driven biorefinery or energy-driven biorefinery. On the basis of targeted products, type of biomass and processes used the biorefinery is of three types [19]. The first type of biorefinery uses one biomass, one process and produces one product, i.e., there is no flexibility in the entire process. The best example of such a biorefinery is the dry grind ethanol process, wherein the milling of corn is carried out first, and then, saccharification and later fermentation to ethanol. In the second type of biorefinery, more number of products is formed. Corn wet milling process is the best example, wherein lactic acid, ethanol, starch, corn oil, and corn syrup are formed. Since a number of products are formed, the economic performance of the biorefinery is high; also, even if the market for ethanol is weak, the biorefinery can still be profitable by making other products. The third type of biorefinery is more flexible than second type of biorefinery since it utilizes different types of feedstock and processing methods for producing multiple products. As the refinery can utilize different feedstock, it will be independent of the shortage of feedstock supply during the off-season [20].

Figure 13.1 shows the process chain using which the lignocellulose biomass can be converted into a number of products. The biomass initially is pre-treated, and then, hydrolysis is carried out to release the sugars used for conversion to fuels and chemicals. In this stage, lignin is also separated and any proteins if present in the feedstock can be recovered and used as food ingredients. The sugars can be used to produce biofuels or maximum part of them can be switched to convert valuable chemicals to obtain maximum turnover. The lignin can be used to produce power or heat or it can be converted into valuable chemicals. Lastly, the left nutrients or water can be recycled for better utilization of the resources and for closing the cycle [21].

If biofuels and bio-based chemicals are integrated in a biorefinery, the feedstock fraction can be efficiently utilized and diversified outputs can be produced resulting in improvement in financial performance as the market price of chemicals is higher than biofuels. Chemicals are low in volume and high in market price, whereas biofuels volume is high

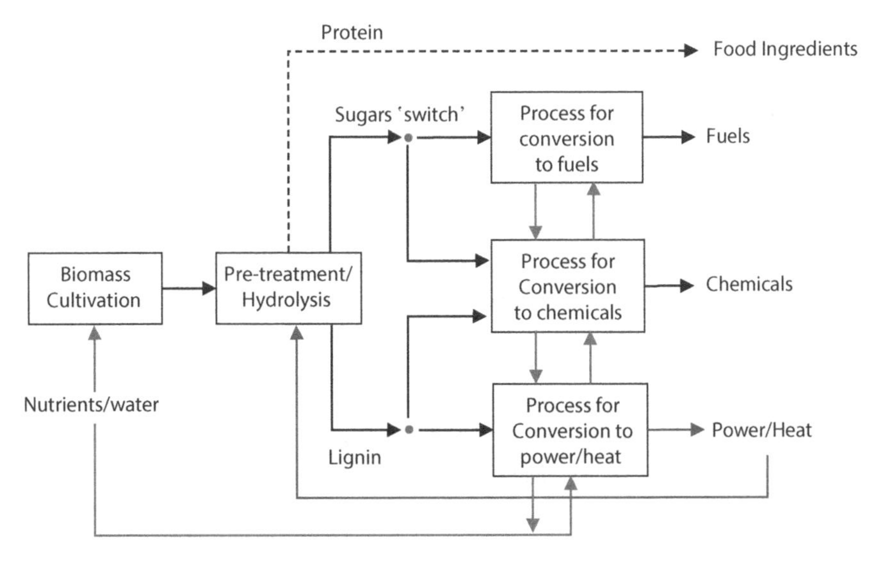

Figure 13.1 Process chain for converting lignocellulosic biomass into different products [21].

but they are cheap [22]. Therefore, revenues can be generated by including chemicals in an energy-driven biorefinery. Greenhouse gas emissions can be reduced to a large extent with biofuels, depending upon the feedstock and the technology used [23]. Chemicals produced from biomass can significantly reduce greenhouse gas emissions. Greenhouse gas saving (t CO_2/t of product) for ethanol, butanol, adipic acid, and succinic acid were estimated to be 2.7, 3.9, 3.3, and 5.0, respectively, if fossil-based chemical is replaced completely with bio-based chemicals [24].

13.2 Bioresources

The biofuels are liquid fuels produced using various bio-resources. They are broadly classified into three generations based upon the source from which the fuel is derived. In this section, different sources having potential to be used in biorefinery is discussed.

13.2.1 First-Generation Feedstock

Edible food crops and vegetable oils are generally considered as the first-generation sources [25]. They are used to produce biodiesel (mono-alkyl esters) and bioethanol which are already being produced in large quantities worldwide and the production technologies for the same are already established. Biodiesel can be substituted for diesel; it is produced by transesterifying the residual oils, fats, and vegetable oils. With minor modifications, it can be used in diesel engines. Gasoline can be substituted with bioethanol derived through fermentation of sugars or starch [26]. Some of the edible vegetable oils used for biofuel production are soybean oil [27], corn oil [27], sunflower oil [27], olive oil [28], palm oil [28], coconut oil [28], rapeseed oil [29], and castor oil [30]. However, the use of edible oils as fuels face challenges, such as pressure on land use and increase in prices of food, making it unsustainable [31]. Therefore, alternate feedstock which can overcome these shortcomings is being developed. Table 13.1 presents the first-generation feedstock, their oil content, and oil yield.

13.2.2 Second-Generation Feedstock

Plant biomass is used to produce biofuels sustainably and they are considered as second-generation biofuels. The fuels production process is carbon neutral or carbon negative [26]. Non-edible feedstock such as non-food crops [31], wood waste, animal fats [45], and waste cooking oil [45], etc., are considered as second-generation feedstock. Table 13.2 shows some second-generation feedstock used in production of biofuels. The cost of the production processes high due to a number of hurdles which needs to be overcome to realize their full potential. Plant biomass is the most underutilized resource and can be a promising source for fuels and raw materials. The plant cell walls contains nearly 75% of polysaccharides, which represents a valuable pool of sugars [46]. For example, the stem of wheat has as much sugar as there is starch in its grains. Many crop residues have so much potential for biofuel production which has not been realized till date. However, the use of agricultural residues for biofuel production can partly satisfy the ever increasing demand for fuels [19].

Table 13.1 First-generation feedstock.

Edible source	Oil content (%)		Oil yield (L/ha)	References
Food crops	**Seed**	**Kernel**		
Potato Waste	20	–	–	[32, 33]
Rice bran oil	15–23	–	825	[34–36]
Barley	2.5–5.0	–	–	[37]
Sugar Beets	40	–	–	[38, 39]
Corn	48	–	172	[35]
Vegetable oil				
Coconut oil	63–65	–	2,689	[34–36, 40]
Rapeseed oil	37–50	–	1,190	[34, 35, 41]
Palm oil	30–60	–	5,950	[34–36, 41]
Soybean oil	15–20	–	446	[35, 36, 41]
Canola oil	43	–	–	[35, 42]
Sesame oil	41	–	–	[43]
Sunflower oil	25–35	45–55	952	[35, 36, 42]
Hemp oil	30–35	–	–	[35, 42]
Moringa oil	35–40	–	250	[34–36]
Mustard oil	30	–	–	[44]
Peanut oil	45–55	–	1,059	[34–36]
Olive oil	45–70	–	1,212	[34–36]
Castor oil	45–50	–	1,413	[34–36]
Cotton seed oil	18–25	–	325	[35, 36]
Linseed oil	40–44	–	–	[34–36]

Lignocellulosic material present in the plant biomass can be converted to advanced biofuels through hydrolysis and fermentation (bioethanol) or through gasification [Fischer-Tropsch (FT) biodiesel, bio-dimethyl ether (DME), bio-synthetic natural gas (SNG)]. Short rotation crops, perennial grasses (red canary grass, miscanthus, and switch grass) and the wood industry, and forest and agriculture residues can be used for producing these biofuels. Bioethanol can be produced by hydrolyzing the lignocellulosic feedstock into sugar, which is later fermented into ethanol. Lignocellulosic biomass can be gasified into syngas, which can be converted into either FT-diesel, bio-DME, or bio-SNG. FT-diesel can fully substitute

Table 13.2 Second-generation feedstock.

Feedstock	Distribution	Plant type	Oil content (%)		Reference
			Seed	Kernel	
Vegetable oils					
Sleichera triguga (Kusum)	Northern India, the western Deccan, Sri-Lanka, Malaysia, Indonesia, Java, etc.	Tree	–	55–70	[48, 49]
Sapium sebifeum L. Roxb (Stillingia)	Native to Japan, China, India and also grows in southern coastal United States	Tree	13–32	53–64	[48, 49]
Ximenia americana (Sea lemon)	India, Africa, South East Asia, New Zealand, Australia, West Indies, Pacific Islands, South America	Tree	–	49–61	[50]
Guizotia abyssinica L. (Niger)	Ethiopia and India	Herbaceous Annual	50–60	–	[51]
Hevea brasiliensis (Rubber seed oil)	South East Asia, India, Nigeria, Brazil, West Africa	Tree	40–60	40–50	[52]
Croton tiglium(Jamaal gota)	China, Ceylon, Amboina Philippines, and Java	Herbaceous Perennial	30–45	50–60	[50]
Jatropha curcas L. (Jatropha)	India, Thailand, Pakistan, Indonesia, Nepal, Malaysia, Philippines	Tree	20–60	40–60	[49, 50]
Mesu ferrera (Cobra's saffron)	Forests in North East India, China, Nepal, Malaysia, Philippines, Sumatra	Herb	35–50	–	[53]
Simmondsia chinesis (Jojoba)	Sonoran and Mojave deserts	Shrub	45–55	–	[49, 54]

(*Continued*)

Table 13.2 Second-generation feedstock. (*Continued*)

Feedstock	Distribution	Plant type	Oil content (%)		Reference
			Seed	Kernel	
Madhuca indica (Mahua)	India	Tree	35–50	50	[49, 50]
Cerbera odollam (Sea mango)	India and some parts of South Asia	Tree	54	6.4	[49]
Terminalia catppa (Bengal almond)	Asia, Brazil, Australia, Africa	Tree	49	–	[54, 55]
Sapindus mukorosso (Soap nut)	America, Asia, Europe	Tree	51.8	–	[56]
Balanites aegyptiaca (Desert date)	Arid regions of Asia and Africa	Tree	–	36–47	[57, 58]
Ricinus communis (Castor)	Brazil, Cuba, China, France, Italy, India, and countries of the former Soviet Union	Tree/Shrub	45–50	–	[48, 49]
Pongamia pinnata (Karanja)	Native of Western Ghats in India, Northern Australia, some regions in Eastern Asia, Fiji	Tree	25–50	30–50	[50, 54]
Calophyllum Inophyllum (Beauty leaf tree)	Australia, East Africa, Southern coastal India to Malaysia	Tree	50	–	[59, 60]
Salvadora oleoides (Vann)	Southern Iran, arid regions of India, Pakistan,	Tree	45	–	[49, 61]
Garcinia indica (Kokum)	Asia, Africa	Tree	45.5	–	[49, 50]
Michela chaampaca (Yellow jade orchid)	China, Burma, eastern Himalayas, Assam, and Western Ghats in India	Tree	45	–	[48, 50]
Azadirachta indica (Neem)	Native to India, Australia, Bangladesh, Burma, Cuba, Malaysia, Pakistan, and Sri Lanka	Tree	20–30	25–45	[48–50, 54]

(*Continued*)

Table 13.2 Second-generation feedstock. (*Continued*)

Feedstock	Distribution	Plant type	Oil content (%)		Reference
			Seed	Kernel	
Linum usitatissimum (Linseed)	Asia, Argentina, Canada, and Europe	Herbaceous annual	35–35	–	[49, 62]
Putranjiva roxburghii (Putranjiva)	Distributed in India	Tree	41–42	–	[48–50]
Ceiba pentandra (Kapok)	Native to Central America, Burma, Mexico, and the Caribbean, northern South America to tropical west Africa, Indonesia, Southern China, Taiwan, and northern Vietnam	Tree	20–40	–	[48, 63]
Milletia pinnata (Pongam oil tree)	Malaysia, Australia, China, India, Pacific Islands, and Japan	Legume Tree	27–39	–	[64]
Cupphea hyssopifolia (Cuphea)	The Eastern and North Central USA to Argentina	Herbaceous annual	20–38	–	[48, 65]
Moringa oleifera (Drumstick tree)	Africa, Asia, Latin America, Oceania	Tree	35	–	[66, 67]
Aphanamixis piolystachya (Pithraj)	Growing in India, China	Tree	–	35	[48]
Eruca sativa (Garden rocket)	Europe, the Mediterranean, America, South Africa, Australia, Northwest of China, New Zealand, tropical regions of Asia	Herbaceous perennial	35	–	[68]
Asclepias syriaca (Milkweed)	Distributed to the Northeast and North-Central United States	Herbaceous perennial	20–25	0.019	[48, 49, 69]

(*Continued*)

Table 13.2 Second-generation feedstock. (*Continued*)

Feedstock	Distribution	Plant type	Oil content (%)		Reference
			Seed	Kernel	
Ricinus communis (Castor oil plant)	Australia, Native to Africa, and Eurasia	Tree	45–50	–	[70]
Crotalaria retusa L. (Fabaceae)	Native in Asia, Coastal Eastern and Africa	Herbaceous annual	15	–	[48]
Waste cookingoil	All countries	–	97.02	–	[71]
Lingo cellulosic biomass					
Bagasse	Almost every country in the world	–	–	–	[72]
Wheat straw	Almost every country in the world	–	–	–	[73]
Rich Straw	Almost every country in the world	–	–	–	[74]
Barley straw	Almost every country in the world	–	–	–	[75]
Animal Fats					
Tallow	Almost every country in the world	–	–	–	[55]
Lard	Almost every country in the world	–	–	–	[55]

diesel in a diesel engine, whereas slight modifications in the engine are required for DME use. Bio-SNG can be used in a gasoline engine with slight modifications [47].

13.2.3 Third-Generation Feedstock

Microorganism-based biomass is the third-generation feedstock for production of biofuels. The biofuel yield from microalgae is nearly ten times that of the best traditional feedstock. Various types of fuels and other valuable chemicals such as bio-ethanol, bio-methane, bio-butanol, biodiesel, jet-fuel, and gasoline can be produced [76]. Algae biomass utilizes environmental carbon dioxide to grow. Thus, they can be grown near sources rich in carbon emission such as power plants, industries, etc., wherein the carbon emissions can be directly converted in to usable fuel and reduce the environmental pollution [77]. They can be grown in places unfit for food crop production, thus reducing the strain on edible crops. However, to satisfy the industry needs, algae should be cultivated on a large scale, for that the requirements of nutrients like phosphorus and nitrogen will be enormous. Thus, the requirement of energy and the greenhouse gas emission in the production of the fertilizer will offset the advantage of the fuel produced from algae [77].

Chlorella vulgaris, *Chamydomonas reinhardtii*, and *Dunaliella salina* are some of the targeted green algae because of around 60%–70%lipid content [78] and high productivity (7.4 g/L/d for C. *vulgaris*) [79]. By transesterification process, the lipids extracted from algae can be converted into biofuels with characteristics similar to second-generation biofuels. Also by hydrogenolysis, kerosene grade alkane, a suitable drop-in aviation fuel, can be produced [80]. Some of the species are tabulated in Table 13.3.

13.3 Chemicals Produced from Biomass

By exploiting all the lignocellulosic components of the biomass wide range of chemicals can be produced which can replace the fossil-based chemicals. US Department of Energy identified 15 chemicals, which can be produced by converting lignocellulosic sugars. The 15 sugar-based building blocks are succinic, fumaric and malic acid, 2,5-furan dicarboxylic acid, 3-hydroxy propionic acid, aspartic acid, glucaric acid, glutamic acid, itaconic acid, levulinic acid, 3-hydroxybutyrolactone, glycerol, sorbitol, and xylitol/arabinitol [101]. This section discusses some chemicals that can be produced from biomass.

13.3.1 Ethylene

During ripening of fruits, ethylene is produced; however, the quantity of ethylene is so less that the process is practically impossible and economically unfeasible. The production of ethylene from bioethanol is very promising [102]. The technology for the conversion has already been established and it comprises of dehydration of ethanol to ethylene. The dehydration takes place at nearly 300°C–600°C using different heterogeneous catalysts such as alumina, activated clays, zeolites, and mordenite [103, 104]. It is expected that in countries like Brazil with cheap access to bioethanol, stand-alone units of bio-ethylene will be economically more attractive. These stand-alone units can severely disrupt the coproduction of C4 olefins, propylene, and B, T, X from naphtha. Production of bio-ethylene can also open

Table 13.3 Microorganisms used for biofuel production.

Microorganism	Oil content per ton of biomass (dry weight %)	Lipid content (weight per dry weight %)	Reference
Microalgae			
Schizochytrium sp.	50–77	35–55	[81, 82]
Chlorella vulgaris	63.2	5–58	[82, 83]
Nitzschia laevis	69.1	–	[82, 84]
Scenedesmus obiquus	35–55	11–55	[82, 85]
Neochloris oleoabundans	35–65	29–65	[82, 86]
Botryococcus braunii	64	25–75	[81, 82, 87, 88]
Parietochloris incise	62	–	[82]
Crypthecodium cohnii	56	11–55	[82, 87, 89]
Nannochloris sp.	–	20–56	[87]
Nannochloropsis oculata	50	22.7–29.7	[82, 90]
Nitzschia sp.	45–47	16–47	[81, 82, 87]
Scenedesmus dimorphus	16–40	–	[82, 91]
Dunaliella primolecta	23	23.1	[78, 87]
Cylindrotheca sp.	16–37	–	[81, 82]
Phaeodactylum tricornutum	20–30	18–57	[87]
Chamydomonas reinhardtii	25.25	–	[78, 92]
Haematococcus pluvialis	25	25	[78, 93]
Monodus subterraneus	39.3	16	[82]
Monallanthus salina	>20	20–22	[87]
Tetraselmis sueica	15–23	8.5–23	[87]
Chlorella sorokiana	22	19–22	[82, 94]
Isochrysis galbana	14.5	7–40	[82]

(Continued)

Table 13.3 Microorganisms used for biofuel production. (*Continued*)

Microorganism	Oil content per ton of biomass (dry weight %)	Lipid content (weight per dry weight %)	Reference
Dunaliella salina	14–20	6–25	[78, 82]
Porphyridium cruentum	19.3	9–18.8	[82]
Spirulina platensis	5–17	4–16.6	[82]
Bacterium			
Arthrobacter sp.	>40	24-31	[81]
Rhodococcus opacus	24–25	–	[81]
Acinetobacter calcoaceticus	27–38	–	[81]
Bacillus alcalophilus	18–24	–	[81]
Yeast			
Rhodotorula glutinis	72	–	[81]
Cryptococcus albidus	65	–	[81]
Rhodosporidium toruloides	48–67.5	–	[95]
Candida curvata	58	–	[81]
Lipomyces starkeyi	64	–	[81]
Fungi			
Aspergillus oryzae	57.2	–	[81, 82]
Mortierella isabellina	86	–	[81, 82, 96]
Humicola lanuginose	75	–	[81, 82]
Mortierella vinacea	66	–	[81, 82]
Cunninghamella echinulata	46	–	[82, 97]
Mortierella ramanniana	54.2	–	[82]
Cunninghamella japonica	50	–	[82, 98]
Mucor rouxii	32	–	[82, 99]

(*Continued*)

Table 13.3 Microorganisms used for biofuel production. (*Continued*)

Microorganism	Oil content per ton of biomass (dry weight %)	Lipid content (weight per dry weight %)	Reference
Mortierella alpine	42	–	[82, 100]
Mucor circinelloides	23	–	[82, 100]
Mucor sp.	3–17	–	[82, 99]

the doors for production of bio-vinyl chloride. This means that two of the most important thermoplastics of the world, namely, polyethylene and polyvinyl chloride to an extent will be bio-based [102].

13.3.2 Propylene

It is used to produce various end user products such as polypropylene (a thermoplastic polymer) used in clothing and automotive industry, acrylonitrile in carpets and nylon, and acrylic acid in adhesives, textiles, and paints [105]. Conventionally, propylene is produced by fluidized catalytic cracking of gas oils and steam cracking of hydrocarbons which account for 30% and 60% of the total propylene capacity, respectively [106]. Propylene can also be produced from biomass. One method for production of propylene is the dehydration of biochemically produced 2-propanol, which is derived from 1,2-propanediol or by acetone reduction which can be produced via ABE (acetone, butanol, ethanol) fermentation process [102]. The ABE process technology with various microorganisms and feedstock is already established [107]. Another method is methanol to olefins/methanol to propylene (MTP) technology wherein syngas derived from any biomass is converted into methanol which is then converted into propylene. The process occurs in two steps: in the first step, methanol is dehydrated to DME over a catalyst, and in the second step, DME is converted into various olefins. However, with ZSM–5 (zeolite-based catalyst), propylene is selectively produced during the process [108].

13.3.3 Propylene Glycol

1,2-Propylene glycol and 1,3-propylene glycol are the two glycols produced by the chemical industry. By opening the ring of propylene oxide in the presence of a basic or an acidic catalyst 1,2-propylene glycol is produced. The glycol is used as abrake fluid or an antifreeze or a humectant or one of the component of alkyd resins and polyesters. 1,3-Propylene glycol is produced using petrochemicals by a process patented by Shell and Degussa. It is used for producing polymers, such as (poly propanediol terephthalate) fibers [102].

Fermentation process can also be used to produce propylene glycols. Altara *et al.* [109] fermented sugars such as xylose, glucose, and galactose using *Thermoanaerobacterium thermosaacharolyticum HG-8* as the microorganism. Under anaerobic conditions and at 60°C, the process was carried out. The authors observed that 12% of 1,2-propanediol could be produced with larger amounts of lactate, ethanol, and CO_2. Propylene glycol can also be produced by thermochemical processes. In this process, glycerol can be hydrogenolyzed using various heterogeneous catalysts [110]. Depending upon specific catalysts and 20%–30% conversion rate of glycerol, the selectivity of 1,2-propanediol or1,3-propanediol is very high. Werpy *et al.* [111] used Ni-Re bimetallic catalyst and observed that 60% of glycerol can be converted with a high selectivity for propylene glycol (78%–88%) along with ethylene glycol (12%) and lactate (8%–10%). Over the years, the production of biodiesel by transesterification process has increased dramatically along with the by-product glycerol. Therefore, the prices of glycerol have reduced, making it an attractive starting material for production of ethylene and propylene glycol.

13.3.4 Butadiene

It can be produced from bio-ethanol which is dehydrogenated to acetaldehyde which is then condensed in presence of aldol and then dehydration in a one-pot process over MgO or SiO_2 catalyst (Lebedew process) resulting in 70% yield of butadiene [112]. The butadiene via a butadiene epoxidation process can be converted into tetrahydrofuran and 1,4-butanediol. It can also act as a base chemical for producing different types of elastomers such as styrene butadiene rubbers [102].

13.3.5 2,3-Butanediol and 2-Butanone Methyl Ethyl Ketone (MEK)

2,3-Butanediol can be the starting chemical in the production of butadiene and MEK. 2,3-Butanediol can be produced by fermentation process wherein glucose and xylose can be efficiently converted into mixtures of 2,3-butanediol and ethanol using the bacterial species such as *Bacillus polymyxa* and *Klebsiella pneumoniae* [113, 114]. Arabinose can also be used as a feedstock for conversion into 2,3-butanediol with conversion efficiency of 40% [115]. A strain of *Lactobacillus brevis* having diol-dehydratase enzyme is used to produce MEK from 2,3-butanediol [116]. Another method to produce MEK is dehydration of 2,3-butanediol by direct reaction with sulphuric acid or using a catalyst such as alumina [117]. Butadiene can also be produced from 2,3-butanediol by dehydrating it over an alumina catalyst. The authors observed that under optimized kinetic reaction conditions nearly 80% selectivity of butadiene can be achieved [118].

13.3.6 Acrylic Acid

Gas phase oxidation of propylene can be used to produce acrylic acid. Polyacrylates can be produced from the acid which are used as super adsorbent materials and water borne-coatings and adhesives [102]. Acrylic acid can be produced from glycerol by dehydrating glycerol to acrolein, then purifying the acrolein, and lastly oxidizing the acrolein to acrylic acid. The glycerol was dehydrated with a WO_3 catalyst supported on ZrO_2 at 320°C. At low temperature the acrolein was purified and its oxidation was carried out in a second reactor using W-Mo-V-O–based catalyst at 245°C [119]. 3-Hydroxypropionic acid can also be used to produce acylric acid which is obtained by sugarfermentation [120]. Craciun *et al.* [121] dehydrated 20% aqueous solution of 3-hydroxypropionic acid over SiO_2 resulting in 97% yield of acrylic acid at 250°C and also over γ-alumina resulting in 97%–98% yield of acrylic acid. Chu *et al.* [122] genetically engineered *E. coli* for fermenting glucose into acrylic acid. These processes however cannot compete with petroleum-based processes due to high energy costs for purification and catalyst separation [123].

13.3.7 Aromatic Compounds

Lignin found in trees and other lignocellulosic plants is the best material for the preparation of aromatic compounds due to their chemical nature as they have large amounts of aromatic structure [124]. Thus, CO_2 emissions are reduced and use of non-renewable fossil resources are also reduced. The lignin is cracked using a high temperature thermal process resulting in the formation of a complex mixture of polyhydroxylated and alkylated phenol compounds along with char and other volatile compounds [125]. This creates a challenge in increasing the content of phenol and further processing of these compounds so that the phenolic-like compounds can be separated. Only 5%–10% of pure phenols can be derived; therefore, it is necessary that all the side products that are available are also developed. Major technology breakthroughs are required for isolating pure phenols from lignin due to the complex structure of lignin [102].

13.4 Production Technologies

Biofuels and other value added products can be produced from biomass by using thermochemical processes or biochemical processes. The biomass is decayed thermally and chemically reformed in the thermochemical processes so that a range of products can be produced. In biochemical processes, the biomass is biologically decayed and chemically reformed. The main difference between the two processes is the complete conversion of organic components of biomass in thermochemical process, whereas in biochemical process, the focus is mainly on polysaccharides [126].

13.4.1 Pre-Treatment

Pre-treating the biomass is an important step for converting it into biofuels and other bio-based products. Pre-treatment modifies the particle size, structure, and chemical composition of the biomass. Studies show that pre-treatment deconstructs the biomass thereby providing the enzymes, the access to cellulose for effectively separating the components from hemicellulose and cellulose [127]. Perfect pre-treatment strategy provides a number of advantages, which are decrease in particle size, quick hydrolysis by enzymes thereby enhancing the yield of monosaccharide, decrease in energy needs, and inhibition to enzymatic action [128, 129]. Generally, the conversion of biomass into saccharides without pre-treatment can yield less than 20% of sugars. Whereas, pre-treatment with methods such as biological, physical, chemical, and bio-physical can result in sugar yields of 90% [130]. However, in the overall conversion process of biomass, pre-treatment is the most expensive step. If the entire process of converting biomass into products is considered, then the major costs (nearly 40%) occurs in recovering sugars (including pre-treatment, enzyme production, and enzymatic hydrolysis) wherein pre-treatment alone costs 18% of the total costs [17]. In biorefineries, all the fractions of the biomass can be isolated by pre-treating the lignin, thus obtaining high purity unadulterated feedstock.

Pre-treatment of lignocellulosic biomass using biological methods is gaining attention from many researchers due to its favorable characteristics such as cost effective, eco-friendly, and high inclination for degradation of plants cell. A wide number of microorganisms are available for degradation of lignin by either enzymatic or chemical pathway, thus making the biomass more responsive toward saccharification [131]. Various fungi can also degrade the biomass by secreting biomass degrading enzymes. White and red-rot fungi were used for degradation of biomass by Sun and Cheng [132].

The biomass can be physically pre-treated by mechanical (grinding, chipping, scissor cutting, knife milling, and milling), ultrasound irradiation (low frequency), microwave irradiation, hot water (liquid), and steam explosion treatments. These methods reduce the particle size which decreases the crystalline structure of cellulose, thus increasing the pore size and the surface area of the biomass substrate. Microwave irradiation pre-treatment is efficient in solubilizing the cellulose and lignin. However, there are demerits in using this technique such as high energy use and increase in formation of phenolic acids like by-products [130].

The structure of biomass can be positively affected by using chemicals for pre-treatment such as organic solvents (methanol, ethanol, hexane, cyclohexane, ether, and benzene), alkalis (NaOH, KOH, $Ca(OH)_2$, ammonia, and hydrazine), acids (HNO_3, H_2SO_4, and HCl), ionic liquids, and ozone. Acids being corrosive and costly are not favored for pre-treatment. However, hemicellulose can be removed by acid pre-treatment which breaks the ether bonds in the lignin complex without dissolving the lignin [130]. Alkali pre-treatment breaks the linkages of the lignin and other carbohydrate fragments, thus allowing easy access to the starch in the matrix. The method is excellent for agricultural residues having low lignin content like [133]. Ozone pre-treatment increases the digestibility of the biomass without producing any toxic residues. Ionic liquids are becoming popular in fractionation of biomass. Ionic liquids have less environmental impact due to their high solvent power, low vapor pressure, low-volatility, non-flammability, eco-friendly, and non-toxic solvent characteristics as compared to other organic solvents. Thus, reducing the cost of solvents used and also they can be recycled [132].

Studies show that combining two pre-treatment methods such as biological with physical or chemical method is more effective than single treatment methods. Kadimaliev *et al.* [134] used ammonia, sulphuric acid, or ultrasound treatment along with a fungi for the conversion of birch and pine sawdust. The authors observed that ultrasound pre-treatment loosened the wood structure by weakening the bonds between and among the polysaccharides and lignin molecules, thus providing more accessibility to the fungal enzymes. The authors observed that alkali delignified the substrate and it became more accessible to the enzymes. Also, the presence of nitrogen in the alkali further accelerated the growth of fungi which was visually more abundant. With sulphuric acid, the lignin content increased and the polysaccharides were partially hydrolyzed, also visually, the growth of the fungus was weak.

13.4.2 Hydrolysis

The cellulose and hemicellulose produced during pre-treatment can be broken down in to simple sugars by hydrolysis. Glucose can be obtained from cellulose, whereas pentose and hexose are obtained from hemicellulose. Mineral acids or enzymes can be used to carry out the hydrolysis. Sulphuric acid or hydrochloric acid is used to hydrolyse the biomass for some specific period at a particular temperature for breaking the cellulose and hemicellulose [135]. Enzymes act as biocatalyst for the hydrolysis of cellulose and hemicellulose [136].

White-rot fungi have three groups of enzymes for converting the cellulose into glucose. First is endo-1,4-β-glucanases which attacks the amorphous region of cellulose resulting in the formation of chain free ends. Second is cellobiohydrolases enzyme which is an exocellulase enzyme and it hydrolyzes the β-1,4-glycosidic bonds from the chain ends, resulting in the formation of cellobiose. Lastly, β-glucosidases enzyme hydrolyzes the cellobiose and cellodextrins to glucose [137].

Hemicellulose consists of xylose and arabinose, mannose, glucose, and galactose, D-glucuronic acid, acetic acid, and D-galacturonic acid [138]. Hemicellulose can be converted into ethanol, butanol, and acetone, and chemicals such as furfural, lactic acid, xylo-oligosaccharides, succinic acid, and xylitol [21]. If hemicellulose is separated from the biomass, then the revenue generation of the biorefinery will be more as value added chemicals can be produced. Castro [139] conducted a techno-economic analysis of rice straw-based biorefinery. In the first scenario, both cellulose and hemicellulose were converted to ethanol, whereas, lignin was used to produce power and phenolic acids. The total revenue of the refinery was US$ 295.1 million per year and the net revenue per same year was negative US$ 37.9 million. In the second scenario, only cellulose was converted to ethanol, whereas hemicellulose was used to produce xylitol and lignin was used to produce the same products. The total revenue of the refinery per year was US$ 409.3 million and the net revenue changed from negative to positive, i.e., US$ 73.5 million per year. This shows that the biomass should be converted not only into fuels but other high value chemicals so that the biorefinery is self-sustainable and is independent of the government subsidies.

13.4.3 Fermentation

The treated biomass can be fermented with microorganisms to produce ethanol or value added chemicals. In the fermentation process, the hydrolyzed cellulose and hemicellulose is added to a water bath containing microorganisms. These organisms ferments the sugar into ethanol *via* Embden-Meyerhof (EM) pathway in anaerobic conditions or ABE process in controlled environment [135]. Yeast, *Saccharomyces cerevisiae* (used in wine and brewery industry, converts glucose by EM pathway), and the bacterium, *Zymomonas mobilis* [converts glucose through Entner-Doudoroff (ED) pathway], are the best known ethanol-producing microorganisms [140]. A major problem in the ethanol production is the suppression of fermentation caused by undesirable side products and the microorganisms unable to consume pentose. The yeast cannot metabolize pentoses since a pentose phosphate pathway is not present. Some non-conventional yeasts can metabolize pentose include *Yarrowia lipolytica*, *Scheffersomyces stipitis* (formerly known as *Pichia stipitis*), *Debaromyces hansenii*, *Kluyveromyces marxianus*, *Spathaspora passalidarum*, and *Candida shehatea* strains. The ethanol from second-generation feedstock can be produced using a widely studied yeast namely *Scheffersomyces stipites* as it can metabolize both pentose and hexose sugars (galactose, glucose, rhamnose, xylose, arabinose, mannose, and cellobiose) [141].

In conventional methods, hydrolysis and fermentation are carried out separately known as SHF. In simultaneous saccharification and fermentation (SSF) method, the hydrolysis of the cellulose is carried out by a fermentative microorganism. In another method, saccharification of both cellulose and hemicellulose is carried out, and co-fermentation of the glucose obtained from cellulose and xylose obtained from hemicellulose is carried out (SSCF). The microbes are genetically engineered for carrying out SSF and SSCF methods. The advantage of these methods is reduction in costs as single tank can be used for both hydrolysis and fermentation. The recovery of ethanol from fermented broth can be carried out either by distillation or distillation combined with adsorption process. The unreacted products like hemicellulose, cellulose, lignin, microorganisms, enzymes, and ash left in the distillation column can be collected, concentrated, and used for burning or converted to other products [142].

Chemicals such as succinic acid and aroma compounds such as vanillin are being produced commercially using biomass and some robust microbial strains. Companies, namely, BioAmber, Reverdia, Myriant, and Succinity are producing bio-based succinic acid. *E. coli* is used by BioAmber with corn as feedstock. Reverdia ferments corn feedstock using recombinant *Saccharomyces cerevisiae* microorganism. Genetically modified *E. coli* is used by Myriant for converting lignocellulose feedstock to sugars and grain sorghum to glucose. A strain of *Basfia succiniciproducens* is used by Succinity for fermentation [143]. Evolva is producing vanillin commercially by using recombinant *Schizosaccharomyces pombe* strain [144]. Engineered *Saccharomyces cerevisiae* is used by Amyris to produce isoprenoids [145].

13.4.4 Pyrolysis

Biomass degradation by heat and in the absence of oxygen is known as pyrolysis. The products formed from the degradation are bio-char, bio-oil, and syngas. Pyrolysis is carried out in the temperature range of 400°C–1200°C. The products formed by the pyrolysis of biomass are dependent upon the temperature and residence time, for example, low temperature and high residence time will result in high amount of char formation [146]. The process on the basis of these two variables can be classified as conventional pyrolysis, fast pyrolysis, and flash pyrolysis.

Massive wood pieces can be pyrolyzed using conventional pyrolysis method wherein the heating rate is 0.1–1 K/s and the residence time is 45–550 s. Initially, the biomass decomposes and internal rearrangement takes place such as elimination of water, breakage of bonds, and formation of hydroperoxide, carboxyl, and carbonyl group. Then, the decomposition rate increases and the products of pyrolysis are formed. Finally, the decomposition rate slows down and the carbon rich residue is formed. Fine particles, with size less than 1 mm, can be fast pyrolyzed with high heating rates (10–200 K/s) at high temperatures (850–1,250 K) and low residence time 0.5–10s [147]. Depending upon the biomass nature and pyrolysis conditions, 60%–75% of liquid product, 15%–25% bio-char (pure carbon and inert materials), and 10%–20% non-condensable gases (H_2, CH_4, CO, CO_2, etc.) are formed [148]. Flash pyrolysis is an improvement of fast pyrolysis. The process is carried out for very fine particles (less than 0.2 mm) in the temperature range of 1,050–1,300 K, with heating rates greater than 1,000 K/s and residence time less than 0.5 s [149]. Flash pyrolysis is typically used to produce bio-oil. The oil and the char produced can be mixed, and the slurry formed can be fed to a gasifier for conversion to syngas [150].

13.4.5 Gasification

The biomass can be gasified using air, oxygen, or steam to produce a mixture of carbon dioxide, carbon monoxide, methane, nitrogen, and hydrogen which can be either called

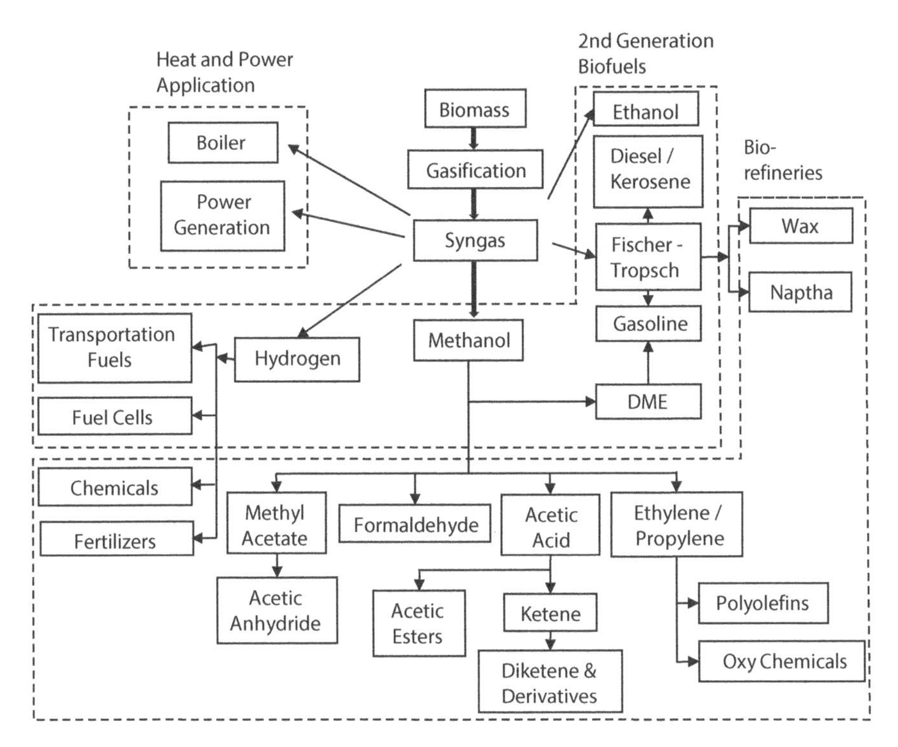

Figure 13.2 Applications of syngas [152].

synthesis gas (syngas) or producer gas based on the proportions of the components in the gas [151]. Stationary engines can be operated with producer gas for power generation. Whereas, a range of fuels and chemical intermediates can be produced from syngas [47]. By water-gas shift reaction, hydrogen can be formed from syngas or by F-T synthesis or methanol synthesis syngas can be converted to hydrocarbons [152]. Figure 13.2 shows some applications of syngas. Syngas can be produced either with catalyst or without catalyst. Nearly 1,300°C temperature is required for non-catalytic gasification, whereas lower temperature is required for catalytic gasification [153].

13.4.6 Supercritical Water

Biomass can also be converted to chemicals by supercritical fluid processing, which is an alternate path to enzymatic hydrolysis and acidic hydrolysis of cellulose to sugars. Enzymatic hydrolysis requires the biomass to be pre-treated, whereas acid recovery after acidic hydrolysis is both costly and polluting. Whereas, supercritical water can quickly and easily convert the biomass in to a mix of methane, organic acids, alcohols, and oil [154]. A fluid is called supercritical fluid when its pressure and temperature are above its vapor liquid critical point. A supercritical fluid is neither liquid nor gas since it cannot be boiled by lowering the pressure at constant temperature and it won't condense by cooling at constant pressure [155]. In supercritical state (300–644 K, 200–250 bar) and near critical state (523–573K), water is converted into acid (H^+) and basic components (OH^-), which dissolves in the biomass. These dissolved components breaks the biomass rapidly into small sugar molecules, xylose, glucose, and oligosaccharide [154]. This makes the supercritical water an attractive alternate approach for converting the biomass into value added products without using a catalyst.

13.4.7 Algae Biomass

The biomass available from various algae can be combusted directly or processed to form gaseous or liquid fuel by thermochemical or biochemical processes [156]. For direct combustion, the algae is dried and then burnt for heat or energy recovery [157]. Previously described thermochemical processes such as gasification or pyrolysis can convert the biomass into gaseous or oil biofuels. The algal biomass can also be fermented, and after anaerobic digestion of the ferment bioethanol/biobutanol, biomethane and biohydrogen is produced [158, 159]. The lipids extracted from biomass can be converted to biodiesel by transesterification process which has a higher yield in comparison to oil seeds [160].

13.5 Conclusion

Biorefinery can become a sustainable platform for the production of biofuels and other value added chemicals like bioplastics, organic acids, etc. While designing, the biorefinery all the steps and various aspects should be evaluated so that the biorefinery can be easily scaled up to the commercial level from lab-level. Some of the aspects are selection and supply of the feedstock, also, the environmental, economic, and social evaluation needs to be considered.

Firstly, a compositional database of the different types of feedstock should be developed. The gathered data will help in forming the guidelines for designing the process and selection of final products. Also, many feedstocks are generally seasonal and difficult to store which is a challenge for operating the refinery round the year. Therefore, the process in the refinery should be designed such that the refinery is independent of feedstock. Secondly, the energy density of biomass is low and they are bulkier to carry which makes their transportation to a central place expensive. Therefore, localized small biorefineries can be developed instead of a large production facility, so that biomass with high water content does not deteriorate. However, these small facilities cannot compete with integrated process massive facilities. Thirdly, the processes should be designed on the basis of the products obtained from the biorefinery, which includes not only biofuels but other bio-based chemicals as well. However, this integrated biorefinery model in which many feedstock can be used to produce many products is not matured enough as compared to single feedstock, single product model and needs to be developed.

Lastly, most of the research and implementation of biorefinery has been carried out in the US or Europe. For developing countries like China and India, which has high amount of biomass, the relevant research is not available. Adopting the processes, technologies for conversion and policies of the developed countries are not recommended. Therefore, studies that target a particular country are the need of the hour.

References

1. British Petroleum, BP Statistical Review of World Energy 2018. *British Petroleum*, 66, 1–52, 2018. https://www.bp.com/content/dam/bp/business-sites/en/global/corporate/pdfs/energy-economics/statistical-review/bp-stats-review-2018-full-report.pdf, [accessed on 10 November 2018].
2. British Petroleum, BP Statistical Review of World Energy June 2017, 2017. https://www.connaissancedesenergies.org/sites/default/files/pdf-actualites/bp-statistical-review-of-world-energy-2017-full-report.pdf, [accessed on 10 November 2018].
3. Thiyagarajan, S., Edwin Geo, V., Martin, L.J., Nagalingam, B., Carbon dioxide (CO2) capture and sequestration using biofuels and an exhaust catalytic carbon capture system in a single-cylinder CI engine: an experimental study. *Biofuels*, 9, 6, 659-668, 1–10, 2017.
4. Nigam, P.S. and Singh, A., Production of liquid biofuels from renewable resources. *Prog. Energy Combust. Sci.*, 37, 52–68, 2011.
5. GoI, P., Cabinet approves National Policy on Biofuels - 2018, 2018. https://pib.gov.in/Pressreleaseshare.aspx?PRID=1532265#:~:text=The%20Union%20Cabinet%2C%20chaired%20by,as%20%22Basic%20Biofuels%22%20viz.&text=5000%20crore%20in%206%20years,as%20compared%20to%201G%20biofuels.
6. Kircher, M., The transition to a bio-economy: Emerging from the oil age. *Biofuels Bioprod. Biorefin.*, 6, 369–375, 2012.
7. Kircher, Sustainability of biofuels and renewable chemicals production from biomass. *Curr. Opin. Chem. Biol.*, 29, 26–31, 2015.
8. Carus, M., Eder, A., Beckmann, J., Green Premium prices along the value chain of biobased products. *Ind. Biotechnol.*, 10, 83–89, 2014.
9. Dammer, L., Carus, M., Raschka, A., Scholz, L., Market Developments of and Opportunities for Biobased Products and Chemicals, 2013. http://bio-based.eu/downloads/market-developments-opportunities-biobased-products-chemicals/, [accessed on 10 November 2018].
10. Becker, J. and Wittmann, C., Advanced biotechnology: Metabolically engineered cells for the bio-based production of chemicals and fuels, materials, and health-care products. *Angew. Chem. Int. Ed.*, 54, 3328–3350, 2015.
11. Festel, G. and Rammer, C., Importance of venture capital investors for the industrial biotechnology industry. *J. Commer. Botechnol.*, 21, 2, 31–42, 2015.
12. Dammer, L. and Carus, M., RED Reform: European Parliament Agrees to Cap the Use of Traditional Biofuels, 2015. http://news.bio-based.eu/media/2015/05/15-05-11_PR_RED_Reform_Material_Sector_nova.pdf
13. Liguori, R. and Faraco, V., Biological processes for advancing lignocellulosic waste biorefinery by advocating circular economy. *Bioresour. Technol.*, 215, 13–20, 2016.
14. Velis, C.A., Circular economy and global secondary material supply chains. *Waste Manage. Resour.*, 33, 5, 389–391, 2015.
15. Velis, C.A. and Vrancken, K.C., Which material ownership and responsibility in a circular economy? *Waste Manage. Resour.*, 33, 9, 773–774, 2015.
16. Zorpas, A.A., Lasaridi, K., Abeliotis, K., Voukkali, I., Loizia, P., Fitiri, L. *et al.*, Waste prevention campaign regarding the Waste Framework Directive. *Fresenius Environ. Bull.*, 23, 11a, 2876–2883, 2014.
17. Mussatto, S.I. and Dragone, G.M., Biomass pre-treatment, biorefineries and potential products for a bioeconomy development, in: *Chapter 1. Biomass Fractionation Technologies for a Lignocellulosic Feedstock Based Biorefinery*, Elsevier Inc, Waltham, MA, 2016.
18. Ohara, H., Biorefinery. *Appl. Microbial Biotechnol.*, 62, 474–477, 2003.
19. Kamm, B. and Kamm, M., Principles of biorefineries. *Appl. Microbiol. Biotechnol.*, 64, 2, 137–145, 2004.

20. Jin, Q., Yang, L., Poe, N., Huang, H., Integrated processing of plant-derived waste to produce value-added products based on the biorefinery concept. *Trends Food Sci. Technol.*, 74, 119–131, 2018.
21. Yamakawa, C.K., Qin, F., Mussatto, S.I., Advances and opportunities in biomass conversion technologies and biorefineries for the development of a bio-based economy. *Biomass Bioenergy*, 119, 54–60, 2018.
22. Budzianowski, W.M., High-value low-volume bioproducts coupled to bioenergies with potential to enhance business development of sustainable biorefineries. *Renewable Sustainable Energy Rev.*, 70, 793–804, 2017.
23. FAO, The State of Food and Agriculture - Biofuels: Prospects, Risks and Opportunities. *Food Agric. Organ. U. N.*, 1–128, 2008. http://www.fao.org/3/a-i0100e.pdf, [accessed on 14 November 2018].
24. Jong, E., Higson, A., Walsh, P., Wellisch, M., Bio-based chemicals - value added products from biorefineries. *IEA Bioenergy Report - Task 42 Biorefinery.*, 1–34 2011. https://www.ieabioenergy.com/wp-content/uploads/2013/10/Task-42-Biobased-Chemicals-value-added-products-from-biorefineries.pdf
25. Lee RA, L.J.-M., From first-to third-generation biofuels: challenges of producing a commodity from a biomass of increasing complexity. *Anim. Front.*, 3, 2, 6–11, 2013.
26. Naik, S.N., Goud, V.V., Rout, P.K., Dalai, A.K., Production of first and second generation biofuels: A comprehensive review. *Renewable Sustainable Energy Rev.*, 14, 2, 578–597, 2010.
27. Ma, F. and Hanna, M.A., Biodiesel production: A review. *Bioresour. Technol.*, 70, 1, 1–15, 1999.
28. Demirbas, A., Biodiesel production from vegetable oils *via* catalytic and non-catalytic supercritical methanol transesterification methods. *Prog. Energy Combust. Sci.*, 31, 5–6, 466–487, 2005.
29. Saka, S. and Kusdiana, D., Biodiesel fuel from rapeseed oil as prepared in supercritical methanol. *Fuel*, 80, 2, 225–231, 2001.
30. Meneghetti, S.M.P., Meneghetti, M.R., Serra, T.M., Barbosa, D.C., Wolf, C.R., Biodiesel Production from Vegetable Oil Mixtures: Cottonseed, Soybean, and Castor Oils. *Energy Fuels*, 21, 6, 3746-3747, 2007.
31. Azad, A.K., Rasul, M.G., Khan, M.M.K., Sharma, S.C., Hazrat, M.A., Prospect of biofuels as an alternative transport fuel in Australia. *Renewable Sustainable Energy Rev.*, 43, 331–351, 2015.
32. Arapoglou, D., Varzakas, T., Vlyssides, A., Israilides, C., Ethanol production from potato peel waste (PPW). *Waste Manage.*, 30, 10, 1898–1902, 2010.
33. Liimatainen, H., Kuokkanen, T., Kääriäinen, J., Development of bio-ethanol production from waste potatoes. *Proceedings of the Waste Minimization and Resources Use Optimization Conference*, Oulu University Press, Oulu, 123–129, 2004.
34. Barnwal, B.K. and Sharma, M.P., Prospects of biodiesel production from vegetable oils in India. *Renewable Sustainable Energy Rev.*, 9, 4, 363–378, 2005.
35. Karmakar, A., Karmakar, S., Mukherjee, S., Properties of various plants and animals feedstocks for biodiesel production. *Bioresour. Technol.*, 101, 19, 7201–7210, 2010.
36. Ramos, M.J., Fernández, C.M., Casas, A., Rodríguez, L., Pérez, Á., Influence of fatty acid composition of raw materials on biodiesel properties. *Bioresour. Technol.*, 100, 1, 261–268, 2009.
37. Hahn-Hägerdal, B., Galbe, M., Gorwa-Grauslund, M.F., Lidén, G., Zacchi, G., Bio-ethanol – the fuel of tomorrow from the residues of today. *Trends Biotechnol.*, 24, 12, 549–556, 2006.
38. Halleux, H., Lassaux, S., Renzoni, R., Germain, A., Comparative life cycle assessment of two biofuels ethanol from sugar beet and rapeseed methyl ester. *Int. J. Life Cycle Assess.*, 13, 3, 184, 2008.
39. Mussatto, S.I., Dragone, G., Guimarães, P.M.R., Silva, J.P.A., Carneiro, L.M., Roberto, I.C. *et al.*, Technological trends, global market, and challenges of bio-ethanol production. *Biotechnol. Adv.*, 28, 6, 817–830, 2010.

40. Bunyakiat, K., Makmee, S., Sawangkeaw, R., Ngamprasertsith, S., Continuous Production of Biodiesel *via* Transesterification from Vegetable Oils in Supercritical Methanol. *Energy Fuels*, 20, 2, 812–817, 2006.
41. Gui, M.M., Lee, K.T., Bhatia, S., Feasibility of Edible Oil vs. Non-Edible Oil vs. Waste Edible Oil as Biodiesel Feedstock. *Energy*, 33, 1646–1653, 2008.
42. Hoekman, S.K., Broch, A., Robbins, C., Ceniceros, E., Natarajan, M., Review of biodiesel composition, properties, and specifications. *Renewable Sustainable Energy Rev.*, 16, 1, 143–169, 2012.
43. Banapurmath, N.R., Tewari, P.G., Hosmath, R.S., Performance and emission characteristics of a DI compression ignition engine operated on Honge, Jatropha and sesame oil methyl esters. *Renewable Energy*, 33, 9, 1982–1988, 2008.
44. Thompson, J. and He, B., Characterization of crude glycerol from biodiesel production from multi plefeedstocks. *Appl. Eng. Agric.*, 22, 2, 261–265, 2006.
45. Alcantara, R., Amores, J., Canoira, L., Fidalgo, E., Franco, M.J., Navarro, A., Catalytic production of biodiesel from soy-bean oil, used frying oil and tallow. *Biomass Bioenergy*, 18, 6, 515–527, 2000.
46. Pauly, M. and Keegstra, K., Cell-wall carbohydrates and their modification as a resource for biofuels. *Plant J.*, 54, 4, 559–568, 2008.
47. Balat, M., Sustainable transportation fuels from biomass materials. *Energy Educ. Sci. Technol.*, 17, 83–103, 2006.
48. Atabani, A.E., Silitonga, A.S., Ong, H.C., Mahlia, T.M.I., Masjuki, H.H., Badruddin, I.A. *et al.*, Non-edible vegetable oils: A critical evaluation of oil extraction, fatty acid compositions, biodiesel production, characteristics, engine performance and emissions production. *Renewable Sustainable Energy Rev.*, 18, 211–245, 2013.
49. No, S.-Y., Inedible vegetable oils and their derivatives for alternative diesel fuels in CI engines: A review. *Renewable Sustainable Energy Rev.*, 15, 1, 131–149, 2011.
50. Mohibbe Azam, M., Waris, A., Nahar, N.M., Prospects and potential of fatty acid methyl esters of some non-traditional seed oils for use as biodiesel in India. *Biomass Bioenergy*, 29, 4, 293–302, 2005.
51. Sarin, R., Sharma, M., Khan, A.A., Studies on Guizotia abyssinica L. oil: Biodiesel synthesis and process optimization. *Bioresour. Technol.*, 100, 18, 4187–4192, 2009.
52. Ramadhas, A.S., Jayaraj, S., Muraleedharan, C., Biodiesel production from high FFA rubber seed oil. *Fuel*, 84, 4, 335–340, 2005.
53. Kushwah, Y.S., Mahanta, P., Mishra, S.C., Some Studies on Fuel Characteristics of Mesua Ferrea. *Heat Transfer Eng.*, 29, 4, 405–409, 2008.
54. Singh, S.P. and Singh, D., Biodiesel production through the use of different sources and characterization of oils and their esters as the substitute of diesel: A review. *Renewable Sustainable Energy Rev.*, 14, 1, 200–216, 2010.
55. McCarthy, P., Rasul, M.G., Moazzem, S., Analysis and comparison of performance and emissions of an internal combustion engine fuelled with petroleum diesel and different bio-diesels. *Fuel*, 90, 6, 2147–2157, 2011.
56. Chhetri, A.B. and Watts, K.C., Densities of canola, jatropha and soapnut biodiesel at elevated temperatures and pressures. *Fuel*, 99, 210–216, 2012.
57. Deshmukh, S.J. and Bhuyar, L.B., Transesterified Hingan (Balanites) oil as a fuel for compression ignition engines. *Biomass Bioenergy*, 33, 1, 108–112, 2009.
58. Chapagain, B.P., Yehoshua, Y., Wiesman, Z., Desert date (Balanites aegyptiaca) as an arid lands sustainable bioresource for biodiesel. *Bioresour. Technol.*, 100, 3, 1221–1226, 2009.
59. Venkanna, B.K. and Venkataramana Reddy, C., Biodiesel production and optimization from Calophyllum inophyllum linn oil (honne oil) – A three stage method. *Bioresour. Technol.*, 100, 21, 5122–5125, 2009.

60. Ong, H.C., Mahlia, T.M.I., Masjuki, H.H., Norhasyima, R.S., Comparison of palm oil, Jatropha curcas and Calophyllum inophyllum for biodiesel: A review. *Renewable Sustainable Energy Rev.*, 15, 8, 3501–3515, 2011.
61. Al-Sohaibani, S. and Murugan, K., Anti-biofilm activity of Salvadora persica on cariogenic isolates of Streptococcus mutans: *in vitro* and molecular docking studies. *Biofouling*, 28, 1, 29–38, 2012.
62. Demirbas, A., Production of biodiesel fuels from linseed oil using methanol and ethanol in non-catalytic SCF conditions. *Biomass Bioenergy*, 33, 1, 113–118, 2009.
63. Silitonga, A.S., Ong, H.C., Mahlia, T.M.I., Masjuki, H.H., Chong, W.T., Characterization and production of Ceiba pentandra biodiesel and its blends. *Fuel*, 108, 855–858, 2013.
64. Subramanian, K.A., Singal, S.K., Saxena, M., Singhal, S., Utilization of liquid biofuels in automotive diesel engines: An Indian perspective. *Biomass Bioenergy*, 29, 1, 65–72, 2005.
65. Knothe, G., Cermak, S.C., Evangelista, R.L., Cuphea Oil as Source of Biodiesel with Improved Fuel Properties Caused by High Content of Methyl Decanoate. *Energy Fuels*, 23, 3, 1743–1747, 2009.
66. Kibazohi, O. and Sangwan, R.S., Vegetable oil production potential from Jatropha curcas, Croton megalocarpus, Aleurites moluccana, Moringa oleifera and Pachira glabra: Assessment of renewable energy resources for bio-energy production in Africa. *Biomass Bioenergy*, 35, 3, 1352–1356, 2011.
67. Kafuku, G., Lam, M.K., Kansedo, J., Lee, K.T., Mbarawa, M., Heterogeneous catalyzed biodiesel production from Moringa oleifera oil. *Fuel Process. Technol.*, 91, 11, 1525–1529, 2010.
68. Li, S., Wang, Y., Dong, S., Chen, Y., Cao, F., Chai, F. *et al.*, Biodiesel production from Eruca Sativa Gars vegetable oil and motor, emissions properties. *Renewable Energy*, 34, 7, 1871–1876, 2009.
69. Holser, R.A. and Harry-O'Kuru, R., Transesterified milkweed (Asclepias) seed oil as a biodiesel fuel. *Fuel*, 85, 14–15, 2106–2110, 2006.
70. Berman, P., Nizri, S., Wiesman, Z., Castor oil biodiesel and its blends as alternative fuel. *Biomass Bioenergy*, 35, 7, 2861–2866, 2011.
71. McCarthy, P., Rasul, M.G., Moazzem, S., Comparison of the performance and emissions of different biodiesel blends against petroleum diesel. *Int. J. Low-Carbon Technol.*, 6, 4, 255–260, 2011.
72. Rasul, M.G., Rudolph, V., Carsky, M., Physical properties of bagasse. *Fuel*, 78, 8, 905–910, 1999.
73. Qureshi, N., Saha, B.C., Hector, R.E., Hughes, S.R., Cotta, M.A., Butanol production from wheat straw by simultaneous saccharification and fermentation using Clostridium beijerinckii: Part I—Batch fermentation. *Biomass Bioenergy*, 32, 2, 168–175, 2008.
74. Amiri, H., Karimi, K., Zilouei, H., Organosolv pre-treatment of rice straw for efficient acetone, butanol, and ethanol production. *Bioresour. Technol.*, 152, 450–456, 2014.
75. Qureshi, N., Saha, B.C., Dien, B., Hector, R.E., Cotta, M.A., Production of butanol (a biofuel) from agricultural residues: Part I - Use of barley straw hydrolysate. *Biomass Bioenergy*, 34, 4, 559–565, 2010.
76. Brennan, L. and Owende, P., Biofuels from microalgae-A review of technologies for production, processing, and extractions of biofuels and co-products. *Renewable Sustainable Energy Rev.*, 14, 2, 557–577, 2010.
77. Maity, J.P., Bundschuh, J., Chen, C.-Y., Bhattacharya, P., Microalgae for third generation biofuel production, mitigation of greenhouse gas emissions and wastewater treatment: Present and future perspectives - A mini review. *Energy*, 78, 104–113, 2014.
78. Scott, S.A., Davey, M.P., Dennis, J.S., Horst, I., Howe, C.J., Lea-Smith, D.J. *et al.*, Biodiesel from algae: challenges and prospects. *Curr. Opin. Biotechnol.*, 21, 3, 277–286, 2010.
79. Chen, C.-Y., Yeh, K.-L., Aisyah, R., Lee, D.-J., Chang, J.-S., Cultivation, photobioreactor design and harvesting of microalgae for biodiesel production: A critical review. *Bioresour. Technol.*, 102, 1, 71–81, 2011.

80. Tran, N.H., Bartlett, J.R., Kannangara, G.S.K., Milev, A.S., Volk, H., Wilson, M.A., Catalytic upgrading of biorefinery oil from micro-algae. *Fuel*, 89, 2, 265–274, 2010.
81. Meng, X., Yang, J., Xu, X., Zhang, L., Nie, Q., Xian, M., Biodiesel production from oleaginous microorganisms. *Renewable Energy*, 34, 1, 1–5, 2009.
82. Yan, Y., Li, X., Wang, G., Gui, X., Li, G., Su, F. *et al.*, Biotechnological preparation of biodiesel and its high-valued derivatives: A review. *Appl. Energy*, 113, 1614–1631, 2014.
83. Tran, D.-T., Chen, C.-L., Chang, J.-S., Effect of solvents and oil content on direct transesterification of wet oil-bearing microalgal biomass of Chlorella vulgaris ESP-31 for biodiesel synthesis using immobilized lipase as the biocatalyst. *Bioresour. Technol.*, 135, 213–221, 2013.
84. Chen, G.-Q., Jiang, Y., Chen, F., Variation of lipid class composition in Nitzschia laevis as a response to growth temperature change. *Food Chemistry*, 109, 1, 88–94, 2008.
85. Lardon, L., Sialve, B., Steyer, J.-P., Bernard, O., Life-cycle assessment of biodiesel production from microalgae. *Environ. Sci. Technol.*, 43, 17, 6475–6481, 2009.
86. Mata, T.M., Martins, A.A., Caetano, N.S., Microalgae for biodiesel production and other applications: A review. *Renewable Sustainable Energy Rev.*, 14, 1, 217–232, 2010.
87. Balat, M., Potential alternatives to edible oils for biodiesel production - A review of current work. *Energy Convers. Manage.*, 52, 2, 1479–1492, 2011.
88. Rao, A.R., Dayananda, C., Sarada, R., Shamala, T., Ravishankar, G., Effect of salinity on growth of green alga Botryococcus braunii and its constituents. *Bioresour. Technol.*, 98, 3, 560–564, 2007.
89. Jiang, Y., Chen, F., Liang, S.-Z., Production potential of docosahexaenoic acid by the heterotrophic marine dinoflagellate Crypthecodinium cohnii. *Process Biochem.*, 34, 6, 633–637, 1999.
90. Khozin-Goldberg, I. and Boussiba, S., Concerns over the reporting of inconsistent data on fatty acid composition for microalgae of the genus Nannochloropsis (Eustigmatophyceae). *J. Appl. Phycol.*, 23, 5, 933–934, 2011.
91. Gouveia, L. and Oliveira, A.C., Microalgae as a raw material for biofuels production. *J. Ind. Microbiol. Biotechnol.*, 36, 2, 269–274, 2009.
92. Kong, Q.L., Martinez, B., Chen, P., Ruan, R., Culture of microalgae Chlamydomonas reinhardtii in wastewater for biomass feedstock production. *Appl. Biochem. Biotechnol.*, 160, 1, 9–18, 2010.
93. Razon, L.F. and T, R.R., Net energy analysis of the production of biodiesel and biogas from the microalgae: Haematococcus pluvialis and Nannochloropsis. *Appl. Energy*, 88, 10, 3507–3514, 2011.
94. Wan, M., Liu, P., Xia, J., Rosenberg, J.N., Oyler, G.A., Betenbaugh, M.J. *et al.*, The effect of mixotrophy on microalgal growth, lipid content, and expression levels of three pathway genes in Chlorella sorokiniana. *Appl. Microbiol. Biotechnol.*, 91, 3, 835–844, 2011.
95. Hu, C., Zhao, X., Zhao, J., Wu, S., Zhao, Z.K., Effects of biomass hydrolysis byproducts on oleaginous yeast Rhodosporidium toruloides. *Bioresour. Technol.*, 100, 20, 4843–4847, 2009.
96. Chatzifragkou, A., Fakas, S., Galiotou-Panayotou, M., Komaitis, M., Aggelis, G., Papanikolaou, S., Commercial sugars as substrates for lipid accumulation in Cunninghamella echinulata and Mortierella isabellina fungi. *Eur. J. Lipid Sci. Technol.*, 112, 9, 1048–1057, 2010.
97. Fakas, S., Papanikolaou, S., Galiotou-Panayotou, M., Komaitis, M., Aggelis, G., Organic nitrogen of tomato waste hydrolysate enhances glucose uptake and lipid accumulation in Cunninghamella echinulata. *J. Appl. Microbiol.*, 105, 4, 1062–1070, 2008.
98. Sergeeva, Y.E., Galanina, L., Andrianova, D., Feofilova, E., Lipids of filamentous fungi as a material for producing biodiesel fuel. *Appl. Biochem. Microbiol.*, 44, 5, 523–527, 2008.
99. Somashekar, D., Venkateshwaran, G., Sambaiah, K., Lokesh, B., Effect of culture conditions on lipid and gamma-linolenic acid production by mucoraceous fungi. *Process Biochem.*, 38, 12, 1719–1724, 2003.

100. Wynn, J.P., Hamid, A.A., Li, Y., Ratledge, C., Biochemical events leading to the diversion of carbon into storage lipids in the oleaginous fungi Mucor circinelloides and Mortierella alpina. *Microbiology*, 147, 10, 2857–2864, 2001.
101. Werpy, T. and Petersen, G., Top Value Added Chemicals from Biomass. Volume I - Results of Screening for Potential Candidates from Sugars and Synthesis Gas, 2004. https://www.nrel.gov/docs/fy04osti/35523.pdf, [accessed on 30 November 2018].
102. Haveren, J., Scott, E.L., Sanders, J., Bulk chemicals from biomass. *Biofuels, Bioprod. Biorefin.*, 2, 41–57, 2008.
103. Takahara, I., Saito, M., Inaba, M., Murata, K., Dehydration of Ethanol into Ethylene over Solid Acid Catalysts. *Catal. Lett.*, 105, 3, 249–252, 2005.
104. Arenamnart, S., Trakarnpruk, W., Science, P., Polymer, F., Ethanol Conversion to Ethylene Using Metal-Mordenite Catalysts. *Int. J. App. Sci. Eng.*, 4, 1, 21–32, 2006.
105. Vasiliadou, E.S. and Lemonidou, A.A., Production of Biopropylene Using Biomass-Derived Sources. *Encycl. Inorg. Bioinorganic Chem.*, R.A. Scott (Ed.), 1–12, 2016.
106. Ren, T., Patel, M., Blok, K., Olefins from conventional and heavy feedstocks: Energy use in steam cracking and alternative processes. *Energy*, 31, 4, 425–451, 2006.
107. Ezeji, T., Qureshi, N., Blaschek, H.P., Production of acetone–butanol–ethanol (ABE) in a continuous flow bioreactor using degermed corn and Clostridium beijerinckii. *Process Biochem.*, 42, 1, 34–39, 2007.
108. Zhang, S., Gong, Y., Zhang, L., Liu, Y., Dou, T., Xu, J. *et al.*, Hydrothermal treatment on ZSM-5 extrudates catalyst for methanol to propylene reaction: Finely tuning the acidic property. *Fuel Process. Technol.*, 129, 130–138, 2015.
109. Altaras, N.E., Etzel, M.R., Cameron, D.C., Conversion of Sugars to 1,2-Propanediol by Thermoanaerobacterium thermosaccharolyticum HG-8. *Biotechnol. Progr.*, 17, 1, 52–56, 2001.
110. Chaminand, J., Djakovitch, L., Gallezot, P., Marion, P., Pinel, C., Rosier, C., Glycerol hydrogenolysis on heterogeneous catalysts. *Green Chem.*, 6, 8, 359–361, 2004.
111. Werphy, T., Frye, J.J.G., Zacher, A.H., Miller, D.J., *Hydrogenolysis of 6-carbon sugars and other organic compounds*, vol. 6, p. 085 B2, US Patent 6841085B2.
112. Weissermel, K. and Arpe, H.J., *Industrial Organic Chemistry*, Fourth Wiley VCH, Weinheim, 2008.
113. Perego, P., Converti, A., Del Borghi, M., Effects of temperature, inoculum size and starch hydrolyzate concentration on butanediol production by Bacillus licheniformis. *Bioresour. Technol.*, 89, 2, 125–131, 2003.
114. Syu, M.-J., Biological production of 2,3-butanediol. *Appl. Microbiol. Biotechnol.*, 55, 1, 10–18, 2001.
115. Saha, B.C. and Bothast, R.J., Production of 2,3-butanediol by newly isolated Enterobacter cloacae. *Appl. Microbiol. Biotechnol.*, 52, 3, 321–326, 1999.
116. Manitto, P., Speranza, G., Fontana, G., Galli, A., Stereochemistry and fate of hydrogen atoms in the diol -dehydratase catalysed dehydration of meso-butane-2,3-diol. *Helvetica Chemica Acta*, 81, 11, 2005–2016, 1998.
117. Emerson, R.R., Flickinger, M.C., Tsao, G.T., Kinetics of dehydration of aqueous 2,3-butanediol to methyl ethyl ketone. *Ind. Eng. Chem. Prod. Res. Dev.*, 21, 3, 473–477, 1982.
118. Liu, X., Fabos, V., Taylor, S., Knight, D.W., Whiston, K., Hutchings, G.J., One-Step Production of 1,3-Butadiene from 2,3-Butanediol Dehydration. *Chem. - A Eur. J.*, 22, 35, 12290–12294, 2016.
119. Dubois, J.L. and Patience, G.H., 2013. Method for the reactive vaporization of glycerol, US8530697B2, 2013.
120. Jiang, X., Meng, X., Xian, M., Biosynthetic pathways for 3-hydroxypropionic acid production. *Appl. Microbiol. Biotechnol.*, 82, 6, 995–1003, 2009.

121. Craciun, L., Benn, G.P., Dewing, J., Schriver, G.W., Peer, W., Preparation of acrylic acid derivatives from alpha- or betahydroxy carboxylic acids. US7538247B2, 2009.
122. Chu, H.S., Ahn, J.-H., Yun, J., Choi, I.S., Nam, T.-W., Cho, K.M., Direct fermentation route for the production of acrylic acid. *Metab. Eng.*, 32, 23–29, 2015.
123. Grasselli, R.K. and Trifirò, F., Acrolein and acrylic acid from biomass. *Rendiconti Lincei*, 28, 59–67, 2017.
124. Chakar, F.S. and Ragauskas, A.J., Review of current and future softwood kraft lignin process chemistry. *Ind. Crops Prod.*, 20, 2, 131–141, 2004.
125. Amen-Chen, C., Pakdel, H., Roy, C., Production of monomeric phenols by thermochemical conversion of biomass: a review. *Bioresour. Technol.*, 79, 3, 277–299, 2001.
126. Gomez, L.D. and Clare GS, M.-M.J., Sustainable liquid biofuels from biomass: the writing's on the walls. *New Phytol.*, 178, 473–485, 2008.
127. Li, J., Lin, J., Xiao, W., Gong, Y., Wang, M., Zhou, P. *et al.*, Solvent extraction of antioxidants from steam exploded sugarcane bagasse and enzymatic convertibility of the solid fraction. *Bioresour. Technol.*, 130, 8–15, 2013.
128. Duque, A., Manzanares, P., Ballesteros, I., Ballesteros, M., Steam Explosion as Lignocellulosic Biomass Pre-treatment, Chapter 15, in: *Biomass Fractionation Technologies for a Lignocellulosic Feedstock Based Biorefinery*, Elsevier Inc, Amsterdam, 2016.
129. Kumar, P., Barrett, D.M., Delwiche, M.J., Stroeve, P., Methods for Pre-treatment of Lignocellulosic Biomass for Efficient Hydrolysis and Biofuel Production. *Ind. Eng. Chem. Res.*, 48, 8, 3713–3729, 2009.
130. Ponnusamy, V.K., Nguyen, D.D., Dharmaraja, J., Shobana, S., Banu, J.R., Saratale, R.G. *et al.*, A review on lignin structure, pre-treatments, fermentation reactions and biorefinery potential. *Bioresour. Technol.*, 271, 462–472, 2019.
131. Valášková, V., Šnajdr, J., Bittner, B., Cajthaml, T., Merhautová, V., Hofrichter, M. *et al.*, Production of lignocellulose-degrading enzymes and degradation of leaf litter by saprotrophic basidiomycetes isolated from a Quercus petraea forest. *Soil Biol. Biochem.*, 39, 10, 2651–2660, 2007.
132. Agbor, V.B., Cicek, N., Sparling, R., Berlin, A., Levin, D.B., Biomass pre-treatment: Fundamentals toward application. *Biotechnol. Adv.*, 29, 6, 675–685, 2011.
133. Saratale, G.D. and Oh, M.-K., Improving alkaline pre-treatment method for preparation of whole rice waste biomass feedstock and bioethanol production. *RSC Adv.*, 5, 118, 97171–97179, 2015.
134. Kadimaliev, D.A., Revin, V.V., Atykyan, N.A., Samuilov, V.D., Effect of Wood Modification on Lignin Consumption and Synthesis of Lignolytic Enzymes by the Fungus Panus (Lentinus) tigrinus. 39, 5, 488–492, 2003.AU: Please provide journal title.
135. Dalena, F., Senatore, A., Tursi, A., Basile, A., *Bioenergy production from second- and third-generation feedstocks*, Elsevier Ltd, Cambridge, 2017.
136. Brethauer, S. and Wyman, C.E., Review: Continuous hydrolysis and fermentation for cellulosic ethanol production. *Bioresour. Technol.*, 101, 13, 4862–4874, 2010.
137. Dashtban, M., Schraft, H., Qin, W., Fungal bioconversion of lignocellulosic residues; opportunities & perspectives. *Int. J. Biol. Sci.*, 5, 6, 578–595, 2009.
138. FitzPatrick, M., Champagne, P., Cunningham, M.F., Whitney, R.A., A biorefinery processing perspective: Treatment of lignocellulosic materials for the production of value-added products. *Bioresour. Technol.*, 101, 23, 8915–8922, 2010.
139. Castro, R.C.A., *Produção de etanol a partir da palha de arroz por sacarificação e fermentação simultânea empregando um reator agitado não convencional*, Universidade de São Paul Lorena, São Paulo, Brazil, 20162016.

140. Claassen, P.A.M., van Lier, J.B., Lopez Contreras, A.M., van Niel, E.W.J., Sijtsma, L., Stams, A.J.M. *et al.*, Utilisation of biomass for the supply of energy carriers. *Appl. Microbiol. Biotechnol.*, 52, 6, 741–755, 1999.
141. Ceccato-Antonini, S.R., Codato, C.B., Martini, C., Bastos, R.G., Tauk-Tornisielo, S.M., Yeast for pentose fermentation: isolation, screening, performance, manipulation, and prospects, in: *Adv. Basic Sci. Second Gener. Bioethanol from Sugarcane*, M.S. Buckeridge and A.P. de Souza (Eds.), pp. 133157–, Springer International Publishing, New York, 2017.
142. Mosier, N., Wyman, C., Dale, B., Elander, R., Lee, Y.Y., Holtzapple, M. *et al.*, Features of promising technologies for pre-treatment of lignocellulosic biomass. *Bioresour. Technol.*, 96, 6, 673–686, 2005.
143. Nghiem, N.P., Kleff, S., Schwegmann, S., Succinic Acid: Technology Development and Commercialization. *Fermentation*, 3, 26, 1–14, 2017.
144. Gallage, N.J. and Møller, B.L., Vanillin - Bioconversion and Bioengineering of the Most Popular Plant Flavor and Its De Novo Biosynthesis in the Vanilla Orchid. *Mol. Plant*, 8, 1, 40–57, 2015.
145. Meadows, A.L., Hawkins, K.M., Tsegaye, Y., Antipov, E., Kim, Y., Raetz, L. *et al.*, Rewriting yeast central carbon metabolism for industrial isoprenoid production. *Nature*, 537 694–697, 2016.
146. Tripathi, M., Sahu, J.N., Ganesan, P., Effect of process parameters on production of biochar from biomass waste through pyrolysis: A review. *Renewable Sustainable Energy Rev.*, 55, 467–481, 2016.
147. Shafizadeh, F., Introduction to pyrolysis of biomass. *J. Anal. Appl. Pyrolysis*, 3, 283–305, 1982.
148. Ni, M., Leung, D.Y.C., Leung, M.K.H., Sumathy, K., An overview of hydrogen production from biomass. *Fuel Process. Technol.*, 87, 5, 461–472, 2006.
149. Current technologies for thermo-conversion of biomass into fuels and chemicals. *Energy Sources*, 26, 715–730, 2004.
150. Mohan, D., Pitman, C.U., Steele, P., Pyrolysis of wood/biomass for bio-oil: A critical review. *Energy Fuels*, 20, 848–889, 2006.
151. Rowlands, W.N., Masters, A., Maschmeyer, T., The biorefinery-challenges, opportunities, and an Australian perspective. *Bulletin Sci. Technol. Soc.*, 28, 2, 149–158, 2008.
152. Steen EV, C.M., Fischer–Tropsch catalysts for the biomass-to-liquid (BTL) process. *Chem. Eng. Technol.*, 31, 5, 650–660, 2008.
153. Lee, S. and Speight JG, L.S., *Hand book of alternative fuel technologies*, CRC Taylor and Francis Group, USA, 2007.
154. Sasaki, M., Kabyemela, B., Malaluan, R., Hirose, S., Takeda, N., Adschiri, T., A.K., Cellulose hydrolysis in subcritical and supercritical water. *J. Supercrit. Fluid*, 13, 261–268, 1998.
155. Saka, S., Ehara, K., Sakaguchi, S., Yoshida, K., Useful products from lignocellulosics by supercritical water technologies., *in: The Second Joint International Conference on Sustainable Energy and Environment*, 485–489, Bangkok, 2006.
156. Rittmann, B., Opportunities for renewable bioenergy using microorganisms. *Biotechnol. Bioeng.*, 100, 203–212, 2008.
157. Kadam, K., Environmental implications of power generation *via* coal- microalgae cofiring. *Energy*, 27, 905–922, 2002.
158. McKendry, P., Energy production from biomass (part 1): Overview of biomass. *Bioresour. Technol.*, 83, 37–46, 2002.
159. McKendry, P., Energy production from biomass (part 2): Conversion technologies. *Bioresour. Technol.*, 83, 47–54, 2002.
160. Chisti, Y., Biodiesel from microalgae. *Biotechnol. Adv.*, 25, 294–306, 2007.

14

Sustainable Production of Biofuels Through Synthetic Biology Approach

Dulam Sandhya[1], Phanikanth Jogam[1], Lokesh Kumar Narnoliya[2*], Archana Srivastava[3] and Jyoti Singh Jadaun[3†]

[1]Department of Biotechnology, Kakatiya University, Warangal, India
[2]Advanced Technology Platform Centre-Regional Centre for Biotechnology, Faridabad, India
[3]Department of Botany, Dayanand Girls Postgraduate College, Kanpur, India

Abstract
Biofuels are termed to those kinds of fuel, which are obtained either from biological sources or produced through the deployment of any biological process. Biofuels have several advancements over the petroleum/fossil fuels as these are the sustainable sources and their production is independent on the geographical distribution. Another gross motivation for affection toward the biofuel production is that excessive use of fossil fuels will deplete all the natural resources in nearby future, and in this racing world, we can't imagine our life which is devoid of fuels because we are utterly dependent on it from kitchen upto transport level globally. Even though, it will not be wrong to say that we are unable to move our vehicles due to lack of fuel. Therefore, biofuel production is the only alternative of fossil fuels, which can accomplish the demand of growing population all over the world. So, we have to search for more resources, which can serve as better platforms for biofuel production. Upgradation of biofuel level in the existing sources or generation of genetically modified organism (GMO) for superior production of biofuels can be a hope of research community to cope up with the crises of fossil fuels, and it is only possible by blending the knowledge of basic biological science and with system biology techniques. In this modern era, we can achieve this goal in lesser cost, time, and space with the aid of synthetic biology. In this book chapter, we unveil diverse sources of biofuels and implications of synthetic biology in attaining the desired level of biofuels in organisms by fluxing the metabolic pathways.

Keywords: Bioethanol, biodiesel, synthetic biology, metabolic engineering, algae

14.1 Introduction

Biofuel is derived by bioenergy and considered as a major source of energy in future. Biofuels might be liquid (ethanol made from sugarcane and corn), gaseous (biogas-methane made from digested material and animal manure), and solid (fuelwood). Other biofuels like biodiesel are produced from plant oils and animal fats, while green diesel is produced from algal and plant species [1]. Currently, research is focusing on production of biofuels

**Corresponding author*: dr.lknarnoliya@gmail.com
†*Corresponding author*: jsjdgpg2017@gmail.com

Arindam Kuila and Mainak Mukhopadhyay (eds.) Biorefinery Production Technologies for Chemicals and Energy, (289–312) © 2020 Scrivener Publishing LLC

because of its potential applications in biofuel-powered vehicles, backup systems, fuel cells, as well as for removing paints and in cleaning of metals. Biofuel is proven to be less toxic than petroleum fuel due to its low volatility; therefore, it can easily and safely used in pure form; thus, it is designated as eco-friendly [2].

Biofuel is considered as an alternative source of fossil fuel but both differs in the nature of their raw materials as biofuels are derived from living biomaterials while fossil fuels are derived from nonliving materials. Biofuel can be produced directly from plants or indirectly from household and agriculture wastes. Several plant sources like wheat, corn, soybean, rapeseed and canola oil, palm oil, switchgrass, and sugarcane are used as a raw material for the production of biofuels, and other resources are bacteria, algal, fungal, and animal fats (Table 14.1) [3]. These resources of biofuels are cheap and ecofriendly; thus, biofuels can overcome the pollution related limitations of fossil fuel [1]. Pollution and increasing amount of wastes is becoming the challenging problem worldwide so researchers have started to develop the technology for managing the household and agricultural waste in sustainable manner. Recycling of renewable wastes helps to reduce waste in environment and simultaneously products can be supplied to society at low price [4]. As per reports, around 26 Mt nutrients can be recycled from animal and crop wastes. This article updates the potentials of various sources for the production of biofuel, their types, and uses. Based on latest estimations world population will become 9 billion by 2050 so it leads to increase in demand of renewable energy sources. Therefore, there is need to enhance the production of non-fossil fuels to meet the global demands as well as use of biofuels will reduce the hazardous impacts of fossil fuels, so it will be helpful to control the climate changes [5–7].

The first-generation biofuels are produced from agricultural products and second-generation biofuels are produced from cellulosic biomass while third- and fourth-generation biofuel are associated with algal species [8]. In recent decades, genetic engineering is one of the promising technology for the production of biofuels to solve the current problem of fossil fuel. Further, research in genetic engineering with synthetic biology may extend the use of sustainable biofuel instead of fossil fuel. Metabolic engineering and gene editing technology like CRISPR/Cas will drive the cost efficient system to produce green biofuels for transportation and other applications [9, 10]. Genetic engineering can be explored for

Table 14.1 Sources of first-generation biofuels.

Source	Application	Reference
Sugarcane	Green diesel	[22]
Corn	Vegetable oil	[94]
Soybean	Bio ethers	[94]
Safflower	Biogas	[16]
Camelia	Syngas	[15]
Sunflower	Solid biofuels	[96]
Peanut	Biodiesel	[153]
Canola	Bioalcohol	[154]

enhancing the production of biofuel levels and to enhance sugar levels and oil content in plants to drive the green biofuels. Genetically altered cyanobacteria and *E. coli* are used for the manufacturing of biodiesel from simple sugars and for the production of important value added products [11]. Biofuel producing plant sources are cultivated separately worldwide such as jatropha in India; soybeans, switchgrass, and corn in USA; sugar cane in Brazil; miscanthus and palm oil in Asia; wheat and sugar beet in Europe; sorghum, tapioca, and cassava in China. Out of these states, USA and Brazil produce up to 87% ethanol fuel in the world. Majorly ethanol is produced through fermentation process from sugars [12]. International Energy Agency (IEA) projects that 30% biofuels should be used in transportation by 2050, so that we can reduce the 2.1 gigatonnes of CO_2 emission per year in the environment. According to Energy Independence and Security Act of 2007, increase in the biofuel up to 36 billion gallons from 9 billion gallons can be estimated by 2022 [13, 14].

14.2 Types of Biofuel

Based on the source of material and emission of CO_2, biofuels are classified into four different categories.

14.2.1 First-Generation Biofuels (Conventional Biofuels)

Biofuel produced from feedstock (sustainable for human consumption) is called first-generation biofuel. They have been derived from different plant oils, sugar beet, animal fats, and starch. Biogas, bio-ether, biodiesel, and bio-alcohols fall under the category of first-generation biofuels. Unlike second generation, first-generation biofuels are used directly in an unprocessed form.

14.2.1.1 Biogas

Biogas is produced from the decomposition of organic matter in the presence of anaerobic organisms (anaerobic reaction). During these reactions, different types of biogases are produced like methane, carbon dioxide, nitrogen, hydrogen, ethers, and sulfurs. Biogas is a source of electricity in the form of heat, and it is also used in vehicle engines for internal combustion to move vehicles [8].

14.2.1.2 Biodiesel and Bioethanol

Biodiesel and bioethanol are produced from animal fat or vegetable oils (Figure 14.1). Biodiesel and bioethanol function as an important liquid transport biofuels and these can be considered as an alternative to diesel and petrol, respectively. Biodiesel cannot be used directly in internal combustion engines but it can be used with petroleum fuel while bioethanol can be used directly or in a blend. Biodiesel is also produced from oil crops, such as sunflower oil, soybean oil, and oilseed rape [15, 16]. It can also be produced from the gunky stuff and restaurant fat fryers. The oil seeds undergo some chemical reactions for the production of biodiesel. In Europe, Tanzania, the nuts of croton tree (*Croton megalocarpus*) are used to produce biodiesel, while in India, biodiesel is produced from *Jatropha curcas* oil.

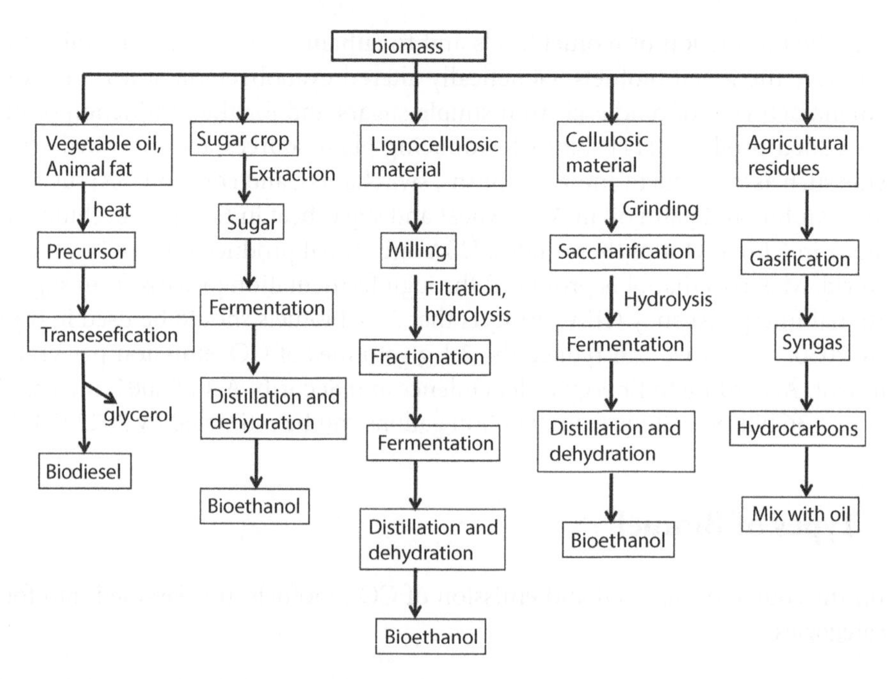

Figure 14.1 Different feedstocks like vegetable oil, animal fats, and agriculture residues used to produce various renewable biofuel through various methods.

Bioethanol is produced from crop plants having high sugar content or with the breakdown of plant oils with bacteria [8, 17–20].

14.2.2 Second-Generation Biofuels

Second-generation biofuel is also known as advanced biofuel and it is produced from processing of feedstock from ligno-cellulosic materials, biomass, municipal residues, forest, and crop residues (Table 14.2). Generally, second-generation biofuels are used in transportation. Mainly, food crops are not considered as second-generation biofuels but food crops and vegetable oils (non-food materials) can be used in second-generation biofuels when they have already fulfilled their food purpose and no longer fit for human consumption.

Table 14.2 Sources of second-generation biofuel.

Sources	Application	Reference
Castor	Bio-hydrogen	[155]
Jatropha	Bio-DME (Dimethylether)	[16]
Polanga	Bio-methanol	[156]
Coconut	DMF (Dimethylformamide)	[55]
Oil palm	HTU diesel HydroThermalUpgrading	[15]
	Fischer-Tropsch diesel	
	Wood diesel	
	Bio-alcohols	
	Green diesel	

The second-generation biofuels act more significantly, to balance the greenhouse gases than first-generation biofuels. Second-generation biofuels play vital role in the control of environmental carbon and it reduces the air contaminants as compared to fossil fuels. Biofuel from feedstock leads to development of viable biofuel to achieve energy security. Different types of processing technologies such as gasification and fermentation are used to produce second-generation biofuels [21].

14.2.2.1 Cellulosic Ethanol

Cellulosic ethanol was produced from crop residues and vegetable wastes, which have high cellulose content. It is commercially produced from grasses and sugar cane residues [22].

14.2.2.2 Biomethanol

Boimethanol produced from synthetic gas processed from biomass. It is an alternative as starting oil in vehicles. Without any modifications, biomethanol can be used in vehicles by mixing with oil [23].

14.2.2.3 Dimethylformamide

DMF or dimethylformamide is a colorless liquid compound used as the dissolvent in reactions, production of pesticide and fibers [8, 18, 19].

14.2.3 Third-Generation Biofuels

Biofuel produced from algae and seaweed is known as third-generation biofuels. Previously, algal resource considered as a second-generation biofuel, but later, it is categorized into the third-generation due to its high yielding with low inputs. Cyanobacteria, seaweed, and green algae used as a source for third-generation biofuels because these species contain high amount of lipids to produce biofuels. Genetically modified algae can produce almost 10-fold higher biofuels than traditional feedstocks. Algal species grew almost all arable lands without affecting the freshwater resources and it consumes carbon dioxide so it acts as a sink for excess CO_2 in the environment. It can grow in wastewater without affecting the environment. Algae are used to produce biodiesel, bioethanol, vegetable oils, bio-butanol, bio-gasoline, and various biofuels [18, 19, 24, 25].

14.2.4 Fourth-Generation Biofuels

It is an extension of third-generation biofuels and it is associated with recent technologies like advanced synthetic biology, metabolic engineering of algal biomass and genetic engineering technique. Genetically modified crops and microbes used as a feedstock for fourth-generation biofuel. Engineered aerobic microbes are used in both third- and fourth-generation biofuels to produce renewable biofuels. Recent research is focusing on the engineering techniques, which can be used to produce alternative biofuel [12, 25].

Algal butanol, ethanol, and gasoline produced from algal species through fermentation are directly used in engines without any alteration due to the presence of high octane molecules.

Various techniques such as bioengineering and recombinant DNA technology are applied to modify cellular and molecular properties of algae [26]. The CRISPR genome editing technology may also be applied to modify the metabolic properties of algae to produce biofuels in the required amount to meet global demands. Biofuel production from algae is value added resources because it will take fewer steps to produce biofuel with less expenditure and easier processing method. Genetically altered microbes and plants can grasp more carbon dioxide and solar energy with enhanced photosynthesis so that gradually decrease the effect of greenhouse gases and also enhance biofuel production in the fourth generation [18, 19, 27].

14.2.5 Advantages of Biofuels

- Biofuel from renewable sources is biodegradable and recyclable in nature and not releases harmful gases into the environment so good to reduce global warming.
- Unlike fossil fuels, biofuels are renewable sources with low cost.
- Biofuels can be used in the transportation of various vehicles.
- Biofuels can serve as cleaning agent from the waste stream.
- Biofuels are used in houses to generate heat in the buildings.
- Biofuels establish economic and energy security with energy balancing potential.
- Biofuels can be used as a power source for electronics.
- Biofuels can act as lubricating for the proper working of engines.

14.2.6 Disadvantages of Biofuels

- Plant-based corn burning might be responsible for release of greenhouse gases into the environment.
- Occupies irrigation land, consume high water, and compete with crop plants.
- The natural habitat of animals might be lost in the ecosystem [28, 29].

14.3 Sources of Biofuel

Several sources are used to produce advanced biofuels to solve the problem of fossil fuels. Oleaginous microorganisms, fungal species, microalgae species, bacterial species, and yeast attracted great attention for producing sustainable biofuels in the world (Table 14.3). Not only microorganism, plant oils and animal fats are also used to produce biofuels [28–30].

14.3.1 Bacterial Source

Bacterial species and yeast are examples of ideal hosts for the production of various renewable biofuels due to their faster growth, genetic amenable, incredible malleability, and well-studied organism. Genetically engineered *E. coli* and yeast generate biofuels from simple sugars through fermentation. Recent research is focusing on metabolically

Table 14.3 Resources of biofuel production.

Resource	Reference
Corn	[31]
Switch grass	[32]
Waste banana	[33]
Maize	[34]
Wheat	[35]
Cotton	[36]
Sugarcane (lignocellulosic material)	[34, 37]
Sweet sorghum	[34]
Grasses (Miscanthus species)	[34, 38]
Edible vegetable (e.g., Canola oil, Sunflower oil, etc.)	[39]
Non-edible vegetable oils (e.g., Jatropha oil, Pongamia oil, etc.)	[40] [41] [42]
Non-edible trees with oilseeds • Jatropha • Karanja • Edible oilseed plants such as Oleaginous, Leguminosae, Brassicaceae, and Euphorbiaceae	 [43] [44] [45]
Animal fats	[46]
Aquatic plants • Cattails (*Typha* species) • Duckweed or Lemnaceae • Water lettuce (*Pistiastratiotes L.*) • Water hyacinth (*Eichhorniacrassipes*)	 [47] [48] [49] [50]
Cassava, sweet potato	[51]
Mahua flower	[52]
Rapeseed oil	[53]
Watermelon	[54]
Algal species • *Chlorella* • *Dunaliella tertiolecta* • *Spirulina platensis* • *Nannochloropsis* sp. • *Scenedesmus* sp. • *Nannochloropsis salina* • *Porphyridium cruentum* • *Tetraselmis suecica* • ·*Desmodesmus* sp.	 [55] [56] [57] [58] [59] [60] [61] [62] [63]

(*Continued*)

Table 14.3 Resources of biofuel production. (*Continued*)

Resource	Reference
Plants	
• *Madhuca indica*	[64]
• *Pongamia pinnata*	[65]
• *Azadirachta indica*	[66]
• Soybean	[67]

engineered cyanobacteria and *E. coli* to encounter the problem with plant-based biofuels. Based on some reasons, bacterial species are well deserve for the production of biofuels such as they require less space and time for growth and they can be easily modified and easier for the extraction of the compound from bacteria [68]. Cyanobacteria need CO_2 and sunlight for its photosynthesis process like plants so these are considered as metabolic factories for the production of large-scale biofuels [69, 70, 152]. In the context of metabolic engineering, *E. coli* is the most prone for attempting the pathway modulation for optimizing the production of biofuels [71]. Earlier engineered cyanobacteria produce sugars, which are further catalyzed by *E. coli* to produce biofuel precursors. Further, extensive research on cyanobacteria provides a way that it can produce products for biofuels precursors without aid of *E. coli* [72, 73].

14.3.2 Algal Source

Photosynthetic microalgae, transform 9–10% of solar energy into biomass with a yield of 77 g/biomass/m^2/day [74, 75]. Algal species are the best alternative sources for the production of biofuels (biohydrogen, biodiesel, bioethanol, biogasoline, and biogas) because of fewer hurdles in their cultivation as it needs no land and they can produce huge biomass in lesser time than oil crop plants. Till date, more than 40,000 microalgae species are identified [30, 76, 77]. Naturally, algal species are found in fresh or marine water but they can be cultivated in wastewater ponds attached with any biorefinery. Biofuel production from algae defined as algae farming and it belongs to third-generation biofuels (Figure 14.2). Further genetic modification has been demonstrated in various green algal species like *Chlorella*, *Chlamydomonas reinhardtii*, *Nannochloropsis*, and *Scenedesmus* to produce biofuels [78–80]. Rapidly growing and photosynthetic microalgae have high lipid content in its body about 20–50% of their biomass, although it can be increased by inserting the desired genes to produce huge amount of biofuels [81, 82].

14.3.3 Fungal Source

Oleaginous fungal species are vital source for biofuel production. Fungal species are used in fermentation process of plant materials such as grape juice and wine. Fungi can accumulate 70% lipids of their biomass during metabolic stress, so fungal lipids may be used as a feedstock for biofuel production [83]. Moreover, various genetic engineering techniques can be applied to optimize the conditions for higher lipid production in fungi. Although,

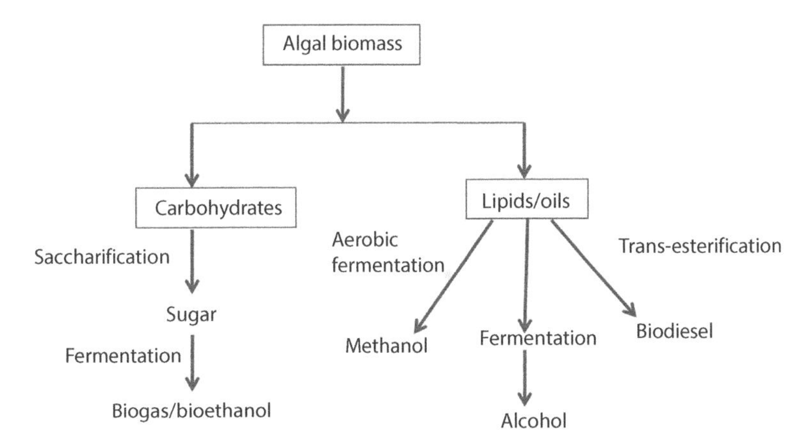

Figure 14.2 Raw material like lipids and carbohydrates from algal biomass used as a source for the production of different biofuels through various methods.

exploitation of fungi for biofuel production is an eco-friendly technique but it should associate with some pre-treatment steps [84]. Yeast is also another model host for biofuel production and it can produce 18% ethanol from the fermentation broth and it can grow on simple sugars and metabolize glucose [85].

14.3.4 Plant Source

Recent trends is focusing on plant material as feedstock for production of various biofuel such as sweet sorghum is used as a source for bio-hydrogen production [86, 87]. One of the challenges faced in biofuel production is with its pre-treatment steps. Therefore, to resolve this problem, the genetically engineering microbe/plants can be used to avoid these pre-treatment steps. Genetic engineering in plants used to enhance polysaccharide content and decrease the lignin biosynthesis to produce advanced biofuels like butanol and ethanol. Aquatic plants also used as feedstock in biofuel production, such as duckweed, water lettuce, water hyacinth, and cattails. Aquatic plants are used in the photo remediation process, sewage treatment, and to clean pollutants in water [47]. Leguminosae family plants such as Indian rosewood, *Acacia nilotica*, and yellow flame tree (*Vachelliani lotica* L.), locust bean, and *Delonix regia*, etc., used as a feedstock for carbohydrate. Various plants like corn, sunflower, coconut, palm, rice bran, canola, and soybean are used as a source for the production of biodiesel. In addition to algal fats, fish oil and animal fats are also used to produce nontoxic biodiesel, which would gradually reduce the dependency on fossil fuel [88].

However, plants are considered as a part of the ecosystem, so biofuel from plants associated with some negative impact on agriculture, economy, and food security. Therefore, non-edible plant oils, animal fats, and kitchen residues are used for the production of biofuels like biodiesel. Some examples of non-edible plant source are *Azadirachta indica, Pongamia pinnata, Jatropha curcas, Madhuca spp., Diploknema butyracea, Camelina sativa, Ricinus communis* L., and Simarouba. Nowadays, various genetic engineering techniques are applying to enhance the desired properties and oil yield in plants for biofuel production

[89, 90]. Plants with high lignin become a barrier to produce biofuels. Therefore, lignin pathway in plants are modified by synthetic biology and genetic engineering techniques [91].

Plant-derived oils are unsaturated fatty acids and liquid (non-volatile) at room temperature while animal-derived fats are saturated fatty acids and solids (volatile) at room temperature. So, we need to modify some properties with genetic engineering to obtain desired biofuel products. These modified saturated fats are used as raw material to produce biodiesel. Sunflower oil, coconut, soybean, jatropha, and cottonseed oil are used as biofuel sources in USA and Europe. Liquid biofuels like ethanol meant for transportation and produced by fermentation of simple sugars. Huge amount of ethanol was produced from corn and sugarcane in the United States and Brazil [3, 92]. Plants like wheat, corn, soybean, rapeseed, canola, palm, switchgrass, and sugarcane can be used for the production of first-generation biofuels such as production of biodiesel from vegetable oil and bioethanol.

14.3.4.1 Plant Materials Utilized for the Production of Biofuels

14.3.4.1.1 Cellulose

Cellulose is a tremendous feedstock and 1.3 billion tons of cellulose has been consumed for various biofuel productions in the US. Cellulose is harvested from various plant materials and processed to produce biofuels [93].

14.3.4.1.2 Corn

Corn is also biggest biomass source for ethanol-based biofuel production in the United States and it is more sustainable source than petroleum. Corn is processed through a fermentation to produce ethanol and it blended with gasoline [94].

14.3.4.1.3 Soy

Soy is one of the biggest seed-based biomass source for biofuel production, and it is processed by transesterification to produce jet fuel and biodiesel, but it is more expensive than other sources [95].

14.3.4.1.4 Sugar Cane

Sugar cane is also one of the important sources for bio-ethanol production. Nowadays, ethanol from sugarcane is used in most vehicles, and it is cheaper than corn [22].

14.3.4.1.5 Camelina and Jatropha

Camelina and Jatropha are flowering plants which serve as feedstock for biofuels, and these are found all over the world. These plants can grow in poor soil with more fertility [16].

14.3.4.1.6 Rapeseed and Canola Oil

Rapeseed and canola oil are vital sources for biodiesel production, and these are used in US and Canada. It is an easy and cheap way, but it needs more land to harvest [96].

14.3.4.1.7 Palm Oil

Palm oil is derived from palm fruit, it is efficient as biodiesel fuel, and it is mostly found in Indonesia and Malaysia [15].

14.3.4.1.8 Switchgrass

Switchgrass containing cellulose is processed to produce more ethanol through fermentation. Culture plants produce 15 tons of biomass per acre, which can produce 4,350 L of ethanol [97].

14.3.5 Animal Source

Raw materials like animal fat, pork lard, fish oil, beef tallow, and chicken fat is used as feed-stock for high-quality biodiesel production by various methods like blending, rendering, trans-esterification, micro-emulsions, and pyrolysis [98]. Non-edible animal fat is more beneficial than vegetable oil as a source due to their low cost and one-third of the useful oils are produced from animal fat in United States. For effective use of animal fat as biodiesel could be possible after blending with biodiesel or by using in warm climatic conditions because animal fats are solids at room temperature for being saturated nature. Biodiesel from animal fat and vegetable fat can be produced by biological or chemical methods. In biological methods, enzymes like lipase are utilized while in chemical method catalysts are used [99, 100]. Pyrolysis method is used to process animal wastes to produce biodiesel [101–103].

14.4 Possible Routes of Biofuel Production Through Synthetic Biology

Synthetic biology caters the opportunity to modify the gene and then its addition in the genome of targeted organisms, which provide the desire metabolites in targeted organism or alter the existent metabolic flux [104–106]. Moreover, synthetic biology is also used to study regulation of cell metabolism in the desired flow by controlling the pathway gene expression [105]. There are some valuable routes, which can be successfully applied for the production of biofuels.

14.4.1 Metabolic Engineering

Metabolic engineering enables to transform native pathways in to new pathways to produce various biofuel products such as *E. coli* is used to produce fatty acid methyl esters from fatty acids through transesterification to produce biodiesel and petroleum-derived products [107]. Alternatively, metabolic engineering also alter pathways which enable to produce different value-added products and molecules through cell factories [104, 105]. Metabolic engineering technology is also used to down regulate the mobilization of substrates in undesired pathways and simultaneously the committed steps can be upregulated [108]. Metabolic engineering technology has the potential to produce various products not only from microbes but also from plants, for example, there are possibilities of genetic improvement of jatropha for high quality biodiesel production [109]. Similarly, microalgae can be modified by metabolic engineering for biofuel production by enhancing the photosynthetic efficiency, biomass growth, oil percentage, stress tolerance, and oxidative stability [110].

14.4.2 Tissue Culture/Genetic Engineering

Tissue culture is an important technique to enhance the production of desired metabolites or products in minimum time and space [111, 112]. Genetic engineering approach is one of the potential ways to produce biofuel like bioethanol, biohydrogen, biodiesel, etc. [76, 91]. Moreover, it is also used to enhance oil content, improves stress resistance, and lowers the toxic molecules in a plant or microalgae. This technology is demonstrated in non-edible plants like *Jatropha* and *Camelina* to enhance the quality and quantity of oil [113]. Genetic engineering tools like RNAi technology is used to alter the expression of genes, which involves in the committed step of the desired pathway. Since last few years, researchers are focusing on synthetic biology to develop GMOs for production of renewable fuels. Genetic engineering is a successful technique to produce advanced biofuels with improved carbohydrate and lipid percentage through modified pathways in a host organism [82, 91].

14.4.3 CRISPR-Cas

Clustered regularly interspaced short palindromic repeats-CRISPR-associated protein (CRISPR/Cas9) is a tool for genome manipulation, has been utilized for genome editing in microbes, plants and animals [114]. Zinc-finger nucleases (ZFN) and transcription activator-like effectors (TALEs) are the other efficient genome editing tools but CRISPR/Cas9 is most efficient technique. CRISPR-Cas9 is the most widely used technique in various host systems including microbes, algae, monocot, and dicot plants [114, 115]. These systems are used to posses, unknown metabolic pathways, targeted genes, modify promoters for the overexpression/repression of genes to modify the host system to produce advanced biofuels as desired products. Recent metabolic engineering strategies with CRISPR-Cas were applied to produce sustainable biofuels from microalgae [116]. Species with gene editing improve the biofuel production than native species [114, 117]. This technology is applied in microalgae to improve the lipid content and enhance the conversion of carbon into lipids to produce renewable biofuels. Interestingly, more than 20 factors are identified which can control the lipid production. Out of them, knock out of 18 factors will almost double the lipid content in engineered microalgae without affecting the growth rate. Modified microalgae are able to convert 40–55% of carbon into lipids while normal algae able to convert only ~20% of carbon to lipids [118].

Recent trends are focused on the biofuels from modified metabolic pathways via CRISPR-Cas technology. Inserting the genes from one organism to another can improve the yield of fatty acids and quality of oils. Thus, various bacterial species are used as biofuel generators enable to produce straight long chain fatty acid ethanol [119]. Robust CRISPR-Cas technology with engineering tools was applied in various microbes to obtain desired traits like stress tolerance and improved yield of biofuels. The CRISPR-Cas system is already reported in yeast, which converts sugars into alcohols with high yield [120]. CRISPR with dCas9 has been established to knockout the gene or regulate the gene expression in yeast [114, 121]. This technique is also developed in non-conventional yeasts and enhances the production of muconic acid precursors for biofuel production by inserting six genes at respective targeted sites in its genome [122]. CRISPR technology is also used to either activates or deactivates the promoter gene for fatty acid synthesis for renewable biofuel production [151].

14.5 Synthetic Biology and Its Application for Biofuels Production

Synthetic biology is a combination of biology and engineering discipline, which includes several branches of science such as metabolic engineering, genetic engineering, biotechnology, systems biology, computational biology, etc. [105, 106]. Synthetic biology is used to alter metabolic pathways via genetic engineering techniques, which facilitates the reprogramming of metabolic pathways in the organism for production of desirable products [104, 117]. Nowadays, system biology approach provides genomic, transcriptomics, and complete pathway information of an organism, which will provide the desired pathway or genes or regulatory elements information [4, 105, 123–125] and could be used for metabolic engineering and genetic engineering to produce enhanced biofuels. Synthetic biology overcomes the drawbacks of traditional methods; hence, a large amount of biofuels can be produced by using genetically engineered organisms [126]. Various microorganisms are used to produce huge amount of biofuels such as *E. coli*, yeast, photosynthetic bacteria and gram negative and gram positive bacteria, etc., [70, 127]. With the aid of synthetic biology, mevalonate pathway gets altered and redesign in *E. coli* to produce terpenoid in required quantity [105]. Synthetic biology is applied successfully to increase the resistance against adverse environmental conditions and to enhance the yield of metabolites, so it can be potentially applied in biofuel industry [117].

Synthetic biology is a promising tool for creating genetic modifications in organisms for the efficient production of biofuels. Currently, yeast and *E. coli* act as model organisms to produce different types of biofuels. Gene addition and gene deletion approach is applied for engineering of microbes as through engineered yeast different sugars were produced by efficient xylose fermentation [128]. Strategies of synthetic biology can be applied on waste material for production of biofuels by applying engineered microbes [4, 105].

Synthetic biology approach is successfully applied in *Zymomonas mobilis*, algae, yeast, *E. coli*, and nonconventional yeast to improve alcohol production [129, 130]. Genetic engineering is used to increase the CO_2 fixation [131] and develop stress resistance in algae to produce enhanced biofuels [132].

14.5.1 Case Study 1: Production of Isobutanol by Engineered *Saccharomyces cerevisiae*

Isobutanol is a branched alcohol and it is considered as superior fuel than ethanol. Isobutanol is generally produced by the yeast cells in lower concentration so many attempts have been performed to increase the production of isobutanol by yeast cells. Isobutanol producing strains of *Saccharomyces cerevisiae* was produced by overexpressing the enzyme 2-keto acid decarboxylase (KDC) from *L. lactis* subsp. lactis KACC 13877 (Kivd) and valine biosynthetic enzymes [133]. Strain having *Kivd* gene produced approximately 93 mg L^{-1} of isobutanol from glucose as a carbon source which was fourfold in comparison to control strains and this production level was further extended up to 151.93 mg L^{-1} by cytosolic overexpression of other enzymes like acetolactate synthase, acetohydroxyacid reductoisomerase, and dihydroxyacid dehydratase.

Manufacturing the cell factories which can utilize pentose sugar like xylose will be more beneficial for production of biofuels because xylose being the main component of lignocellulosic biomass of plants. Hence, its fermentation can overcome the problem of sustainable management of lignocellulosic waste. Brat and Boles [134] demonstrated the conversion of xylose into isobutanol by overexpressing the xylose isomerase in combination with the xylokinase and transaldolase along with the overexpression of isobutanol production pathway.

14.5.2 Case Study 2: Generation of Biofuel From Ionic Liquid Pretreated Plant Biomass Using Engineered *E. coli*

Bokinsky and his coworkers developed a technique which may be fruitful in reduction of total time for production of biofuels from lignocellulosic biomass [135]. They modified the *E. coli* strain in such a way that it was able to digest the ionic liquid treated or untreated lignocellulosic biomass directly and produce hydrocarbons of fuel grade. These strains were genetically modified for expressing the cellulase, xylanase, beta glucosidase, and xylobiosidase enzymes, and for solubilization of lignocellulosic biomass, they were applied either separately or in combination of co-cultures. Pathways for production of alcohols and linear or branched hydrocarbons were integrated in these modified strains. A plasmid consisting of six genes of fatty acid synthesis pathway was introduced into *E. coli* MG1655 (deprived of acy-CoA dehydrogenase gene) and the resultant strain was able to produce up to 405 ± 27 mg L^{-1} fatty acid ethyl ester. Use of ionic liquid was successfully applied by other group of researchers for production of biofuel precursor molecules [136]. Pre-treatment of switchgrass biomass by using ionic liquid 1-ethyl-3-methylimidazolium acetate ([C_2C_1Im][OAc]) resulted to deconstruction of lignocellulosic components. This cellulosic biomass is further utilized by an [C_2C_1Im]-tolerant *E. coli* strain for the production of bio-jet fuel precursor D-limonene [136].

14.5.3 Case Study 3: CRISPRi-Mediated Metabolic Pathway Modulation for Isopentenol Production in *E. coli*

Repression of multiple genomic DNA loci by applying CRISPR interference (CRISPRi) technology leads to production of isopentenol in *E. coli* [137]. CRISPRi machinery functions via a target guide RNA (gRNA) and a deactivated Cas9 (dCas9) protein. Expression of genes related to competitive pathways was knocked down by gRNA of the target sites for increasing the availability of precursor for isopentenol pathway which was heterologously expressed in host.

14.6 Current Status of Biofuel

Renewable Fuels Standard (RFSII) estimated the production of renewable fuels up to 36 billion gallons by 2022. US government is focusing to reduce the dependence on foreign oil up to

30% by using agriculture feedstock to produce biofuels by 2030. US department has decided to produce biofuel from algae up to 5 billion gallons/year at reasonable cost by 2030 [138, 139]. Up to 90% of ethanol is produced from corn-starch and cane-sugar in USA and Brazil, respectively. India, Malaysia, China, and Indonesia successfully produce biodiesel from plant-based feedstocks. According to the RFSII, approximately, 16 billion gallons of lignocellulosic mass is required for production of 1 billion gallons biodiesel/year and till the year of 2022, the target is the production of 58% of advanced biofuels, which can reduce 50% emission of the greenhouse gases. The United States and Europe are the largest producers of biodiesel, 0.89 billion gallons and 6 billion gallons, respectively. The USA is focused to increase the production of renewable fuel includes increase the ethanol production from 4 billion gallons (2005) to 14 billion gallons (2014) and also increase the production of biodiesel from 112 million gallons (2005) to 1.3 billion gallons (2014) [140, 141].

Currently, 85% of fossil fuels are consumed in diverse sectors such as power industry and transportation but this dependency can be reduced up to 75% till the year 2040 because of emergence of non-fossil fuel production [142]. About 468 MW of electricity was generated from renewable fuel in 2016 and this may increase up to 1,000 MW by the year 2023. By the usage of biofuels, emission of CO_2 is stabilized gradually which was 2.2% during previous decade [143]. The European Union (EU) set the guidelines to maintain 10% of green biofuel in final transportation by 2020 and maintain the 10 GW of solar energy and 16 GW of wind energy by 2030 [141, 144]. The current research is focused on biomass-derived biofuels production through genetic engineering technology in the various host to increase the quantity and quality of the desired molecule. The US Department of Energy's Aquatic Species Program investigated and identified around 3,000 different microalgal species to produce biofuel through genetic engineering. Genetic modification of microalgae is a good alternative source for fossil fuels [78, 109]. The US Department of Energy has set the aim to decrease dependence on fossil petroleum-based fuels up to 30% by 2030 with Metabolic engineering of *Escherichia coli* to produce cellulosic ethanol [71, 117].

14.7 Future Aspects

World population is gradually increasing and reaches up to 1.7 billion by 2050. The usage of fuel also increases due to high population, so there is urgent need to pay attention toward renewable energy sources for biofuel production. Gradually, demand for biofuel will be increased from $82.7 billion to $185.3 billion by 2021 [143]. Therefore, researcher pays attention for renewable biofuel production through various methods like genetic engineering, metabolic engineering, and synthetic biology. Currently, research is focused on advanced biofuel production through fungal, algal, and plant species to produce bioethanol [145]. Use of high lipid accumulating algal species (*Euglena sanguine*) for the production of biodiesel is also eco-friendly [146].

Metabolic engineering is used to modify the pathways in microalgae and enable to produce long chain fatty acids, alcohols, and other polymers for biofuel production, which, ultimately, enhance the yield of precursors in the pathway of interest to produce a large number of biofuels in near future. Recent technology regarding biofuel production evaluates the limitation and strategies to fill the gap between first- and fourth-generation biofuels to meet the global demands. Metabolic engineering with synthetic biology should be focus to enhance metabolic flux and destroy the competing metabolic flux in the respected host system to increase the production of advanced biofuels for future aspects. For example, enhanced the expression of thioesterases gene and knockout the fadD and fadL genes in the fatty acid biosynthetic pathway is a successful story of accumulating high amount of fatty acids [147, 148]. Overexpression of aldehyde reductases gene enhances the production of n-butanol from butyraldehyde in a host system. Not only algal species but also cyanobacteria (enhance propane yield), yeast, and *E. coli* also serve as host system in biofuel production [149, 150].

Genetic engineering in microalgae offers high yield, rapid growth, and high lipid percentage to produce biofuels, value-added products, and other proteins. In future, expanded understanding about genetic engineering will lead to achieve novel renewable energy source and face the challenges regarding advanced biofuels. Nowadays, biofuels have great attention for the replacement of fossil fuels in near future. At present, new innovative methods are developed to produce advanced biofuels for future generations. Currently, few species are available on the platform with fully annotated genomes, so there is need to annotate more genome information with advance, concentrated on the reconstruction of pathways, alter the traits with novel functions and construct novel biological systems. With these new methods, we need to address some challenges for green biofuel production in the near future [68].

14.8 Conclusion

Advanced biofuel attracts great attention globally to replace fossil fuels so main aim of this chapter is to present the recent developments in genetic engineering, metabolic engineering, and synthetic biology regarding biofuel production to address the present problem. This study also discussed all generations of biofuels with examples and various sources of biofuel like bacterial, algal, fungal, plants, and animal fat. This chapter is focused on synthetic biology role in biofuel production, its current status, and its application in biofuel production. Synthetic biology in association with genetic engineering, metabolic engineering, and CRISPR genome editing technology offers possible routes to improve the yield of advanced biofuels to replace fossil fuel in near future. Research is continuously progressive to explore new opportunities for biofuel production to make them economically viable.

References

1. Buhroo, Z.I., Bhat, M.A., Ganai, N.A., Lone, J.A., Biofuels and their future perspectives. *J. Med. Plants*, 4, 6, 256–264, 2016.
2. Zahan, K. and Kano, M., Biodiesel production from palm oil, its by-products, and mill effluent: A review. *Energies*, 11, 8, 2132, 2018.
3. Singh, S.P. and Singh, D., Biodiesel production through the use of different sources and characterization of oils and their esters as the substitute of diesel: A review. *Renew. Sustain. Energy Rev.*, 14, 1, 200–216, 2010.
4. Narnoliya, L.K., Jadaun, J.S., Singh, S.P., Management of agro-industrial wastes with the aid of synthetic biology, in: *Book: Biosynthetic technology and environmental challenges*, pp. 11–28, Springer Nature Book, Singapore, 2017.

5. Yue, D., You, F., Snyder, S.W., Biomass-to-bioenergy and biofuel supply chain optimization: Overview, key issues and challenges. *Comput. Chem. Eng.*, 66, 36–56, 2014.
6. Marchal, V., Dellink, R., van Vuuren, D., Clapp, C., Château, J., Lanzi, E., van Vliet, J., OECD, environmental outlook to 2050: The consequences of inaction, *OECD*, 1, 1–353, 2012.
7. Bharathidasan, A.K., *Production of Biobutanol from inulin-rich biomass and industrial food processing wastes*, Doctoral dissertation, The Ohio State University, Ohio, 2013.
8. Aro, E.M., From first generation biofuels to advanced solar biofuels. *Ambio*, 45, 1, 24–31, 2016.
9. Mussatto, S.I., Dragone, G., Guimarães, P.M., Silva, J.P.A., Carneiro, L.M., Roberto, I.C., Teixeira, J.A., Technological trends, global market, and challenges of bio-ethanol production. *Biotechnol. Adv.*, 28, 6, 817–830, 2010.
10. Fortman, J.L., Chhabra, S., Mukhopadhyay, A., Chou, H., Lee, T.S., Steen, E., Keasling, J.D., Biofuel alternatives to ethanol: Pumping the microbial well. *Trends Biotechnol.*, 26, 7, 375–381, 2008.
11. Oliver, N.J., Rabinovitch-Deere, C.A., Carroll, A.L., Nozzi, N.E., Case, A.E., Atsumi, S., Cyanobacterial metabolic engineering for biofuel and chemical production. *Curr. Opin. Chem. Biol.*, 35, 43–50, 2016.
12. Tharakan, P., Baker, B., Montecalvo, R., Biofuels in Asia: An Analysis of Sustainability Options. *RDM f. A. US Agency Int. Dev.*, 151, 1–27, 2009.
13. Organisation for Economic Co-operation and Development, *Technology roadmap: Biofuels for transport*, OECD Publishing, Paris, 2011.
14. Gehlhar, M., *Effects of Increased Biofuels on the US Economy in 2022*, DIANE Publishing, United States, 2011.
15. Chew, T.L. and Bhatia, S., Catalytic processes towards the production of biofuels in a palm oil and oil palm biomass-based biorefinery. *Bioresour. Technol.*, 99, 17, 7911–7922, 2008.
16. Patil, P.D., Gude, V.G., Deng, S., Biodiesel production from *Jatropha curcas*, waste cooking, and *Camelina sativa* oils. *Ind. Eng. Chem. Res.*, 48, 24, 10850–10856, 2009.
17. Alves, R.M.D.B., do Nascimento, C.A.O., Biscaia, E.C., *10th International Symposium on Process Systems Engineering-PSE2009*, Elsevier Science, Brazil, 2009.
18. Nigam, P.S. and Singh, A., Production of liquid biofuels from renewable resources. *Progr. Energy Combustion Sci.*, 37, 1, 52–68, 2011.
19. Demirbas, A., Competitive liquid biofuels from biomass. *Appl. Energy*, 88, 1, 17–28, 2011.
20. Van Gerpen, J.H., Peterson, C.L., Goering, C.E., *Biodiesel: An alternative fuel for compression ignition engines*, American Society of Agricultural and Biological Engineers, Kentucky, USA, 2007.
21. Facts, G., Liquid Biofuels for Transport Prospects, risks and opportunities. *Greenfacts*, 1, 1–70 2016.
22. Chandel, A.K., da Silva, S.S., Carvalho, W., Singh, O.V., Sugarcane bagasse and leaves: Foreseeable biomass of biofuel and bio-products. *J. Chem. Technol. Biotechnol.*, 87, 1, 11–20, 2012.
23. Shamsul, N.S., Kamarudin, S.K., Rahman, N.A., Kofli, N.T., An overview on the production of bio-methanol as potential renewable energy. *Renew. Sustain. Energy Rev.*, 33, 578–588, 2014.
24. Dismukes, G.C., Carrieri, D., Bennette, N., Ananyev, G.M., Posewitz, M.C., Aquatic phototrophs: Efficient alternatives to land-based crops for biofuels. *Curr. Opin. Biotechnol.*, 19, 3, 235–240, 2008.
25. Kagan, J., Third and Fourth Generation Biofuels: Technologies. *Markets Econom. Through*, 6, 2010, 2015.
26. Lee, S.Y. and Papoutsakis, E.T. (Eds.), *Metabolic engineering*, vol. 24, CRC Press, Florida, United States, 1999.
27. Lü, J., Sheahan, C., Fu, P., Metabolic engineering of algae for fourth generation biofuels production. *Energy Environ. Sci.*, 4, 7, 2451–2466, 2011.

28. Rasool, U. and Hemalatha, S., A review on bioenergy and biofuels: Sources and their production. *Braz. J. Biol. Sci.*, 3, 5, 3–22, 2016.
29. Sameera, V., Sameera, C., Ravi Teja, Y., Current strategies involved in biofuel production from plants and algae. *J. Microb. Biochem. Technol.*, 1, 002, 2011.
30. Behera, S., Singh, R., Arora, R., Sharma, N.K., Shukla, M., Kumar, S., Scope of algae as third generation biofuels. *Front. Bioeng. Biotechnol.*, 2, 90, 2015.
31. Klingenfeld, D. and Kennedy, H., *Corn stover as a bioenergy feedstock: Identifying and overcoming barriers for corn stover harvest, storage, and transport*, John F. Kennedy School of Government, Massachusetts, United States, 2008.
32. Tao, L., Aden, A., Elander, R.T., Pallapolu, V.R., Lee, Y.Y., Garlock, R.J., Ladisch, M.R., Process and technoeconomic analysis of leading pre-treatment technologies for lignocellulosic ethanol production using switchgrass. *Bioresour. Technol.*, 102, 24, 11105–11114, 2011.
33. Alshammari, A.M., Adnan, F.M., Mustafa, H., Hammad, N., Bioethanol fuel production from rotten banana as an environmental waste management and sustainable energy. *Afr. J. Microbiol. Res.*, 5, 6, 586–598, 2011.
34. Weijde, T.V.D., Alvim Kamei, C.L., Torres, A.F., Vermerris, W., Dolstra, O., Visser, R.G.F., Trindade, L.M., The potential of C4 grasses for cellulosic biofuel production. *Front. Pant Sci.*, 4, 107, 2013.
35. McKendry, P., Energy production from biomass (part 1): Overview of biomass. *Bioresour. Technol.*, 83, 1, 37–46, 2002.
36. Arthe, R., Rajesh, R., Rajesh, E.M., Rajendran, R., Jeyachandran, S., Production of bio-ethanol from cellulosic cotton waste through microbial extracellular enzymatic hydrolysis and fermentation. *Electron. J. Environ. Agricult. Food Chem.*, 7, 6, 2948–2958, 2008.
37. Elbehri, A., Segerstedt, A., Liu, P., *Biofuels and the sustainability challenge: A global assessment of sustainability issues, trends and policies for biofuels and related feedstocks*, Food and Agriculture Organization of the United Nations (FAO), Rome, 2013.
38. Robson, P., Jensen, E., Hawkins, S., White, S.R., Kenobi, K., Clifton-Brown, J., Farrar, K., Accelerating the domestication of a bioenergy crop: identifying and modelling morphological targets for sustainable yield increase in Miscanthus. *J. Exp. Botany*, 64, 14, 4143–4155, 2013.
39. Rosillo-Calle, F., Pelkmans, L., Walter, A., A global overview of vegetable oils, with reference to biodiesel. *A report for the IEA Bioenergy Task*, 40, 1–89, 2009.
40. Shikha, K. and Chauhan, Y.R., Biodiesel production from non edible-oils: A review. *J. Chem. Pharm. Res.*, 4, 9, 4219–4230, 2012.
41. Liaquat, A.M., Masjuki, H.H., Kalam, M.A., Varman, M., Hazrat, M.A., Shahabuddin, M., Mofijur, M., Application of blend fuels in a diesel engine. *Energy Proc.*, 14, 1124–1133, 2012.
42. Ahmad, M., Teong, L.K., Sultana, S., Zafar, M., Biodiesel production from non-food crops: A step towards self reliance in energy, in: *Materials and processes for energy: Communicating current research and technological developments*, pp. 239–243, Formatex Research Center, Badajoz, Spain, 2013.
43. Ahmad, M., Khan, M.A., Zafar, M., Sultana, S., Biodiesel from non edible oil seeds: A renewable source of bioenergy, in: *Economic effects of biofuel production*, InTech, London, UK, 2011.
44. Padhi, S.K. and Singh, R.K., Non-edible oils as the potential source for the production of biodiesel in India: a review. *J. Chem. Pharm. Res.*, 3, 2, 39–49, 2011.
45. Atabani, A.E., Silitonga, A.S., Badruddin, I.A., Mahlia, T.M.I., Masjuki, H.H., Mekhilef, S., A comprehensive review on biodiesel as an alternative energy resource and its characteristics. *Renew. Sustain. Energy Rev.*, 16, 4, 2070–2093, 2012.
46. Banković-Ilić, I.B., Stojković, I.J., Stamenković, O.S., Veljkovic, V.B., Hung, Y.T., Waste animal fats as feedstocks for biodiesel production. *Renew. Sustain. Energy Rev.*, 32, 238–254, 2014.

47. Zhang, B., Xiu, S., Shahbazi, A., Aquatic plants: Is it a viable source for biofuel production?, *Advances in Energy Research.*, 11, 203–216, 2012.
48. Suda, K., Shahbazi, A., Li, Y., The feasibility of using cattails from constructed wetlands to produce bioethanol, in: *Proceedings of the 2007 National Conference on Environmental Science and Technology*, pp. 9–15, Springer, New York, NY, 2009.
49. Xiu, S.N., Shahbazi, A., Croonenberghs, J., Wang, L.J., Oil production from duckweed by thermochemical liquefaction. *Energy Sourc Part A: Recover. Util. Environ. Eff.*, 32, 14, 1293–1300, 2010.
50. Mishima, D., Tateda, M., Ike, M., Fujita, M., Comparative study on chemical pre-treatments to accelerate enzymatic hydrolysis of aquatic macrophyte biomass used in water purification processes. *Bioresour. Technol.*, 97, 16, 2166–2172, 2006.
51. Ziska, L.H., Runion, G.B., Tomecek, M., Prior, S.A., Torbet, H.A., Sicher, R., An evaluation of cassava, sweet potato and field corn as potential carbohydrate sources for bioethanol production in Alabama and Maryland. *Biomass Bioenergy*, 33, 11, 1503–1508, 2009.
52. Benerji, D.S.N., Ayyanna, C., Rajini, K., Rao, B.S., Banerjee, D.R.N., Rani, K.S., Rajkumar, G., Studies on physico-chemical and nutritional parameters for the production of ethanol from mahua flower (Madhuca indica) using *Saccharomyces cerevisiae*-3090 Through Submerged Fermentation (smf). *J. Microb. Biochem. Technol.*, 2, 2, 046–050, 2010.
53. Simacek, P., Kubička, D., Šebor, G., Pospíšil, M., Hydroprocessed rapeseed oil as a source of hydrocarbon-based biodiesel. *Fuel*, 88, 3, 456–460, 2009.
54. Fish, W.W., Bruton, B.D., Russo, V.M., Watermelon juice: A promising feedstock supplement, diluent, and nitrogen supplement for ethanol biofuel production. *Biotechnol. Biofuels*, 2, 1, 18, 2009.
55. Iyovo, G.D., Du, G., Chen, J., Poultry manure digestate enhancement of *Chlorella Vulgaris* biomass under mixotrophic condition for biofuel production. *J. Microb. Biochem. Technol.*, 2, 051–057, 2010.
56. Rismani-Yazdi, H., Haznedaroglu, B.Z., Bibby, K., Peccia, J., Transcriptome sequencing and annotation of the microalgae *Dunaliella tertiolecta*: Pathway description and gene discovery for production of next-generation biofuels. *BMC Genom.*, 12, 1, 148, 2011.
57. Nautiyal, P., Subramanian, K.A., Dastidar, M.G., Kinetic and thermodynamic studies on biodiesel production from *Spirulinaplatensis* algae biomass using single stage extraction-transesterification process. *Fuel*, 135, 228–234, 2014.
58. Susilaningsih, D., Djohan, A.C., Widyaningrum, D.N., Anam, K., Biodiesel from indigenous Indonesian marine microalgae *Nannochloropsis* sp. *J. Biotechnol. Res. Trop. Reg.*, 2, 1–4, 2009.
59. Choi, W.Y., Kim, G.V., Lee, S.Y., Lee, H.Y., Biodiesel production from Scenedesmus sp. through optimized *in situ* acidic transesterification process. *Chem. Biochem. Eng. Q.*, 28, 3, 367–374, 2014.
60. Muthukumar, A., Elayaraja, S., Ajithkumar, T.T., Kumaresan, S., Balasubramanian, T., Biodiesel production from marine microalgae *Chlorella marina* and *Nannochloropsis salina*. *J. Petrol. Technol. Alternat. Fuels*, 3, 5, 58–62, 2012.
61. Kim, H.M., Oh, C.H., Bae, H.J., Comparison of red microalgae *Porphyridium cruentum* culture conditions for bioethanol production. *Bioresour. Technol.*, 233, 44–50, 2017.
62. Reyimu, Z. and Ízšimen, D., Batch cultivation of marine microalgae *Nannochloropsis oculata* and *Tetraselmis suecica* in treated municipal wastewater toward bioethanol production. *J. Cleaner Prod.*, 150, 40–46, 2017.
63. Rizza, L.S., Smachetti, M.E.S., Do Nascimento, M., Salerno, G.L., Curatti, L., Bioprospecting for native microalgae as an alternative source of sugars for the production of bioethanol. *Algal Res.*, 22, 140–147, 2017.

64. Ghadge, S.V. and Raheman, H., Biodiesel production from mahua (*Madhuca indica*) oil having high free fatty acids. *Biomass Bioenergy*, 28, 6, 601–605, 2005.
65. Poojary, S. and Rao, C.V., Process optimisation of pilot scale biodiesel production from pongamia and waste cooking oil feedstocK. *J. Eng. Sci. Technol.*, 13, 9, 2670–2684, 2018.
66. Awolu, O.O. and Layokun, S.K., Optimization of two-step transesterification production of biodiesel from neem (*Azadirachta indica*) oil. *Int. J. Energy Environm. Eng.*, 4, 1, 39, 2013.
67. Martins, M.I., Piresb, R.F., Alvesb, M.J., Horib, C.E., Reisb, M.H., Cardosob, V.L., Transesterification of soybean oil for biodiesel production using hydrotalcite as basic catalyst. *Chem. Eng.*, 32, 817–822, 2013.
68. Peralta-Yahya, P.P., Zhang, F., Del Cardayre, S.B., Keasling, J.D., Microbial engineering for the production of advanced biofuels. *Nature*, 488, 7411, 320, 2012.
69. Ljungdahl, L.G., A life with acetogens, thermophiles, and cellulolytic anaerobes. *Annu. Rev. Microbiol.*, 63, 1–25, 2009.
70. Rabinovitch-Deere, C.A., Oliver, J.W., Rodriguez, G.M., Atsumi, S., Synthetic biology and metabolic engineering approaches to produce biofuels. *Chem. Rev.*, 113, 7, 4611–4632, 2013.
71. Xu, P. and Koffas, M.A., Metabolic engineering of *Escherichia coli* for biofuel production. *Biofuels*, 1, 3, 493–504, 2010.
72. Niederholtmeyer, H., Wolfstädter, B.T., Savage, D.F., Silver, P.A., Way, J.C., Engineering cyanobacteria to synthesize and export hydrophilic products. *Appl. Environ. Microbiol.*, 76, 11, 3462–3466, 2010.
73. Ducat, D.C., Way, J.C., Silver, P.A., Engineering cyanobacteria to generate high-value products. *Trends Biotechnol.*, 29, 2, 95–103, 2011.
74. Formighieri, C., Franck, F., Bassi, R., Regulation of the pigment optical density of an algal cell: Filling the gap between photosynthetic productivity in the laboratory and in mass culture. *J. Biotechnol.*, 162, 1, 115–123, 2012.
75. Melis, A., Solar energy conversion efficiencies in photosynthesis: minimizing the chlorophyll antennae to maximize efficiency. *Plant Sci.*, 177, 4, 272–280, 2009.
76. Nazari, F. and Raheb, J., Genetic engineering of microalgae for enhanced biodiesel production suitable fuel replacement of fossil fuel as a novel energy source. *Am. J. Life Sci.*, 3, 32–41, 2015.
77. Hannon, M., Gimpel, J., Tran, M., Rasala, B., Mayfield, S., Biofuels from algae: Challenges and potential. *Biofuels*, 1, 5, 763–784, 2010.
78. Shuba, E.S. and Kifle, D., Microalgae to biofuels: 'Promising' alternative and renewable energy, review. *Renew. Sustain. Energy Rev.*, 81, 743–755, 2018.
79. Khan, M.I., Shin, J.H., Kim, J.D., The promising future of microalgae: current status, challenges, and optimization of a sustainable and renewable industry for biofuels, feed, and other products. *Microb. Cell Factories*, 17, 1, 36, 2018.
80. Jeon, S., Jeong, B.R., Chang, Y.K., Chemicals and fuels from microalgae, in: *Consequences of microbial interactions with hydrocarbons, oils, and lipids: production of fuels and chemicals*, pp. 1–21, 2017.
81. Ahmad, I., Sharma, A.K., Daniell, H., Kumar, S., Altered lipid composition and enhanced lipid production in green microalga by introduction of brassica diacylglycerol acyltransferase 2. *Plant Biotechnol. J.*, 13, 4, 540–550, 2015.
82. Radakovits, R., Jinkerson, R.E., Darzins, A., Posewitz, M.C., Genetic engineering of algae for enhanced biofuel production. *Eukaryotic Cell*, 9, 4, 486–501, 2010.
83. Ratledge, C. and Wynn, J.P., The biochemistry and molecular biology of lipid accumulation in oleaginous microorganisms. *Adv. Appl. Microbiol.*, 51, 1–52, 2002.
84. Devi, N.M. and Velayutham, P., Biodiesel production from fungi. *IJONS*, 1, 275–281, 2011.
85. Lin, Y. and Tanaka, S., Ethanol fermentation from biomass resources: Current state and prospects. *Appl. Microbiol. Biotechnol.*, 69, 6, 627–642, 2006.

86. Rao, S.P., Rao, S.S., Seetharama, N., Umakath, A.V., Reddy, P.S., Reddy, B.V.S., Gowda, C.L.L., *Sweet sorghum for biofuel and strategies for its improvement*, International Crops Research Institute for the Semi-Arid Tropics, Andhra Pradesh, India, 2009.
87. Ntaikou, I., Gavala, H.N., Kornaros, M., Lyberatos, G., Hydrogen production from sugars and sweet sorghum biomass using *Ruminococcus albus*. *Int. J. Hydrogen Energy*, 33, 4, 1153–1163, 2008.
88. Hossain, A.S., Salleh, A., Boyce, A.N., Chowdhury, P., Naqiuddin, M., Biodiesel fuel production from algae as renewable energy. *Am. J. Biochem. Biotechnol.*, 4, 3, 250–254, 2008.
89. Alaba, P.A., Sani, Y.M., Daud, W.M.A.W., Efficient biodiesel production *via* solid superacid catalysis: a critical review on recent breakthrough. *RSC Adv.*, 6, 82, 78351–78368, 2016.
90. delPilar Rodriguez, M., Brzezinski, R., Faucheux, N., Heitz, M., Enzymatic transesterification of lipids from microalgae into biodiesel: A review, *AIMS Energy*, 4, 6, 817–855, 2016.
91. Paudel, S. and Menze, M.A., Genetic engineering, a hope for sustainable biofuel production. *Int. J. Environ.*, 311, 311–323, 2014.
92. Wojcik, E.Z., Singleton, C., Chapman, L.N., Parker, D.A., Love, J., Plant Biomass as Biofuels. *ELS*, 1, 1–11, 2001.
93. Robertson, G.P., Hamilton, S.K., Barham, B.L., Dale, B.E., Izaurralde, R.C., Jackson, R.D., Landis, D.A., Swinton, S.M., Thelen, K.D., Tiedje, J.M., Cellulosic biofuel contributions to a sustainable energy future: Choices and outcomes. *Science*, 356, 6345, p.eaal2324, 2017.
94. Taheripour, F., Fiegel, J., Tyner, W.E., Development of corn stover biofuel: Impacts on corn and soybean markets and crop rotation. *Sustain. Agricult. Res.*, 5, 526-2016-37746, 1–9, 2015.
95. Lima, M., Skutsch, M., Costa, G., Deforestation and the social impacts of soy for biodiesel: Perspectives of farmers in the South Brazilian Amazon. *Ecol. Soc.*, 16, 4, 1–17, 2011.
96. Gunstone, F.D. (Ed.), *Rapeseed and canola oil: production, processing, properties and uses*, CRC Press, Florida, United States, 2004.
97. Calumpang, L.M., Biofuels in Asia: An Analysis of Sustainability Options. *Pol. Brief Ser.*, 1–2, 2009.
98. Cunha, A., Jr., Feddern, V., Marina, C., Higarashi, M.M., de Abreu, P.G., Coldebella, A., Synthesis and characterization of ethylic biodiesel from animal fat wastes. *Fuel*, 105, pp.228–234, 2013.
99. Yaakob, Z., Mohammad, M., Alherbawi, M., Alam, Z., Sopian, K., Overview of the production of biodiesel from waste cooking oil. *Renew. Sustain. Energy Rev.*, 18, 184–193, 2013.
100. Gog, A., Roman, M., Toşa, M., Paizs, C., Irimie, F.D., Biodiesel production using enzymatic transesterification–current state and perspectives. *Renew. Energy*, 39, 1, 10–16, 2012.
101. Guru, M., Artukoğlu, B.D., Keskin, A., Koca, A., Biodiesel production from waste animal fat and improvement of its characteristics by synthesized nickel and magnesium additive. *Energy Convers. Manage.*, 50, 3, 498–502, 2009.
102. Ito, T., Sakurai, Y., Kakuta, Y., Sugano, M., Hirano, K., Biodiesel production from waste animal fats using pyrolysis method. *Fuel Process. Technol.*, 94, 1, 47–52, 2012.
103. Mata, T.M., Martins, A.A., Caetano, N.S., Valorization of waste frying oils and animal fats for biodiesel production, in: *Advanced biofuels and bioproducts*, pp. 671–693, Springer, New York, NY, 2013.
104. Sangwan, N.S., Jadaun, J.S., Tripathi, S., Mishra, B., Narnoliya, L.K., Sangwan, R.S., Plant metabolic engineering, in: *Book: Omics Technologies and Bio-engineering*, pp. 143–175, Elsevier Book, Massachusetts, United States, 2018.
105. Narnoliya, L.K., Sangwan, R.S., Singh, S.P., Transcriptome mining and in silico structural and functional analysis of ascorbic acid and tartaric acid biosynthesis pathway enzymes in rose-scanted geranium. *Molec. Biol. Rep.*, 45, 3, 315–326, 2018.
106. Narnoliya, L.K. and Jadaun, J.S., Biotechnological avenues for fruit juices debittering, in: *Book: Green Bio-Processes*, pp. 119–149, Springer Nature Book, Singapore, 2019.

107. Wheeldon, I., Christopher, P., Blanch, H., Integration of heterogeneous and biochemical catalysis for production of fuels and chemicals from biomass. *Curr. Opin. Biotechnol.*, 45, 127–135, 2017.
108. Jullesson, D., David, F., Pfleger, B., Nielsen, J., Impact of synthetic biology and metabolic engineering on industrial production of fine chemicals. *Biotechnol. Adv.*, 33, 7, 1395–1402, 2015.
109. Kumar, N., Singh, A.S., Kumari, S., Reddy, M.P., Biotechnological approaches for the genetic improvement of *Jatropha curcas* L.: A biodiesel plant. *Ind Crops Prod*, 76, 817–828, 2015.
110. Dunlop, M.J., Engineering microbes for tolerance to next-generation biofuels. *Biotechnol. Biofuels*, 4, 1, 32, 2011.
111. Jadaun, J.S., Sangwan, N.S., Narnoliya, L.K., Tripathi, S., Sangwan, R.S., *Withania coagulans* tryptophan decarboxylase gene cloning, heterologous expression, and catalytic characteristics of the recombinant enzyme. *Protoplasma*, 254, 1, 181–192, 2017.
112. Jadaun, J.S., Sangwan, N.S., Narnoliya, L.K., Singh, N., Bansal, S., Mishra, B., Sangwan, R.S., Over expression of DXS gene enhances terpenoidal secondary metabolite accumulation in geranium and *Withania somnifera*: Active involvement of plastid isoprenogenic pathway in their biosynthesis. *Physiol. Plantar.*, 159, 4, 381–400, 2017.
113. SalehiJouzani, G., Sharafi, R., Soheilivand, S., Fueling the future; plant genetic engineering for sustainable biodiesel production. *Biofuel Res. J.*, 5, 3, 829–845, 2018.
114. Kumar, J., Narnoliya, L.K., Alok, A., A CRISPR technology and biomolecule production by synthetic biology approach, *Current Developments in Biotechnology and Bioengineering.*, 6, 143–161, 2018.
115. Nymark, M., Sharma, A.K., Sparstad, T., Bones, A.M., Winge, P., A CRISPR/Cas9 system adapted for gene editing in marine algae. *Sci. Rep.*, 6, 24951, 2016.
116. Banerjee, C., Dubey, K.K., Shukla, P., Metabolic engineering of microalgal based biofuel production: prospects and challenges. *Front. Microbiol.*, 7, 432, 2016.
117. Jagadevan, S., Banerjee, A., Banerjee, C., Guria, C., Tiwari, R., Baweja, M., Shukla, P., Recent developments in synthetic biology and metabolic engineering in microalgae towards biofuel production. *Biotechnol. Biofuels*, 11, 1, 185, 2018.
118. Banerjee, A., Banerjee, C., Negi, S., Chang, J.S., Shukla, P., Improvements in algal lipid production: A systems biology and gene editing approach. *Crit. Rev. Biotechnol.*, 38, 3, 369–385, 2018.
119. Jiang, W., Qiao, J.B., Bentley, G.J., Liu, D., Zhang, F., Modular pathway engineering for the microbial production of branched-chain fatty alcohols. *Biotechnol. Biofuels*, 10, 1, 244, 2017.
120. Papapetridis, I., Dijk, M., Dobbe, A.P., Metz, B., Pronk, J.T., Maris, A.J., Improving ethanol yield in acetate-reducing *Saccharomyces cerevisiae* by cofactor engineering of 6-phosphogluconate dehydrogenase and deletion of ALD6. *Microb. Cell Factories*, 15, 1, 67, 2016.
121. Bao, Z., Xiao, H., Liang, J., Zhang, L., Xiong, X., Sun, N., Zhao, H., Homology-integrated CRISPR-Cas (HI-CRISPR) system for one-step multigene disruption in *Saccharomyces cerevisiae. ACS Synth. Biol.*, 4, 5, 585–594, 2014.
122. Madhavan, A., Jose, A.A., Binod, P., Sindhu, R., Sukumaran, R.K., Pandey, A., Castro, G.E., Synthetic biology and metabolic engineering approaches and its impact on non-conventional yeast and biofuel production. *Front. Energy Res.*, 5, 8, 2017.
123. Narnoliya, L.K., Rajakani, R., Sangwan, N.S., Gupta, V., Sangwan, R.S., Comparative transcripts profiling of fruit mesocarp and endocarp relevant to secondary metabolism by suppression subtractive hybridization in *Azadirachta indica* (Neem). *Molec. Biol. Rep.*, 41, 5, 3147–3162, 2014.
124. Sangwan, R.S., Tripathi, S., Singh, J., Narnoliya, L.K., Sangwan, N.S., De novo sequencing and assembly of *Centella asiatica* leaf transcriptome for mapping of structural, functional and regulatory genes with special reference to secondary metabolism. *Gene*, 525, 1, 58–76, 2013.

125. Rajakani, R., Narnoliya, L., Sangwan, N.S., Sangwan, R.S., Gupta, V., Subtractive transcriptomes of fruit and leaf reveal differential representation of transcripts in *Azadirachta indica. Tree Genet. Genom.*, 10, 5, 1331–1351, 2014.
126. Rodriguez, M.G. and Atsumi, S., Synthetic biology approaches to produce C3-C6 alcohols from microorganisms. *Curr. Chem. Biol.*, 6, 1, 32–41, 2012.
127. Wackett, L.P., Engineering microbes to produce biofuels. *Curr. Opin. Biotechnol.*, 22, 3, 388–393, 2011.
128. Kim, S.R., Skerker, J.M., Kang, W., Lesmana, A., Wei, N., Arkin, A.P., Jin, Y.S., Rational and evolutionary engineering approaches uncover a small set of genetic changes efficient for rapid xylose fermentation in *Saccharomyces cerevisiae. PloS One*, 8, 2, e57048, 2013.
129. Chubukov, V., Mukhopadhyay, A., Petzold, C.J., Keasling, J.D., Martín, H.G., Synthetic and systems biology for microbial production of commodity chemicals. *NPJ Syst. Biol. Applic.*, 2, 16009, 2016.
130. Tsai, C.S., Kwak, S., Turner, T.L., Jin, Y.S., Yeast synthetic biology toolbox and applications for biofuel production. *FEMS Yeast Res.*, 15, 1, 1–15, 2015.
131. Yang, B., Liu, J., Ma, X., Guo, B., Liu, B., Wu, T., Chen, F., Genetic engineering of the Calvin cycle toward enhanced photosynthetic CO_2 fixation in microalgae. *Biotechnol. Biofuels*, 10, 1, 229, 2017.
132. Mukhopadhyay, A., Tolerance engineering in bacteria for the production of advanced biofuels and chemicals. *Trends Biotechnol.*, 23, 8, 498–508, 2015.
133. Lee, W.H., Seo, S.O., Bae, Y.H., Nan, H., Jin, Y.S., Seo, J.H., Isobutanol production in engineered *Saccharomyces cerevisiae* by overexpression of 2-ketoisovalerate decarboxylase and valine biosynthetic enzymes. *Bioprocess Biosyst. Eng.*, 35, 9, 1467–1475, 2012.
134. Brat, D. and Boles, E., Isobutanol production from D-xylose by recombinant *Saccharomyces cerevisiae. FEMS Yeast Res.*, 13, 2, pp.241–244, 2013.
135. Bokinsky, G., Peralta-Yahya, P.P., George, A., Holmes, B.M., Steen, E.J., Dietrich, J., Lee, T.S., Tullman-Ercek, D., Voigt, C.A., Simmons, B.A., Keasling, J.D., Synthesis of three advanced biofuels from ionic liquid-pretreated switchgrass using engineered *Escherichia coli. Proc. Natl. Acad. Sci.*, 108, 50, pp.19949–19954, 2011.
136. Frederix, M., Mingardon, F., Hu, M., Sun, N., Pray, T., Singh, S., Simmons, B.A., Keasling, J.D., Mukhopadhyay, A., Development of an *E. coli* strain for one-pot biofuel production from ionic liquid pretreated cellulose and switchgrass. *Green Chem.*, 18, 15, pp.4189–4197, 2016.
137. Tian, T., Kang, J.W., Kang, A., Lee, T.S., Redirecting metabolic flux *via* combinatorial multiplex CRISPRi-mediated repression for isopentenol production in *E. coli*. ACS synthetic biology, *ACS Synth. Biol.*, 8, 391–402, 2019.
138. Brownstein, A.M., *Renewable motor fuels: The past, the present and the uncertain future*, Butterworth-Heinemann, United Kingdom, 2014.
139. Herrera, S., Bonkers about biofuels. *Nat. Biotechnol.*, 24, 7, 755, 2006.
140. Balan, V., Current challenges in commercially producing biofuels from lignocellulosic biomass. *ISRN Biotechnol.*, 2014, 1–31, 2014.
141. Hood, E.E., Plant-based biofuels. *F1000Research*, 5, 1–9, 2016.
142. Gaurav, N., Sivasankari, S., Kiran, G.S., Ninawe, A., Selvin, J., Utilization of bioresources for sustainable biofuels: a review. *Renew. Sustain. Energy Rev.*, 73, 205–214, 2017.
143. REN21, R., *Global Status Report*, REN21 Secretariat, *In Tech. Rep.*, Paris, France, 2017.
144. Uyanik, S., Sucu, Y., Zaimoglu, Z., Potential of Biofuel Usage in Turkey's Energy Supply, in: *Biofuels-State of Development*, IntechOpen, 2018.
145. Narnoliya, L.K., Jadaun, J.S., Singh, S.P., Synthetic Biology Advances for Enrichment of Bioactive Molecules in Plants, in: *Recent Trends and Techniques in Plant Metabolic Engineering*, pp. 117–145, Springer, Singapore, 2018.

146. Kings, A.J., Raj, R.E., Miriam, L.M., Visvanathan, M.A., Cultivation, extraction and optimization of biodiesel production from potential microalgae *Euglena sanguinea* using eco-friendly natural catalyst. *Energy Convers. Manage.*, 141, 224–235, 2017.
147. Liu, H., Yu, C., Feng, D., Cheng, T., Meng, X., Liu, W., Xian, M., Production of extracellular fatty acid using engineered *Escherichia coli. Microb. Cell Factories*, 11, 1, 41, 2012.
148. Janben, H.J. and Steinbüchel, A., Fatty acid synthesis in *Escherichia coli* and its applications towards the production of fatty acid based biofuels. *Biotechnol. Biofuels*, 7, 1, 7, 2014.
149. Anfelt, J., *Metabolic engineering strategies to increase n-butanol production from cyanobacteria*, Doctoral dissertation, KTH Royal Institute of Technology, Stockholm, Sweden, 2016.
150. Pásztor, A., Advanced biofuel production: Engineering metabolic pathways for butanol and propane biosynthesis, University of Turku., 1–93, 2015.
151. Jeon, S., Lim, J.M., Lee, H.G., Shin, S.E., Kang, N.K., Park, Y.I., Chang, Y.K., Current status and perspectives of genome editing technology for microalgae. *Biotechnol. Biofuels*, 10, 1, 267, 2017.
152. Machado, I.M. and Atsumi, S., Cyanobacterial biofuel production. *J. Biotechnol.*, 162, 1, 50–56, 2012.
153. Pérez, Á., Casas, A., Fernández, C.M., Ramos, M.J., Rodríguez, L., Winterization of peanut biodiesel to improve the cold flow properties. *Bioresour. Technol.*, 101, 19, 7375–7381, 2010.
154. Şensöz, S. and Angın, D., Pyrolysis of safflower (Charthamus tinctorius L.) seed press cake: Part 1. The effects of pyrolysis parameters on the product yields. *Bioresour. Technol.*, 99, 13, 5492–5497, 2008.
155. do Nascimento Garritano, A., de Sá, L.R.V., Aguieiras, É.C.G., Freire, D.M.G., Ferreira-Leitão, V.S., Efficient biohydrogen production via dark fermentation from hydrolyzed palm oil mill effluent by non-commercial enzyme preparation. *Int. J. Hydrog. Energy.*, 42, 49, 29166–29174, 2017.
156. Milano, J., Ong, H.C., Masjuki, H.H., Chong, W.T., Lam, M.K., Loh, P.K., Vellayan, V., Microalgae biofuels as an alternative to fossil fuel for power generation. *Renew. Sustain. Energy Rev.*, 58, 180–197, 2016.

15

Biorefinery Approach for Bioethanol Production

Rituparna Saha[1,2], Debalina Bhattacharya[3] and Mainak Mukhopadhyay[1*]

[1]Department of Biotechnology, JIS University, Agarpara, West Bengal, India
[2]Department of Biochemistry, University of Calcutta, Ballygunge, West Bengal, India
[3]Department of Microbiology, Maulana Azad College, Kolkata, West Bengal, India

Abstract
The growth in population and condition of the economy has increased the dependency on fossil fuels to meet energy demands. Extensive use of fossil fuels has led to increased emissions of CO_2, rising atmospheric temperatures, deteriorating air quality, thus ultimately leading to climate change. This has led countries and governments to seek out more renewable and sustainable alternatives that can replace fossil fuels. Bioethanol is one of the most promising biofuels to have successfully become one of the best alternatives to fossil fuels. Ethanol can be produced both from agri crops like barley, oats, rice, wheat, etc., as well as from lignocellulosic biomass feedstock and other wastes. All the different kinds of biomass constitute sugars in high concentrations which when fermented produce bioethanol. The quantity and quality are determined through different types of methods like pre-treatment, hydrolysis, fermentation, and distillation.

This chapter will provide a comprehensive study of the biorefinery approach to bioethanol production. The consecutive steps leading from the type of biorefinery to further purification of ethanol will also be discussed.

Keywords: Bioethanol, biorefineries, pre-treatments, lignocellulosic wastes, feasibility

15.1 Introduction

The last century has witnessed continuous growth in the world population. Projections by the United Nations show that, with this rapid rise, the population will approximately reach up eleven billion inhabitants by 2050. This will require the use of basic resources like food, living space, and energy [1]. But overpopulation has caused increased pressure on the existing natural resources, such that it has caused an increase in energy demand, with a decline in food security, population capacity, and the ecosystem [2]. Over time, the whole population will have to deal with two major problems, which are energy supply and environmental resources, as well as the management and disposal of the wastes produced by humans.

**Corresponding author*: m.mukhopadhyay85@gmail.com

Arindam Kuila and Mainak Mukhopadhyay (eds.) Biorefinery Production Technologies for Chemicals and Energy, (313–334) © 2020 Scrivener Publishing LLC

Currently, the energy supply, as well as the world economy, is dependent on fossil fuel sources such as coal, oil and natural gas, etc. These are mostly used for providing electricity, fuels and other useful chemicals for use in day-to-day life [3]. However, this extensive use of fossil fuels and its combustion has led to increased emissions of greenhouse gases, leading to the accumulation of CO_2 in the atmosphere and causing global warming [4]. This, along with the continuous exhaustion of non-renewable fossil fuel sources, has led the world into looking for an alternative, less greenhouse emitting, and renewable sources that can meet the international energy requirements.

Using fossil fuels has become completely unreliable due to their unsustainability and relation, in effect, with air, water, and soil pollution [5]. Moreover, the dwindling of fossil fuels harms the economy. Therefore, to prevent this and limit the amount of greenhouse gas emissions, it is necessary to reduce the usage of fossil fuels and increase the consumption of renewable sources. One such form of a crucial source is plant biomass which can meet the requirement for energy, in comparison to fossil fuels. For a long time, several types of solid agri residues have been used as an energy source for heating, cooking, and also for power generation. In abundance and highly renewable, plant biomass is rich in carbohydrates, which can be converted, both biologically and chemically into biofuels and different bio-based products [6]. Biofuels are generally referred to as the liquid or gaseous fuels that have been synthesized from plant matter and other agri residues, for use in, mainly, the transportation sector. Due to this, biofuels have become an interesting alternative to fossil fuels and has mostly been used in the transportation industry [7]. Although these advantages have led to an increase in the production and usage of biofuels over the years, renewable energy accounts for only 10% of the total energy demands [8].

Shifting energy consumption from fossil fuels to bioenergy could lead to the growth of a sustainable form of economy. It is because the renewable plant feedstock used for these processes gives rise to a broad range of wastes and other by-products from which a range of biofuels and useful chemicals can be synthesized [9]. These agri residues undergo photosynthesis and have the ability to fix CO_2 in the atmosphere, thus reducing greenhouse emissions. Mostly, they are either used for animal feedstock, disposed, or incinerated in landfills. As a result, these wastes can be alternatively used for the manufacturing of different value-added products, proving to be advantageous for the economy and the environment, proving to be a highly inexpensive biorefinery [10]. The organizations International Energy Agency (IEA) and National Renewable Energy Laboratory (NREL) define biorefinery as "A facility that integrates biomass conversion processes and equipment to produce fuels, power, and chemicals from biomass". The development of bio-refineries is one of the most advantageous commitments for building a pro-active biofuel industry.

Studies have proved that bio-refineries are highly effective as they mostly use waste products and other non-utilizable plant matter and use the substrates sustainably for the production of chemicals and energy. Whereas, microorganisms are also utilized which uses these raw materials as substrates to produce different products with no wastes, thus making the process more sustainable and environmentally useful [11].

Mostly, biofuels generated from biomass are classified into two types: first-generation biofuels and second-generation biofuels. The first-generation biofuels are produced through conventional methods from food and feedstock, which includes starch, sugar, vegetable oils, wheat, corn and rapeseed, etc. These fuels are mainly bioethanol, biodiesel, biomethanol, and bioethers and starch-derived biogas. But, in addition to their simple conversion techniques and reduced impact on various environmental aspects, their productions compete with food crops in terms of fertile land and total yield, which are partly overcome through the production of second-generation biofuels. In comparison to first-generation biofuels, second-generation biofuels are derived from the whole plant, non-food crops, and lignocellulosic wastes produced from various sources like agriculture, forestry, and other industries. Fuels like bioethanol and Fisher-Tropsch (FT) diesel synthesized from lignocellulosics are more beneficial in terms of land use, environmental safety, and economic growth and could be further developed through improved biorefineries for better sustainability [12, 13].

15.2 Bioethanol

Bioethanol is one of the most commonly used renewable fuels and its production currently surpasses 4 billion gallons in the United States alone. Its most common raw materials are corn starch and sugarcane bagasse, but due to their limitations of use, production of this fuel has shifted to lignocellulosic biomass [14]. The fuel has generated a lot of interest in both the national and international levels, which is why many countries across the world have released policies to increase the utilization of bioethanol as a means to decrease the dependability on fossil fuels [15]. It has the capability to become a valuable replacement of gasoline due to the broad range of properties like higher octane number, wider flammability limits, higher heats of vaporization, and flame speeds. These provide ethanol with a shorter burn time and higher compression ratio, with higher oxygen content that reduces hydrocarbon and CO_2 emissibility, proving to be quite useful for its role in the transportation sector as combustion fuels [16–18].

The quality of bioethanol is dependent on the production processes involved. Firstly, the source from which it is produced, the types of substrates used, pre-treatments, hydrolysis, followed by fermentation, and finally, the distillation. Each of these production routes contains several subdivisions; the combinations of which are directly related to the conversion efficiency of substrates and production quality. Every method has its own advantage, disadvantage, and final outcome [19, 20]. Therefore, it is necessary for the governments looking to develop the bioethanol sector to select specific methods and viable technologies for implementation in the production of highly efficient and superior quality bioethanol for increasing renewable energy consumption.

15.3 Classification of Biorefineries

Recently, the utilization of renewable fuels and improvement in policies to do the same has led to an overwhelming effort in the development of new technologies, which has led to commercializing biofuels especially bioethanol [21]. The production of bioethanol from edible crops has turned out to less advantageous in terms of energy input and farmland requirement, proving to be costlier than utilizing lignocellulosic biomass as a substrate. Exploitation of lignocellulosic biomass and wastes, completely, proves to be more cost-effective because with spending equal amount of energy, the biomass not only generates bioethanol but also co-produces several types of by-products in the form of useful chemicals or thermochemical fuels to generate power and energy, thus reducing the application of fossil fuel sources [12, 22].

Brazil is one of the countries which produces the most bioethanol and utilizes the lignocellulosic biomass sugarcane bagasse as the substrate. Through ordinary methods, most of the biomass and its residue are used for the fermentation and distillation of ethanol, while the waste generated is thrown away [20]. To maximize the application of every aspect of biomass, biorefining technologies have been introduced and developed in order to make ethanol production more environmental and economic friendly.

The biorefinery process involves the complete conversion of biomass into fuels, chemicals, and other useful by-products and is a complete analog of a petroleum refinery [23]. The method and its outcome are dependent on the type of components that make up the biomass and be profitable by producing end-products according to market needs and availability of biomass [24]. Mostly lignocellulosic biomass is made up of three components: cellulose, hemicellulose, and the recalcitrant polymer lignin, making it difficult for separation and synthesis of any type of products [25]. This is why the production of second-generation bioethanol includes multi-step process—starting with pre-treatment to make the cellulose more accessible, hydrolysis with the help of bio or chemical catalysts for release of the monomeric sugars, followed by fermentation to convert the sugars into ethanol, and finally, distillation of the end product for recovery and use [26].

Despite the many advantages, biorefinery technologies have also created many challenges like optimization of the production of biofuel and other by-products, along with increasing the efficiency of the conversion processes [27]. Another challenge is the large amounts of wastewater generated in the process. An economical solution would be removing the residual organic matter from the effluents and use them for converting to biogas and use the remaining wastes as fertilizers for agricultural purposes [28].

The development of biorefinery technologies is still undergoing. Several studies involving pre-treatment and genetic engineering to increase the efficiency of production, and other downstream processing techniques have been made for making the technology more economically viable in comparison to fossil fuels.

15.3.1 Agricultural Biorefinery

The sources which make up agricultural biorefinery are mostly agri residues and wastes generated from food crops, which become a direct source for the production of bioethanol. Agri residues are available during harvesting or processing of food crops or both and may be solid, semi-solid, or liquid [29]. Two types of agri residues are produced: primary residues, that are collected after the harvesting of the main agro product from the crop and mainly constitutes plant parts like plant materials, leaves, stalks, etc., whereas secondary residues are components produced during the processing of the main agri product, for example, maize cob, rice husk, etc. [30].

The biomass waste generated from agricultural and forestry products and other sectors is increasing due to the growing population and rapid industrialization. Waste such as rice straw, rice husk, corn straw, wheat straw, and sugarcane bagasse is produced in tons, as does from the food industry and causes environmental damage when thrown away. Thus, using these wastes in biorefineries could solve the disposal problems effectively [31, 32].

Agri and forestry residues have shown immense potential in biorefineries due to their composition which is mostly rich in cellulose and hemicellulose, which, when fermented, produce ethanol. Food industry wastes like fruit peels and citrus fruit pomaces also form ideal substrates for fermentation, to synthesize bioethanol [33]. Investments have also been made for the extraction of ethanol directly from grains, like maize, sorghum, barley, cane sugar, sugar beet, peas, and also almond hulls. Corn-based biorefineries account for about 95% of the total bioethanol supply. The USA uses milling operation for the dry corn grinding process in order to produce ethanol. Different starch enriched grains like sorghum has been genetically modified for making it ideal toward enzymatic hydrolysis before fermentation, in order to reduce production costs. Crops like barley have become an important stepping stone toward the generation of cellulosic ethanol, which upon pre-treatment and enzymatic hydrolysis produces fermentable sugars with much more ease [34, 35]. Governments of certain countries like Canada and Mexico have taken the initiative to develop suitable technology and commercialize cellulosic ethanol, in order to fulfill their clean energy initiative and reduce greenhouse emissions. Cellulosic feedstocks are produced in abundance and measure about 500 million tons per year; therefore, research and development into the availability of biorefinery technologies commercially will help to establish the feasibility and business perspective into producing large-scale cellulosic ethanol [36, 37].

In the last few years, research has also identified some new renewable feedstocks which can be used for biorefinery purposes. Coffee pulp was recognized to be one of the underutilized waste, which has the potential to become a resourceful feedstock if the processing methods are made economically viable [38]. Another such example is the cassava food crop, which has high starch content and is one of the third largest sources of carbohydrate used widely as a food source for humans and also as animal feedstock. Waste residues generated from cassava-based industries are also a cause for environmental concern. The abundant availability and the low cost has made cassava a very important source of starch for biorefinery purposes to produce bioethanol and other value-added products, providing it with the ability to make a successful industry [39, 40].

Integrating biorefineries with agri crops, residues, and other wastes has their own challenges. To implement agricultural refineries in the market, more reliable and developed processing machinery is required. Research and developments in the future should also include improvement in cultivation by selecting genotypes that have maximum production, including the underutilized biomass [41, 42].

15.3.2 Lignocellulosic Biorefinery

Lignocellulosic biorefineries are economically and environmentally more useful than agriculture biorefineries, in terms of its use with lignocellulosic biomass that are abundant and are produced mainly as waste from various industries. This is why more efforts have been made in the last decades, with respect to investments and developments to technologies for the production of bioethanol and useful chemicals from lignocellulosic wastes [43]. Lignocellulosic biomass as residue or wastes is mostly produced from herbaceous plant species and woody plants and is one of the most easily available biomass on the planet. It mainly consists of cellulose, hemicellulose, and the recalcitrant polymer lignin, enmeshed with the carbohydrate polymers, and is mostly found in the cell walls of plants. Their individual percentage, properties, and appearance are dependent on the type of plant species [44, 45].

Lignocellulosic biomass is non-edible plants and arises in the form of agricultural residues, wood residues, energy crops, and residues from saw-mill and paper mill; and this has the capability to sustainably support the production of biofuels in the near future [46]. One of the main barriers in using this biomass comes in the form of the polymer lignin, which is difficult to digest by normal methods, and even after partial digestion due to chemical treatments gives rise to half-digested end products which are environmentally harmful [47]. To overcome this challenge, several biological pre-treatments have been developed with a wide range of microorganisms, which are capable of completely breakdown the complex polymer lignin [48].

One of the most common lignocellulosic biomass available for the production of second-generation ethanol is sugarcane bagasse. Countries like Brazil produce a surplus of this underutilized stock to be used in biorefineries for producing bioethanol and bioelectricity [49]. Other likely examples of lignocellulosics used in biorefineries are wheat straw, castor seeds, and also fiber sorghum, etc. [50–52]. Studies have also explored other potential lignocellulosic feedstocks, like the moringa tree and vetiver grass, which considering the lignocellulosic biomass content, makes them highly suitable for the bioenergy industry, especially bioethanol [53]. Among the agricultural lignocellulosics, the perennial species including the short rotation woody crops including the likes of poplar, black locust, and willow, have also proved to be an essential feedstock for the production of bioethanol in integrated lignocellulosic biorefineries [54]. These short rotation and fast-growing sources of biomass are rich in carbohydrate polymers and therefore provides a higher yield of bioethanol, per hectare of its cultivation [55].

Lignocellulosic biorefineries are more advantageous than agricultural biorefineries, in terms of availability of biomass, competition with food crops with respect to utilization and land requirement, and the amount of bioethanol production [56, 57]. Although the quantity in which lignocellulosic biomass is available is a lot more in comparison with the agricultural residues and wastes, still more development is required in reducing the overall cost of the conversion technologies for upgrading and increasing the efficiency of lignocellulosic biorefinery [58].

15.4 Types of Pre-Treatments

The abundance, sustainability, and cost-effectiveness of lignocellulosic biomass have made it a highly valuable source for the production of bioethanol through biorefineries. But, the recalcitrant nature of the polymer lignin is what makes a hindrance in the utilization and application for commercial purposes [59]. Pre-treatments are methods applied for removing the biological barrier of polymer lignin to ease the conversion process and are mainly classified into physical, chemical, mechanical, physicochemical, and biological methods of pre-treatment [60]. Though the feasibility of any pre-treatment process is biomass dependent, still characteristics like high saccharification efficiency, minimum production of waste products, minimal degradation of fermentable carbohydrates and lignin, and destruction of potential inhibitors [61].

15.4.1 Physical Pre-Treatments

Effective physical pre-treatments mostly include mechanical splintered treatment, milling, microwave treatment, ultrasound treatment, pyrolysis, and using a pulsed electric field. Although these methods are relatively simple and cause less environmental pollution, they require the usage of high energy and power, thus effectively increasing the production cost [62].

The method of mechanical splintering includes dry crushing, wet crushing, mill grinding, and compression and is used to make the subsequent processes easier for production. However, this pre-treatment process fails to remove lignin and hemicelluloses completely, even when high energy is used [63]. So, many times researchers have tried a more synergistic approach, i.e., a combination of pre-treatments together to achieve much more desired results [62].

The milling process or mechanical grinding mainly reduces the crystallinity of cellulose, by chipping, grinding, and milling [62]. This results in reduced biomass size and particle size which further aides in enzymatic hydrolysis. The carbohydrate yields especially of glucose and xylose, reduction percentage in biomass, followed by effective hydrolysis, all depends on the type and extent to which the milling process was attempted. One advantage of this pre-treatment process is the zero production of toxic compounds like hydroxymethyl furfuraldehyde (HMF) and levulinic acid [64].

Pre-treatment with microwave irradiation is a more acceptable approach due to its easy operation, requires low energy, reaches a high temperature under a short duration of time, generates a low number of inhibitors, and heavily degrades the lignocellulosic structure. A combination of mild-alkali treatment and microwave irradiation has proven to be more useful in getting a yield close to 70%–90% sugars. Correlation also exists between the percentage of lignin present in the biomass and the type of chemical pre-treatment done either before or along with microwave irradiation [65].

Ultrasound pre-treatment technique is relatively new as compared to other physical pre-treatment techniques and has resulted to be a more feasible option. The ultrasound waves alter the morphology of the lignocellulosic biomass by producing a combinatorial physical and chemical effect [66]. The treatment increases the subsequent accessibility of the cellulose-degrading enzymes to produce reducing sugars, by rupturing the cellulose and hemicellulose fractions through the formation of small cavitation bubbles, depending upon the duration and frequency of treatment. Factors like the characteristics of the biomass, configuration of the reactor, and the temperature under which the pre-treatment is carried out play a role in the sugar yield as well as the delignification percentage [67].

The pyrolysis pre-treatment process takes place at a high temperature of 300°C, which swiftly decomposes the cellulose and results in the release of gaseous products and other unwanted residues. The rate of decomposition is directly proportional to the rise of temperature and additives like metal ions and the presence of oxygen, which greatly increases the reaction rate and the conversion process [62, 68].

Another form of physical pre-treatment process is the pulsed-electric field (PEF) pre-treatment, under which the biomass gets exposed to high voltage pulses for short durations. The PEF creates pores in the membrane, which enables the entry of molecules that exposes the cellulose underneath and breaks down to its constituent sugars. One of the major advantages of using this treatment method is its low energy requirement and the ability of the process to take place in any conditions [69, 70].

15.4.2 Chemical Pre-Treatments

The various methods of chemical pre-treatments are more environmentally friendly and cost-effective in comparison to physical treatments, due to the minimal requirement of heat and energy. The chemicals involved removes lignin and hemicellulose and increases the biodegradability of cellulose by decreasing the crystallinity of the cellulosic portions and their degree of polymerization. This form of pre-treatment is mainly utilized in industries where lignocellulosic biomass is a preliminary source and its digestibility in the form of delignification is of utmost importance [71, 72]. Research has shown that certain chemicals, which react under ambient temperature and pressure, have a significant effect of delignification on the biomass and produce minimal to zero production of toxic residues, as a result of treatment [73].

Some common chemical agents used for this purpose are strong alkali's, acids including organic acids, various salts, and oxidizing agents, which are used to degrade the lignin and hemicelluloses, in order to make the digestion of cellulose more effective and feasible [74].

One of the most extensively studied chemical pre-treatment method is the alkaline pre-treatment method employing compounds like sodium hydroxide, potassium hydroxide, calcium hydroxide, and hydrogen peroxide, etc. [72] These alkaline solutions mainly disposes off the acetyl groups and other group substitutions to increase the availability of hemicellulose and cellulose to enzymatic hydrolysis and fermentation [73]. But, the accumulation of salts onto the biomass and the cellulosic fibers during digestion causes a major hindrance for further downstream processes.

Pre-treatment with strong acids is another very powerful method for obtaining a suitable structure of lignocellulosic biomass, for obtaining bioethanol. The most common steps include acidic hydrolysis followed by saccharification with the enzyme cellulase. Using strong acids like hydrochloric acid, sulfuric acid, and nitric acid requires high temperature and pressure, where the process of using sulfuric acid for treatment purposes has been studied in detail. Dilute acid treatment also provides better hydrolysis of biomass, but one of the greatest disadvantages is the production of toxic, inhibitory compounds, whose removal increases the overall cost of the treatment process [74–76].

Pre-treatment in the form of wet oxidation takes place at a temperature of 195°C and takes place after the drying and milling of the biomass. This is followed by the addition of water and Na_2CO_3 which reduces the byproducts produced during fermentation. This is followed by the oxidation of the dissolved compounds in the water, by passing air through it. This method gives an added advantage in that after lignin is removed and hemicellulose gets solubilized, the lignocellulosic components get fractionated. Studies of the wet oxidation method in combination with other chemical and/or physicochemical methods have produced results with increasing sugar yields [75, 76].

Other forms of crucial but expensive chemical pre-treatments involve the treatment of the biomass with ionic liquid solvents, which tends to precipitate the biomass by solubilizing lignin and cellulose. Another method involves the treatment of biomass with oxidizing agents like ozone, hydrogen peroxide, or oxygen and is limited to oxidation of lignin only. Treatment with organic solvents like ethanol, acetone, methanol, and ethylene glycol is also used for removal of lignin and hydrolysis of hemicellulose, in order to make the biomass more prone to enzymatic hydrolysis [76, 77].

Chemical pre-treatment methods are mostly performed to achieve higher sugar yields, by focusing on lignin removal, dissolving hemicellulose and subsequently enhancing the effect of enzymatic hydrolysis on the biomass to aid in the production of bioethanol [78].

15.4.3 Physico-Chemical Pre-Treatments

Physico-chemical pre-treatments are mostly a combination or mixture of physical and chemical methods [79]. Through this type of treatment, the lignin and hemicellulose are removed, and cellulose gets disrupted, depending on conditions like temperature and pressure, both in the presence or absence of a chemical. Some common methods are steam explosion, liquid hot-water treatment, ammonia fiber explosion, and carbon dioxide explosion [80].

Steam explosion is one of the most commonly used methods for the treatment of lignocellulosics and agricultural residues. In this process, the raw material gets exposed to high-pressure saturated steam at a temperature of about 160°–240°C, for some seconds to a few minutes, which is then followed with the release of pressure. This mainly solubilizes hemicellulose which is found dissociated into oligomeric and monomeric sugars in the liquid phase, which results in the exposure of cellulose to the hydrolysis of enzymes. The yield of sugars from hemicellulose and cellulose in the liquid and solid phases depends upon the severity of the pre-treatment process and addition of an acid catalyst, like sulfuric acid or sulfur dioxide. A combination of pre-treatment with either acid gives an increase in the sugar yield from hemicellulose while making the cellulose more vulnerable to enzymatic hydrolysis [80–82].

Liquid hot-water treatment is another treatment that includes treatment of biomass with water at high temperature, but lower than that used for the steam explosion, which increases the overall yield of oligosaccharides and polysaccharides. The addition of catalysts like acid increases the sugar yield, but the higher amount of water dilutes it eventually [83, 84].

The ammonia fiber explosion (AFEX) treatment is ultimately an alkaline method but, similar to the steam explosion treatment, requires high pressure. In this method, the biomass gets treated for about 10–60 min with liquid ammonia, at a temperature below that of 100°C and in presence of pressure greater than 3 MPa. A large amount of ammonia is required for the process, which can later be recycled after reducing the total pressure. During the process, no lignin or hemicellulose gets removed, but the hemicellulose gets degraded and deacetylated, thus making the hemicellulose insoluble and increasing the digestibility of the overall biomass. This treatment works more efficiently on agricultural biomass but has proven to be less effective on lignocellulosic biomass due to its lignin content [81, 85, 86].

Pre-treatment with carbon dioxide explosion is similar to steam explosion and AFEX, though requiring lower temperatures than steam explosion and proving to be more profitable than AFEX. Though like other physicochemical methods, this process also requires high pressure and does not greatly remove or modifies lignin and hemicellulose in the biomass; it provides a much larger surface area and produces no subsequent inhibitory compounds as a byproduct. CO_2, when dissolved in water, forms carbonic acid which helps in acid hydrolysis of the biomass. The molecules of CO_2, due to their size, are able to penetrate into pores of the raw material and disrupt the structures of hemicellulose and cellulose, to yield sugars [79–81].

Recent studies have also proposed other combinations of physicochemical pre-treatment methods, whose main objective is to change the physical and chemical conditions of biomass and maintain the required treatment conditions like duration of treatment, temperature, pressure, and types of chemical compounds [81].

15.4.4 Biological Pre-Treatments

The biological pre-treatment methods, in comparison to ones mentioned till now, are more promising due to its eco-friendly nature and energy-efficient process. These treatments are mild in nature and produce no inhibitory compounds as byproducts and mainly employ microorganisms like white-rot and soft-rot fungi, actinomycetes, and bacteria, which helps to degrade lignin—one of the most recalcitrant polymers present in the biomass, with the help of lignin-degrading enzymes like the peroxidases and laccases [87, 88].

Several studies have been reported, where researchers have tried and achieved biological pre-treatments on different types of agricultural and lignocellulosic biomass, to prepare them for the successive saccharification and fermentation processes. A suitable example is sugarcane biomass, which forms an important cash crop in countries like Brazil and India. Microbial pre-treatment into sugarcane waste with a consortium of fungi and bacteria decreased the C:N ratio by 61%, which varied along with the different microorganisms, rendering easy accessibility of cellulose to enzymatic hydrolysis, for the production of ethanol [89]. Another readily available lignocellulosic material is sugarcane bagasse, which when treated with basidiomycetes fungi, degrades the lignocellulosic substrate, exposing the sugars underneath for its conversion to ethanol [90].

The main objective behind developing biological pre-treatment methods is because it is economically viable and environmentally useful and also helps in asserting the facilitation in saccharification of cellulose and its subsequent conversion to bioethanol. Some model examples of fungi used for this purpose are *Phanerochaete chryosporium*, *Trichoderma reesei*, *Lentinus edodes*, *Pleurotus* spp., *Penicillium camemberti*, *Cerporiopsis subvermispora*, and *Phlebia subserialis*; whereas, certain bacterial laccases which have been characterized from *Azospirillum lipoferum* and *Bacillus subtilis* have also proved useful for this purpose [88, 91, 92].

A major disadvantage of this particular method is the requirement of a prolonged pre-treatment time, as observed with *Trametes hirsuta* yj9 in treating the biomass corn stover, which degraded lignin as high as 71.49% after a 42-day pre-treatment. Sugar yields were also high with enzymatic digestibility reaching 73.99%, with effective structural changes taking place in the biomass, making it coarser in nature [93]. The white-rot fungus *Irpex lacteus* was also used to carry out saccharification in agricultural residues like wheat straw, corn stover, barley straw, and corncob, whose treatment reduced the lignin content by half and exponentially increased the sugar yield, as compared to other physical and chemical treatment methods [94, 95].

A novel approach that has been undertaken with this treatment method is a combination of fungal treatment with a physical or chemical method. A likewise example is of the fungi *Lenzites betulina* C5617 that enhanced its enzymatic hydrolysis in addition to liquid hot water treatment at 200°C, producing the highest hemicellulose removal and a two-fold increase in sugar yield [96]. Another combination is that of using either of ligninolytic fungi (*Irpexlacteus* and *Pleurotus eryngii*) in solid-state fermentation of wheat straw along with a mild alkali treatment, under optimum conditions. This produced the highest digestibility of cellulose and hemicellulose, giving a maximum glucose yield of 84% after 14 days of treatment, along with producing an overall bioethanol yield of 74% [97].

Although biological pre-treatment methods have proved invaluable in enhancing the saccharification process and increasing the final yield of bioethanol, still the longer treatment time required for delignification, the presence of lignin and hemicellulose on the cellulosic surface, and the limited accessibility to their low surface area prove to be quite a hurdle in order to achieve a complete microbial degradation [98]. To overcome this problem, new strategies should be developed for enzyme production or generation of genetically modified organisms to make the process more efficient [99].

15.5 Enzymatic Hydrolysis of Biomass

Enzyme hydrolysis is one of the most significant steps in the process of bioethanol production and involves cellulolytic enzymes which cause depolymerization of complex carbohydrates into simple monomeric sugars. This step requires less energy and takes place under mild conditions [100]. The productivity and efficiency of this step depend heavily on the type of pre-treatment methods, for their alteration capability involving the lignocellulosic structure, and increasing the accessibility to the cellulose underneath. The choice of pre-treatment methods and enzyme hydrolysis are inter-dependent on factors like sugar yield and the enzymes required. Therefore, starting from the composition of the biomass, type of pre-treatment, ability, and efficiency of the enzymes used in hydrolysis, all play an important role in the digestibility process, as well as the yield of ethanol [101, 102].

Demonstration into the combinatorial effect of pre-treatment and hydrolysis has been established by researchers on agricultural biomass like a corncob, rice straw, and banana plant waste for efficient bioethanol production. Pre-treatment was done by chemical and steam explosion along with microwave treatment followed by enzyme hydrolysis with *Trichoderma reesei* cellulases, giving a higher yield of sugar [103]. Different types of biomass like sorghum fibers and sugarcane bagasse were pretreated with dilute ammonium hydroxide at a high temperature and pressure, leading to the removal of lignin and hemicellulose almost close to 50%. This was followed by hydrolysis by commercial enzymes providing about 85% digestibility of cellulose [104, 105]. Organosolv pre-treatment with compounds like ethanol, acetone, methanol, and acetic acid also effectively dissolves lignin to enhance the hydrolysis process by the enzymes [106, 107]. Other pre-treatment methods that have positive effect on the final yield of pentose and hexose sugars after enzyme hydrolysis are alkaline peroxide treatment performed on rice hulls giving a yield of about 70%; in case of sugarcane bagasse pretreated with alkaline hydrogen peroxide also led to a higher yield of 691 mg g^{-1} of glucose [108, 109].

Enzymatic hydrolysis is carried out by cellulases that are produced by a wide range of bacteria and fungi and belongs to three major groups: endoglucanase, exoglucanase, and ß- glucosidase, giving rise to reducing sugars as a result. The process takes place under mild conditions, and the utility cost is low as well, making the process economically viable and environmentally friendly as it hardly produces any inhibitory compounds as byproducts [110, 111]. Challenges faced for this system are the low yield and high titer value of the enzyme which is required for hydrolysis. So, to improve the total yield and rate of reaction of the enzymes, researchers have focused on optimizing the process by increasing the enzyme activity and factoring the substrate concentration and reaction conditions like temperature, pH, and pressure, etc. [112].

There are certain factors which various studies have identified that aids in cellulose conversion. One is the particle size of the biomass and the effect that it has on enzyme hydrolysis. Experiments with *Miscanthus* after the samples were grounded and passed through screens having different size of sieve openings and was subjected to hot water, dilute acid, or ammonium hydroxide pre-treatment. When enzymatic hydrolysis was performed, the concentrations of the sugars were monitored, with the highest percent of total conversion observed in the biomass with the smallest size. The reason behind the increased conversion rate could be attributed to the fact that the availability of the external surface area and porosity making the biomass readily available to the cellulolytic enzymes, but it requires a more detailed investigation [113, 114]. Another factor that aids in cellulose conversion depends on the selection of microorganisms that produces both complex and non-complexed cellulolytic enzymes or genetically engineered microorganisms which produce robust enzymes that accounts for rapid conversion of the cellulose in the biomass to monomeric sugars [100, 115].

15.6 Fermentation

Over the years, fermentation of biomass has proved to be quite challenging, due to the necessarily required conversion of the pentose and the hexose sugars produced after enzyme hydrolysis. For this reason, the fermentation process has undergone continuous improvements with the help of genetic engineering as well as biophysical manipulation [102]. The ethanol fermentation process is ultimately a biological process which utilizes microorganisms for converting the fermentable sugars to ethanol and CO_2, by producing enzymes that catalyze the chemical reaction [116].

Commonly, the main substrates which are taken up for fermentation can be categorized into sugars, starch, and celluloses. The most preferred microorganism for ethanol fermentation has always been yeast, especially *Saccharomyces cerevisiae* which produces ethanol at a concentration of 18% of the fermentation broth, and can grow on both simple and complex sugars, by metabolizing it through the Embden-Meyerhof (EM) pathway. Bacteria like the *Zymomonas* have also received special attention due to its unique physiology which has the ability to metabolize glucose. Other prominent members of microorganisms include the gram-negative bacteria *Zymomonas mobilis*, genetically engineered *Escherichia coli*, the thermophilic bacteria *Clostridium thermocellum*, filamentous fungi *Neurospora crassa*, *Aspergillus* sp., and *Trichoderma viride* [116–118].

Studies have revealed that the development of fermentation technologies depends on the pre-treatment methods utilized for any raw materials, followed by the enzymatic hydrolysis that takes place subsequently. For example, pre-treatment of corn stalk with dilute sulfuric acid, followed with hydrolysis increased ethanol production, as compared to other pre-treatment strategies [119]. While corn stover pretreated with a combination of steam explosion and dilute sulfuric acid, and organosolv underwent hydrolysis followed by simultaneous saccharification and fermentation (SSF) process, produced 78.3% and 62.3% ethanol [120]. Another energy crop used for biorefinery purpose is the Jerusalem artichoke for its high biomass yield. When treated and optimized with alkaline hydrogen peroxide using response surface methodology, the sugar yield was improved without the biomass undergoing the hydrolysis step and with SSF produced 36.5% more ethanol [121]. Treatment with different kinds of ionic liquids on lignocellulosic biomass such as cedar, eucalyptus, and sugarcane bagasse followed by direct ethanol fermentation gives a yield of about 75% after fermenting the biomass for 96 h [122].

Investigations into more cost-effective and environmentally friendly strategies for efficient fermentation have led to the development of processes, which has shown potential advantages like reducing the overall costs. One such innovation is the continuous hydrolysis and fermentation processes for converting cellulosic biomass to produce ethanol. This method is generally performed within a bioreactor that keeps the reactions going continuously and has been used commercially for the production of ethanol from agricultural biomasses like corn and cane sugar, utilized for their voluminous amounts of biomass, cutting the labor costs, and diminishing the time required for cleaning and filling. But challenges like contamination of the reactor with microorganisms require more sophisticated development, and more investigations are required to find the alternative usage of expensive enzymes for the process [123–125].

Using *Saccharomyces cerevisiae* for fermentation for ethanol production has several limitations, which affect the ethanol yield. Utilization of cellulose as the raw material for SSF undoubtedly produces a high amount of bioethanol but requires optimum pH, temperature, and substrate concentration for reaching compatible SSF conditions [126]. Other factors involved in the process include the concentration of sugar to ethanol production ratio and the condition of the medium used, which in turn is related to the rate of growth, the viability of the cell populations, and the rate of fermentation [127].

To solve the regular limitations involved with the fermentation process, several new techniques have been developed. Like the genetically engineered thermophilic acetogen, *Moorella thermoacetica* which was constructed for successfully fermenting sugars in hydrolysate prepared after acid pre-treatment, to produce ethanol [128]. Other biological approaches involve the use of the enzyme laccase, to enhance the hydrolysability and fermentation capability of wheat straw pretreated with steam-explosion [129]. Laccase has also been used in the treatment of the pretreated biomass to remove the phenolic lignin and other inhibitory compounds, finally increasing the fermentation capability of the thermotolerant yeast *Kluyveromyces marxianus* CECT 10875 for producing ethanol [130].

Ethanol fermentation has become a very important part of the biorefinery industry and has put forth a range of emerging technologies, for the successful production of ethanol. Though a gap exists between academia and industry research, many of the scientific approaches like thermotolerant yeasts, genetically engineered microorganisms, bioreactors, yeast cell immobilization, and others could be used to develop technologies from laboratory scale to industrial scale [131, 132].

15.7 Future Prospects for the Production of Bioethanol Through Biorefineries

Production of bioethanol from both agricultural and lignocellulosic biomass has become one of the most sought-after green alternatives to fossil fuels. In recent times, innovation and development of new strategies and technologies have rendered the ability to generate bioethanol from edible crops as well as lignocellulosic biomass, through biorefinery. But, the conversion efficiency and ethanol yield are highly dependent on the source and nature of the biomass, primarily because of their lignocellulosic content. The major polymers in the biomass, i.e., lignin and hemicellulose remain tightly linked and form a mesh-like network with the cellulosic structure, thus reducing the accessibility to hydrolytic enzymes for its conversion to reducing sugars [133, 134]. Thus, this type of biomass requires a robust and highly efficient conversion process, to completely utilize this promising feedstock.

Another problem faced with the production process is the low production of ethanol due to the hindrance of the recalcitrant polymer lignin and technical difficulties involved in the process like the high viscosity, deficiency in free water, temperature, pH, and the increased levels of inhibitors, all contributing to the obstruction in enzymatic hydrolysis. More research is required to overcome these restrictions along with developing advanced enzyme systems that are cost-effective and can be utilized on a commercial scale [135].

The generation of ethanol from biomass is a very important challenge that depends upon factors like the type, source, enzyme titer, fermentation strategies, and process conditions. It is very difficult to find the optimum combination of these characteristics for individual biomasses, for the extent to which these implicate on the production yield of ethanol [136].

15.8 Conclusion

Biorefinery approach for the production of bioethanol from agricultural and lignocellulosic biomass can become the alternative technology for using as fuel and generating energy. Though the process has many limitations and hindrances, restricting its production yield, more research and investigations into the development and further development could make the processes more efficient and robust in the future. It would be highly beneficial for the researchers and the industry to work hand-in-hand, to commercialize the process, and develop a large-scale technology.

References

1. Pimentel, D. and Pimentel, M., Global environmental resources versus world population growth. *Ecol. Econ.*, 59, 2, 195–198, 2006.
2. Escobar, J.C., Lora, E.S., Venturini, O.J., Yanez, E.E., Castillo, E.F., Almazan, O., Biofuels: Environment, technology and food security. *Renew. Sust. Energ. Rev.*, 13, 6–7, 1275–1287, 2009.
3. Alaswad, A., Dassisti, M., Prescott, T., Olabi, A.G., Technologies and development of third generation biofuel production. *Renew. Sust. Energ. Rev.*, 51, 1446–1460, 2015.
4. Lankoski, J. and Ollikainen, M., Biofuel policies and the environment: Do climate benefits warrant increased production from biofuel feedstocks? *Ecol. Econ.*, 70, 4, 676–687, 2011.
5. Tilman, D., Socolow, R., Foley, J.A., Hill, J., Larson, E., Lynd, L., Pacala, S., Reilly, J., Searchinger, T., Somerville, C., Williams, R., Beneficial Biofuels – The food, energy and environmental trilemma. *Science*, 325, 5938, 270–271, 2009.
6. Antoni, D., Zverlov, V.V., Schwarz, W.W., Biofuels from microbes. *Appl. Microbiol. Biotechnol.*, 77, 1, 23–35, 2007.
7. Lee, R.A. and Lavoie, J., From first- to third-generation biofuels: Challenges of producing a commodity from a biomass of increasing complexity. *Anim. Front.*, 3, 2, 6–11, 2013.
8. Rodriguez-Couto, S., Green Nanotechnology for Biofuel Production, in: *Sustainable Approaches for Biofuels Production Technologies*, pp. 73–82, 2018.
9. Somerville, C., Youngs, H., Taylor, C., Davis, S.C., Long, S.P., Feedstocks for Lignocellulosic Biofuels. *Science*, 329, 5993, 790–792, 2010.
10. Ravindranath, N.H., Sita Lakshmi, C., Manuvie, R., Balachandra, P., Biofuel production and implications for land use, food production and environment in India. *Energy Policy*, 39, 10, 5737–5745, 2011.
11. Cherubini, F., The biorefinery concept: Using biomass instead of oil for producing energy and chemicals. *Energ. Convers. Manage.*, 51, 7, 1412–1421, 2010.
12. Taylor, G., Biofuels and the biorefinery concept. *Energy Policy*, 36, 12, 4406–4409, 2008.

13. Menon, V. and Rao, M., Trends in bioconversion of lignocellulose: Biofuels, platform chemicals & biorefinery concept. *Prog. Energy Combust. Sci.*, 38, 4, 522–550, 2012.
14. Gray, K.A., Zhao, L., Emptage, M., Bioethanol. *Curr. Opin. Chem. Biol.*, 10, 2, 141–146, 2006.
15. Sarkar, N., Ghosh, S.K., Bannerjee, S., Aikat, K., Bioethanol production from agricultural wastes: An overview. *Renew. Energy*, 37, 1, 19–27, 2012.
16. Balat, M., Balat, H., Oz, C., Progress in bioethanol processing. *Prog. Energy Combust. Sci.*, 34, 5, 551–573, 2008.
17. Vohra, M., Manwar, J., Manmode, J., Padgilwar, S., Patil, S., Bioethanol production: Feedstock and current technologies. *J. Environ. Chem. Eng.*, 2, 1, 573–584, 2014.
18. Chandel, A.K., ES, C., Rudravaram, R., Narasu, M.L., Rao, L.V., Ravindra, P., Economics and environmental impact of bioethanol production technologies: An appraisal. *Afr. J. Environ. Econ. Manage.*, 1, 5, 126–143, 2013.
19. Aditiya, H.B., Mahlia, T.M.I., Chong, W.T., Nur, H., Sebayang, A.H., Second generation bioethanol production: A critical review. *Renew. Sust. Energ. Rev.*, 66, 631–653, 2016.
20. Rabelo, S.C., Carrere, H., Maciel Filho, R., Costa, A.C., Production of bioethanol, methane and heat from sugarcane bagasse in a biorefinery concept. *Bioresour. Technol.*, 102, 17, 7887–7895, 2011.
21. Kim, S. and Dale, B.E., Global potential bioethanol production fromwasted crops and crop residues. *Biomass Bioenergy*, 26, 4, 361–375, 2004.
22. Zhang, Y., Feng, J., Xiao, Z., Liu, Y., Ma, H., Wang, Z., Pan, H., Highly efficient and selective fractionation strategy for lignocellulosic biomass with recyclable dioxane/ethylene glycol binary solvent. *Ind. Crops Prod.*, 144, 112038, 2020.
23. Kumar, B., Bhardwaj, N., Agrawal, K., Chaturvedi, V., Verma, P., Current perspective on pre-treatment technologies using lignocellulosic biomass: An emerging biorefinery concept. *Fuel Process. Technol.*, 199, 106244, 2020.
24. Bhatia, S.K., Jagtap, S.S., Bedekar, A.A., Bhatia, R.K., Patel, A.K., Pant, D., Banu, J.R., Rao, C.V., Kim, Y., Yang, Y., Recent developments in pre-treatment technologies on lignocellulosic biomass: Effect of key parameters, technological improvements, and challenges. *Bioresour. Technol.*, 122724, 2020.
25. Siqueira, J.G.W., Rodrigues, C., Vandenberghe, L.P., Woiciechowski, A.L., Soccol, C.R., Current advances in on-site cellulase production and application on lignocellulosic biomass conversion to biofuels: A review. *Biomass Bioenergy*, 132, 105419, 2020.
26. Kuhad, R.C., Gupta, R., Khasa, Y.P., Singh, A., Zhang, Y.-H.P., Bioethanol production from pentose sugars: Current status and future prospects. *Renew. Sust. Energ. Rev.*, 15, 9, 4950–4962, 2011.
27. Amore, A., Ciesielski, P.N., Lin, C., Salvachua, D., Nogue, V.S., Development of Lignocellulosic Biorefinery Technologies: Recent Advances and Current Challenges. *Aust. J. Chem.*, 69, 11, 1201–1218, 2016.
28. Kaparaju, P., Serrano, M., Thomsen, A.B., Kongjan, P., Angelidaki, I., Bioethanol, biohydrogen and biogas production from wheat straw in a biorefinery concept. *Bioresour. Technol.*, 100, 9, 2562–2568, 2009.
29. Muktham, R., Bhargava, S.K., Bankupalli, S., Ball, A.S., A review on 1st and 2nd generation bioethanol production - Recent progress. *J. Sustain. Bioenergy Syst.*, 6, 3, 72–92, 2016.
30. Cho, E.J., Trinh, L.T.P., Song, Y., Lee, Y.G., Bae, H., Bioconversion of biomass waste into high value chemicals. *Bioresour. Technol.*, 298, 122386, 2019.
31. Ubando, A.T., Felix, C.B., Chen, W., Biorefineries in circular bioeconomy: A comprehensive review. *Bioresour. Technol.*, 299, 122585, 2020.
32. Ramesh, D., Muniraj, I.K., Thangavelu, K., Karthikeyan, S., Chemicals and fuels production from agro residues: A biorefinery approach, in: *Sustainable Approaches for Biofuels Production Technologies*, pp. 47–71, 2018.

33. Ankush, Khusboo, Dubey, K.K., Food industry waste biorefineries: Future energy, valuable recovery, and waste treatment, in: *Refining Biomass Residues for Sustainable Energy and Bioproducts*, pp. 391–406, 2020.
34. Orts, W.J. and McMahan, C.M., Biorefinery developments for advanced biofuels from a sustainable array of biomass feedstocks: Survey of recent biomass conversion research from agricultural research service. *BioEnergy Res.*, 9, 2, 430–446, 2016.
35. Kluts, I.N., Brinkman, M.L.J., de Jong, S.A., Junginger, H.M., Biomass Resources: Agriculture, in: *Advances in Biochemical Engineering/Biotechnology*, 2017.
36. Mupondwa, E., Li, X., Tabil, L., Large-scale commercial production of cellulosic ethanol from agricultural residues: A case study of wheat straw in the Canadian Prairies. *Biofuel Bioprod. Bior.*, 11, 6, 955–970, 2017.
37. Hernandez, C., Escamilla-Alvarado, C., Sanchez, A., Alarcon, E., Ziarelli, F., Musule, R., Valdez-Vazquez, I., Wheat straw, corn stover, sugarcane, and *Agave* biomasses: Chemical properties, availability, and cellulosic-bioethanol production potential in Mexico. *Biofuel Bioprod. Bior.*, 13, 1143–115, 2019.
38. Gurram, R., Al-Shannag, M., Knapp, S., Das, T., Singsaas, E., Alkasrawi, M., Technical possibilities of bioethanol production from coffee pulp: A renewable feedstock. *Clean Technol. Envir.*, 18, 1, 269–278, 2015.
39. Zhang, M., Xie, L., Yin, Z., Khanal, S., Zhou, Q., Biorefinery approach for cassava-based industrial wastes: Current status and opportunities. *Bioresour. Technol.*, 215, 50–62, 2016.
40. Nair, R.B., Lennartsson, P.R., Taherzadeh, M.J., Bioethanol production from agricultural and municipal wastes, in: *Current Developments in Biotechnology and Bioengineering*, pp. 157–190, 2017.
41. Nguyen, Q.A., Yang, J., Bae, H., Bioethanol production from individual and mixed agricultural biomass residues. *Ind. Crops Prod.*, 95, 718–725, 2017.
42. De Corato, U., De Bari, I., Viola, E., Pugliese, M., Assessing the main opportunities if integrated biorefining from agro-bioenergy co/by-products and agroindustrial residues into high-value added products associated to some emerging markets: A review. *Renew. Sust. Energ. Rev.*, 88, 326–346, 2018.
43. Ghosh, D., Dasgupta, D., Agrawal, D., Kaul, S., Adhikari, D.K., Kurmi, A.K., Arya, P.K., Bangwal, D., Negi, M.S., Fuels and chemicals from lignocellulosic biomass: An integrated biorefinery approach. *Energ. Fuel*, 29, 5, 3149–3157, 2015.
44. De Bhowmick, G., Sarmah, A.K., Sen, R., Lignocellulosic biorefinery as a model for sustainable development of biofuels and value added products. *Bioresour. Technol.*, 247, 1144–1154, 2018.
45. Arevalo-Gallegos, A., Ahmad, Z., Asgher, M., Parra-Saldivar, R., Iqbal, H.M.N., Lignocellulose: A sustainable material to produce value-added products with a zero waste approach – A review. *Int. J. Biol. Macromol.*, 99, 308–318, 2017.
46. Nanda, S., Azargohar, R., Dalai, A.K., Kozinski, J.A., An assessment on the sustainability of lignocellulosic biomass for biorefining. *Renew. Sust. Energ. Rev.*, 50, 925–941, 2015.
47. Moreno, A.D., Ibarra, D., Alvira, P., Tomas-Pejo, E., Ballesteros, M., A review of biological delignification and detoxification methods for lignocellulosic bioethanol production. *Crit. Rev. Biotechnol.*, 35, 3, 342–354, 2014.
48. Rambo, M.K.D., Schmidt, F.L., Ferreira, M.M.C., Analysis of the lignocellulosic components of biomass residues for biorefinery opportunities. *Talanta*, 144, 696–703, 2015.
49. Khatiwada, D., Leduc, S., Silveira, S., McCallum, I., Optimizing ethanol and bioelectricity production in sugarcane biorefineries in Brazil. *Renew. Energy*, 85, 371–386, 2016.
50. Zucaro, A., Forte, A., Fierro, A., Life cycle assessment of wheat straw lignocellulosic bio-ethanol fuel in a local biorefinery prospective. *J. Clean. Prod.*, 194, 138–149, 2018.

51. Khoshnevisan, B., Rafiee, S., Tabatabaei, M., Ghanavati, H., Mohtasebi, S.S., Rahimi, V., Shafiei, M., Angelidaki, I., Karimi, K., Life cycle assessment of castor-based biorefinery: A well to wheel LCA. *Int. J. Life Cycle Ass.*, 23, 9, 1788–1805, 2017.
52. Forte, A., Zucaro, A., Fagnano, M., Fierro, A., Potential environmental impact of bioethanol production chain from fiber sorghum to be used in passenger cars. *Sci. Total Environ.*, 598, 365–376, 2017.
53. Raman, J.K., Alves, C.M., Gnansounou, E., A review on moringa tree and vetiver grass – Potential biorefinery feedstocks. *Bioresour. Technol.*, 249, 1044–1051, 2018.
54. Stolarski, M.J., Krzyzaniak, M., Luczynski, M., Zaluski, D., Szczukowski, S., Tworkowski, J., Golaszewski, J., Lignocellulosic biomass from short rotation woody crops as a feedstock for second-generation bioethanol production. *Ind. Crops Prod.*, 75, 66–75, 2015.
55. Dominiguez, E., Romani, A., Domingues, L., Garrote, G., Evaluation of strategies for second generation bioethanol production from fast growing biomass Paulownia within a biorefinery scheme. *Appl. Energy*, 187, 777–789, 2017.
56. Liguori, R. and Faraco, V., Biological processes for advancing lignocellulosic waste biorefinery by advocating circular economy. *Bioresour. Technol.*, 215, 13–20, 2016.
57. Chandel, A.K., Garlapati, V.K., Singh, A.K., Antunes, F.A.F., da Silva, S.S., The path forward for lignocellulose biorefineries: Bottlenecks, solutions, and perspective on commercialization. *Bioresour. Technol.*, 264, 370–381, 2018.
58. Cheali, P., Posada, J.A., Gernaey, K.V., Sin, G., Upgrading of lignocellulosic biorefinery to value-added chemicals: Sustainability and economics of bioethanol-derivatives. *Biomass Bioenergy*, 75, 282–300, 2015.
59. Shafiei, M., Kumar, R., Karimi, K., Pre-treatment of lignocellulosic biomass, in: *Biofuel and Biorefinery Technologies*, 85–154, 2015.
60. Maurya, D.P., Singla, A., Negi, S., An overview of key pre-treatment processes for biological conversion of lignocellulosic biomass to bioethanol. *3 Biotech*, 5, 5, 597–609, 2015.
61. Akhtar, N., Gupta, K., Goyal, D., Goyal, A., Recent advances in pre-treatment technologies for efficient hydrolysis of lignocellulosic biomass. *Environ. Prog. Sustain. Energy*, 35, 2, 489–511, 2015.
62. Chen, H., Liu, J., Chang, X., Chen, D., Xue, Y., Liu, P., Lin, H., Han, S., A review on the pre-treatment of lignocellulose for high-value chemicals. *Fuel Process. Technol.*, 160, 196–206, 2017.
63. Saini, J.K., Saini, S., Tewari, L., Lignocellulosic agriculture wastes as biomass feedstocks for second-generation bioethanol production: Concepts and recent developments. *3 Biotech*, 5, 4, 337–353, 2014.
64. Kumari, D. and Singh, R., Pre-treatment of lignocellulosic wastes for biofuel production: A critical review. *Renew. Sust. Energ. Rev.*, 90, 877–891, 2018.
65. Shirkavand, E., Baroutian, S., Gapes, D.J., Young, B.R., Combination of fungal and physico-chemical processes for lignocellulosic biomass pre-treatment – A review. *Renew. Sust. Energ. Rev.*, 54, 217–234, 2016.
66. Bahmani, M.A., Shafiei, M., Karimi, K., Anaerobic digestion as a pre-treatment to enhance ethanol yield from lignocelluloses. *Process Biochem.*, 51, 9, 1256–1263, 2016.
67. Perrone, O.M., Colombari, F.M., Rossi, J.S., Moretti, M.M.S., Bordignon, S.E., da C.C. Nunes, C., Gomes, E., Boscolo, M., Da-Silva, R., Ozonolysis combined with ultrasound as a pre-treatment of sugarcane bagasse: Effect on the enzyme saccharification and the physical and chemical characteristics of the substrate. *Bioresour. Technol.*, 218, 69–76, 2016.
68. Kumar, A.K. and Sharma, S., Recent updates on different methods of pre-treatment of lignocellulosic feedstocks: A review. *Bioresour. Bioprocess.*, 4, 7, 1–9, 2017.
69. Singh, R., Krishna, B.B., Kumar, J., Bhaskar, T., Opportunities for utilization of non-conventional energy sources for biomass pre-treatment. *Bioresour. Technol.*, 199, 398–407, 2016.

70. Rastogi, M. and Shrivastava, S., Recent advances in second generation bioethanol production: An insight to pre-treatment, saccharification and fermentation processes. *Renew. Sust. Energ. Rev.*, 80, 330–340, 2017.
71. Bensah, E.C. and Mensah, M., Chemical pre-treatment methods for the production of cellulosic ethanol: Technologies and Innovations. *Int. J. Chem. Eng.*, 2013, 2013.
72. Behera, S., Arora, R., Nandhagopal, N., Kumar, S., Importance of chemical pre-treatment for bioconversion of lignocellulosic biomass. *Renew. Sust. Energ. Rev.*, 36, 91–106, 2014.
73. Zhu, J.Y. and Pan, X.J., Woody biomass pre-treatment for cellulosic ethanol production: Technology and energy consumption evaluation. *Bioresour. Technol.*, 101, 4992–5002, 2010.
74. Zhao, X., Zhang, L., Liu, D., Comparative study on chemical pre-treatment methods for improving enzymatic digestibility of crofton weed stem. *Bioresour. Technol.*, 99, 3729–3736, 2008.
75. Karimi, K., Shafiei, M., Kumar, R., Progress in physical and chemical pre-treatment of lignocellulosic biomass. *Biofuel Technol.*, 53–96, 2013.
76. Badiei, M., Asim, N., Jahim, J.M., Sopian, K., Comparison of chemical pre-treatment methods for cellulosic biomass. *APCBEE Procedia*, 9, 170–174, 2014.
77. Akhtar, N., Gupta, K., Goyal, D., Goyal, A., Recent advances in pre-treatment technologies for efficient hydrolysis of lignocellulosic biomass. *Environ. Prog. Sustain. Energy*, 35, 2, 489–511, 2015.
78. Yu, J., Zhang, J., He, J., Liu, Z., Yu, Z., Combinations of mild physical or chemical pre-treatment with biological pre-treatment for enzymatic hydrolysis of rice hull. *Bioresour. Technol.*, 100, 903–908, 2009.
79. Galbe, M. and Zacchi, G., Pre-treatment of lignocellulosic materials for efficient bioethanol production. *Adv. Biochem. Eng. Biotechnol.*, 108, 41–65, 2007.
80. Shirkavand, E., Baroutian, S., Gapes, D.J., Young, B.R., Combination of fungal and physico-chemical process for lignocellulosic biomass pre-treatment – A review. *Renew. Sust. Energ. Rev.*, 54, 217–234, 2016.
81. Agbor, V.B., Cicek, N., Sparling, R., Berlin, A., Levin, D.B., Biomass pre-treatment: Fundamentals toward application. *Biotechnology Advances*, 29, 6, 675–685, 2011.
82. Banoth, C., Sunkar, B., Tondamanati, P.R., Bhukya, B., Improved physicochemical pre-treatment and enzymatic hydrolysis of rice straw for bioethanol production by yeast fermentation. *3 Biotech*, 7, 5, 675–685, 2017.
83. Xu, Z. and Huang, F., Pre-treatment methods for bioethanol production. *Appl. Biochem. Biotechnol.*, 174, 1, 43–62, 2014.
84. Ortiz, P.S. and de Oliveira, S., Jr., Exergy analysis of pre-treatment processes of bioethanol production based on sugarcane bagasse. *Energy*, 76, 130–138, 2014.
85. Brodeur, G., Yau, E., Badal, K., Collier, J., Ramachandran, K.B., Ramakrishnan, S., Chemical and physicochemical pre-treatment of lignocellulosic biomass: A review. *Enzyme Res.*, 2011, 787532, 2011.
86. Tomas-Pejo, E., Alvira, P., Ballesteros, M., Negro, M.J., Pre-treatment technologies for lignocellulose-to-bioethanol conversion. *Biofuels*, 149–176, 2011.
87. Sindhu, R., Binod, P., Pandey, A., Biological pre-treatment of lignocellulosic biomass – A review. *Bioresour. Technol.*, 199, 76–82, 2016.
88. Saritha, M., Arora, A., Lata, Biological pre-treatment of lignocellulosic substrates for enhanced delignification and enzyme digestibility. *Indian J. Microbiol.*, 52, 2, 122–130, 2011.
89. Singh, P., Suman, A., Tiwari, P., Arya, N., Gaur, A., Shrivastava, A.K., Biological pre-treatment of sugarcane trash for its conversion to fermentable sugars. *World J. Microbiol. Biotechnol.*, 24, 5, 667–673, 2007.
90. Cardona, C.A., Quintero, J.A., Paz, I.C., Production of bioethanol from sugarcane bagasse: Status and perspectives. *Bioresour. Technol.*, 101, 12, 4754–4766, 2010.

91. Chen, S., Zhang, X., Singh, D., Yu, H., Yang, X., Biological pre-treatment of lignocellulosics: potential, progress and challenges. *Biofuels*, 1, 1, 177–199, 2010.
92. Isroi, Millati, R., Syamsiah, S., Niklasson, C., Cahyanto, M.N., Lundquist, K., Taherzadeh, M.J., Biological pre-treatment of lignocelluloses with white-rot fungi and its applications: A review. *BioResources*, 6, 4, 5224–5259, 2011.
93. Sun, F., Li, J., Yuan, Y., Yan, Z., Liu, X., Effect of biological pre-treatment with *Trametes hirsuta* yj9 on enzymatic hydrolysis of corn stover. *Int. Biodeterioration Biodegradation*, 65, 7, 931–938, 2011.
94. Lopez-Abelairas, M., Lu-Chau, T.A., Lema, J.M., Enhanced saccharification of biologically pretreated wheat straw for ethanol production. *Appl. Biochem. Biotechnol.*, 169, 4, 1147–1159, 2013.
95. Garcia-Torreiro, M., Lopez-Abelairas, M., Lu-Chau, T.A., Lema, J.M., Fungal pre-treatment of agricultural residues for bioethanol production. *Ind. Crops Prod.*, 89, 486–492, 2016.
96. Wang, W., Yuan, T., Wang, K., Cui, B., Dai, Y., Combination of biological pre-treatment with liquid hot water pre-treatment to enhance enzymatic hydrolysis of *Populus tomentosa*. *Bioresour. Technol.*, 107, 282–286, 2012.
97. Lopez-Abelairas, M., Alvarez Pallin, M., Salvachua, D., Lu-Chau, T.A., Martinez, M.J., Lema, J.M., Optimisation of the biological pre-treatment of wheat straw with white-rot fungi for ethanol production. *Bioprocess Biosyst. Eng.*, 36, 9, 1251–1260, 2012.
98. Sharma, H.K., Xu, C., Qin, W., Biological pre-treatment of lignocellulosic biomass for biofuels and bioproducts: An overview. *Waste Biomass Valorization*, 10, 235–251, 2017.
99. Moreno, A.D., Ibarra, D., Alvira, P., Tomas-Pejo, E., Ballesteros, M., A review of biological delignification and detoxification methods for lignocellulosic bioethanol production. *Crit. Rev. Biotechnol.*, 35, 3, 342–354, 2014.
100. Gupta, A. and Verma, J.P., Sustainable bio-ethanol production from agro-residues: A review. *Renew. Sust. Energ. Rev.*, 41, 550–567, 2015.
101. Alvira, P., Tomas-Pejo, E., Ballesteros, M., Negro, M.J., Pre-treatment technologies for an efficient bioethanol production process based on enzymatic hydrolysis: A review. *Bioresour. Technol.*, 101, 13, 4851–4861, 2010.
102. Mielenz, J.R., Ethanol production from biomass: Technology and commercialization status. *Curr. Opin. Microbiol.*, 4, 3, 324–329, 2001.
103. El-Zawawy, W.K., Ibrahim, M.M., Abdel-Fattah, Y.R., Soliman, N.A., Mahmoud, M.M., Acid and enzyme hydrolysis to convert pretreated lignocellulosic materials into glucose for ethanol production. *Carbohydr. Polym.*, 84, 3, 865–871, 2011.
104. Salvi, D.A., Aita, G.M., Robert, D., Bazan, V., Dilute ammonia pre-treatment of sorghum and its effectiveness on enzyme hydrolysis and ethanol fermentation. *Appl. Biochem. Biotechnol.*, 161, 1–8, 67–74, 2010.
105. Aita, G.A., Salvi, D.A., Walker, M.S., Enzyme hydrolysis and ethanol fermentation of dilute ammonia pretreated energy cane. *Bioresour. Technol.*, 102, 6, 4444–4448, 2011.
106. Zhao, X., Cheng, K., Liu, D., Organosolv pre-treatment of lignocellulosic biomass for enzymatic hydrolysis. *Appl. Microbiol. Biotechnol.*, 82, 5, 815–827, 2009.
107. Sannigrahi, P., Miller, S.J., Ragauskas, A.J., Effects of organosolv pre-treatment and enzymatic hydrolysis on cellulose structure and crystallinity in loblolly pine. *Carbohydr. Res.*, 345, 7, 965–970, 2010.
108. Diaz, A., Le Toullec, J., Blandino, A., de Ory, I., Caro, I., Pre-treatment of rice hulls with alkaline peroxide to enhance enzyme hydrolysis for ethanol production. *Chem. Eng. Trans.*, 32, 949–954, 2013.
109. Rabelo, S.C., Amezquita Fonseca, N.A., Andrade, R.R., Maciel Filho, R., Costa, A.C., Ethanol production from enzymatic hydrolysis of sugarcane bagasse pretreated with lime and alkaline hydrogen peroxide. *Biomass Bioenergy*, 35, 7, 2600–2607, 2011.

110. Kumar, S., Singh, S.P., Mishra, I.M., Adhikari, D.K., Recent advances in production of bioethanol from lignocellulosic biomass. *Chem. Eng. Technol.*, 32, 4, 517–526, 2009.
111. Wan, C. and Li, Y., Microbial pre-treatment of corn stover with *Ceriporiopsis subvermispora* for enzymatic hydrolysis and ethanol production. *Bioresour. Technol.*, 101, 16, 6398–6403, 2010.
112. Sun, Y. and Cheng, J., Hydrolysis of lignocellulosic materials for ethanol production: A review. *Bioresour. Technol.*, 83, 1, 1–11, 2002.
113. Khullar, E., Dien, B.S., Rausch, K.D., Tumbleson, M.E., Singh, V., Effect of particle size on enzymatic hydrolysis of pretreated *Miscanthus*. *Ind. Crops Prod.*, 44, 11–17, 2013.
114. Narayanaswamy, N., Faik, A., Goetz, D.J., Gu, T., Supercritical carbon dioxide pre-treatment of corn stover and switchgrass for lignocellulosic ethanol production. *Bioresour. Technol.*, 102, 13, 6995–7000, 2011.
115. Binod, P., Sindhu, R., Singhania, R.R., Vikram, S., Devi, L., Nagalakshmi, S., Kurien, N., Sukumaran, R.K., Pandey, A., Bioethanol production from rice straw: An overview. *Bioresour. Technol.*, 101, 13, 4767–4774, 2010.
116. Lin, Y. and Tanaka, S., Ethanol fermentation from biomass resources: Current state and prospects. *Appl. Microbiol. Biotechnol.*, 69, 6, 627–642, 2005.
117. Sukumaran, R.K., Singhania, R.R., Mathew, G.M., Pandey, A., Cellulase production using biomass feed stock and its application in lignocellulose saccharification for bio-ethanol production. *Renew. Energy*, 34, 2, 421–424, 2009.
118. Geddes, C.C., Nieves, I.U., Ingram, L.O., Advances in ethanol production. *Curr. Opin. Biotechnol.*, 22, 3, 312–319, 2011.
119. Li, P., Luo, Z., Qin, P., Chen, C., Wang, Y., Zhang, C., Wang, Z., Tan, T., Effect of acid pretreatment on different parts of corn stalk for second generation ethanol production. *Bioresour. Technol.*, 206, 86–92, 2016.
120. Katsimpouras, C., Zacharopoulou, M., Matsakas, L., Rova, U., Christakopoulos, P., Topakas, E., Sequential high gravity ethanol fermentation and anaerobic digestion of steam explosion and organosolv pretreated corn stover. *Bioresour. Technol.*, 244, 1129–1136, 2017.
121. Li, K., Qin, J., Liu, C., Bai, F., Optimization of pre-treatment, enzymatic hydrolysis and fermentation for more efficient ethanol production by Jerusalem artichoke stalk. *Bioresour. Technol.*, 221, 188–194, 2016.
122. Yamada, R., Nakashima, K., Asai-Nakashima, N., Tokuhara, W., Ishida, N., Katahira, S., Kamiya, N., Ogino, C., Kondo, A., Direct ethanol production from ionic liquid-pretreated lignocellulosic biomass by cellulase-displaying yeasts. *Appl. Biochem. Biotechnol.*, 182, 1, 229–237, 2016.
123. Brethauer, S. and Wyman, C.E., Review: Continuous hydrolysis and fermentation for cellulosic ethanol production. *Bioresour. Technol.*, 101, 13, 4862–4874, 2010.
124. Dasgupta, D., Ghosh, D., Bandhu, S., Adhikari, D.K., Lignocellulosic sugar management for xylitol and ethanol fermentation with multiple cell recycling by *Kluyveromyces marxianus* IIPE453. *Microbiol. Res.*, 200, 64–72, 2017.
125. Tang, Y., An, M., Liu, K., Nagai, S., Shigematsu, T., Morimura, S., Kida, K., Ethanol production from acid hydrolysate of wood biomass using the flocculating yeast *Saccharomyces cerevisiae* KF-7. *Process Biochem.*, 41, 4, 909–914, 2006.
126. Lin, Y., Zhang, W., Li, C., Sakakibara, K., Tanaka, S., Kong, H., Factors affecting ethanol fermentation using *Saccharomyces cerevisiae* BY4742. *Biomass Bioenergy*, 47, 395–401, 2012.
127. Klinke, H.B., Thomsen, A.B., Ahring, B.K., Inhibition of ethanol-producing yeast and bacteria by degradation products produced during pre-treatment of biomass. *Appl. Microbiol. Biotechnol.*, 66, 1, 10–26, 2004.
128. Rahayu, F., Kawai, Y., Iwasaki, Y., Yoshida, K., Kita, A., Tajima, T., Kato, J., Murakami, K., Hoshino, T., Nakashimada, Y., Thermophilic ethanol fermentation from lignocellulose

hydrolysate by genetically engineered *Moorella thermoacetica*. *Bioresour. Technol.*, 245, 1393–1399, 2017.
129. Moreno, A.D., Ibarra, D., Mialon, A., Ballesteros, M., A bacterial laccase for enhancing saccharification and ethanol fermentation of steam-pretreated biomass. *Fermentation*, 2, 4, 11, 2016.
130. Moreno, A.D., Ibarra, D., Ballesteros, I., Gonzalez, A., Ballesteros, M., Comparing cell viability and ethanol fermentation of the thermotolerant yeast *Kluyveromyces marxianus* and *Saccharomyces cerevisiae* on steam-exploded biomass treated with laccase. *Bioresour. Technol.*, 135, 239–245, 2013.
131. Bai, F.W., Anderson, W.A., Moo-Young, M., Ethanol fermentation technologies from sugar and starch feedstocks. *Biotechnol. Adv.*, 26, 1, 89–105, 2008.
132. Chen, X., Zhai, R., Shi, K., Yuan, Y., Dale, B.E., Gao, Z., Jin, M., Mixing alkali pretreated and acid pretreated biomass for cellulosic ethanol production featuring reduced chemical use and decreased inhibitory effect. *Ind. Crops Prod.*, 124, 719–725, 2018.
133. Zabed, H., Sahu, J.N., Boyce, A.N., Faruq, G., Fuel ethanol production from lignocellulosic biomass: An overview on feedstocks and technological approaches. *Renew. Sust. Energ. Rev.*, 66, 751–774, 2016.
134. Liu, C., Xiao, Y., Xia, X., Zhao, X., Peng, L., Srinophakun, P., Bai, F., Cellulosic ethanol production: Progress, challenges and strategies for solutions. *Biotechnol. Adv.*, 37, 491–504, 2019.
135. Matano, Y., Hasunuma, T., Kondo, A., Display of cellulases on the cell surface of *Saccharomyces cerevisiae* for high yield ethanol production from high-solid lignocellulosic biomass. *Bioresour. Technol.*, 108, 128–133, 2012.
136. Manochio, C., Andrade, B.R., Rodriguez, R.P., Moraes, B.S., Ethanol from biomass: A comparative overview. *Renew. Sust. Energ. Rev.*, 80, 743–755, 2017.

16

Biorefinery Approach for Production of Biofuel From Algal Biomass

Bhasati Uzir and Amrita Saha*

Department of Environmental Sciences, Amity Institute of Environmental Sciences, Amity University Kolkata, Kolkata, India

Abstract

In the recent years, with the growing interest in socio-economic awareness and the adverse impact of excessive fossil fuel utilization, the research interests shifted toward bio-resources such as algal-based biomass and for meeting the current bio-economy challenges, new integrated processes in research and development are necessary.

Fuel required for transportation, that is obtained from biological matter, also known as bio-fuel, is an alternative renewable source to fossil fuel and algal biomass is a raw material for the manufacturing of bio-fuel.

For the cultivation of algal biomass, it does not involve convoluted treatment method as compared to lignocellulose-enriched biomass and its cultivation does not compete for cropland with agricultural growing of food crop for biofuel. Algal biomass proceeds third-generation feedstock for liquid transportation fuel that does not compete with food crops for cropland. For the production of different biofuel like biodiesel, biohydrogen, bio-oil, and biosyngas, microalgae have been explored which are a diverse group of prokaryotic and eukaryotic photosynthetic that grows promptly due to their non-complex structure.

The biorefinery approach provides different technologies which are capable of converting this biomass into value added chemicals, products (food and fuel), and biofuels (biodiesel, bioethanol, biohydrogen). Biorefinery is a process that assimilates biomass conversion processes and equipments to produce value added chemicals, biofuel, etc., from biomass. The most efficient producers of biomass of the potential sources of biofuel are the photosynthetic microalgae and cyanobacteria.

This article presents comprehensive information on the production of biofuel from algal-biomass using biorefinery approach.

Keywords: Biofuel, algal biomass, biorefinery, biomass

16.1 Introduction

Algae have been concluded to have exclusive potential to yield a variety of biofuels concomitantly with the generation of value added products and also phycoremediation of

**Corresponding author*: asaha@kol.amity.edu

Arindam Kuila and Mainak Mukhopadhyay (eds.) Biorefinery Production Technologies for Chemicals and Energy, (335–346) © 2020 Scrivener Publishing LLC

wastewater and some algal strains like *Chlamydomonas, Chlorella, Scenedesmus,* and *Btryococcus braunii* have been reported to produce biofuel.

Algae are the kind of organisms which can be cultivated in aquatic environment and light and carbon dioxide is used to create biomass. There are two classifications of algae, namely, microalgae and macroalgae. Microalgae are tiny, unicellular algae that grow in suspension within a body of water and are measured in micrometers. Macroalgae are multicellular and large algae that often grow in ponds and are measured in inches. Seaweeds are the largest multicellular algae and Giant Kelp plant is an example which can be more than 100 feet long.

Microalgae have been recognized as potentially good sources for biofuel production because they have relatively higher oil content as compared to macroalgae and rapid biomass production. Also, microalgae grow very quickly, and the algal mass culture practice can be performed on non-arable lands using wastewater and non-potable saline water. Thus, using microalgae for the production of biofuel is gaining interest from researchers, general public, and the entrepreneurs.

The biorefinery approach has been reorganized as the most promising way to discover a biomass-based industry and is determined as the sustainable biomass process to obtain biofuel and high-valued products through processes for biomass. Biofuel can be produced from microalgae biomass such as biohydrogen, bioelectricity, bioethanol, and biodiesel.

Microalgae have the ability to perform photosynthesis where carbon dioxide is taken from the atmosphere and oxygen is produced during the process. Also, microalgae have the potential to produce high-value compound structures such as neutraceuticals and compound for human consumption and low-value added products like biofertilisers, bioenergy, etc. It also helps in wastewater treatment, nitrogen fixation, and carbon dioxide mitigation.

Microalgae have been considered as a feedstock for renewable biofuels production, like bioelectricity, methane produced by anaerobic digestion of the algal biomass, biohydrogen produces under anaerobic condition, and bioethanol and biodiesel derived from microalgal oil [10].

Because microalgae offer higher growth rate as compared to renewable resources and require less water as compared to terrestrial crops, therefore, it is considered as a sustainable or raw material for the production of biofuel. Also, microalgae have a very short harvesting cycle (1–10 days) as compared to other land-based feedstock and do not compete with food or arable lands for algae cultures.

Biorefinery is a process of fractionation of products from biomass for different applications. A biorefinery is "the sustainable processing of biomass into a spectrum of bio-based products (food, fuel, chemicals and materials) and bioenergy (biofuels, power, and/or heat)" [12]. The biorefinery of the algae is all about extracting value from bioactive proteins, carbohydrates, pigments, lipids, and all other different metabolites produced by algae during its real-time growth.

16.2 Algal Biomass: The Third-Generation Biofuel

Algae are considered as raw material for production of biofuel. Microalgae biofuels [1] are recognized as third-generation type of biofuel that are considered as an alternative source for fossil fuels without the disadvantages linked with the first- and the second-generation biofuel.

Table 16.1 Biofuels yields from algae [2–9].

Species	Product	Yield
Chlamydomonas reinhardtii (CC124)	**Biohydrogen**	102 ml/1.2 L
		0.58 ml/hl
		0.30 mol/m^2
		0.6 ml/L h
Chlamydomonas reinhardtii (Dang 137C mt+)		175 ml/L
		4.5 mmol/L
		71 mL/L
Chlorella vulgaris MSU 01		26 ml/0.5L
Scenedesmus obliquus		3.6 ml/µg Chl a
Platymonas subcordiformis		11,720 nl/h
		7.20 ml /h
		0.339 ml/hl
Dunaliella tertiolecta	**Bio-oil**	43.8%, 34 MJ/Kg
		42.6%, 37.8 MJ/Kg
		25.8%, 30.74 MJ/Kg
Chlorella protothecoides		52%
		57.9
Chlorella sp.		28.6%
Chlorela vulgaris		35.83%
Nannochloropsis sp.		31.1%
Chlorella vulgaris	**Biogas**	0.63–0.79 LCH4 /gVS
Dunaliella salina		0.68 LCH4 /gVS
Euglena graciis		0.53 LCH4 /gVS
Scenedesmus		140 LCH4 /KgVS
Scenedesmus (Biogas from lipid-free biomass)		212 LCH4 /KgVS
Scenedesmus (Biogas from amino acidsfree biomass)		272 LCH4 /KgVS

(*Continued*)

Table 16.1 Biofuels yields from algae [2–9]. (*Continued*)

Species	Product	Yield
Botryococcus braunii	**Lipid content for Biodiesel**	25%–75%
Chlorella sp.		28%–32%
Chlorella vulgaris		56%
Crypthecodinium cohnii		20%
Monallanthus salina		20%–70%
Nannochlorisis sp.		20%–35%
Nannochloropsis sp.		31%–68%
Neochloris oleoabundans		35%-54%
Nitzschia sp.		45%–47%
Scenedesmus dimorphus		6%–40%
Scenedesmus obliquus		11%–55%
Schizochytrium sp.		77%
Chlorella pyrenoidosa	**Carbohydrates content for Bioethanol**	26%
Chlorella vulgaris		12%–17%
Dunaliella salina		32%
Scenedesmus obliquus		10%–17%
Porphyridium cruentum		40%–57%
Euglena gracilis		14%–18%

The first-generation biofuels are generally obtained from crop plants like soybean, corn, maize, sugarcanes, and sugar beet; rapeseed oil, vegetable oil, palm oil, and animal fats. But this type of biofuels has created a lot of negative impacts on food security, water variety, global food markets, and deforestation. The second-generation biofuel is obtained from non-edible oils (*Jatropha curcas, Pongamia pinnata,* etc.), lignocellulosic biomass, and forest residues and requires huge areas of land. Currently, the second-generation biofuel production lacks efficient technology and inventive ideas for commercial exploitation of wastes as source of biofuel generation.

As mentioned above about the drawbacks associated with first-generation biofuel and second-generation biofuel, the third-generation biofuel, that is, microalgae biofuels seem to be a viable alternative source of energy to replace or supplement fossil fuels.

16.2.1 Algae as a Raw Material for Biofuels Production

Algae biofuel (Table 16.1) is commercially feasible because it is competitive regarding cost as compared to fossil-based fuels. It also does not require extra lands, and it consumes atmospheric carbon dioxide thereby improving air quality. Also, it requires minimal water for its growth. However, it also have some disadvantages such as low biomass production, low lipid content in the cells, and the size of the cell is small so harvesting process becomes very costly.

But these restrictions can be overpowered by using advanced technologies for harvesting, drying, and genetic engineering of metabolic pathways for higher growth rate and increased lipid content. The microalgae were evaluated as a potential source for production in 1970 but were temporarily suspended due to technical and economic problems. The subsequent studies about the potential of the microalgae biofuel began from 1980.

An algae-based industrial process includes two types: the first type, that is, high-volume low value which includes biofuels, carbon dioxide bioremediation, and wastewater treatment. The second type, that is, low-volume high value that consists of neutraceuticals, cosmoceutical, and pharmaceutical applications biomass consists of variety of algae components like proteins, lipids, etc., which makes them feasible feedstock for biofuel production. But, because the production cost is higher for algae cultivation, it is a challenge to commercialize algae biofuels. Hence, to enhance the economic feasibility of algae biofuel production, extraction of other high-value added products are necessary to maximize the economic returns on the high-value added products as compared to the low-value products to be burned for energy.

16.2.2 Algae as Best Feedstock for Biorefinery

Algae, being the photosynthetic microorganisms, show tremendously enormous biological diversity and metabolic plasticity as contrasted to terrestrial plants, that is, they can more widely adapt to biochemical metabolic pathways.

When it comes to the capability to produce fuel, no feedstock can be compared with the potential of algae in terms of quality or diversity. The algae microorganism has two characteristics which help in the manufacture of different types of fuels: first, algae can produce an oil that can be easily refined to provide diesel or even certain components of gasoline. Secondly, it can be genetically manipulated to produce all from ethanol and butanol to even gasoline and diesel fuel precisely. Algae can produce biodiesel, butanol, gasoline, methane, ethanol, vegetable oil, and jet fuel.

Not only algae can produce different type of fuels, it is also capable of producing phenomenal yields. In fact, algae have been utilized for the production of up to 9,000 gallons biofuel per acre as indicated in Table 16.2, which is 10-fold to what the best traditional feedstock

Table 16.2 Yield in oil/ha of selected biomass [24].

Culture	Yield of oil (L/ha)
Microalgae	58.700
Palm	5.950
Coconut	2.689
Canola	1.190
Soy	446
Corn	172

have been able to provide. According to the U.S. Department of Energy, yields that are 10 times greater than second-generation biofuels mean that only 0.42% of the U.S. land area would be required to generate biofuel that is enough to meet all the needs of U.S., given that U.S. is the largest consumer of fuel in the world [14]. Some of the advantages of microalgae over any other biomass are given below:

1. Microalgae have the ability to synthesize and acquire huge quantities of neutral lipids/oils (20%–50% dry cell weight). They also grow in high rate (example, 1–3 doublings per day) [13].
2. Oil yield per area of microalgae overtakes the oil yield of best oilseed crops.
3. Microalgae can be cultivated in marginal lands (desert, arid, and semi-arid land) which are not suitable for conventional agriculture.
4. Microalgae can also be grown or cultivated in saline/brackish water/coastal seawater on non-arable lands and do not clash for resources with conventional agriculture.
5. Microalgae can sequester carbon dioxide from flue gases emitted from fossil fuel-fired power plants and other sources and reduces greenhouse gas emission (1 kg of algal biomass requires about 1.8 kg of carbon dioxide) [15].
6. It also develops value added coproducts or byproducts and does not involve pesticide or herbicide for its growth (example, biopolymers, proteins, fertilizers, etc.).

In addition to microalgae, there are also myriad of macroalgae species like seaweeds that can be developed as biomass feedstock as well as many potential pathways to produce bioenergy or bio-based products. AD Technologies is among the most researched methods for converting seaweeds to biogas. However, AD Technologies approached for macroalgae may prove to be problematic over time because of its capability for high salinity and sand accumulation. It is also unlikely to harvest seaweed at a scale that would be sufficient to provide significant quantities for transportation fuel or to meet the unceasing supply needs for biofuel-focused biorefinery. Seaweed-based production for bioenergy products is in the initial stage of development and is still yet unexplored for which species will be more efficient for bioenergy production. Various parameters, including the method of cultivation, species of seaweed, seaweed yield hectare, time of harvest, suitability of seaweed to ensiling, gross and net energy yields of biogas, carbon balance, cost of the harvested seaweed, cost of the harvested seaweed, cost of the produced biofuel, etc., have not yet been sufficiently evaluated [16]. To understand the potential for microalgae-based production [11], more investigation or research is required.

16.3 Microalgal Biomass Cultivation/Production

Most algal species are photosynthetic in nature and requires light and carbon dioxide as energy and carbon sources and this form is called as photoautotrophic, whereas some algae species can grow in darkness using organic carbons such as glucose or acetate as energy and carbon sources and are called heterotrophic. Heterotrophic algal culture has high capital and operational cost which is difficult to justify for biodiesel production, and in order to minimize cost, photoautotrophic culture that uses sunlight as source of energy is usually reliable for algal biomass production.

Phototrophic microalgae require water, carbon dioxide, light, and inorganic salts for its growth. For optimal growth of this kind of algae, the culture temperature should be between 15°C and 30°C (~60°F–80°F). The growth medium must be provided with inorganic elements like nitrogen, phosphorus, iron, etc., which helps in making up the algal cell. For large-scale production of microalgae, to prevent the algal biomass from settling, the algal cells should be continuously mixed and nutrients should be provided during daylight hours when the algae are reproducing but through respiration, during the night, up to one-quarter of algal biomass produced during the day can be lost.

There are different photoautotrophic culture systems are available having their own advantages and disadvantages. The commonly used system for algal biofuel production is the suspension-based open ponds and enclosed photobioreactors (PBRs).

16.3.1 Open Pond Production

These microalgae cultivating methods have been practiced since 1950s [17] and are the oldest and simplest system of all and are commonly used for microalgae production. Currently, approximately 98% of commercial algae are produced in this system [18].

There are distinctive types of open pond systems depending on the size, shape, and material that are used for construction, type of agitation, and inclination. Raceways are the most regularly used artificial system. Open pond system is cheaper contrasted to closed PBRs for large cultivation of microalgae and does not challenge agricultural crops for land as they need minimal crop production areas. Open ponds are rectangular grid having closed loops with recirculation channel and usually operates when water depth is 15–20 cm. And for mixing and circulating the algal biomass, paddlewheel is present. The maintenance and cleaning of these ponds are easier, require comparatively low energy, and are more scalable.

The drawbacks of these kinds of system are the contamination from the air and ground for algae cultivation, and therefore, many species that are cultured in this system are provided with alkalinity and salinity environment. Also these are limited by abiotic growth factors like temperature, pH, light intensity, and dissolved oxygen concentration.

16.3.2 Closed Bioreactors/Enclosed PBRs

Closed bioreactors support up to five-fold higher productivity as compared to reactor volume and on yield basis, they have a smaller "footprint". Not only it saves water, energy, and chemical, it also has many other advantages, which makes them the preferred reactor for biofuel production as they reduce the cost.

Enclosed PBRs are mostly accessible as tubes, bags, or plates that are made up of glass, plastic, or other transparent materials. Here, algae are cultivated by providing light, nutrients, and carbon dioxide. Some PBR designs include annular, flat-panel reactors, and tubular with large surface area.

16.3.3 Hybrid Systems

This is the combination of both open ponds as well as close bioreactor system to get optimum results. Open ponds are very efficient method of cultivating algae but they have the possibility of becoming contaminated with superfluous species very rapidly, and therefore, the combination of both the system is more profitable for cultivation of high yielding strains for biofuels. In this system, the inocula, that is free from contamination is obtained from PBRs and transferred to open ponds to yield maximum biomass. But, this system is more expensive and involves batch culture system process than a continuous culture system; therefore, it is not much relevant for biofuel production.

16.4 Strain Selection and Microalgae Genetic Engineering Method Strain Selection Process for Biorefining of Microalgae

Strain selection is the key to fundamental issues for any microalgae biorefinery process. Polycultures give rise to low-value products, whereas monocultures give higher-value products with control over better quality product and also undertake extensive environmental optimization for increased productivity and performance. Another advantage of monocultures is the capacity to undertake extensive environmental optimization for increased productivity and performance. In recent years, with the advent of genomic technologies, genetic manipulation has redefined the metabolic potential of microalgae by offering the ability to enhance existing, and even insert entirely new, strain properties [19].

The few microalgae strains that are developed successfully at commercially relevant industrial scale are: *Haematococcus pluvialis* for production of the red pigment and potent antioxidant, *astaxanthin*; *Dunaliella salina* or bardawil for β-carotene; the cyanobacteria *Arthrospira platensis* or maxima (known as *Spirulina* sp.) for phycocyanin and biomass for food applications; *Chlorella vulgaris* for food applications and more [20].

The genetic engineering process is facilitated by the knowledge of the genome of the interested microalgae and *Chlamydomonas reinhardtii* is the first microalgae to have its genome sequenced. Since then, many strains have been sequenced with the new generation sequencing technologies.

To facilitate the insertion of the foreign DNA into microalgae, different protocols have been developed of which the most common ones are electroporation, glass beads, bolistic and *Agrobacterium*-mediated transformations. Also with the development of new improved genetic engineering tools, chloroplast transformation protocols are also being developed to deal with complex sequence manipulation. The CRISPR-Cas9 system permits scar-free gene modification and can be utilized to better understand the phenotype after knockout of specific genes or to upgrade the manufacturing of specific molecules blocking competitive pathways [21, 22].

The limitation of this type of genetically modified technologies to microalgae is the higher cost that is associated with strain development. Other than this, the other drawbacks like non-sterile condition of microalgae to allow industry-controlled distribution like with genetically modified plant varieties and the protection of properties of strains that can go airborne and be "isolated naturally" from surrounding sites poses a bigger threat to commercial risk. But provided, the returns being sufficient, genetically modified microalgae can change scenario to process economics. Currently, the genetically modified microalgae regulations are evolving rapidly like gene editing, which includes changes in native genes.

16.5 Harvesting Methods

The most challenging aspect of the biorefinery process is the harvesting process of microalgae because of its large operational cost at industrial scale (20%–30% approx. of total downstream processing cost). The type of harvesting method depends on the strain and culturing conditions and on the utilization of biomass and the type of product that needs to be derived. Harvesting of microalgae is removing a huge amount of water from a little amount of suspended biomass. The overall cost of harvesting biomass per unit volume gets reduced with the increase in culture density.

To concentrate biomass before centrifugation, flocculation is a process that reduces the cost, but in the harvested biomass and also in the spent medium, the residues of flocculation can be found, which can create complications for commercial application. For harvesting microalgae, magnetic nanoparticle (Fe_2SO_3) have been researched which adsorb on the outside of the cells and works better in the presence of cationic polymer coating, by increasing the adsorption rate on the negatively charged microalgal cell surface. The application of magnetic field helps in achieving two processes in one step, that is, flocculation and separation and also helps in recovering the nanoparticles that can be reused again after harvesting. Genetic manipulation of *Chlamydomonas* has been utilized to discover magnetic cells through overexpression of the iron binding protein ferritin in the chloroplast that, associated with high iron level in the medium right before harvesting, permitted magnetic separation of the biomass, facilitating lower energy cost dewatering [23].

16.6 Cellular Disruption

To begin the isolation or fractionation of components of importance, the extraction process in a biorefinery depends on cellular disruption, which is an integral step and can be divided into four distinct approaches: mechanical, physical, biochemical, and biological. However, these processes are not mutually exclusive.

The bead milling and high pressure homogenization are included in mechanical approaches. However, these processes are energy expensive and dispersion of cell debris is created which can complicate the purification of different fractions.

Pulse Electric Field (PEF) or Ultrasound Treatment (UT) as well as microwave treatment are included in physical cell disruption approaches. PEF generally has low efficiency of microalgae because of the presence of a complex which protects them from lysis and is a low shear and low temperature method. UT confers high shear stress and is performed in low temperatures, but the cost is significant in case of controlling the temperature of the biomass during the process.

Biochemical approaches include enzyme digestion and alkali/acid treatments that breaks microalgal cells. Alkali/acid treatment can degrade valuable components, whereas enzyme digestion is a process that promotes the weakening and /or degradation of the cell wall and is generally expensive when applied at industrial scale. The permeability of some microalgal cells can be increased with the mixing of different degradation enzymes and when coupled with PEF, the approach becomes effective.

Biological disruption for certain microalgal strains can only be achieved where virus has been described.

16.7 Extraction

The most profitable compounds present in microalgae are carbohydrates, proteins, pigments, and oils (sterol, fatty acids), unless a distinct high-value molecule is produced by that algae. In the first step, the polar and non-polar molecules are separated by using organic solvents and affect the quality of the proteins. In the second step, using HPCL methods, components of two fractions can be purified.

In a biorefinery proposition, the various components of a biomass are extracted without hampering the other components. Cascade extractions are often necessary where the drivers for the cascade order are yield, stability, and value.

To approach the biorefinery concept, biomass valorization is required to be handled as a full process to be optimized by avoiding multistep process. After the completion of the process of extraction of valuable compounds, the waste biomass that remain behind can be used for the production of gas, bio char, and bio oil through the process of Hydrothermal Liquification (HTL).

16.8 Conclusion

As the cost of petroleum and petrochemicals are escalating rapidly, it is accelerating the people to shift from conventional sources of energy toward more environment-friendly, low cost chemical products derived from renewable sources of energy, and microalgae have the potential of becoming sustainable renewable energy feedstock and could meet the global demand for energy. Despite the advantages of microalgae biofuel, they also have some disadvantages like the small cell size property which makes the harvesting process costly. But, this can be overcome by developing new technologies that are efficient for biomass harvesting, drying, and oil extraction at low cost and also by designing advanced PBRs.

The biorefinery approach with regard to manufacturing of a high-valued product from microalgae and the harvesting and the extracting processes of primary and secondary components of microalgae can become costly for any industrial scale. But this limitation could be overcome by advancing new technologies, by upgrading algal biology (in terms of biomass yield and oil content) and through culture system engineering. Besides, production of high-valued products from microalgae along with low-valued products through biorefinery process is a way to reduce the algal biofuel production cost. Therefore, further research in the advancement of technologies and in ways of exploiting both microalgae and macroalgae can benefit the economic production of biofuels.

References

1. Jones, C.S and Mayfield, S.P., Algae biofuels: versatility for the future of bioenergy. *Curr. Opin. Biotech.*, 23, 3, 346–351, 2012.
2. Singh, A. and Olsen, S.I., A critical review of biochemical conversion, sustainability and life cycle assessment of algal biofuels. *Appl. Energy*, 88, 3548–3555, 2011. 7.
3. Chisti, Y., Biodiesel from microalgae. 25, 294–306, 2007.

4. Das, D. and Veziroglu, T., Advances in biological hydrogen production processes. *Int. J. Hydrogen Energy*, 2008.
5. Gouveia, L. and Oliveira, A.C., Microalgae as a raw material for biofuels production. *J. Ind. Microbiol. Biotechnol.*, 36, 269–74, 2009.
6. Marcilla, A., Catalá, L., García-Quesada, J.C., Valdés, F.J., Hernández, M.R., A review of thermochemical conversion of microalgae. *Renew. Sustain. Energy Rev.*, 27, 11–19, 2013.
7. Oncel, S.S., Microalgae for a macroenergy world. *Renew. Sustain. Energy Rev.*, 26, 241–264, 2013.
8. Ramos-Suárez, J.L. and Carreras, N., Use of microalgae residues for biogas production. *Chem. Eng. J.*, 242, 86–95, 2014.
9. Demirbas, M.F., Biofuels from algae for sustainable development. *Appl. Energy*, 88, 3473–3480, 2011.
10. Yen, H.W., Hu, I.C., Chen, C.-Y., Ho, S.-H., Lee, D.-J., Chang, J.-S., Microalgae-based biorefinery – From biofuels to natural products. *Bioresour. Technol.*, 135, 166–174, 2013.
11. https://www.sciencedirect.com/science/article/pii/S096085241201601X
12. Jungmeier, G., Ree, R.V., de Jong, E., Stichnothe, H., de Bari, I., Wellisch, M., Bell, G., Spaeth, J., Torr, K., Kimura, S., The "Biorefinery Fact Sheet" and its Application to Wood Based Biorefining-Case Studies of IEA Bioenergy Task 42 "Biorefining". *Biorefining Circ. Econ.*, 1–6, 2015.
13. https://www.academia.edu/9384659/Algal_Biorefinery_A_Road_towards_Energy_Independence_and_Sustainable_Future?auto=download
14. http://biofuel.org.uk/third-generation-biofuels.html
15. Rodolfi, L., Zittelli, C.G., Bassi, N., Padovani, G., Biondi, N., Biondietal, G., Microalgae for oil: strain selection, induction oflipid synthesis and outdoor ma.ss cultivation in a low costphotobioreactor. *Biotechnol. Bioeng.*, 102, l, 100–I 12, 2009.
16. https://www.osti.gov/servlets/purl/1358333
17. Borowitzka, M.A., Commercial production of microalgae: ponds, tanks, tubes and fermenters. *J. Biotechnol.*, vol. 70, no. 1–3, pp. 313–321, 1999. View at: Publisher Site | Google Scholar.
18. Sheehan, J., Camobreco, V., Duffield, J., Graboski, M., Shapouri, H., An overview of biodiesel and petroleum diesel life cycles, in: *US Department of Agriculture and Energy Report*, 1998, View at: Google Scholar.
19. Singh, P., Kumari, S., Guldhe, A., Misra, R., Rawat, I., Bux, F., Trends and novel strategies for enhancing lipid accumulation and quality in microalgae. *Renew. Sustain. Energy Rev.*, 55, 1–16, 2016. [CrossRef].
20. Available online: https://www.algenuity.com/algenuitys-chlorella-colours-platform-launches-at-this-yearsvitafoods-europe (accessed on 19 August 2019).
21. Nomura, T., Inoue, K., Uehara-Yamaguchi, Y., Yamada, K., Iwata, O., Suzuki, K., Mochida, K., Highly efficient transgene-free targeted mutagenesis and single-stranded oligodeoxynucleotide-mediated precise knock-in in the industrial microalga Euglena gracilis using Cas9 ribonucleoproteins. *Plant Biotechnol. J.*, 17, 2032–2034, 2019. [CrossRef] [PubMed].
22. Naduthodi, M.I.S., Mohanraju, P., Südfeld, C., D'Adamo, S., Barbosa, M.J., van der Oost, J., CRISPR–Cas ribonucleoprotein mediated homology-directed repair for efficient targeted genome editing in microalgae Nannochloropsis oceanica IMET1. *Biotechnol. Biofuels*, 12, 66, 2019. [CrossRef] [PubMed].
23. Sayre, R. and Postier, B., Modification of microalgae for magnetic properties. *US Patent App.*, 13/808, 746, 2013.
24. https://www.intechopen.com/books/biofuels-state-of-development/cultivation-systems-of-microalgae-for-the-production-of-biofuels

17

Biogas Production and Uses

Anirudha Paul[1], Saptarshi Konar[2], Sampad Ghosh[3] and Anirban Ray[4*]

[1]India Dairy Products Limited, Dankuni, West Bengal, India
[2]Molecular Biology and Biotechnology Department, University of Kalyani, Kalyani, West Bengal, Kalyani, India
[3]Department of Chemistry, Nalanda College of Engineering, Nalanda, Bihar, India
[4] Department of Microbiology, R.G. Kar Medical College, Kolkata, West Bengal, India

Abtract
Fossil fuels are the main non-renewable energy sources used to sustain human civilization. Biogas is an eco-friendly, renewable energy source which can be produced from low cost substrates by anaerobic digestion. In this regard, the concept of biorefinery is based on utilization of bio-based raw materials for the production of different value added chemicals and bio fuel, energy, and heat. Now a days, state of art green-technology offers maximum utilization of resources with minimum wastage and pollution, although there are still some aspects which should be improved. The major objective of this paper is to give an idea on different energy sources, props and techniques applied for the industrial production of biogas, along with future perspective.

Keywords: Biogas, fossil fuel, biorefinery, green technology, non-renewable energy sources, renewable energy sources, process & purification technology, biowaste

17.1 Introduction

Fossil fuels such as oil, coal, and natural gas are the main non-renewable energy sources till now used for civilization, development, and industrialization. Most of the fossil fuels are used either to produced power and heat or as fuel for transportation. Very low amount is being used for the production of value added chemicals [1]. Apart from this, many key factors such as fossil fuel depletion and its global localization, high amount of green house gas emission during purification, fatal political issues associated with fossil fuel reserving countries and vulnerability in global economy regarding oil price hike has draw attention of many countries toward new alternatives [2]. Biogas is an eco-friendly, renewable energy source, that can be produced from low cost various substrates by anaerobic digestion.

Biorefinery concept is based on utilization of bio-based raw materials for the production different value added chemicals along with bio fuel, energy, and heat. Such flexible multi technology offers many advantages including reduction in waste management, decrease in

**Corresponding author*: anirbanrayiitkgp@gmail.com

Arindam Kuila and Mainak Mukhopadhyay (eds.) Biorefinery Production Technologies for Chemicals and Energy, (347–356) © 2020 Scrivener Publishing LLC

green house gas emission, availability of low-cost renewable material worldwide, and most importantly eco friendly, green technology. Different substrates such as energy crops, straw, food waste, wood and wood waste, algae, and fish wastes are of major interest, and different studies are still running for proper improvement and implementation as large-scale production. The major objective of this paper is to give an idea on different energy sources, processes, and purification technologies that have been applied for the industrial production of biogas, along with future perspective.

17.2 Potential Use of Biogas

The non-upgraded form of biogas is widely used for heat and electricity. Cogeneration is a process where biogas is utilized for Central Heat and Power (CHP) production. Trigeneration is another process where the heat produced from CHP engine is being used for different cooling purposes (combined cooling, heat, and power) [3]. The upgraded form of biogas (by removing CO_2, water, and H_2S) mainly consists of methane and high energy content. Such clean biomethane mainly used for transportation of bus, cars, and trucks. Through national transmission network, upgraded biomethane is easily available to different customers and industries [4]. Upgraded biogas is very clean and odor free and can therefore be used in food and bakery industries for continuous heat and energy supply. In a variety of distiller, brewery, and creameries industries where natural gas is the primary choice of energy for evaporation, biogas can be used. Biogas can also be used in roasting of coffee, chips, and biscuits. For fast and efficient air drying in carpentries and laundries, biogas can be a great choice.

17.2.1 Anarobic Digestion

Under anaerobic condition, organic matters are subjected to biological degradation in anaerobic digestion technique to produce biogas [5]. The first step in this process is hydrolysis of organic material, where different microbes are applied to convert polymeric carbohydrate, protein, and lipid into its simple forms such as sugar, amino acid, fatty acid, and glycerol. In the next fermentation step, acidogenic bacteria are utilized to produce alcohol and volatile fatty acids. The third step is known as acetogenesis, where acetogenic bacteria (*Acetobacterium woodii* and *Clostridium aceticum*) acted on volatile fatty acid to yield acetate, carbon dioxide, and hydrogen. In the final step, methanogenic bacteria (*Methanosarcina barkeri, Methanococcus mazei*, and *Methanotrix soehngenii*) are utilized for methanogenesis [6]. Apart from such variation, a wide range of diverse bacterial special also play crucial role in this complex biochemical process. For the first two steps, anaerobic (*Clostridia, Bacteriocides*, and *Bifidobacteria*) and facultative anaerobic bacteria (*Streptococci and Enterobacteriaceae*) are the main choice [7]. In the final step, methane is produced either by cleavage of two acetic acid molecule into methane and carbon dioxide or by reduction of carbon dioxide. Anaerobic digestion is generally carried out in to two different temperature conditions, *viz.*, mesophillic temperature (35°C–42°C) and thermophillic temperature (45°C–60°C) [7]. However, the temperature optimum varies according to the composition of feedstocks and types of digester used. The retention time in anaerobic digestion is depended on the temperature condition type. For mesophillic condition, it is 15–20 days, where as in case of thermophillic condition, it is 12–14 days. Anaerobic digestion is also a pH-dependent process. Different pH optima are required for acedogenesis and methanogenesis to operate successfully. The optimum ranges between pH 7 and 8. Fluctuation in pH range strongly inhibited the process (below pH 6 and above pH 8.5). Accumulation of ammonia increases the pH, while accumulation of volatile fatty acid decreases the pH value. Therefore, effective control parameters are required. Generally, the microbial growth rate is very low in anaerobic digestion; therefore, key nutrient ratio should be maintained. For effective biogas production, the nutrient ratio is C:N:P:S = 600:15:5:1. Additionally, different micronutrients such as cobalt, iron, molybdenum, selenium, and nickel are also added in trace amount to promote better microbial growth [7].

17.2.2 Biogas from Energy Crops and Straw

Various sugar (sugar beet) and starch (corn, potato) crops are mainly used for the commercial production of biogas. Other main sources are the lignocellulosic plant materials, however, due to complex cell wall structure such substances, are far more resistant to microbial fermentation, and therefore, various pre-treatment processes are required for large-scale production and implementation of biogas [6, 8]. Such lignocellulosic materials are mainly composed of cellulose, hemicelluloses, and lignin, which make them more resistant to enzymatic degradation. After pre-treatment, lignocellulosic compounds are separated into solid and liquid fractions. The liquid fraction mainly contains hemicelluloses, a pentose sugar, while solid fraction consists of cellulose, a polymer of glucose [9, 10]. As lignocellulosic materials are rich in carbon but deficient in nutrient content, therefore, digestion with other substrate is required for high and progressive biogas yield [11].

17.2.3 Biogas from Fish Waste

All over the world, fish processing industrial waste generally contains different fish parts such as scales, viscera, and fish head. Such waste materials are rich in high amount of proteins and fats and thus further processing (after extraction of fish oil, enzymes, essential fatty acids, etc.) can give rise to feedstock for anaerobic digestion. According to Nges *et al.* [12], salmon fish waste can be used for biogas production. Similarly sardine, tuna, needle fish, and mackerel fish waste can be used for biogas production [13]. Although fish waste can be used as biogas production but risk factors are associated with accumulation of long chain fatty acid and ammonia in the bioreactor. For stable digestion, addition of more carbohydrate rich co-inoculam is of better choice.

17.2.4 Biogas from Food Waste

Wasted food materials can also be used as the feed stocks for biogas production. According to Food and Agricultural Organization (FAO) of United Nation (UN), one third of the food produced for human consumption is mistreated globally. According to Al Seadi *et al.* [14], an average of 1.3 billion tons per year, food is lost. Many countries, especially Sweden, have taken strong initiative to manufacture biogas by anaerobic digestion from such food waste. Reports showed that 32% of food waste was treated for the recovery of nutrient and energy as biogas.

17.2.5 Biogas from Sewage Sludge

The sewage treatment plants in Sweden are able to yield biogas from the anaerobic digestion of both primary and secondary sludge. Such sludge generally contains microbial biomass and certain extra cellular polymeric substances (EPS). Carrère *et al.* [15], suggested that only 30%–50% of the EPS can be digested for the biogas production. Different advanced pre-treatment (physical, chemical, and biological) can lead to high amount of biogas production.

17.2.6 Biogas from Algae

Both microalgae and macroalgae have high potential for biogas production. According to Debrowski *et al.* [16], suitable nutrient supply can lead to high biomass production with minimum effort and also beneficial from economical point of view. On the other hand, absence of lignin in algal biomass makes it suitable for degradation and digestion. However, the protein content varies from species to species [17].

17.2.7 Some Biogas Biorefinery

BioGTS is a Finland based (founded in 2011) biorefinery company, which uses different biodegradable waste to produce cost effective biogas. It is accompanied with biogas upgradation unit for better quality of biogas. Through various Research and Development (R&D). It has managed to produce new generation biogas with high energy efficiency at low investment cost (www.biogts.com).

It is a Norway-based company. This biogas plant utilizes food waste to produce green or climate-friendly biogas. Households of Eastern Norway are the main source of food waste. Currently, the total amount of biogas produced here can replace 6.8 million liters of diesel per year.

17.3 Pre-Treatment

Prior to anaerobic digestion, different solid materials such as lignocellulosic substances, food waste, and waste from sewage sludge might need different pre-treatment for better biogas yield along with less waste production. Different pre-treatment processes and parameters were studied for proper application [18, 19]. Different pre-treatment processes includes i) physical; ii) physiochemical; iii) chemical; and iv) biological process, which are described below.

17.3.1 Physical Pre-Treatment

Milling, irradiation, grinding, and chipping are the major physical pre-treatment processes used in biogas biorefinery. Milling generally used to reduce substrate size and making it appropriate for digestion and degradation. Both dry and wet materials are subjected to milling; however, excess milling may cause lose in carbohydrate concentration and require high energy consumption [20]. Irradiations with gamma rays, electron beam, or microwave are also applied for physical pre-treatment but the actual results were somehow fluctuating.

Some results indicated positive impact [21], whereas other results are associated with less importance in biogas production [22]. Application of high pressure is also sometime associated with enhancement in biogas yield [23]. Application of ultra-sonication has been reported to be associated with inclination in biogas yield [15].

17.3.2 Physiochemical Pre-Treatment

Sometimes, lignocellulosic substances are subjected to hot water pre-treatment to remove excess hemicelluloses [8]. Such pre-treatment is effective for further digestion. Substances with low carbohydrate content such as slaughterhouse waste showed good impact when are subjected to pasteurization [24]. Application of saturated steam at high pressure, alone or with different chemicals (SO_2, H_2SO_4, NH_3) and enzymes on lignocellulosic substances, had shown positive result on biogas production by anaerobic digestion. In such processes, the temperature varies from 160°C–260°C with different time exposures [8, 18].

17.3.3 Chemical Pre-Treatment

Chemical pre-treatments have been done to improve biogas yield. Concentrated acid treatment at low temperature or dilute acid treatment at high temperature have shown to improve anaerobic digestion. However, due to corrosive action and complex recovery process in concentrated acid treatment at low temperature, most pre-treatments are carried out with low acid concentration at high temperature [25]. In another pre-treatment process, lignocellulosic materials are subjected to alkali treatment along with H_2O_2 at room temperature and had kept for 6 to 12 h. Such pre-treatment processes have shown to remove lignin and thereby making it comfortable for anaerobic digestion [8]. Ozonolysis and other alkali treatments [8] such as concentrated NaOH, NH_3, or $Ca(OH)_2$ have also been associated with effective biogas production.

17.3.4 Biological Pre-Treatment

In biological pre-treatment process, different types of microbes are applied prior to anaerobic digestion. Such microorganisms are able to degrade lignocellulosic materials such as lignin, cellulose, hemicelluloses, etc. Brown and soft rot fungi are the most suitable choice and have been associated with good results [8].

17.4 Process and Technology

Biogas production is mainly dependent on dry fermentation or wet fermentation technology. In dry fermentation the total solid content maintained in fermenter, ranges between 15% and 35%. Most of the dry fermentation processes are operated by batch culture system or continuous culture system. In wet fermentation process, the total solid content maintained below 10% and mostly operated by continuous culture system in stirred tank bioreactor.

In the wet fermentation process, mostly continuous stirred tank bioreactor is used, which is of vertical type [26]. To achieve active stirring, pneumatic, mechanical, and hydraulic mixing were done. Such stirring is essential to maintain constant temperature in bioreactor and for proper mixing of microorganisms with feedstock. Most of the stirring is done by mechanical stirrer and depending on the rotational speed, they are subdivided into fast and slow mixers. Slow mixers are operated continuously, whereas fast mixers are operated with particular time interval. In most of the biogas plants, the temperature required for wet fermentation process is maintained between 38°C and 42°C with few exceptions, where the optimum temperature is maintained at 55°C. In such thermophillic condition, lowering the temperature may affect decrease biogas production due to low microbial growth but ammonia toxicity may be reduced [27].

In dry fermentation technique, generally, several batch processes are applied. Mechanical mixing processes are not required here. Depending on the type of solid substrate, preceding batch inoculam is mixed with solid substrate. Sprinkled water is added throughout the digestion period for proper inoculam mixing and to maintain temperature and moisture content in the closed digester. As most of the digestion process is exothermic, imbalance in water content may inhibit the process. After digestion (generally a period of 3–4 weeks), new set of batch is initiated. Continuous process is applied rarely in dry fermentation, where total solid content is more than 25%, and generally, vertical plug flow fermenters or horizontal mechanically mixed fermenters are used [28, 29].

17.5 Biogas Purification and Upgradation

Biogas thus produced contains different impurities including CO_2, N_2, O_2, NH_3, water vapor, hydrocarbon, siloxanes, H_2S, and some other compounds [30, 31]. The main artifacts are CO_2, H_2S, and water. Such compounds are corrosive and also reduce the concentration of biogas leading to decrease in calorific value. Thus, purification is needed for proper use of biogas.

17.5.1 Removal of CO_2

Removal of CO_2 is necessary to improve calorific value and density of biogas. Different technologies are applied in industrial processes. In physical adsorption process, liquid scrubbing is applied. In this technology, water is used as selective solvent, and depending on the solubility, different gases are absorbed at different extent. CO_2 is more polar than methane (CH_4), and therefore, the solubility of methane is less than CO_2. At room temperature (25°C), CO_2 is 26 times more soluble than CH_4 [32]. Application of organic substances (dimethyl ethers of polyethylene glycol and methanol) as solvent has shown to remove more CO_2 than water in industrial application [33]. In chemical adsorption process, different chemical compounds are used to remove CO_2, where selective reaction take place between solvent and CO_2. Different chemical such as solutions of K_2CO_3, KOH, NaOH, $FeCl_2$ and $Fe(OH)_3$, alkanol amines, or di-methyl ethanol amine (DMEA) are effectively used [34]. Among them, amine solutions are widely used because methane loss is reduced up to nil. In cryogenic separation process, the gas mixture is subjected to cooling at elevated pressure. As the boiling temperature of CO_2 is −78°C and of CH_4 is −160°C, application of such technology can effectively remove CO_2 [35], leading to pure liquid methane. In another technique, membrane separation process, selective permeability is applied to separate CO_2. In such case, either gas-gas or gas-liquid separation is applied. Hydrophobic porous selective membranes are used with pressure variation (high pressure or low pressure) and up to 97% methane can be produced [36].

17.5.2 Removal of H_2S

Removal of H_2S is essential because conversion of H_2S to SO_2 and H_2SO_4 is very common, and the final products are of corrosive in nature that can cause damage to bioreactor, engine, pipes, and pumps [35]. Liquid scrubbing is also applied here to remove H_2S either by using water or organic solvent [37]. Different chemical solutions are also used to convert H_2S to elemental sulfur or to metal sulphides [38]. In another technology, activated carbon is used to remove H_2S. In such process, activated carbon may cause oxidation of H_2S to elemental sulfur [35]. Microorganisms belonging to the genera Thiomonas, Thiobacillus, Sulfurimonas, Acidithiobacillus, Halothiobacillus, and Paracoccus are also applied in such circumstances. Such sulfur oxidizing microorganisms can convert H_2S to elemental sulfur [39]. In membrane separation process, alkaline solutions are used to remove H_2S up to 98% [37, 40].

17.5.3 Removal of Water

Biogas produced in the digester always is saturated with water depending on the release temperature. More the temperature, more amount of water is present. Such water can be removed by physical treatment such as condensation utilizing centrifugal force or by expansion [30, 31]. In chemical treatment process, adsorption principle is mostly applied. Different drying agents such as alumina, activated carbon, silica gel, glycol, and magnesium oxide [30] are used in such cases. Sometimes, hygroscopic salts are also applied to remove water content [41].

17.6 Conclusion

Due to economically attractiveness and as eco-friendly renewable energy source, biogas will become more popular in next few decades. Fast utilization of biogas in different public and industrial sector increases market value worldwide. Application and utilization of low-cost waste materials as feedstock will enhance the production. Different materials such as energy crops, straw, food waste, wood, fish waste, algae, and other lost cost materials are widely used for biogas production. Further researches and improvements are required for better monitoring, pre-treatment, and purification. In case of anaerobic digestion, microbial community plays crucial role. Further extensive studies also required to improve microbial community as well as isolation and identification of new beneficial microbes. Improving microbial community will smooth the process control parameters.

References

1. Bhaskar, T., Bhavya, B., Singh, R., Naik, D.V., Kumar, A., Goyal, H.B., Thermochemical conversion of biomass to biofuels, in: *Biofuels – Alternative feedstocks and conversion processes*, A. Pandey, C. Larroche, S.C. Ricke, C.G. Dussap, E. Gnansounou (Eds.), pp. 51–77, Academic Press;, Oxford, UK, 2011.

2. Cherubini, F. and Strømman, A.H., Principles of biorefining, in: *Biofuels - Alternative feedstocks and conversion processes*, A. Pandey, C. Larroche, S.C. Ricke, C.G. Dussap, E. Gnansounou (Eds.), pp. 3–24, Academic Press;, Oxford, UK, 2011.
3. Fagerström, A., Al Seadi, T., Rasi, S., Briseid, T., *The role of Anaerobic Digestion and Biogas in the Circular Economy*, vol. 37, J.D. Murphy (Ed.), p. 2018: 8, IEA Bioenergy Task, Ireland, 2018.
4. Wall, D., Dumont, M., Murphy, J.D., Green gas: Facilitating a future green gas grid through the production of renewable gas. *Int. Energy Agency (IEA) Bioenergy: Task*, 37, 2 2018, 2018.
5. Frigon, J.C. and Guiot, S.R., Biomethane production from starch and lignocellulosic crops: A comparative review. *Biofuels Bioprod Bioref.*, 4, 4, 447–58, 2010.
6. Zheng, Y., Zhao, J., Xu, F., Li, Y., Pre-treatment of lignocellulosic biomass for enhanced biogas production. *Prog Energy Combust Sci.*, 42, 35–53, 2014.
7. Weiland, P., Biogas production: Current state and perspectives. *Appl. Microbiol. Biotechnol.*, 85, 4, 849–60, 2010.
8. Taherzadeh, M.J. and Karimi, K., Pre-treatment of lignocellulosic wastes to improve ethanol and biogas production: A review. *Int. J. Mol. Sci.*, 9, 1621–1651, 2008.
9. Klinke, H., Ahring, B., Schmidt, A., Thomsen, A., Characterization of degradation products from alkaline wet oxidation of wheat straw. *Bioresour. Technol.*, 82, 1, 15–26, 2002.
10. Bercier, A., Plantier-Royon, R., Portella, C., Convenient conversion of wheat hemicelluloses pentoses (D-xylose and L-arabinose) into a common intermediate. *Carbohydr. Res.*, 342, 16, 2450–2455, 2007.
11. Mata-Alvarez, J., Dosta, J., Romero-Güiza, M.S., Fonoll, X., Peces, M., Astals, S., A critical review on anaerobic co-digestion achievements between 2010 and 2013. *Renew. Sustain. Energy Rev.*, 36, 0, 412–427, 2014.
12. Nges, I.A., Mbatia, B., Björnsson, L., Improved utilization of fish waste by anaerobic digestion following omega-3 fatty acids extraction. *J. Environ. Manage.*, 110, 159–165, 2012.
13. Eiroa, M., Costa, J.C., Alves, M.M., Kennes, C., Veiga, M.C., Evaluation of the biomethane potential of solid fish waste. *Waste Manage.*, 32, 1347–1352, 2012.
14. Al Seadi, T., Owen, N., Hellström, H., Kang, H., Source separation of MSW. An overview of the source separation and separate collection of the digestible fraction of household waste, and of other similar wastes from municipalities, aimed to be used as feedstock for anaerobic digestion in biogas plants. *IEA Bioenergy*, 2013. Ireland, Task 37 Technical Brochure, pp. 4–50.
15. Carrère, H., Dumas, C., Battimelli, A., Batstone, D.J., Delgenès, J.P., Steyer, J.P., Ferrer, I., Pre-treatment methods to improve sludge anaerobic degradability: A review. *J. Hazard Mater.*, 183, 1–15, 2010.
16. Dębowski, M., Zieliński, M., Grala, A., Dudek, M., Algae biomass as an alternative substrate in biogas production technologies - Review. *Renew. Sustain. Energy Rev.*, 27, 596–604, 2013.
17. Fleurence, J., Seaweed proteins: Biochemical, nutritional aspects and potential uses. *Trends Food Sci. Tech.*, 10, 25–28, 1999.
18. Menon, V. and Rao, M., Trends in bioconversion of lignocellulose: Biofuels, platform chemicals & biorefinery concept. *Prog. Energ. Combust.*, 38, 522–550, 2012.
19. Carlsson, M., Lagerkvist, A., Morgan-Sagastume, F., The effects of substrate pre-treatment on anaerobic digestion systems: A review. *Waste Manage.*, 32, 1634–1650, 2012.
20. Monlau, F., Sambusiti, C., Barakat, A., Guo, X.M., Latrille, E., Trably, E., Steyer, J.-, Carrere, H., Predictive models of biohydrogen and biomethane production based on the compositional and structural features of lignocellulosic materials. *Environ. Sci. Technol.*, 46, 12217–12225, 2012.
21. Kuglarz, M., Karakashev, D., Angelidaki, I., Microwave and thermal pre-treatment as methods for increasing the biogas potential of secondary sludge from municipal wastewater treatment plants. *Bioresour. Technol.*, 134, 290–297, 2013.

22. Mehdizadeh, S.N., Eskicioglu, C., Bobowski, J., Johnson, T., Conductive heating and microwave hydrolysis under identical heating profiles for advanced anaerobic digestion of municipal sludge. *Water Res.*, 47, 5040–5051, 2013.
23. Ferrer, I., Serrano, E., Ponsá, S., Vázquez, F., Font, X., Enhancement of thermophilic anaerobic sludge digestion by 70° pre-treatment: Energy considerations. *Jl Res. Sci. Technol.*, 6, 11–18, 2009.
24. Rodríguez-Abalde, A., Fernández, B., Silvestre, G., Flotats, X., Effects of thermal pre-treatments on solid slaughterhouse waste methane potential. *Waste Manage.*, 31, 1488–1493, 2011.
25. Galbe, M. and Zacchi, G., Pre-treatment of lignocellulosic materials for efficient bioethanol production. *Adv. Biochem. Eng. Biotech.*, 108, 41–65, 2007.
26. Gemmeke, B., Rieger, C., Weiland, P., *Biogas-Messprogramm II, 61 Biogasanlagen im Vergleich*, FNR, Gülzow, 2009.
27. Angelidaki I Ellegaard, L. and Ahring, B., Application of the anaerobic digestion process, in: *Biomethanation II*, Adv. Biochem Eng/Biotechnol, pp. 2–33, Springer, 2003.
28. De Baere, L. and Mattheeuws, B., State-of-the-art 2008—Anaerobic digestion of solid waste. *Waste Manage. World*, 9, 1–8, 2008.
29. Weiland, P., Verstraete, W., van Haandel, A., Biomass digestion to methane in agriculture: A successful pathway for the energy production and waste treatment worldwide, in: *Biofuels*, W. Soetaert and E.J. Vandamme (Eds.), pp. 171–195, Wiley, 2009.
30. Persson, M., Jonsson, O., Wellinger, A., Biogas upgrading to vehicle fuel standards and grid. *IEA Bioenerg.*, Ireland, Task 37, 1–32, 2007.
31. Bauer, F., Hulteberg, C., Persson, T., Tamm, D., *Biogas upgrading-review of commercial technologies*, vol. 83, SGC Rapport, Sweden, 2013:270. SGC Rapp, 2013.
32. Gavin, T. and Sinnott, R., *Chemical engineering design: Principles, practice and economics of plant and process design*, Elsevier Butterworth-Heinemann, Oxford, 2013.
33. Tock, L., Gassner, M., Maréchal, F., Thermochemical production of liquid fuels from biomass: Thermo-economic modeling, process design and process integration analysis. *Biomass Bioenergy*, 34, 1838–1854, 2010.
34. Lasocki, J., Kodziejczyk, K., Matuszewska, A., Laboratoryscale investigation of biogas treatment by removal of hydrogen sulfide and Carbon Dioxide. *Polish J. Environ. Stud.*, 24, 1427–1434, 2015.
35. Hosseini, S.E. and Wahid, M.A., Development of biogas combustion in combined heat and power generation. *Renew. Sustain. Energy Rev.*, 40, 868–875, 2014.
36. Bauer, F., Persson, T., Hulteberg, C., Tamm, D., Biogas upgrading—Technology overview, comparison and perspectives for the future. *Biofuels Bioprod. Biorefining*, 7, 499–511, 2013.
37. Ryckebosch, E., Drouillon, M., Vervaeren, H., Techniques for transformation of biogas to biomethane. *Biomass Bioenerg.*, 35, 1633–1645, 2011.
38. Schiavon, D.C., Cardoso, F.H., Frare, L.M., Gimenes, M.L., Pereira, N.C., Purification of biogas for energy use. *Chemical Engineering Transactions*, 37, 643–648, 2014.
39. Montebello, A.M., Aerobic biotrickling filtration for Andrea Monzón Montebello. *J. Hazard. Mater.*, 280, 200–208, 2013.
40. Scholz, M., Melin, T., Wessling, M., Transforming biogas into biomethane using membrane technology. *Renew. Sustain. Energy Rev.*, 17, 199–212, 2013.
41. Sun, Q., Li, H., Yan, J., Liu, L., Yu, Z., Yu, X., Selection of appropriate biogas upgrading technology-a review of biogas cleaning, upgrading and utilisation. *Renew. Sustain. Energy Rev.*, 51, 521–532, 2015.

18

Use of Different Enzymes in Biorefinery Systems

A.N. Anoopkumar[1,2], Sharrel Rebello[2], Embalil Mathachan Aneesh[2], Raveendran Sindhu[3]*, Parameswaran Binod[3], Ashok Pandey[4] and Edgard Gnansounou[5]

[1]*Department of Zoology, Christ College Irinjalakuda, University of Calicut, Kerala, India*
[2]*Communicable Disease Research Laboratory (CDRL), St. Joseph's College, Irinjalakuda, Kerala, India*
[3]*Microbial Processes and Technology Division, CSIR-National Institute of Interdisciplinary Science and Technology (CSIR-NIIST), Trivandrum, India*
[4]*Center for Innovation and Translational Research, CSIR- Indian Institute for Toxicology Research (CSIR-IITR), 31 MG Marg, Lucknow, India*
[5]*Bioenergy and Energy Planning Research Group, Ecole Polytechnique Fédérale de Lausanne (EPFL), Lausanne, Switzerland*

Abstract
Biorefineries are renowned as an integrative concept that merges essential technologies between the economically feasible biogenic raw material and final products. There is an aggregating trend toward the use of easily available, cheap, and renewable biomass in the production of several chemicals in bulk quantities in different biorefineries. A wide variety of enzymes from the microbial cells attained much more important in the biorefineries to convert biomass into final products. Most of the processes involved in biorefineries necessitate thermostable enzymes for effective substrate solubility, high transfer rate, together with the lowered risk of contamination. The thermostable enzymes can be beneficial as industrial catalysts since they can able to tolerate the harsh conditions during the industrial production processing. Here, we discuss the potential applications of different enzymes used in biorefinery systems.

Keywords: Enzymes, biorefinery, thermostable, bioproducts

18.1 Introduction

The sustainable uses of diverse resources are considered as the fundamental political concerns for the advanced developments in the 21st century. During the last two decades, the thermostable enzymes have been gained much more significance in the biorefineries than any other bioproducts [1]. In addition, several microorganisms are also used in this field since they can tolerate high temperature from below 20°C (psychrophiles) to above 55°C (thermophiles) [2]. The advanced developments in the molecular biology approach allowing

**Corresponding author*: sindhurgcb@gmail.com; sindhufax@yahoo.co.in

Arindam Kuila and Mainak Mukhopadhyay (eds.) Biorefinery Production Technologies for Chemicals and Energy, (357–368) © 2020 Scrivener Publishing LLC

Table 18.1 Production of enzymes from microbes.

Microbial sp.	Glucosidases	Cellulases	Hemicellulases	Lytic polysaccharide monooxygenases	Pectinases	Lipases	Proteases
Clostridium sp.		x	x				
Phanerochaete chrysosporium				x			
Pseudomonas sp.		x				x	x
Aspergillus sp.	x			x	x	x	x
Penicillium sp.					x	x	x
Trichoderma reesei		x	x	x			
Bacillus sp.			x		x	x	x

gene transfer and genetic analysis for recombinant production have dramatically stimulated the various activities in the area of thermostable enzymes [3]. This has also encouraged the researchers to isolate the microorganisms from thermal surroundings to access the numerous enzymes that might prominently open the frame for enzymatic bioprocess actions (Table 18.1) [4, 5]. The applications of enzymes as industrial catalysts in very large scale can be expedient since they creating the possibility to practice environment-based processing [5]. These industrial enzymes (Table 18.2) as alternatives for several robust catalysts can also be able to endure harsh conditions during industrial processing. The biorefinery (Figure 18.1) has currently grown into a key concept that links the essential molecular technologies for industrial production using renewable raw materials and industrial intermediates [6]. The principal goal of this key concept is to produce high value-low volume yields together with low value-high volume products. Based on the feedstock accessible from the different nations over the world, different kinds of raw materials (Figure 18.2) have been recognized and they include sugar cane [7], sorgo, cassava, cotton [8], wheat [9], lignocelluloses [10], and corn [11]. The bio-catalysis using different enzymes might generate dual task including carbon-carbon bond formations and oxidation-reduction reactions. For instance, monooxygenases have been primarily used for Baeyer-Villiger oxidation and hydroxylation reactions [12]; the formation of C-C bonds in reactions can be achieved using lyases [13]. The use of oxidoreductases and lipases in epoxide synthesis has significant perspective for the synthesis of various chemicals including enzymatic reactions to replace some toxic elements [14]. The polymer-hydrolyzing enzymes may offer a significant contribution to the

Table 18.2 Enzymes for biomass conversion.

Enzyme classification	Subclasss	Substrates	Products
Cellulases	Endoglucanases	Amorphous cellulose	Cellooligosaccharides
	Exoglucanases	Reducing and non-reducing ends of cellulose	Cellobiose
Hemicellulases	Glucosidases	Cellobiose	Glucose
	Endoxylanases	Xylans	Xylooligosaccharides
	Xylosidases	Xylooligosaccharides	Xylose
Monooxygenases	AA9 family	Crystalline cellulose, chitin	Cellooligosaccharides
Proteases		Proteins	Amino acids
Ligninases	lignin peroxidases	Lignin	Phenolic compounds,
Pectinase	Pectin methylesterases,	Pectin polymers	Galacturonic acids
Amylases	α-amylase	Starch	Maltose and glucose
Lipases		Triglycerides	Glycerol

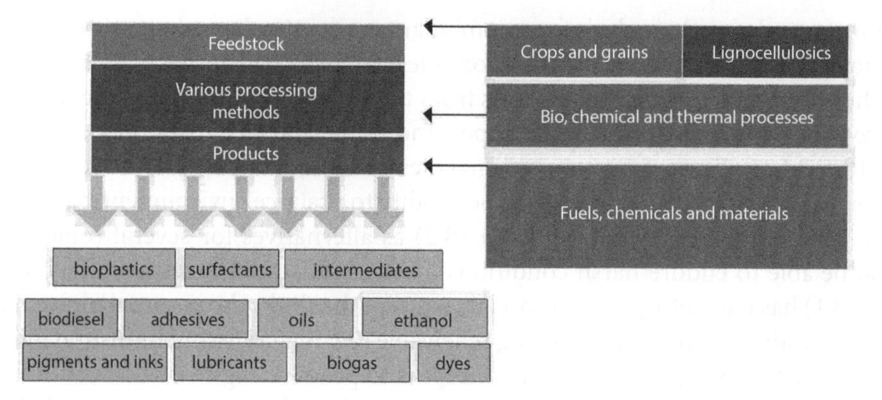

Figure 18.1 Perspectives on biorefinery principles.

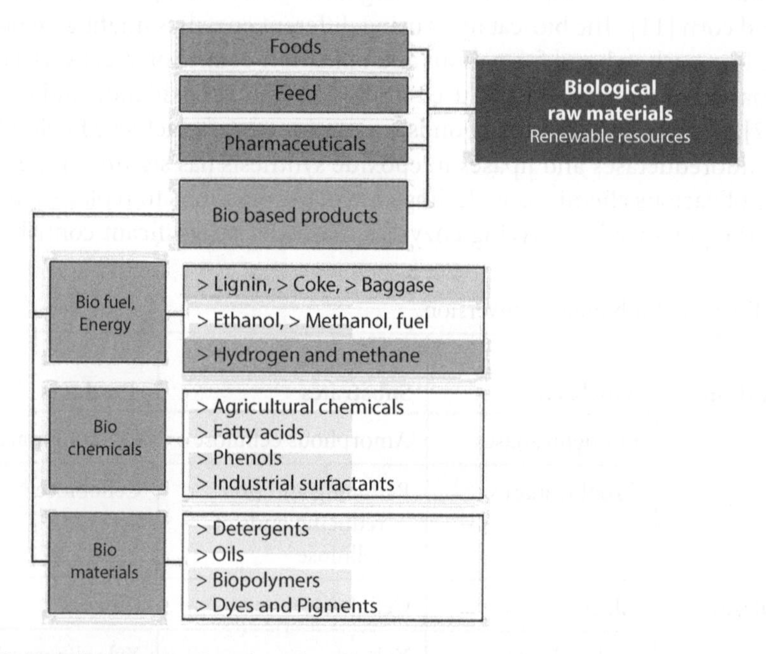

Figure 18.2 Biological raw materials and products.

biorefineries by increasing the substrate availability. The glycoside hydrolases commonly used in feed processing can able to degrade the building materials of herbs into monosaccharide that are easier for microbes to metabolize.

The process of catalysis at high temperature is beneficial in bioconversion of lignocellulosic materials into xylitol [15].

18.2 Perspectives of the Biorefinery Concept

As per the statistical analysis from the Business Communications Company Inc., the global market was assessed to $2 billion in 2004 for industrial enzymes. In addition, its annual growth rate is found to be between 4 and 5% since the prices of enzymes in the global market has lower due to the increased competition between the companies. Based on the distinct applications of industrial enzymes, they are divided into three separate sections such as animal feed enzymes, technical enzymes, and food enzymes. In which technical enzymes section is considered as the largest space where enzymes have been used for the preparation of detergents and paper and it constitutes about 52% whole world market [16–18]. The hydrolytic enzymes such as amylases and proteases comprise the 25 and 20% of the whole market, respectively.

18.3 Starch Degradation

The starch refined from trees is one of the most prominent and easily available energy forms in the world. The wheat, tapioca, potato, and corn are the different sources of starch for processing at industrial level; however, rice, sweet potato, mung beans, arrowroot, and sorghum are used to a reduced level [19]. Either catalytic or biological transformation of the aforementioned resources might leads to the formation of propionic acid, poly(3-hydroxyalkanoate acid), 1,3-propane diol (PDO), and fuels (ethanol and butanol) [20]. The processed starch is chiefly used for maltose and glucose production; however, a number of intermediate products have been produced by cyclodextrins. For instance, high-fructose syrups were formed from glucose by the enzyme glucose isomerase. During the liquefaction of starch, the starch was exposed to thermal treatment needing around 70–90°C temperature (for corn), to assure the lipid-amylose complex removal [21].

During retrogradation of starch processing, the hydrogen bonds make interactions between amylose chains and the chains become hydrated. A thermostable α-amylase is required before the high temperature (105–110°C for 5–7 min) treatment processes. The various enzymes refined from *Bacillus stearothermophilus* and *B. licheniformis* have been used for the aforementioned processes. The reduced level of water content in the starch could be more beneficial in economic aspects and this can be accomplished by the elevated formation of isomaltose [22, 23]. β-amylase and glucoamylase are the two important enzymes that are required for the hydrolysis of oligosaccharides during the saccharification of starch processing [24]. Pullulanase, a debranching enzyme, can be also added during the refining processes to upsurge the efficiency of saccharification. Amylomaltases offer a significant contribution in the synthesis of cyclic dextrins, especially cycloamyloses [25]. These cycloamyloses are principally used as the adhesive material for biodegradable plastics; however, they are also able to solubilize larger compounds including Buckminsterfullerene [26, 27].

18.4 Biodegradation and Modification of Lignocellulose and Hemicellulose

Lignin is an important element in the papermaking process using wood. The papermaking at industrial level comprises the mechanical and chemical separation of the cellulosic fibers from lignified plant parts (Figure 18.3). Edward John Bevan (1856–1921) and Charles Frederick Cross (1855–1935) have introduced the term "lignocellulose"; it is a combination

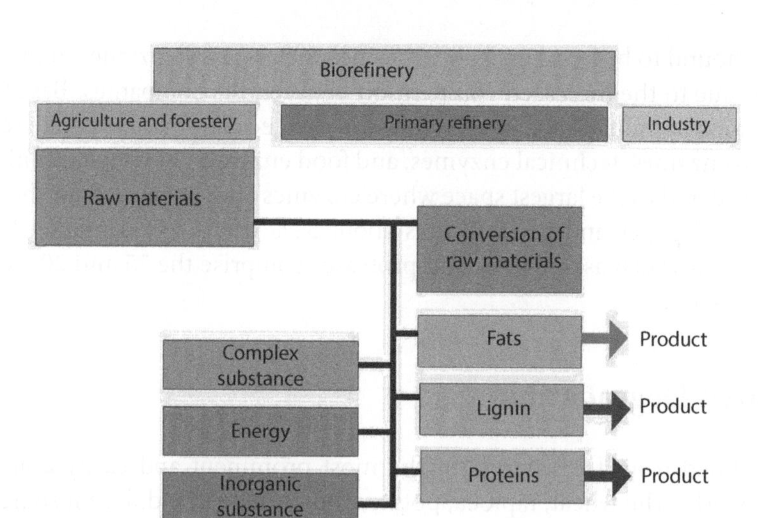

Figure 18.3 Schematic diagram of biodegradation raw materials.

of hemicellulose and polymeric carbohydrates hemicelluloses [28]. The industrial exploitation of lignocellulose-comprising raw materials was predominantly focused on several aspects including the production of vanillin from lignin, paper production from wood, wood liquification, cellulose-based synthetic fibers, and wood-based sugar production [29, 30]. Effective degradation of lignocellulose necessitates complex synergistic activities of different enzymes including hemicellulases, auxiliary enzymes, and cellulases [31, 32]. However, owing to the recalcitrant character of lignocellulose, a preheat treatment (steam pre-treatment, hydrothermal pre-treatment, and steam explosion) is required to upsurge the effectiveness of the enzyme allied hydrolysis [33]. Most of all the cellulases (CBH and EG) comprises the catalytic domain (CD) and carbohydrate binding module (CBM) structure. The correct orientation and adsorption of cellulases over the surface of cellulose are efficiently mediated by CBM part [34]. The significance of lytic polysaccharide monoxygenases (LPMOs) has been increased in recent years since LPMO-based deconstruction of cellulose exhibits effective results. These enzymes use oxidative cleavage in the cellulose backbone. The C4 oxidation yields geminaldiol; whereas, the oxidative action at the C1 position forms an aldonic acid [35]. The LPMOs enhance the hydrolysis of cellulose, and they are an essential element in the present state of the cellulase cocktails. The high action of LPMOs can eventually lead to the formation of substantial amounts of gluconic acids [36]. The agricultural waste usually has substantial lignocellulose content. The commercially available enzymes can be used to explore the applications of lignocellulosic elements for the degradation of different biomass resources [37]. The endoglucanases (EG), β-glucosidases (BG), and cellobiohydrolases (CBH) are the different kinds of cellulose hydrolyzing enzymes that are reacting with β-1,4-glycosidic bonds [38]. An important application of cellulases in industrial production during the 1990s is the detergent, textile, and paper industry. The second most abundant renewable biomass is the hemicellulose (25–35% of total mass). They are made up of pentoses, sugar acids, and hexoses [39]. The degradation of hemicellulose is accomplished by means of using several enzymes; endo-1,4-β-xylanase, β-xylosidase, α-glucuronidase, α-L-arabinofuranosidase, and acetylxylan esterase are responsible for xylan degradation; whereas, β-mannanase and β-mannosidase are responsible for the degradation of glucomannan. The enzymes that come under the category of thermophiles exhibit intrinsic thermostability together with maximum activity at high temperature. For instance, xylanase Xyn10A refined from *R. marinus* was found to be enhancing the brightness during the bleaching process of softwood and hardwood. Xylanasesare also showing significant applications at industrial level; since they are used as additives to upsurge the quality of baked products [15, 40]. The combination of xylanase with mannanases enlightens potential applications in food and feed industry since this combination can induce certain actions that are needed for decreasing the viscosity of coffee extracts [41, 42]. The stability of the enzymes during the processing is have vital importance; in this respect, it is significant to remember that cellulsae preparation at the industrial level is generally found to be a combination of enzymes with different stability [43]. Previous studies have investigated the temeperature stability of three commercial cellulase preparations of Accellerase 1000, spezyme, and celluclast at varying temperatures [44]. The optimum pH ranging from 4.5 to 5.5 has significant importance enzyme stability. Otter et al. investigated that the pH values greater than 11.5 can be used to desorb the 90% of cellulase protein [45]. Sammond *et al.* sophisticatedly reported how the surface hydrophobicity of seven hemicellulases and cellulases has been interrelated with their affinity toward the lignin along with Bovine Serum Albumine (BSA), and thereby offered evidence for the role of aforementioned interactions in the adsorption [46]. However, other studies have reported that the significant modulation in the pH might directly influence the adsorption [47].

18.5 Conversion of Pectins

Pectin has been found to exhibit potential applications in industrial production of biogas [48], soluble sugars, and ethanol [48] from common waste materials; furthermore, they can also be used for the preparation of biodegradable films [49]. The microbial pectinases are the significant group of enzymes that are broadly used for juice extraction, refinement of vegetable fibers, pectin free starch preparation, fruit juice clarification, and wastewater treatment [50]. Pectin depolymerase, pectinase, exopolygalacturanosidase, and exopolygalacturonase are the most plentiful among all the pectin degrading enzymes as for the reason that they are able to make hydrolysis reactions [51]. The α-L-rhamnosidases can have the ability to hydrolyze the rhamnogalacturonan exists in the pectic backbones, whereas α-L-arabinofuranosidases can hydrolyze the endo-arabinase on the arabinan side chains of pectin [52]. The synergistic action of these aforementioned enzymes in lysing branched arabinan has leads to the formation of L-arabinose. The methyl ester linkages in the pectin backbone could be de-esterified using the pectinesterase/pectin methyl esterase [53].

18.6 Microbial Fermentation and Biofuel and Biodiesel Aimed Biorefinery

It is significant to enlighten that biorefineries are envisioned to have versatility to accomplish several objectives at national and global level. Therefore, many studies has been focused on the industrial production of bioethanol corn cobs, wheat straw, and sugarcane bagasse by converting the wastes and vinasses to methane [54]. In addition, the biomass can also be used for the industrial production of methane and hydrogen [55]. Research on the lipid production of hydrolysates has gained much more significance in the production of biodiesel [56, 57]. The microbial fermentations involve the formation of biofuels (butanol, lipids, acetone, and ethanol from lignocellulosic biomass) or biochemical. This can be accomplished using the enzymatic hydrolysis together with separate hydrolysis and fermentation (SHF) or consolidated bioprocessing (CBP). The production of butanol was usually accompanied by ABE (acetone, butanol, and ethanol) fermentation with the help of anaerobic bacterium *Clostridium acetobutylicum* [58]. The recent research has been attempts to produce butanol at industrial level using the genetically modified *Escherichia coli, S. cerevisiae*, and *C. Acetobutylicum* strains. The oleaginous yeasts have been renowned for their ability to accumulate lipids up to 75% of their total body weight by consuming both xylose and glucose [59, 60]. These stored lipids can provide significant contribution in biodiesel making processes using transesterification process. Based on the cost and economic aspects, cofermentation is assumed to be better than any other similar fermentation approaches.

Many of interesting efforts have been developed to genetically modify *Zymomonas mobilis* and *Saccharomyces cerevisiae* to facilitate and induce the fermentation processes of xylose. The two important fermenting pathways in relation with xylose have been explored and engineering into *S. cerevisiae* are i) XR and XDH pathway and ii) XI pathway. The former one exists in the fungi and the later one was exists in bacteria [61, 62]. Traditional distillation can be used to isolate the 95% pure ethanol from the fermentation broth. The researchers in this field have watching at different types of biofuels to avoid distillation procedure. The practice of fermentation residues with anaerobic digestion aspects has improved the economic affordability of bioethanol fermentation [63]. The biogas formed can be explored to generate electricity and heat, consequently enhancing the global energy balance. Rabelo *et al.* established alike scenarios for biogas and sugarcane bagasse bioethanol production at industrial level [56]. Another fascinating approach to produce methanol at industrial level includes the use of 1-d-germinated rye grains [64]. This approach has employed the inherent amylase production to produce intermediate product amylose for its transformation to bioethanol. This system has produced two streams: a liquid fraction which is able to produce biogas and a protein-rich solid fraction which can be used as feed for animals. Sewages from the anaerobic digestion have offer significant contribution in the agriculture since it can be used as fertilizer [63, 64]. The production of methane from anaerobic digestion of biomass is recognized as another significant perspective of biogas intended biorefinery. The use of enzymes in this aspect to produce biogas is relatively considered as scarce. The laccase system can provide elevated biomethane production up to 14% increase when compared with control groups without the addition of enzymes.

These aforementioned results have evidenced that laccase induced production have offer biogas production at industrial level; but the increase was not adequate to replace the economic concerns including the over cost executed enzymes. Due to the economic concerns and expensive nature of the approaches in biogas production, several other strategies might be also used for the same purpose. Schroyen et al. recommended an anaerobic effluent for ethanol production [65]. Conventional strategies regarding the biodiesel-producing processes have never used enzymes as catalysts for the first instance. The substrates for biodiesel are primarily microbial biomass and oleaginous seeds; however, industrial waste and animal fat together with cooking oil are also used. Gong et al. have developed a remarkable approach for the production of biodiesel in biorefinery aspect [63]. He increased the production of lipids from corn stover by assimilating certain chemical reactions such as enzymatic hydrolysis using cellobiase, xylanase, and cellulose together with lipid biosynthesis with the help of curvatus. The various enzyme companies in recent years have prominently upgraded the enzyme preparations and their mechanisms to ensure better hydrolysis within short time with special reference to economic aspects.

18.7 Conclusion

The material production and formation of energy from biomass is increased to extreme level when the raw materials are processed using biorefinery methods, where various advanced technological approaches has been united together to develop various bioproducts from different biomass. The applications of enzymes in biorefineries have to date received a prodigious deal of interest as biocatalysts. The cellulases, hemicellulases, peroxidases, pectinases, laccases, amylases, and proteases produced by fungi, yeast, and bacteria have currently observed in the scenario as significant element for the industrial production of food, paper, pharmaceutical, chemicals, cosmetic, and textile industries together with biofuels. In which, cellulases, hemicellulases, and ligninases are primarily renowned as most widely employed enzymes for the industrial production of biogas, biodiesel, and bioethanol. However, based on the type of feedstock, conversion strategies and final products different biorefinery pathways have been established to form bioproducts from biomass. The volume of enzyme production has dramatically increased in recent years together with low cost since its market was relatively raised. In addition to this, the need of various catalysts of biological/microbial origin is predicted to upsurge, and definitely, there will be a sustained and increased need of specific biocatalysts of enzyme origin in the future.

Acknowledgement

Raveendran Sindhu acknowledges the Department of Science and Technology for sanctioning this project under DST WOS-B scheme.

References

1. Brock, T.D. and Freeze, H., Thermus aquaticus gen. n. and sp. n., a nonsporulating extreme thermophile. *J. Bacteriol.*, 98, 289–297, 1969.
2. Brock, T.D., Thermophiles: General, molecular and applied microbiology. *N. Y.*, 1986.
3. Rebello, S., Anoopkumar, A., Puthur, S., Sindhu, R., Binod, P., Pandey, A. *et al.*, Zinc oxide phytase nanocomposites as contributory tools to improved thermostability and shelflife. *Bioresour. Technol. Rep.*, 3, 3, 4–5, 1–6, 2018.
4. Satyanarayana, T., Raghukumar, C., Shivaji, S., Extremophilic microbes: Diversity and perspectives. *Current Science*, 89, 1, 78–90, 2005.
5. Podar, M. and Reysenbach, A.-L., New opportunities revealed by biotechnological explorations of extremophiles. *Curr. Opin. Biotechnol.*, 17, 250–255, 2006.

6. Kamm, B. and Kamm, M., Principles of biorefineries. *Appl. Microbiol. Biotechnol.*, 64, 137–145, 2004.
7. Edye, L., Doherty, W., Blinco, J.-A., Bullock, G., The sugarcane biorefinery: Energy crops and processes for the production of liquid fuels and renewable commodity chemicals, in: *Proceedings of the 2005 Conference of the Australian Society of Sugar Cane Technologists held at Bundaberg, Queensland, Australia, 3–6 May 2005*, pp. 9–22, 2005.
8. Enze, M., Developing biorefinery by utilizing agriculture and forestry biomass resources: Striding forward the carbohydrate era, in: *Progress in Chemistry*, vol. 18, p. 131, CNKI, Beijing, 2006.
9. Koutinas, A., Wang, R., Webb, C., Restructuring upstream bioprocessing: Technological and economical aspects for production of a generic microbial feedstock from wheat. *Biotechnol. Bioeng.*, 85, 524–538, 2004.
10. Pan, X., Kadla, J.F., Ehara, K., Gilkes, N., Saddler, J.N., Organosolv ethanol lignin from hybrid poplar as a radical scavenger: Relationship between lignin structure, extraction conditions, and antioxidant activity. *J. Agric. Food Chem.*, 54, 5806–5813, 2006.
11. Gáspár, M., Juhász, T., Szengyel, Z., Réczey, K., Fractionation and utilisation of corn fibre carbohydrates. *Process Biochem.*, 40, 1183–1188, 2005.
12. Mihovilovic, M.D., Müller, B., Stanetty, P., Monooxygenase-Mediated Baeyer–Villiger Oxidations. *Eur. J. Org. Chem.*, 2002, 3711–3730, 2002.
13. Fessner, W.-D., Enzyme mediated C-C bond formation. *Curr. Opin. Chem. Biol.*, 2, 85–97, 1998.
14. Hatti-Kaul, R., Törnvall, U., Gustafsson, L., Börjesson, P., Industrial biotechnology for the production of bio-based chemicals–a cradle-to-grave perspective. *Trends Biotechnol.*, 25, 119–124, 2007.
15. Wong, K.K., Application of hemicellulases in the food, feed, and pulp and paper industries. *Hemicelluloses Hemicellulases*, pp. 127–143, Portland Press, London, 1993.
16. Hasan, F., Shah, A.A., Hameed, A., Industrial applications of microbial lipases. *Enzyme Microb. Technol.*, 39, 235–251, 2006.
17. Houde, A., Kademi, A., Leblanc, D., Lipases and their industrial applications. *Appl. Biochem. Biotechnol.*, 118, 155–170, 2004.
18. Jobling, S., Improving starch for food and industrial applications. *Curr. Opin. Plant Biol.*, 7, 210–218, 2004.
19. Wang, T.L., Bogracheva, T.Y., Hedley, C.L., Starch: As simple as A, B, C? *J. Exp. Bot.*, 49, 481–502, 1998.
20. Yang, S.-T. and Yu, M., Integrated biorefinery for sustainable production of fuels, chemicals, and polymers. *Bioprocess. Technol. Biorefin. Sustainable Prod. Fuels Chem. Polym.*, 1, 1–26, 2013.
21. Crabb, W.D. and Mitchinson, C., Enzymes involved in the processing of starch to sugars. *Trends Biotechnol.*, 15, 349–352, 1997.
22. Van der Veen, M., Veelaert, S., Van der Goot, A., Boom, R., Starch hydrolysis under low water conditions: A conceptual process design. *J. Food Eng.*, 75, 178–186, 2006.
23. Souza, R.C. and Andrade, C.T., Investigation of the gelatinization and extrusion processes of corn starch. *Adv. Polym. Technol.: J. Polymer Process. Inst.*, 21, 17–24, 2002.
24. Pandey, A., Glucoamylase research: An overview. *Starch-Stärke*, 47, 439–445, 1995.
25. Strater, N., Przylas, I., Saenger, W., Terada, Y., Fuji, K., Takaha, T., Structural basis of the synthesis of large cycloamyloses by amylomaltase, in: *Biologia, Bratislava*, vol. 57, pp. 93–100, Versita, Bratislava, 2002.
26. Terada, Y., Fujii, K., Takaha, T., Okada, S., Thermus aquaticus ATCC 33923 amylomaltase gene cloning and expression and enzyme characterization: Production of cycloamylose. *Appl. Environ. Microbiol.*, 65, 910–915, 1999.
27. Fujii, K., Minagawa, H., Terada, Y., Takaha, T., Kuriki, T., Shimada, J. *et al.*, Use of random and saturation mutageneses to improve the properties of Thermus aquaticus amylomaltase for efficient production of cycloamyloses. *Appl. Environ. Microbiol.*, 71, 5823–5827, 2005.

28. Cross, C.F., Bevan, E., Beadle, C., Viscose syndicate. *British Patent*, 8700, 1892.
29. Brown, R.C., Biomass refineries based on hybrid thermochemical-biological processing-An overview. *Biorefin. Ind. Processes Prod.: Status Quo Future Directions*, 1, 227–252, 2005.
30. Hayes, D.J., Fitzpatrick, S., Hayes, M.H., Ross, J.R., The biofine process–production of levulinic acid, furfural, and formic acid from lignocellulosic feedstocks. *Biorefin. Ind. Processes Prod.*, 1, 139–164, 2006.
31. Horn, S.J., Vaaje-Kolstad, G., Westereng, B., Eijsink, V., Novel enzymes for the degradation of cellulose. *Biotechnol. Biofuels*, 5, 45, 2012.
32. Gao, D., Uppugundla, N., Chundawat, S.P., Yu, X., Hermanson, S., Gowda, K. *et al.*, Hemicellulases and auxiliary enzymes for improved conversion of lignocellulosic biomass to monosaccharides. *Biotechnol. Biofuels*, 4, 5, 2011.
33. Himmel, M.E., Ding, S.-Y., Johnson, D.K., Adney, W.S., Nimlos, M.R., Brady, J.W. *et al.*, Biomass recalcitrance: Engineering plants and enzymes for biofuels production. *Science*, 315, 804–807, 2007.
34. Várnai, A., Siika-aho, M., Viikari, L., Carbohydrate-binding modules (CBMs) revisited: reduced amount of water counterbalances the need for CBMs. *Biotechnol. Biofuels*, 6, 30, 2013.
35. Srivastava, N., Srivastava, M., Mishra, P., Singh, P., Ramteke, P., Application of cellulases in biofuels industries: An overview. *J. Biofuels Bioenergy*, 1, 55–63, 2015.
36. Cannella, D., Chia-wen, C.H., Felby, C., Jørgensen, H., Production and effect of aldonic acids during enzymatic hydrolysis of lignocellulose at high dry matter content. *Biotechnol. Biofuels*, 5, 26, 2012.
37. Ward, O.P., Moo-Young, M., Venkat, K., Enzymatic degradation of cell wall and related plant polysaccharides. *Crit. Rev. Biotechnol.*, 8, 237–274, 1989.
38. Wood, T.M. and Garcia-Campayo, V., Enzymology of cellulose degradation. *Biodegradation*, 1, 147–161, 1990.
39. Saha, B.C., α-L-Arabinofuranosidases: Biochemistry, molecular biology and application in biotechnology. *Biotechnol. Adv.*, 18, 403–423, 2000.
40. Fagerstrom, R.B., Paloheimo, M., Lantto, R., Lahtinen, T., Suominen, P., Xylanases and uses thereof, ed: Google Patents. 1999.
41. Clarke, J., Davidson, K., Rixon, J., Halstead, J., Fransen, M., Gilbert, H. *et al.*, A comparison of enzyme-aided bleaching of softwood paper pulp using combinations of xylanase, mannanase and α-galactosidase. *Appl. Microbiol. Biotechnol.*, 53, 661–667, 2000.
42. Sachslehner, A., Foidl, G., Foidl, N., Gübitz, G., Haltrich, D., Hydrolysis of isolated coffee mannan and coffee extract by mannanases of Sclerotium rolfsii. *J. Biotechnol.*, 80, 127–134, 2000.
43. Rebello, S., Balakrishnan, D., Anoopkumar, A., Sindhu, R., Binod, P., Pandey, A. *et al.*, Industrial Enzymes as Feed Supplements—Advantages to Nutrition and Global Environment, in: *Green Bio-processes*, pp. 293–304, Springer Nature Singapore Pte Ltd, 2019.
44. Pribowo, A., Arantes, V., Saddler, J.N., The adsorption and enzyme activity profiles of specific Trichoderma reesei cellulase/xylanase components when hydrolyzing steam pretreated corn stover. *Enzyme Microb. Technol.*, 50, 195–203, 2012.
45. Otter, D., Munro, P., Scott, G., Geddes, R., Elution of Trichoderma reesei cellulase from cellulose by pH adjustment with sodium hydroxide. *Biotechnol. Lett.*, 6, 369–374, 1984.
46. Sammond, D.W., Yarbrough, J.M., Mansfield, E., Bomble, Y.J., Hobdey, S.E., Decker, S.R. *et al.*, Predicting enzyme adsorption to lignin films by calculating enzyme surface hydrophobicity. *J. Biol. Chem.*, 289, 30, 20960–20969, 2014.
47. da Silva, V.M., de Souza, A.S., Negrão, D.R., Polikarpov, I., Squina, F.M., de Oliveira Neto, M. *et al.*, Non-productive adsorption of bacterial β-glucosidases on lignins is electrostatically modulated and depends on the presence of fibronection type III-like domain. *Enzyme Microb. Technol.*, 87, 1–8, 2016.

48. Doran, J.B., Cripe, J., Sutton, M., Foster, B., Fermentations of pectin-rich biomass with recombinant bacteria to produce fuel ethanol, in: *Twenty-First Symposium on Biotechnology for Fuels and Chemicals*, pp. 141–152, 2000.
49. Hoagland, P.D. and Parris, N., Chitosan/pectin laminated films. *J. Agric. Food Chem.*, 44, 1915–1919, 1996.
50. Alkorta, I., Garbisu, C., Llama, M.J., Serra, J.L., Industrial applications of pectic enzymes: A review. *Process Biochem.*, 33, 21–28, 1998.
51. Kashyap, D., Vohra, P., Chopra, S., Tewari, R., Applications of pectinases in the commercial sector: A review. *Bioresour. Technol.*, 77, 215–227, 2001.
52. Takao, M., Akiyama, K., Sakai, T., Purification and characterization of thermostable endo-1, 5-α-L-arabinase from a strain of Bacillus thermodenitrificans. *Appl. Environ. Microbiol.*, 68, 1639–1646, 2002.
53. Jayani, R.S., Saxena, S., Gupta, R., Microbial pectinolytic enzymes: A review. *Process Biochem.*, 40, 2931–2944, 2005.
54. Kaparaju, P., Serrano, M., Thomsen, A.B., Kongjan, P., Angelidaki, I., Bioethanol, biohydrogen and biogas production from wheat straw in a biorefinery concept. *Bioresour. Technol.*, 100, 2562–2568, 2009.
55. Escamilla-Alvarado, C., Ríos-Leal, E., Ponce-Noyola, M., Poggi-Varaldo, H., Gas biofuels from solid substrate hydrogenogenic–methanogenic fermentation of the organic fraction of municipal solid waste. *Process Biochem.*, 47, 1572–1587, 2012.
56. Rabelo, S., Carrere, H., Maciel Filho, R., Costa, A., Production of bioethanol, methane and heat from sugarcane bagasse in a biorefinery concept. *Bioresour. Technol.*, 102, 7887–7895, 2011.
57. Demirel, B. and Yenigün, O., Two-phase anaerobic digestion processes: A review. *J. Chem. Technol. Biotechnol.: Int. Res. Process Environ. Clean Technol.*, 77, 743–755, 2002.
58. Ennis, B., Marshall, C., Maddox, I.S., Paterson, A., Continuous product recovery by *in-situ* gas stripping/condensation during solvent production from whey permeate using Clostridiumacetobutylicum. *Biotechnol. Lett.*, 8, 725–730, 1986.
59. Mohagheghi, A., Evans, K., Chou, Y.-C., Zhang, M., Cofermentation of glucose, xylose, and arabinose by genomic DNA-lntegrated xylose/arabinose fermenting strain of zymomonas mobilis ax101, in: *Biotechnology for fuels and chemicals*, pp. 885–898, Springer Springer Nature Singapore Pte Ltd, 2002.
60. Zhang, K., Sawaya, M.R., Eisenberg, D.S., Liao, J.C., Expanding metabolism for biosynthesis of nonnatural alcohols. *Proc. Natl. Acad. Sci.*, 32, 13212-13217, 2011. 2008. p. pnas. 0807157106.
61. Wohlbach, D.J., Kuo, A., Sato, T.K., Potts, K.M., Salamov, A.A., LaButti, K.M. *et al.*, Comparative genomics of xylose-fermenting fungi for enhanced biofuel production. *Proc. Natl. Acad. Sci.*, 108, 32, 13212–13217, 2011.
62. Ha, S.-J., Galazka, J.M., Kim, S.R., Choi, J.-H., Yang, X., Seo, J.-H. *et al.*, Engineered Saccharomyces cerevisiae capable of simultaneous cellobiose and xylose fermentation. *Proc. Natl. Acad. Sci.*, 108, 504–509, 2011.
63. Gong, Z., Shen, H., Wang, Q., Yang, X., Xie, H., Zhao, Z.K., Efficient conversion of biomass into lipids by using the simultaneous saccharification and enhanced lipid production process. *Biotechnol. Biofuels*, 6, 36, 2013.
64. Oleskowicz-Popiel, P., Kádár, Z., Heiske, S., Klein-Marcuschamer, D., Simmons, B.A., Blanch, H.W. *et al.*, Co-production of ethanol, biogas, protein fodder and natural fertilizer in organic farming–evaluation of a concept for a farm-scale biorefinery. *Bioresour. Technol.*, 104, 440–446, 2012.
65. Schroyen, M., Vervaeren, H., Van Hulle, S.W., Raes, K., Impact of enzymatic pre-treatment on corn stover degradation and biogas production. *Bioresour. Technol.*, 173, 59–66, 2014.

Part 4

CONCLUSIONS

19

Wheat Straw Valorization: Material Balance and Biorefinery Approach

Sachin A. Mandavgane[1]* **and Bhaskar D. Kulkarni**[2]

[1]*Department of Chemical Engineering, Visvesvaraya National Institute of Technology, Nagpur, India*
[2]*CSIR-National Chemical Laboratory, Pune, India*

Abstract
The production of wheat grain leads to a considerable quantity of wheat straw (WS) which is generally considered as waste and burnt in farm land. In general, WS has no economic value, even though its composition is rich in cellulose hemicelluloses, lignin, silicate, potassium oxide, and calcium oxide that could form the basis of several industrial processes. In this paper, information was collected on the technologies of valorization of WS, a block diagram of each process is presented and material balance across each unit operation is computed taking basis as processing 1 tonne of WS. Possible applications include edible wax, silica, biochar, syngas, biodiesel, bio-oil, and platform chemicals from bio-oil [levoglucosan (LG) and levoglucosenone]. For sequential utilization of WS, a schematic of biorefinery is proposed. The concept and the inventory data presented here could be useful for evaluating the techno-economic feasibility of the biorefinery as also conduct the life cycle analysis to establish the sustainability.

Keywords: Waste valorisations, waste biorefinery, fruit and vegetable waste, life cycle assessment, sustainability index

19.1 Introduction

In the overall grain production in 2016, wheat amounts for about 93.5 MMT [1]. With a assumed residue to crop ratio of 1.3, the wheat straw (WS) residue produced amounts to 121.5 MMT [2]. This WS can either be burned (combustion) in boiler to produce energy instead of coal or it could be pyrolyzed to obtain bio-oil, biochar, and flue gases. The WS is typically composed of 15% moisture by weight and the dry weight consist of 44.25% silicate and around 28.33% potassium oxide along with 12.64% of calcium oxide [3]. The organic composition could be stated as 33.7% cellulose (by wt), 21% to 26% of hemicellulose, and around 20% of lignin [4]. This WS when burned in boiler produces ash, every 100 kg of WS produces around 2.6 kg of WS ash [5]. This ash can be used to extract silicate and potassium. Also, if the WS is pyrolyzed, the bio-oil produced is a good source of LG (1,6-anhydro-β-D-glucopyranose) and levoglucosenone (LGO;1,6-anhydro-3,4-dideoxy-β-D-glycero-hex-3-enopyranos-2-ulose) [6]. And the biochar obtained on pyrolysis is a

**Corresponding author*: sam@che.vnit.ac.in; mandavgane1@gmail.com

Arindam Kuila and Mainak Mukhopadhyay (eds.) Biorefinery Production Technologies for Chemicals and Energy, (371–382)

good source of activated carbon and has many applications. The flue gases produced here compose mainly of CO, CO_2, and CH_4 and can be reused internally as the energy required for pyrolysis. Apart from this, the waxy cuticle of the straw if extracted can provide us with bio wax which could have vast uses as it could be categorized as edible wax. WS can also prove to be a raw material for biodiesel production.

Herein, a few case studies have been put forth which deal with the applications of WS by performing different unit process onto it. And hence, an overall biorefinery ideology has been put forth.

The biorefinery concept is the development of bio-based industrial products from waste biomass. Comparable to the well-established oil-refinery, the biorefinery will be an integrated system for securing renewable materials to convert into bio-chemicals, bio-energy, and bio-materials in order to maximize the valuable products and minimize the waste.

19.2 Wax Extraction Process

WSs have a cuticle outer layer of wax. This valuable natural wax, when extracted in a sustainable manner, can provide an alternative to the currently used commercial wax as derived from petroleum or other synthetic routes.

Around 1% to 3% of the total weight of WS feed can be extracted as wax [6, 7]. That is, 1 tonne of WS, excluding the moisture content could yield about 25.5-kg wax (by mass balance). Thus, annually, 3.36 MMT of wax could be generated. And the bio wax has great demand in cosmetic industry and for providing waxy layer over fruits, as these waxes are edible.

Wax extraction from WS is carried out using either solvent extraction method or super critical CO_2 extraction method. But, following the traditional solvent extraction method, as shown in Figure 19.1, soxhelete apparatus could be used for the process along with hexane as the solvent. Around eight to nine rounds of the extraction yields the wax-solvent mixture. The wax from this is extracted by distillating the solution in rota evaporator, where the provided hexane (as solvent) is separated from the wax-oil mixture. This wax-oil again undergoes purification to yield 69.2% as pure wax and remaining as oil and resinous material [8]. This is because of the high selectivity of hexane toward wax and not toward the wax esters or triglycerides. This obtained wax is further dried and bleached to form flakes and transport it for further use. The process could be shown as in the flowsheet shown further.

Mass balance for the process:

For every 1000 kg (1 tonne) of WS taken and dried to remove the 15% moisture content, hence left with 850 kg of dry WS feed same is presented schematically in Figure 19.2. This dry feed further undergoes solvent extraction in the soxhlete extraction apparatus where hexane is used as the solvent in the ratio of 13:300, i.e., 300-ml solvent is used for extraction from 13-g feed [8]. Thus, for 850 kg of feed, 19615.4-L hexane is used; 3% of wax is extracted in this process, and hence, leaves back 825 kg [6] of solid residue which can be further pyrolyzed or burnt in air. The solvent-wax mixture formed is distilled in the rota evaporator to separate out all the solvent, which can be reused later. This leaves back 25.5-kg residual mixture of wax and oil out of which 69.2% ispure wax, that is, 17.6 kg, and remaining 7.8 is oil and resinous material [8].

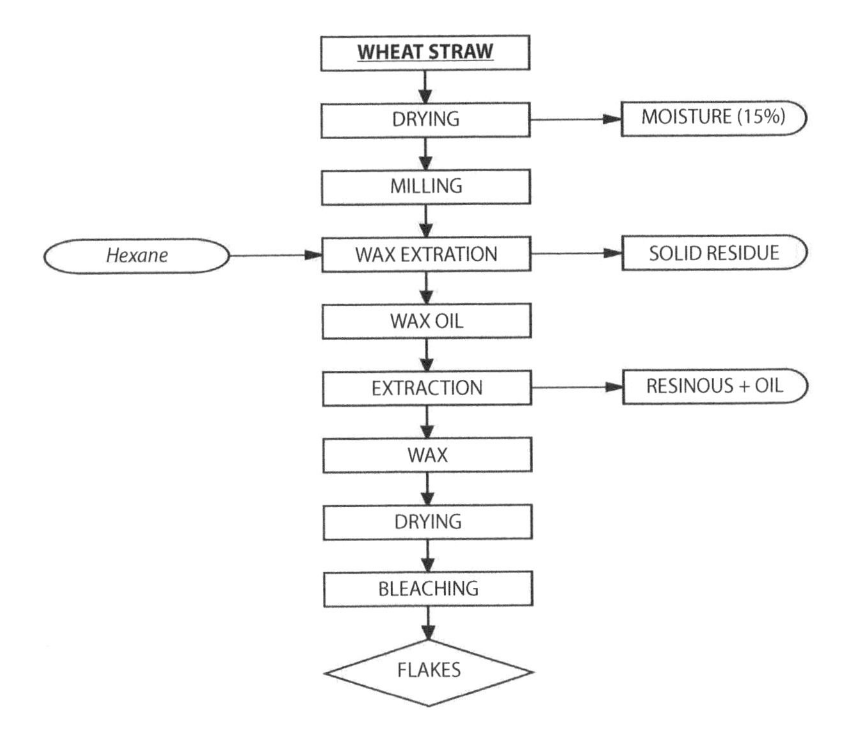

Figure 19.1 Wax extraction from wheat straw.

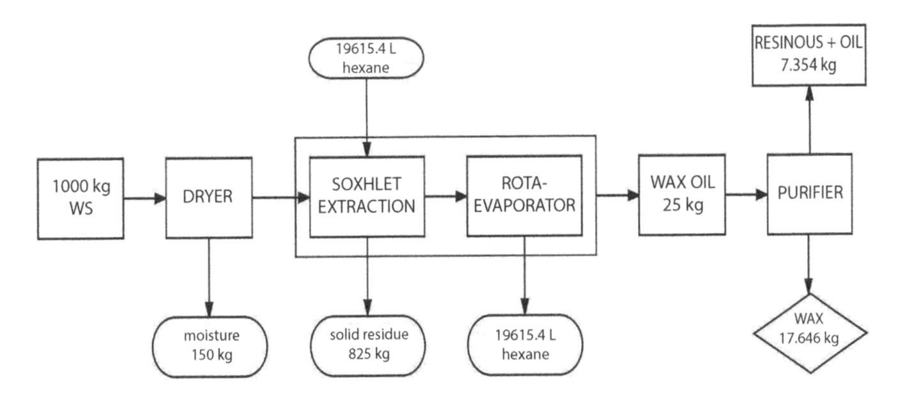

Figure 19.2 Mass balance for wax extraction process.

19.3 Combustion Process

The WS generated every year is mostly used as cattle feed or burned, either in the farms for increasing nutritive value of soil or is used as fuel in boilers. This open air combustion leaves back ash which is about 2.6% of the total straw [5]. As described earlier, WSA has high content of silica and potassium. These components could be extracted and used for further applications of silica and potassium.

Silica and oxides of silica find extensive applications in several industries such as in making electronic circuitry and devices, as composite fillers in ceramic and polymer industries, solar panel, and semiconductor industries, as a thixotropic agent, heat insulator etc. Although naturally available, it is also made synthetically using high temperatures processes. The silica content of the WSA is high [4] and can be used as a natural source for its extraction. The method involves acid treatment to remove metallic impurities [9] and then transforming the remaining ash to sodium silicate and reducing it further to silica. The scheme of silica extraction from WSA is presented in Figure 19.3. The base reactions can be described as:

$$SiO_2 + 2NaOH \rightarrow Na_2SiO_3 + H_2O$$

$$Na_2SiO_3 + H_2SO_4 \rightarrow SiO_2 + Na_2SO_4 + H_2O$$

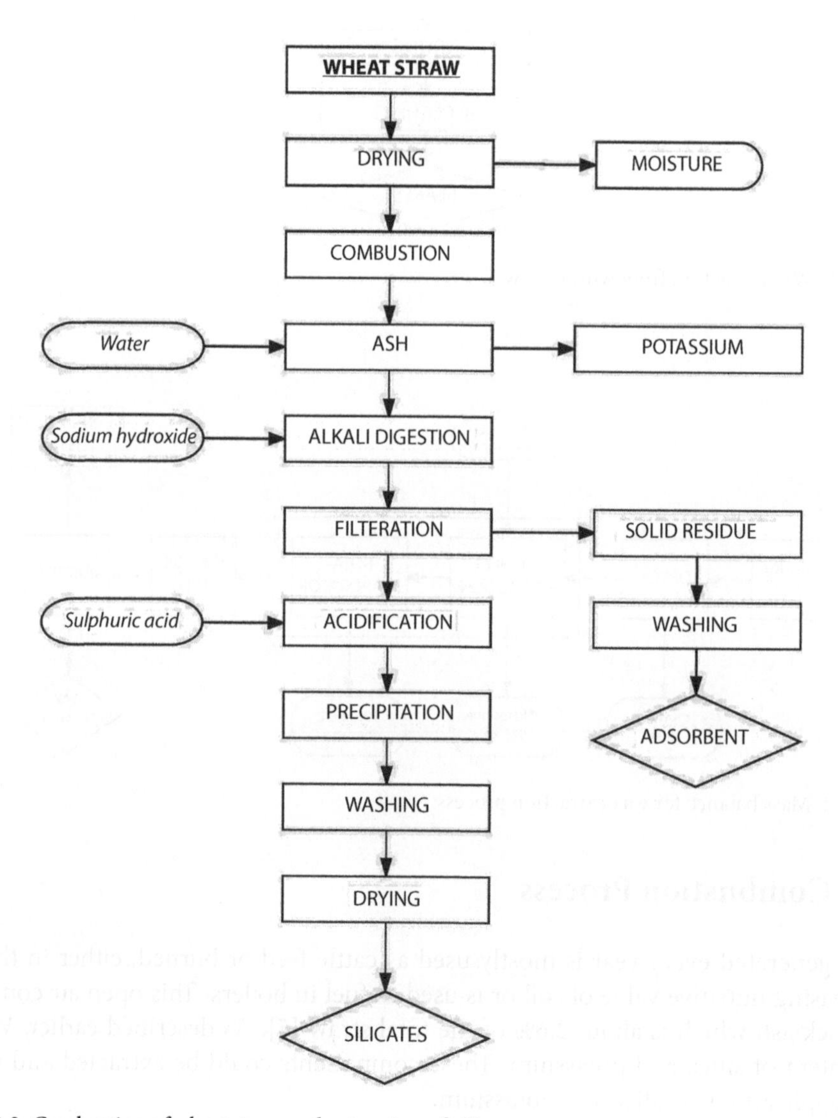

Figure 19.3 Combustion of wheat straw and extraction of silicate.

The WS is dried and milled for the purpose of densification. This feed is then burned in open air (combustion) at the temperature of around 500°C. The ash hence produced is quenched in water so that the potassium oxide gets dissolved in water and can be separated. Around 70% potassium can be extracted successfully [3]. The quenched ash then is reacted with alkali (NaOH) to form sodium silicate. This is then filtered to separate the unreacted solid residue. The sodium silicate is then acidified by mineral acid until pH 7 is reached [9]. The silica gets precipitated at this stage and the Na_2SO_4 and water get separated. The precipitated silica is washed and dried to obtain about 95% of total silica content.

19.4 Mass Balance for Combustion

If a feed of 1,000 kg of WS is dried, the 15% moisture, that is, the 150-kg moisture content is removed. This dried straw is then combusted at around 500°C to produce around 23 kg ash (2.7% of total dry feed). The quenching dissolves the potassium content from the ash, which is around 6.74% by weigh [7]. 70% of this potassium can be extracted [3]. Thus, 1.08 kg of potassium can be obtained from the ash. The remaining undissolved ash undergoes alkali digestion where 80-g NaOH is added to per 70-g ash on experimental scale [9]. Thus, linearly, 25 kg of NaOH is added to the ash. This is then filtered where the unreacted residue gets separated. This solid residue amounts for over 9.68 kg as per the overall system mass balance. This residue is then washed and dried and used as adsorbent [10]. Now, the sodium silicate undergoes acidification till pH 7 is reached [9]. The extraction is 95%. Thus, for total neutralization of sodium silicate, 14.363-kg H_2SO_4 of same normality as NaOH is required. The silica precipitated after neutralization is 12.26 kg, and Na_2SO_4 and water get separated. Figure 19.4 presents the mass balance of combustion of WS.

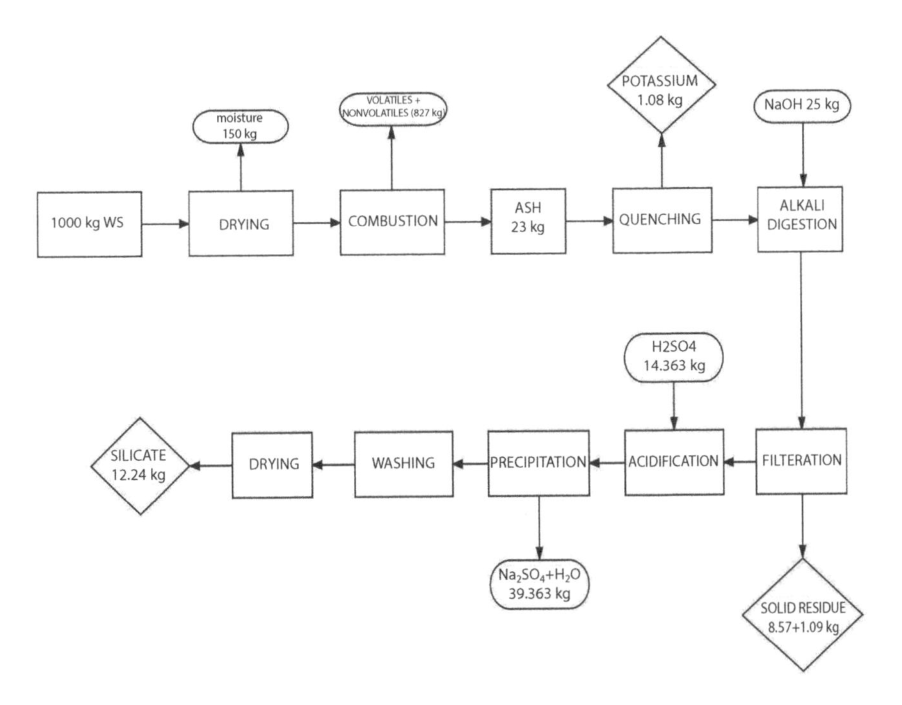

Figure 19.4 Mass balance of combustion of wheat straw and silica extraction.

19.5 Pyrolysis of Wheat Straw

As the WS can be burned directly in open air, so can it be pyrolyzed in closed system to obtain biochar, bio-oil, and syngas. Pyrolysis is one of the most promising thermo-chemical conversion routes for recovering energy from biomass [6]. The scheme of WS pyrolysis is presented in Figure 19.5.

The process has been extensively studied over the years, covering a wide range of temperature from 180° to 900° [11], and gives rise to solid, liquid, and gaseous products along the temperature path. The relative proportions of these products vary with the type of biomass, operating parameters including size of particles, type and mode of reactor operation, heating rate, residence time, etc. Different controlling regimes during operation controls the product distribution.

WS is milled and pyrolyzed at 773 to 1,173 K. Inert nitrogen is provided in the incubator to avoid unwanted reactions. The main products of pyrolysis are biochar (32.15% by weight), syngas (12.5%), and bio-oil (55.3%) [12]. The solid residue, that is, the biochar gets separated at the bottom of the downflow reactor and the fluid (gas + oil) is taken out from the reactor. Further, the gas gets separated from the bio-oil in a hydro cyclone separator.

Maximizing the bio-oil output is often one of the objectives of process parameter optimization, since it can supplement or be an alternative to crude oil for producing fuel or chemicals.

The biochar is a good source of activated carbon and hence can be used for water filteration, detoxification, extraction of heavy metals, etc. It can also be used in building materials and medicines as well. The syngas is directly used as fuel. The composition of syngas at different tempertures is given in Table 19.1 [12].

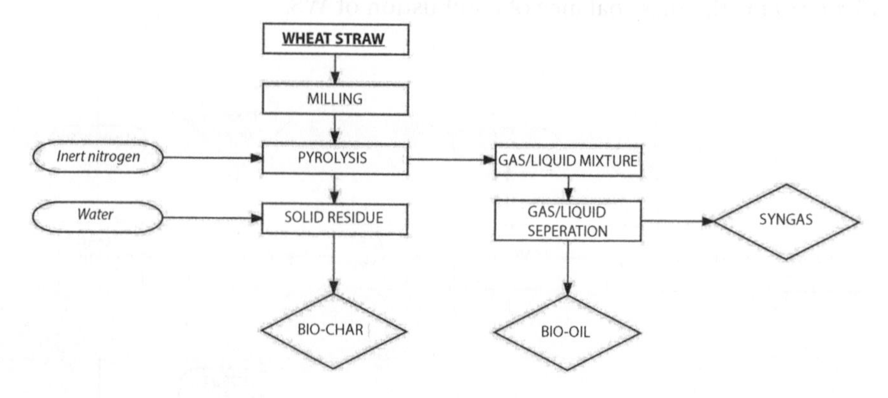

Figure 19.5 Pyrolysis of wheat straw.

Table 19.1 Composition of syngas at different microwave power.

Microwave power (W)	CO_2 (vol%)	CO (vol%)	CH_4 (vol%)	C_{2+} (vol%)	H_2 (vol%)
400	54.27	25.83	0.17	1.9	17.83
600	54.02	25.98	0.24	0.67	21.09
800	42.99	25.48	0.24	1.83	29.47

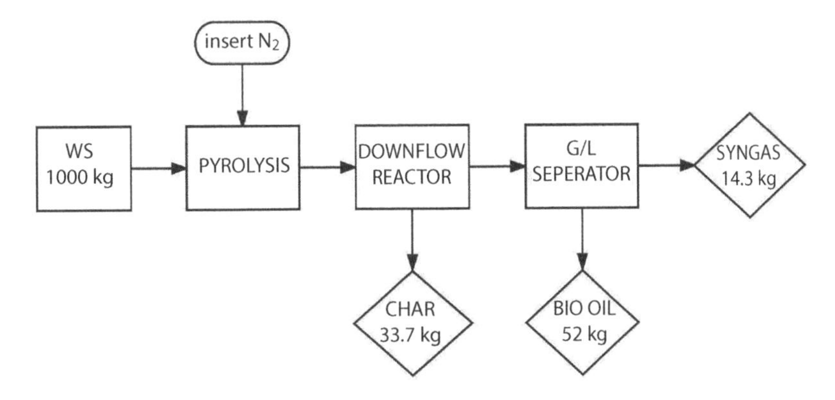

Figure 19.6 Mass balance for pyrolysis of wheat straw.

19.6 Mass Balance of Pyrolysis

For 1,000 kg of WS taken and pyrolyzed in a downflow reactor, 321.5 kg of biochar is obtained along with 125 kg of syngas and 553 kg of bio-oil (percentage mentioned above). The pyrolysis process is demonstrated below along with its mass balance as presented in Figure 19.6.

19.7 Separation of Valuable Chemicals From Bio-Oil

Bio-oil is a rich source of some of the hydrocarbons such as 2,3-dihydrobenzofuron (DHBT) and mono saccharides particularly LG [6]. DHBT and its derived products have found many applications in pharmaceutical industries as therapeutic agents, as inhibitors, and cytoprotective agents. The sugars can be easily extracted in the water and can be further fermented to various other products. The renewable platform allows specific chemistries to alter free hydroxyl groups [13] and provides a base molecule for stereoregular polysaccharides that have bio activity (anti-human immunodeficiency, anticoagulants, etc.). LG itself is a starting raw material for number of products such as resins, surfactants, plasticizers, etc.

For separating these chemicals from bio-oil, the aqueous mixture first is treated with an azeotrope-foaming solvent, after which the resulting mixture is distilled for removing the water by azeotropic distillation. This leaves a water-free mixture of LG and the carbohydrate derived acids. The former is separated from the latter by application of a solvent, particularly methyl isobutyl ketone, in which the LG selectively is soluble. The resulting LG solution then is separated from the resulting insoluble residue of carbohydrate derived acids, after which the LG is crystallized from the solvent medium in which it is contained. This leads to the production of a high yield of white, crystalline LG and solid, carbohydrate-derived acid products which are well suited for use in their various applications. Around 22.6% by weight LG can be obtained by this process presented in Figure 19.7. [13].

LG, derived from bio-oil, is another platform molecule for making 2,3-dideoxyribose, hydroxylactones, and several other molecules used as medicines.

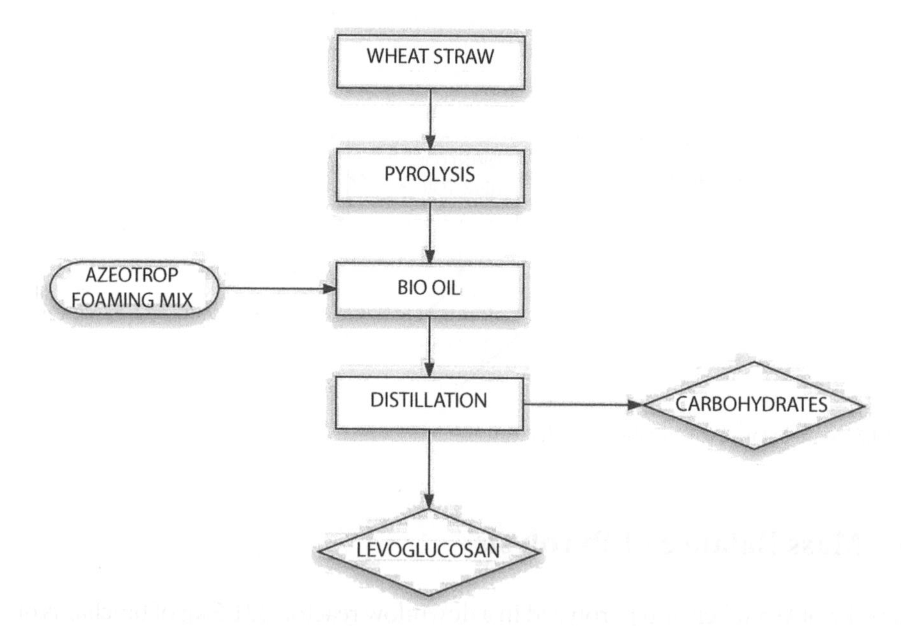

Figure 19.7 Extraction of levoglucosan from bio-oil.

19.8 Production of Biodeisel From Wheat Straw

Lignocellullosic biomass can be converted using thermochemical or biochemical transformation routes. The chemical route employs hydrolysis of biomass followed by reduction of sugars to fuels and chemicals. Bio route converts the stock to bio ethanol and biodiesel. This involves some pre-treatment steps such as steam explosion of biomass, lipid extraction, and trans esterification. Extensive work on this topic has been reported in the literature.

Biorefinery is a facility that integrates biomass conversion processes and equipment to produce fuels, power, heat, and value-added chemicals from biomass. The biorefinery concept is analogous to today's petroleum biorefinery, which produce multiple fuels and products from petroleum.

The International Energy Agency has defined biorefining as the sustainable processing of biomass into a spectrum of bio-based products (food, feed, chemicals, and materials) and bioenergy (biofuels, power, and/or heat). By producing multiple products, a biorefinery takes advantage of the various components in biomass and their intermediates, therefore maximizing the value derived from the biomass feedstock. Some researchers have considered the exploration of a biorefinery as a practical method of improving the economic performance of stand-alone biomass to bioenergy system since biochemicals are produced. A biorefinery could, for example, produce one or several low-volume, but high-value, chemical or nutraceutical products and a low-value, but high-volume liquid fuel such as biodiesel or bioethanol. At the same time, generating electricity and process heat, through combined heat and power (CHP) technology, for its own use and perhaps enough for sale of electricity to the local utility. Figure 19.8 presents one of the schemes of WS bio refinery and production of biodiesel. The high-value products increase profitability, the high-volume fuel helps meet

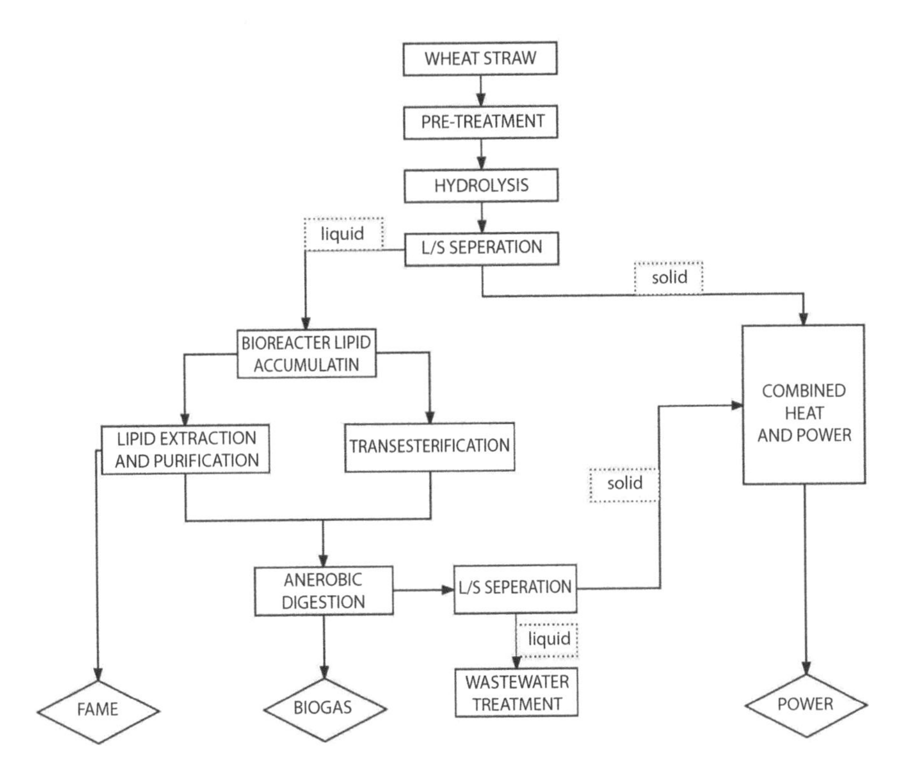

Figure 19.8 Production of biodiesel from wheat straw.

energy needs, and the power production helps to lower energy costs and reduce greenhouse gas emissions from traditional power plant facilities. Although some facilities exist that can be called biorefineries, the biorefinery has yet to be fully realized. Future biorefineries may play a major role in producing chemicals and materials that are traditionally produced from petroleum.

In the present context, considering the compositions of WS and its ash, a biorefinery system could be proposed. As mentioned in above sections, WS ash has high percentage of silica and potassium. And this ash is abundantly available or rather a waste in an agriculture-based country like India. Extraction of silica from WSA is economic and easy. In the similar manner, potassium could be extracted as it is water soluble. Silica further has many uses cosmetics, zeolite, pharmaceutical industry, rubber valorization, adsorption, and in information technology sector, as well for the production of micro chips. The ash remaining after silica extraction could be used for adsorbing heavy pesticides in farms.

Also, the WS if pyrolyzed could produce desirable amount of char, which is a good source of activated carbon and hence can be used in adsorption of heavy metals and detoxification as well. Similarly, the bio-oil produced is proving to be a good replacement for crude oil. Apart from that, bio-oil constitutes of valuable chemicals like LG and levoglucosenone. These two chemicals, though in lesser quantities, have many further uses. The syngas produced is directly used as fuel.

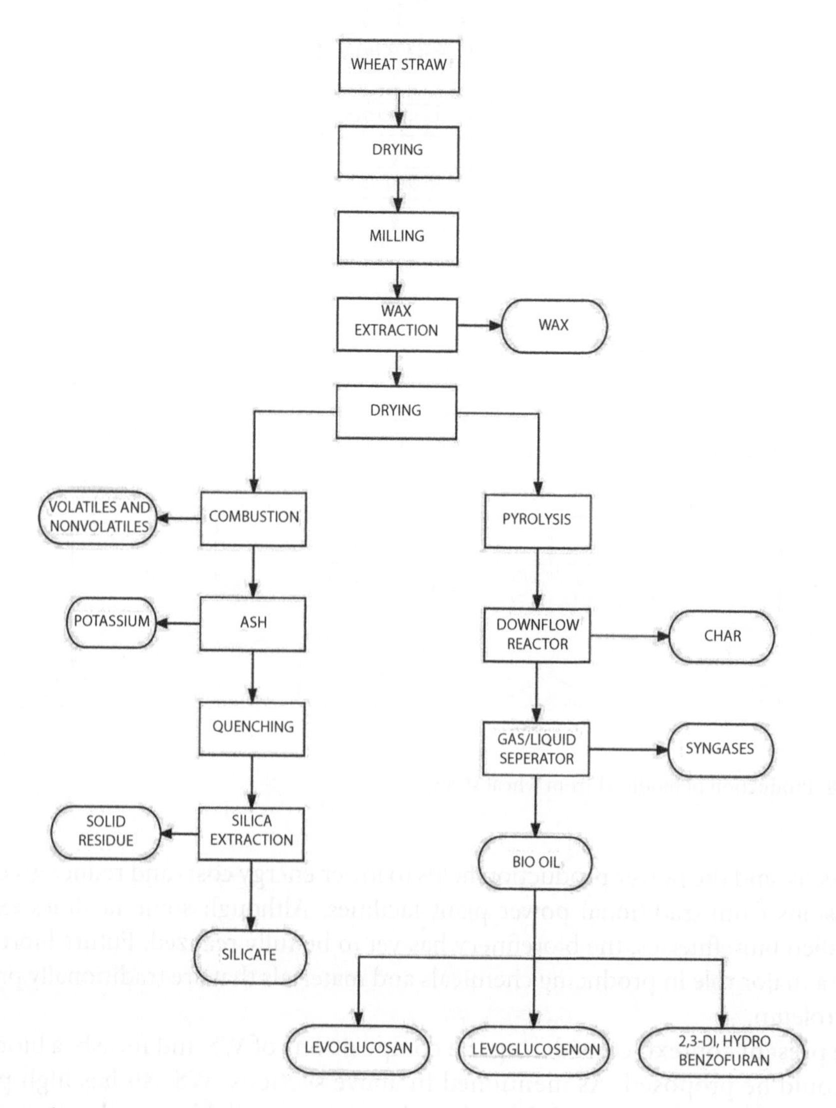

Figure 19.9 Wheat straw biorefinery.

The wax extracted from WS has great demand in cosmetic and food industry. And the straw could be directly used to produce biodiesel by only hydrolysis, fermentation, and transesterification of the straw. This diesel can readily replace the current in use fuels.

Thus, the WS biorefinery could be demonstrated as follows as presented in Figure 19.9:

19.9 Conclusion

WS produced every year stands as a hurdle for disposition. And the ash produced on its burning results in pollution as well. This WS poses many applications due to its chemical and organic composition. The surface chemicals like silica and potassium, which are in high percentage, can be suitably extracted. And the organic composition, like cellulose and hemicellulose, can be degraded by pyrolysis to obtain valuable chemicals.

Wax extraction from the WS prior to combustion or pyrolysis can serve multiple benefits as it provides biowax as well as enhances the further processes on the residue. Biodiesel production from the WS by its fermentation and transesterification is a separate process to deal with and stands as an economic process to replace current fuel scenario.

Thus, WS, which is generally treated as waste, can have many uses from its different fractions extracted or processed.

The concept and the inventory data presented here could be useful for evaluating the techno-economic feasibility of the biorefinery as also conduct the life-cycle analysis to establish the sustainability.

Acknowledgment

The authors are grateful to the Department of Science and Technology, India, for providing the research grant (Grant No. DST/TDT/TDP-02/2017) to undertake this work. BDK acknowledges the support as SERB Distinguished Fellow.

References

1. Food and Agriculture Organisation of United Nations, Wheat grain production, 2016.
2. Talebnia, F., Karakashev, D., Angelidaki, I., Production of bioethanol from wheat straw: An overview on pre-treatment, hydrolysis and fermentation. *Bioresour. Technol.*, 101, 4744–4753, 2010.
3. Dodson, J.R., Hunt, A.J., Budarin, V.L., Matharu, A.S., Clark, J.H., The chemical value of wheat straw combustion residues. *RSC Adv*, 1, 523–530, 2011.
4. Saleem Khan, T. and Mubeen, U., Wheat Straw: A Pragmatic Overview. *Curr. Res. J. Biol. Sci.*, 4, 673–675, 2012.
5. Morissette, R., Savoie, P., Villeneuve, J., Corn Stover and Wheat Straw Combustion in a 176-kw Boiler Adapted for Round Bales. *Energies*, 6, 5760–5774, 2013.
6. Budarin, V.L., Shuttleworth, P.S., Dodson, J.R., Hunt, A.J., Lanigan, B., Marriott, R., Milkowski, K.J., Wilson, A.J., Breeden, S.W., Fan, J., Sin, E.H.K., Clark, J.H., Use of green chemical technologies in an integrated biorefinery. *Energy Environ. Sci.*, 4, 471–479, 2011.
7. Trivedi, N.S., Mandavgane, S.A., Mehetre, S., Kulkarni, B.D., Characterization and valorization of biomass ashes. *Environ. Sci. Pollut. Res.*, 23, 20243–20256, 2016.
8. Sin, E.H.K., The extraction and fractionation of waxes from biomass, Thesis submitted to University of York, 2012.
9. Galassi, C., Bertoni, F., Ardizzone, S., Bianchi, C.L., Water-based Si3N4 suspensions: Part I. Effect of processing routes on the surface chemistry and particle interactions. *J. Mater. Res.*, 15, 155–163, 1999.
10. Kumar, A., Sengupta, B., Dasgupta, D., Mandal, T., Datta, S., Recovery of value added products from rice husk ash to explore an economic way for recycle and reuse of agricultural waste. *Rev. Environ. Sci. Biotechnol.*, 15, 47–65, 2016.
11. Khonde, R.D. and Chaurasia, A.S., Pyrolysis of Sawdust, Rice Husk and Sugarcane Bagasse: Kinetic Modeling and Estimation of Kinetic Parameters using Different Optimization Tools. *J. Inst. Eng. Ser. E.*, 96, 23–30, 2015.

12. Zhao, X., Wang, W., Liu, H., Ma, C., Song, Z., Microwave pyrolysis of wheat straw: Product distribution and generation mechanism. *Bioresour. Technol.*, 158, 278–285, 2014.
13. Esterer, A., Separating levoglucosan and carbohydrate acids from aqueous mixtures containing the same-by solvent extraction, United States Patent 3309356, Application Number: US47150465A, 1965.
14. Karlsson, H., Ahlgren, S., Sandgren, M., Passoth, V., Wallberg, O., Hansson, P.A., A systems analysis of biodiesel production from wheat straw using oleaginous yeast: Process design, mass and energy balances. *Biotechnol. Biofuels*, 9, 1–13, 2016.

Index

Printed in the USA/Agawam, MA
January 3, 2021

767483.121